Astronomy, Weather, and Calendars in the Ancient World

The focus of this book is the interplay between ancient astronomy, meteorology, physics and calendrics. It looks at a set of popular instruments and texts (parapegmata) used in antiquity for astronomical weather prediction and the regulation of day-to-day life. Farmers, doctors, sailors, and others needed to know when the heavens were conducive to various activities, and they developed a set of fairly sophisticated tools and texts for tracking temporal, astronomical, and weather cycles. For the first time the sources are presented in full, with an accompanying translation. A new and comprehensive analysis explores questions such as: what methodologies were used in developing the science of astrometeorology? What kinds of instruments were employed and how did these change over time? How was the material collected and passed on? How did practices and theories vary in the different cultural contexts of Egypt, Mesopotamia, Greece, and Rome?

DARYN LEHOUX is Senior Lecturer in Roman History, Classics and Ancient History at the University of Manchester. He has published numerous articles on ancient science, astronomy, astrology, and epigraphy.

T0174441

Astronomy, Weather, and Calendars in the Ancient World

Parapegmata and Related Texts in Classical and Near Eastern Societies

DARYN LEHOUX

CAMBRIDGE
UNIVERSITY PRESS

CAMBRIDGE UNIVERSITY PRESS
Cambridge, New York, Melbourne, Madrid, Cape Town,
Singapore, São Paulo, Delhi, Mexico City

Cambridge University Press
The Edinburgh Building, Cambridge CB2 8RU, UK

Published in the United States of America by Cambridge University Press, New York

www.cambridge.org
Information on this title: www.cambridge.org/9781107404779

First published 2007
First paperback edition 2011

A catalogue record for this publication is available from the British Library

ISBN 978-0-521-85181-7 Hardback
ISBN 978-1-107-40477-9 Paperback

Contents

List of illustrations

Preface

This book has been in the works for quite some time. It began as a doctoral dissertation project at the University of Toronto in 1996, and the present version has emerged much changed from that original in the decade since. In large part this is due to the very many helpful comments I have received over that time from more people than I can count or thank here by name, although if I had to name a few, I would have to include Alan Bowen, Serafina Cuomo, Leo Depuydt, Paolo Desideri, Jim Evans, Denis Feeney, Karljurgen Feuerherm, Jay Foster, Judith Gilliland, Anthony Grafton, Ian Hacking, Bert Hall, Robert Hannah, James Hoch, Kinch Hockstra, Brad Inwood, Alexander Jones, Csaba La'da, André Leblanc, Marcellus Martyr, Kevin McNamee, Erica Reiner, David Sider, Heinrich von Staden, John Steele, Peter Struck, Noel Swerdlow, Liba Taub, and Katherina Zinn. As anyone who has dealt with parapegmata knows, the material is often difficult, confusing, contradictory, and − say it isn't so! − sometimes a trifle dry. It was in conversation with these readers and others that I was able to give a meaningful shape to these texts and to see how widely their relevance and interest extends.

For particular thanks, I should single out Alexander Jones, who first suggested this project to me and whose supervision of the original dissertation was so outstanding, and Brad Inwood, who played the role almost of a co-supervisor on the dissertation. The two of them made for a crack team of readers on the early versions of this text, and their exceedingly high standards kept me very busy. Jay Foster also contributed in more ways than I can count at virtually every stage of writing. I would also like to thank Michael Sharp and the anonymous referees for Cambridge University Press for their many helpful comments. One referee in particular took a very keen interest in the project from the beginning and his (or her) sharp eye and patient criticism have considerably helped to shape the final form of this book. I only wish I could thank them by name.

For their generous support through various stages of this project, I would like to extend my gratitude to the Institute for Advanced Study, Princeton, where I spent a very productive and stimulating year finishing the manuscript, the Loeb Classical Library Foundation, Harvard University, the Social Sciences and Humanities Research Council of Canada, and the University of King's College, Halifax.

Finally, I would like to thank my wife, Jill Bryant, for her patience and encouragement again, and again. Because of the long genesis of this book, and because of the short tenure of my memory, it is entirely possible (nay, probable!) that I am forgetting someone here, and if so I beg their forgiveness.

Abbreviations

For Greek and Latin works, I have used the standard abbreviations found in Liddell and Scott's *Greek–English Lexicon*, and Lewis and Short's *A Latin Dictionary*. Other abbreviations are as follows:

ACT	*Astronomical Cuneiform Texts* (Neugebauer, 1955)
AfO	*Archiv für Orientforschung*
CCAG	*Catalogus Codicum Astrologorum Graecorum* (Cumont et al., 1898–)
Chronology	al-Bīrūnī, *The Chronology of Ancient Nations*, C. E. Sachau, trans. (London, 1879)
CIL	*Corpus Inscriptionum Latinarum*
CT	*Cuneiform Texts from Babylonian Tablets in the British Museum* (London, 1896–)
d	day(s)
D–K	H. Diels and W. Krantz, *Die Fragmente der Vorsokratiker* (Berlin, 1922)
DSB	*The Dictionary of Scientific Biography* (New York, 1970–)
EAT	*Egyptian Astronomical Texts* (Neugebauer, 1969)
HAMA	*History of Ancient Mathematical Astronomy* (Neugebauer, 1975)
Id.	Ides
IG	*Inscriptiones Graecae* (Berlin, 1863–)
K.	Tablets in the Kouyunjik Collection in the British Museum, or
K.	Kalends
KAR	*Keilschrifttexte aus Assur religiösen Inhalts* (Ebeling, 1915)
LSJ	Liddell and Scott's *Greek-English Lexicon*
m	month(s)
MUL.APIN	*MUL.APIN: An Astronomical Compendium in Cuneiform* (Hunger and Pingree, 1989)
N.	Nones
OCD	*The Oxford Classical Dictionary*, 3rd edn. (Oxford, 1996)
P. Oxy.	*The Oxyrhynchus Papyri* (London, 1898–)

RE	*Paulys Real-Encyclopädie der classischen Altertumswissenschaft* (Stuttgart, 1893–)
τ	tithi(s)
TLG	*Thesaurus Linguae Graecae*
TU	*Tablettes d'Uruk* (Thureau-Dangin, 1922)
y	year(s)

PART I

Parapegmata and astrometeorology

1 | The rain in Attica falls mainly under Sagitta

An' as it blowed an' blowed
I often looked up at the sky
an' assed meself the question
what is the stars, what is the stars?
                                    Sean O'Casey

1. Calendars, weather, and the stars

For several years running now, Canada's national radio broadcaster, the CBC, has run an annual human interest piece in which they interview Gus Wickstrom, a farmer from Saskatchewan who predicts how cold the coming winter will be, how much precipitation we will have in what months, and so on. What is odd is that Wickstrom makes his prediction by looking at the thickness and texture of pig spleens. Occasionally, more sensationalist news sources claim that he also chews on the spleen, raw. He tells us that he learned the technique from his father, who learned it from his father and his father's father. Curiously, Wickstrom also claims to be very accurate, and there are websites full of testimonials from Saskatchewan residents who back him up enthusiastically.

In a similar if a little less gruesome vein, gardeners and farmers everywhere have used and handed down all kinds of indicators for seasonal weather: the widths of the bands on woolly caterpillars, the thickness of onion skins, whether the oak beats the ash into leaf in springtime, and more. A favourite story of mine centres on what are called Aunt Bertha's 'borrowing days'. I am told by my wife's grandmother, Mary McLeod, who grew up on a farm in Huron County, Ontario, that one can predict the weather for a coming season by observing the weather on the three days around the solstice or equinox. The day before the winter solstice, for example, mirrors the weather for January, the day of the solstice mirrors the weather for February, and the day after the solstice that for March. As arbitrary as this method may seem to us (not having tried it, mind you) Mary swears by it, and still prepares for each season (stocking firewood, for example) according to the prediction she gets in this way. She learned the technique from her aunt, Alberta Forrest, who got it from her father, and her father's father, and so on. But this 'and so on' hides an interesting detail for the historian. It turns out that a version of this same method of weather prediction can be found in Pliny's first-century encyclopedia, the *Natural History*, where he ascribes it to Democritus:

Democritus talem futuram hiemem arbitratur qualis fuerit brumae dies et circa eum terni, item solstitio aestatem.

Democritus thinks that the coming winter will be like the day of the winter solstice and the three days around it, and the same with the summer solstice.[1]

Almost 2,000 years later, this same tradition was being passed from farmer to farmer in rural Ontario as a useful rule of thumb. This is a unique kind of transmission, one not paralleled in other genres of classical literature. Aunt Bertha's borrowing days, in essence a kind of unattributed fragment of Democritus, were simply passed by word of mouth from parent to child. How far back this goes is anyone's guess. That it was an uninterrupted string from ancient times onward I rather doubt. Nevertheless, it was thought of by the Forrests as 'the sort of thing you learned at your father's knee', as one elder put it, and it must have come into the family's oral tradition well before the middle of the nineteenth century. In any case, the rule as I encountered it in Huron County did at least come bundled with a guarantee of accuracy that depended on its being time-honoured even back in Aunt Bertha's day. Its great age was meant to underscore its reliability as a method. After all, as Cicero says, 'in everything, great age brings about an extraordinary knowledge by means of continuous observation . . . since what event happens after what event is seen with repeated observation, and also what event is a sign of what thing.'[2]

There are two points I want to make here. The first is the importance of long-term weather prediction to farmers, ancient and modern alike. The second is the significance of agricultural and meteorological topics to some of the most important figures in the classical tradition. I have already mentioned Pliny and Democritus, and in the course of this book I will go on to discuss Hesiod, Homer, Vergil, Varro, Aristotle, Theophrastus, Cicero, Ovid, Aratus, Ptolemy, Diodorus Siculus, Petronius, Sextus Empiricus, Galen, 'Hippocrates', and many more.

---

[1] *HN*, XVIII.231. All translations in the text are my own.

[2] *De div.* I.109: *adfert autem vetustas omnibus in rebus longinqua observatione incredibilem scientiam; . . . cum quid ex quoque eveniat et quid quamque rem significet crebra animadversione perspectum est.*

In classical antiquity there are two more-or-less distinct traditions of weather prediction.[3] One (which I will call *Theophrastan*, after its most famous ancient exemplar, the *De signis* often attributed to Aristotle's successor, Theophrastus of Eresus) uses rules of thumb and day-to-day observations of events like the croaking of frogs, the colour of the sky, and the appearance of haloes around the sun. The observations from which the predictions get made have two salient features. One is that they generally can be seen as fortuitous: the frogs, for example, happen to be croaking today, but we have no way of knowing when they will be croaking next. All we can do is pay attention when we hear it. The other feature is the use of earthly (which includes atmospheric) phenomena, and the exclusion of celestial phenomena (I say this because haloes around the sun, the changing dimness of certain stars, or the apparent colour of the moon are *atmospheric* rather than *astronomical* phenomena in a strict sense). These distinctions are significant because they mark a boundary between Theophrastan weather prediction and the second major ancient tradition of weather prediction, that of *astrometeorology*. This is not to say, however, that no authors brought these two traditions together (I think of Aratus here in particular) but only that the two kinds of weather prediction are conceptually distinct and follow different historical trajectories. Astrometeorology uses the motions of the stars as signs for predicting the weather and for tracking the seasons: when such-and-such a star becomes visible for the first time this year, it marks such-and-such a season, and we will have such-and-such weather.[4] To take an example from one of our earliest classical sources, Hesiod:

ἤματα πεντήκοντα μετὰ τροπὰς ἠελίοιο,
ἐς τέλος ἐλθόντος θέρεος καματώδεος ὥρης,
ὡραῖος πέλεται θνητοῖς πλόος . . . .
τῆμος δ᾽ εὐκρινέες τ᾽ αὖραι καὶ πόντος ἀπήμων

Fifty days after the solstice,
at the arrival of the end of the season of weary heat,
that is the time for mortals to sail . . .
Then are the winds orderly and the sea propitious.[5]

---

[3] See Taub, 2003; Sider, 2002, p. 292–6; Lehoux, 2004a.

[4] There is also a later tradition, attested in book ɪɪ of Ptolemy's *Tetrabiblos*, for example, that looks at the positions and qualities of the planets as indicators of the weather. The cycles involved are considerably more complex than those of fixed-star astrometeorology, and are not tied to the seasons in the same way. *Tetrabiblos*-type astrometeorology is a distinct and later development to the fixed-star kind we will be examining in this book.

[5] Hes. *Op.* 663f. On Hesiodic time reckoning, see West, 1978, p. 376–81.

In other passages, we see the timing of various agricultural activities being determined by the appearances of the fixed stars:

πληιάδων Ἀτλαγενέων ἐπιτελλομενάων
ἄρχεσθ' ἀμήτου, ἀρότοιο δὲ δυσομενάων.

At the rising of the Atlas-born Pleiades,
begin the harvest, and you should plough when they set.[6]

δμωσὶ δ' ἐποτρύνειν Δημήτερος ἱερὸν ἀκτήν
δινέμεν, εὖτ' ἂν πρῶτα φανῇ σθένος Ὠρίωνος

Urge the slaves to thresh Demeter's sacred corn
when strong Orion would first appear.[7]

Here the signs are celestial only, and they are not fortuitous in the same way as the Theophrastan signs are. The general sequence and timing of these stellar events was understood. Unlike the croaking of frogs, stellar risings and settings – also called stellar *phases*, and which will be more fully explained shortly – don't just happen at any time. They happen in a particular order, and they repeat from year to year in that same order. This cyclicality is an important feature of these signs. Because of their rigid cyclicality, they are useful for predicting weather, both in a short-term sense, and also in the larger sense of marking out the seasons of the year.

It is because of their cyclical nature, recurring at the same season, year in and year out, that these signs were so very useful to ancient farmers. Greece, Mesopotamia, Rome, and even Egypt all had calendars that for one reason or another were not adequately tied to the changes of the seasons.[8] In Greece and Mesopotamia, this was because the calendars in question were lunar, which inherently gives them a certain amount of wobble relative to the seasons: where the seasonal year is $365\frac{1}{4}$ days long, a lunar year would be 354 or 384 days, depending on the number of (approximately) $29\frac{1}{2}$-day months there are in a particular year. And so the (lunar) calendrical year and the (solar) seasonal year would shift about relative to each other. Add to this the notorious capriciousness with which days and months were added to the many different Greek civil calendars,[9] and one confronts a situation where over the course of a single lifetime, a Greek farmer may have seen

---

[6] Hes. *Op.* 383–4.   [7] Hes. *Op.* 597–8.

[8] The various ancient calendars may not have been intended to track the seasons at all (this seems particularly possible for the Greek calendars: see Nilsson, 1962), in which case the 'problem' is not really the calendar's. But the farmer is still faced with a challenge: How does he know in advance when the seasons will be changing?

[9] For a fuller discussion, see chapter 4, below. See also Mikalson, 1996; Samuel, 1972; Pritchett and Neugebauer, 1947.

his civil calendar wander relative to the seasons to such an extent that it is not especially useful for the timing of agricultural activities. When would I plant my corn if the month called 'May' didn't always happen at the same time of year? Or what would I do if my 'May' was my neighbour's 'June'?

And it was not only farmers who needed to track the seasons accurately. For sailors as well, as Hesiod showed above, some seasons are more favourable to navigation than others. So Aratus, writing what was in his day (third century BC) an extraordinarily popular poem about weather signs and astronomy, also has sailors watching the fixed stars to determine weather:

καὶ μέν τις καὶ νηὶ πολυκλύστου χειμῶνος
ἐφράσατ᾽ ἢ δεινοῦ μεμνημένος ἀρκτούροιο
ἠέ τεων ἄλλων, οἵ τ᾽ ὠκεανοῦ ἀρύονται
ἀστέρες ἀμφιλύκης, οἵ τε πρώτης ἔτι νυκτός.

And the man on board ship has seen the wavy storm,
remembering dread Arcturus or another star,
which draw themselves from the ocean
in the morning dusk or at the start of night.[10]

The Roman historian Polybius (second century BC), to take another example, tells the dramatic story of what was for him the greatest marine disaster in all history, the complete loss of 284 ships under the command of Marcus Aemilius and Servius Fulvius in a storm in 255 BC, during the first Punic war. After describing this titanic disaster, Polybius says:

ἧς τὴν αἰτίαν οὐχ οὕτως εἰς τὴν τύχην ὡς εἰς τοὺς ἡγεμόνας ἐπανοιστέον·
πολλὰ γὰρ τῶν κυβερνητῶν διαμαρτυραμένων μὴ πλεῖν παρὰ τὴν ἔξω πλευρὰν
τῆς Σικελίας . . . ἅμα δὲ καὶ τὴν μὲν οὐδέπω καταλήγειν ἐπισημασίαν, τὴν
δ᾽ ἐπιφέρεσθαι· μεταξὺ γὰρ ἐποιοῦντο τὸν πλοῦν τῆς Ὠρίωνος καὶ κυνὸς ἐπιτολῆς.

We must lay the blame of this [disaster] not on fortune, so much as on the commanders, for many of the pilots warned them not to sail along the outer coast of Sicily . . . and also warned that a shift in the weather was not yet over, and another one was coming, for they were sailing between the rising of Orion and that of the Dog Star.[11]

So we see references to storms at sea in several of the texts known as *parapegmata* (we will come to a fuller discussion of these texts shortly) and one even marks the date (17 March) on which it becomes safe to sail on the open sea again after the stormy winter.[12]

---

[10] Aratus, *Phaen.* 744–7.   [11] Polybius, I.37.4–5.

[12] See the Clodius Tuscus parapegma in part II: sources, the texts and translations section of this book.

We also know that the astronomical determination of the seasons played an important role in Greek and Roman medicine, where the qualities of the climate could profoundly affect the humoural balance in the body. So in Hippocratic medical works like *Airs, Waters, Places*, and the *Epidemics*, we see physicians paying close attention to seasonal markers, such as the solstices, equinoxes, and the phases of the fixed stars:

φυλάσσεσθαι δὲ χρὴ μάλιστα τὰς μεταβολὰς τῶν ὡρέων τὰς μεγίστας καὶ μήτε φάρμακον διδόναι ἑκόντα μήτε καίειν ὅ τι ἐς κοιλίην μήτε τάμνειν, πρὶν παρέλθωσιν ἡμέραι δέκα ἢ καὶ πλείονες· μέγισται δέ εἰσιν αἵδε καὶ ἐπικινδυνόταται· ἡλίου τροπαὶ ἀμφότεραι καὶ μᾶλλον αἱ θεριναὶ καὶ αἱ ἰσημερίαι νομιζόμεναι εἶναι ἀμφότεραι, μᾶλλον δὲ αἱ μετοπωριναί· δεῖ δὲ καὶ τῶν ἄστρων τὰς ἐπιτολὰς φυλάσσεσθαι καὶ μάλιστα τοῦ κυνός, ἔπειτα ἀρκτούρου, καὶ ἔτι πληιάδων δύσιν.

It is necessary to be especially careful at the most important changes of season, and neither give a purgative drug, nor perform abdominal cautery or surgery until ten or more days have passed. The following are the most important and most dangerous [changes of season]: both of the solstices, especially the summer, and both of the so-called equinoxes, especially the autumnal. It is also necessary to be careful at the risings of stars, especially Sirius, then Arcturus, and again at the setting of the Pleiades.[13]

Knowing when the seasons begin and end is clearly very important in a number of ancient disciplines, and again, the calendars available to the Greeks and Romans before Julius Caesar were not ideal for reckoning this. We also find Thucydides, Plato, Aristotle, and Theophrastus, among others, using astronomical rather than calendrical markers for indicating the time of year.[14] For the Egyptians, who used a rigid 365-day calendar with no leap years, the problems may be a little less pronounced, but they are still a factor.[15]

Since their calendars were at best of limited usefulness for the timing of seasonal activities, Greeks, Romans, Mesopotamians, and Egyptians all turned to the observation of the fixed stars in order to determine the best times for planting, harvesting, pruning, sailing, and more. This is because what are called the *phases of the fixed stars* are very closely tied to the agricultural seasons, and so are good indicators of when those seasons begin and end. Indeed, Greeks and Romans would often link their seasons to particular stellar phases, rather than, as we do it, to just the two solstices and two

---

[13] [Hippocrates] *Aër.* xi. Compare also, e.g., Galen, *In Hipp. lib. prim. epid.*, vol. xviiа, p. 15, l. 8; *De diebus decretoriis*, vol. ix, p. 914, l. 15. The phases of the moon are also important in later medicine. See Aetius Amidenus, *Tetr.* 162.

[14] See Wenskus, 1990; West, 1978, pp. 376–81; Gomme, 1956.   [15] For details, see chapter 6.

equinoxes. We tend to take it for granted that the year is divided into four seasons, each of approximately equal length. But why should it be? In many locations, the schematic division of the seasons according to the solstices and equinoxes does not correspond particularly well to the actual changes in the weather. By the time 'the first day of summer' on or around 21 June arrives, southern Ontario, for example, has typically been enjoying very un-spring-like weather for several weeks. Indeed, June is about as hot as August, and has even less precipitation. But this summer weather starts earlier in the central regions of the continent than it does on the east coast, so why should we not say that the season called 'summer' starts earlier there too?

Along these lines, different ancient cultures used a variety of different schemes for dividing up the seasons. In Egypt there were three seasons in the year: that of 'inundation' (*ȝḥt*, when the Nile flooded its banks and submerged the fields around it), 'emergence' (*prt*, when seedlings began to grow), and 'harvest' (*šmw*). From the time of the earliest Egyptian texts, the beginning of the Nile flood was associated with the heliacal rising of the star Sirius. In classical cultures, on the other hand, there were several different schemes in use, some four-season divisions, some eight, and others. Varro (first century BC), for example, gives us a fourfold division of the year that begins its seasons on the twenty-third day after the sun enters each of Aquarius, Taurus, Leo, and Scorpio. Thus spring starts on 7 February in the Julian calendar, summer on 9 May, autumn on 11 August, and winter on 10 November. He then tells us that if more precision is required, the farmer can use an eightfold scheme that divides the seasons more minutely:

primum a favonio ad aequinoctium vernum dies XLV, *hinc ad vergiliarum exortum dies* XLIV, *ab hoc ad solstitium dies* XLIIX, *inde ad caniculae signum dies* XXVII, *dein ad aequinoctium autumnale dies* LXVII, *exin ad vergiliarum occasum dies* XXXII, *ab hoc ad brumam dies* LVII, *inde ad favonium dies* XLV.

The first [season runs] from [the rising of] the west wind [*favonius*, the same as that called Zephyrus in the Greek sources] to the vernal equinox, forty-five days; from there to the rising of the Pleiades, forty-four days; from this to the solstice, forty-eight days; then to the rising of Sirius, twenty-seven days; next to the autumnal equinox, sixty-seven days; from that to the setting of the Pleiades, thirty-two days; from this to the winter solstice, fifty-seven days; then to the west wind, forty-five days.[16]

We see that the seasons are here divided by the solstices and equinoxes, but also by a regularly occurring annual wind, and by the risings and settings of some of the fixed stars.[17]

---

[16] Varro, *Rust.* I.28.   [17] Compare also Plin. *HN* II.122f.

The Mesopotamians also used the phases of the fixed stars as indicators of weather patterns, and stellar phases played an important role in the so-called *Uruk scheme*, which regularized their lunar calendar with respect to the solar year. As we shall see in chapter 5 of this book, direct connections between the Near Eastern material and the classical sources are often difficult to establish, but we can show that fixed-star astrometeorology was practised in one form or another in classical and Near Eastern cultures alike.

1.i.   The phases of the fixed stars

I have mentioned the annual risings and settings of the fixed stars repeatedly, but since naked-eye astronomy is no longer common knowledge, it will be worth a short digression to explain what these risings and settings, the so-called *phases* of the fixed stars, are.

We all know that the sun has a motion from east to west, which it repeats every day, moving the 360° around the earth in twenty-four hours.[18] But the sun also has another, less obvious motion from west to east. This can be observed as follows: go out just at sunset and watch the sky begin to fill with stars as the brightness of the setting sun recedes. You should take special notice of the stars in the general vicinity of the recently disappeared sun. In particular, remember how far they are from the western horizon a little after sunset. The next day, go out again and observe the same stars. You may notice that they are a little closer to the horizon than they were at this exact time the day before.[19] The next day they will be closer still, until one day they will vanish entirely in the obscuring brightness of the sun. This vanishing is one of the four important 'phases' of a star, called its '*heliacal setting*', its '*evening setting*', or just its '*setting*'. It is said, in Greek terms, to be due to the slow west-to-east motion of the sun (relative to the fixed stars), which it completes in one (sidereal) year, moving at a rate of approximately 1° per day.[20]

After about thirty days (assuming the chosen star is on or near the eclip-tic)[21] the star will rise from the eastern horizon just before sunrise, thus

---

[18] For the purposes of this work, the earth sits still in the centre of the Cosmos and the sun, stars, and planets all move around it.

[19] I say that sunset occurs 'at the same time' each day since it is one of the events which defines 'time' in the ancient world. Hours were counted relative to sunset or sunrise in antiquity, rather than from an artificially determined 'midnight' as we do it today.

[20] The measurement of this motion in degrees only began in the second century BC in Greece. For that matter, it is not even clear in the earliest sources, such as Hesiod, whether the sun or the stars were thought to be moving.

[21] The ecliptic is the path traversed by the sun through the signs of the zodiac over the course of the year. If, following classical practice, we picture the Cosmos as spherical, then the ecliptic is a

ending its period of invisibility. This first appearance is the next significant phase of the star, called its '*heliacal rising*', '*morning rising*', or just its '*rising*'. After this phase, the star will rise earlier and earlier each day until its '*acronychal rising*' (from ἀκρόνυχος, 'at nightfall'), or '*evening rising*', when it rises in the east just after the sun sets on the western horizon. A little while later, the star will set on the western horizon just before the sun rises in the east, making its '*cosmical setting*', or '*morning setting*'. Stars north or south of the ecliptic have some differences in the sequential order of these phases, but the terminology remains the same.[22]

All the phases I have discussed above, the morning and evening risings and settings, are *apparent* insofar as they are observable phenomena. Some astrometeorological texts, however, distinguish the *true* rising and setting from the observed phase.[23] The true heliacal rising, for example, occurs when the star rises at exactly the same time as the sun rather than a little before it, as in the apparent rising. Due to the brightness of the sun, the true phase is never observable, but must be calculated. Examples of the distinction between a true and apparent phase may be found in Geminus:

κζ Εὐκτήμονι Κύων ἐπιτέλλει.

[On the] 27th [of Cancer]: According to Euctemon, Sirius rises.[24]

And four days later, we see:

ἐν μὲν οὖν τῇ α´ ἡμέρᾳ Εὐκτήμονι Κύων μὲν ἐκφανής, πνῖγος δὲ ἐπιγίνεται· ἐπισημαίνει.

On the 1st day [of Leo]: According to Euctemon, Sirius is visible, it becomes very hot: the weather changes.[25]

Here there is a four-day gap between the star's 'rising' and its becoming 'visible', probably referring to the difference between its true rising and its apparent rising.[26] Not all cases are so clearly worded, however, and it is often

---

great circle, inclined at a little over 23° to the celestial equator and intersecting that circle at two points, the equinoxes.

[22] Details can be found in *HAMA*, p. 760f.

[23] Bowen and Goldstein, 1988, p. 54f., disagree, but Ptolemy goes to some length to distinguish true and apparent phases in his introduction to the *Phaseis* (§3). See also Souriban, 1969, p. 208–10.

[24] Geminus, 212.4.   [25] Geminus, 212.16–17.

[26] There is some debate about whether this is really what Geminus was intending, as there are astronomical problems with the precise dating in this instance (see e.g., Bowen and Goldstein, 1988). Nevertheless, alternate explanations seem to me to pose more difficulties than they solve, and the levels of astronomical accuracy we should expect in the writing and transmission of a text of this sort are, as attested in many other examples, not so fine-tuned as to rule out a small dating error here.

difficult to tell whether a given text is referring to the true or the apparent phases.

II.   What is a parapegma?

If the fixed stars are a good way to track the agricultural seasons, how do we then track the phases of the fixed stars? The obvious answer would seem to be 'by observation'. One imagines that ancient farmers would have watched the horizon on a semi-regular basis to see what stars were rising or setting, and then used these observations to time their various activities. Certainly this is what we find in Hesiod and other early authors: a handful of rules of thumb that coordinate observed phases with seasons and activities.

But by the third century BC, and possibly even earlier, these rules of thumb had been developed by the Greeks into a more complex system of weather prediction that accounted for more than just the beginnings and ends of seasons and a handful of weather rules. What we see is the emergence of a detailed set of correlations that tie specific weather phenomena to a host of stellar phases throughout the entire year.[27] In order to keep track of the increasing number of significant phases, something more than just the farmer's memory was needed, and this gap was filled by an instrument – using this word in a very broad sense – called a *parapegma*.

At its most basic, the word *parapegma* (plural: *parapegmata*) describes any instrument that tracked cyclical phenomena by the use of a movable peg or pegs. To take a simple, if late, example, the fragmentary marble parapegma from Pausilipum[28] (see Fig. 1.1) was used to track the seven days of the Roman hebdomadal week alongside the eight days of the nundinal week.[29]

---

[27] On how much of a role *observation* plays in this more detailed tradition, see chapter 3, below.

[28] Excavated in 1891 at Pausilipum, currently in the Johns Hopkins University Archaeological Collection. Published by Fulvio, 1891, p. 238; Degrassi, 1963, p. 304. See also *CIL* I[1], p. 218.

[29] The nundinal day is the market-day for a given Italian town, which occurred, from archaic times onward, every ninth day on the Roman reckoning (every eighth day counted as we would do). The local market-day was a holiday from agricultural work, and farmers could thus come to town to exchange wares and produce, as well as to keep up on local affairs, politics, laws, etc. Macrobius, *Sat.*, lists the nundinae as *feriae* (1.16.5) but points out that there was a divergence of opinion in antiquity on the matter (see 1.16.28–31). On nundinae as news-gathering opportunities, see Macrobius, *Sat.* 1.16.34. On specific nundinal lists, see the astrological and nundinal parapegmata in part II in the Catalogue, below. On the nundinae more generally, see Lehoux, forthcoming; Rüpke, 2000a; Marino, 2000; Andreau, 2000; de Ligt, 1993; Frayn, 1993; Tibiletti, 1976–7; Deman, 1974; MacMullen, 1970; Michels, 1967, p. 23f., and especially 187–90.

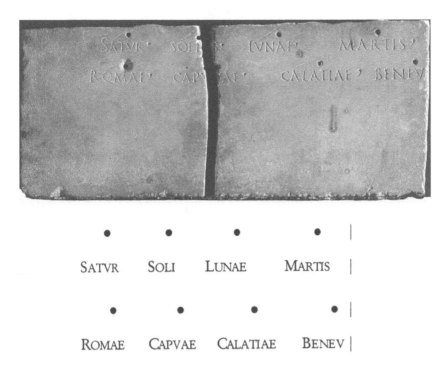

Fig. 1.1 The Pausilipum parapegma.

In the first line we see the planets that give their names to the first four days of the seven-day week: Saturn, the Sun, the Moon, and Mars. Beneath these there are the Cities that lend their names to the local nundinal days: Rome, Capua, Calatia, and Benev(entum). Above each day name, a hole has been drilled in the marble. For each cycle, the seven-day and the eight-day, a single peg would be inserted above one of the day names, and then shifted to the right the next day. By looking at the inscription and locating the pegs, a reader would be able to tell what day of the week it was in either cycle.

Different parapegmata deal with different kinds of cycles. In the Pausilipum parapegma it is the days of the hebdomadal and nundinal weeks. In other examples we see the phases of the fixed stars and seasonal indicators, as in one of the fragmentary parapegmata unearthed in the theatre at Miletus in 1904 (the so-called *Miletus I*, see Figs. 1.2 and 1.3). The fragment consists of part of three columns for tracking the sun's motion through each of three zodiacal signs, probably Sagittarius, Aquarius, and Aries, by means of a single peg that was shifted from day to day.[30] The middle column, that

---

[30] That there was only one peg per cycle, see my argument in section A.ii of the catalogue of parapegmata, in part II below.

Λ	30 (days)
• ἐν Ὑδροχόω[ι ὁ] ἥλιος	• The sun is in Aquarius
• [.....] ἑῶιος ἄρχεται δύνων καὶ Λύρα δύνει.	• .....] begins setting in the morning and Lyra sets.
• •	• •
• Ὄρνις ἀ[κ]ρόνυχος ἄρχεται δύν[ων.]	• Cygnus begins to set acronychally.
• • • • • • • • •	• • • • • • • • •
• Ἀνδρομέδα ἄρχεται ἑῶια ἐ[πι] τέλλειν.	• Andromeda begins rising in the morning.
• •	• •
• Ὑδροχόος μεσοῖ ἀνατέλλων.	• Aquarius is in the middle of rising.
• Ἵππος ἑῶιος ἄ[ρ]χ[ετ]αι ἐπι τέλλειν.	• Pegasus begins to rise in the morning.
•	•
• Κένταυρος ὅλος ἑῶιος δύνει.	• The whole of Centaurus sets in the morning.
• Ὕδρος ὅλος ἑῶιος δύνει.	• The whole of Hydra sets in the morning.
• Κῆτος ἄρχετα[ι] ἀκρόνυχον δύνειν.	• Cetus begins to set acronychally.
• Ὀιστὸς δύνει. ζ[ε]φύρων ὥρα συνεχῶν	• Sagitta sets. The season of the west wind accompanying.
• • • •	• • • •
• Ὄρνις ὅλος ἀκρόνυχος δύνει.	• The whole of Cygnus sets acronychally.
[•]	[•]

Fig. 1.2. Part of a pegged astronomical parapegma (Miletus I, 456B, middle column).[31] A single peg was moved each day from hole to hole. By locating the peg, information about the current astronomical configuration, and hence the season, could be determined easily.

for Aquarius, is almost entirely preserved. The beginning of the west wind (Ζέφυρος), a major seasonal indicator, is indicated for the day of the setting of the constellation Sagitta.

This astronomical parapegma from Miletus uses the same basic technology as the Pausilipum parapegma, but it tracks quite different phenomena. In inscriptional parapegmata, as a general rule, there is a fairly hard-and-fast division between the kinds of information tracked in Greek parapegmata and the kinds tracked in Latin ones. Only one Latin inscriptional parapegma is astrometeorological, and no Greek parapegmata track the days of the hebdomadal or nundinal weeks.

[31] Text and translation based on my new edition of the fragment, Lehoux, 2005.

Fig. 1.3. Miletus 456 B. The text in Figure 1.2, above, is from the middle column of this inscription.[32]

32 Currently in the Pergamon Museum, Berlin (Inv. no. SK 1606). Reproduced with permission of the Staatliche Museen zu Berlin, Preussischer Kulturbesitz

Fig. 1.4 A pegged astrological parapegma (Thermae Traiani). [33]

Some Latin parapegmata also track lunar phenomena, as we can see in the fourth-century *Thermae Traiani parapegma* (Fig. 1.4).[34] Here there are three cycles being tracked. Across the top are the images of the deities presiding over each day of the hebdomadal week: [Saturn], Sol, Luna, Mars, Mercury, [Jupiter], and Venus. In the centre are the signs of the zodiac, with two holes drilled per sign, reading counterclockwise from just to the right of the top: Aries, Taurus, Gemini, Cancer, Leo, Virgo, Libra, Scorpio, Sagittarius, Capricorn, Aquarius, and Pisces. This was probably meant to track the motion of either the sun or the moon through the signs of the zodiac. Finally, down the sides of the inscription are the Roman numerals from i through xv, and xvi through xxx for tracking the days of the moon, counting from either new moon or full moon as day i. The days of the moon, the hebdomadal week, and the motion of the sun and moon through

---

[33] Now lost. Image courtesy of the Thomas Fisher Rare Book Library, University of Toronto.

[34] The inscription itself was a graffito in plaster, originally excavated in 1812 from a house near the baths of Trajan, which was later converted into a shrine to Santa Felicita. It is published in Marulli, 1813; Guatanni, 1817; de Romanis, 1822; Degrassi, 1963, pp. 308–9. The original was destroyed by erosion shortly after de Romanis sketched it in 1822.

the zodiac all had astrological significance for the Romans and were used in timing agricultural and other activities, as we shall see in chapter 2.

Not all parapegmata are inscriptional, however. Some parapegmata survive in literary texts. Because manuscripts cannot physically make use of pegs and holes as inscriptional parapegmata do, a new system of indexing was developed, and this typically linked the stellar phases to dates in some calendar. Ptolemy's *Phaseis*, for example, uses the Alexandrian calendar:

Θώθ

α′. ὡρῶν ιδ′ L′ ὁ ἐπὶ τῆς οὐρᾶς τοῦ Λέοντος ἐπιτέλλει. Ἱππάρχῳ ἐτησίαι παύονται. Εὐδόξῳ ὑετία, βρονταί, ἐτησίαι παύονται.

β′. ὡρῶν ιδ′ ὁ ἐπὶ τῆς οὐρᾶς τοῦ Λέοντος ἐπιτέλλει, καὶ Στάχυς κρύπτεται. Ἱππάρχῳ ἐπισημαίνει.

γ′. ὡρῶν ιγ′ L′ ὁ ἐπὶ τῆς οὐρᾶς τοῦ Λέοντος ἐπιτέλλει. ὡρῶν ιε′ ὁ καλούμενος Αἲξ ἑσπέριος ἀνατέλλει. Αἰγυπτίοις ἐτησίαι παύονται. Εὐδόξῳ ἄνεμοι μεταπίπτοντες. Καίσαρι ἄνεμος, ὑετός, βρονταί. Ἱππάρχῳ ἀπηλιώτης πνεῖ.

δ′. ὡρῶν ιε′ ὁ ἔσχατος τοῦ Ποταμοῦ ἑῷος δύνει. Καλλίππῳ χειμαίνει καὶ ἐτησίαι παύονται.

ε′. ὡρῶν ιγ′ L′ Στάχυς κρύπτεται. ὡρῶν ιε′ L′ ὁ λαμπρὸς τῆς Λύρας ἑῷος δύνει. Μητροδώρῳ δυσαερία. Κόνωνι ἐτησίαι λήγουσιν.

Month of Thoth

1. [For the latitude where the longest day is] $14\frac{1}{2}$ hours: the star on the tail of Leo rises. According to Hipparchus the Etesian winds stop. According to Eudoxus rainy, thundery, the Etesian winds stop.
2. 14 hours: the star on the tail of Leo rises, and Spica disappears. According to Hipparchus there is a change in the weather.
3. $13\frac{1}{2}$ hours: the star on the tail of Leo rises. 15 hours: the star called Capella rises in the evening. According to the Egyptians the Etesian winds stop. According to Eudoxus variable winds. According to Caesar wind, rain, thundery. According to Hipparchus the east wind blows.
4. 15 hours: the rearmost star of Eridanus sets in the morning. According to Callippus it is stormy and the Etesian winds stop.
5. $13\frac{1}{2}$ hours: Spica disappears. $15\frac{1}{2}$ hours: the bright star in Lyra sets in the morning. According to Metrodorus bad air. According to Conon the Etesian winds finish.[35]

---

[35] Ptolemy, *Phas.* xiv.1–16.

Here one looks up the date to see what stellar events are happening, and what weather can be expected.

For the reader's convenience, all the extant parapegmata are catalogued and translated in part II of this book.

II.i.   What kinds of cycles are tracked by parapegmata?

Different parapegmata track different kinds of cycles. We have seen examples so far of parapegmata that track astrometeorological cycles (e.g., Ptolemy's *Phaseis*), astronomical cycles (e.g., *Miletus I*), and ones that track astrological and calendrical cycles (e.g., *Pausilipum* and *Thermae Traiani*).[36] These are the main classes of parapegmata, but there are examples of parapegmata that were used to track calendrical information only, and a few whose use is unclear. Of these different types, the astrometeorological is usually seen as paradigmatic. Partly this stems from ancient usage, where most references to a 'parapegma' are clearly to an astrometeorological parapegma. But the main reason for the canonizing of the astrometeorological parapegmata as the norm is just that modern scholars have known about those for the longest. Indeed, until Wiegand unearthed the first inscriptional parapegma fragments at Miletus in 1902,[37] the only parapegmata the modern world knew were the literary astrometeorological texts like Ptolemy's *Phaseis* and the Geminus parapegma. It was even unclear *why* such texts should have been called parapegmata until these Miletus fragments, with their peg-holes, were found. As a result of this, most modern scholarship on parapegmata tended to ignore the other classes, or else tried to subsume them under the astrometeorological class. For example, Rehm, the most prolific writer on parapegmata to date, simply classed all the astrological parapegmata as 'false (*uneigentliche*) parapegmata'.[38] Since then most scholars have followed suit, and very little of the work on parapegmata has dealt with the Latin inscriptional parapegmata at all.[39] In chapter 2, we will collapse the sharp distinction between astrological and astrometeorological parapegmata.

In some astrometeorological parapegmata, stellar phases are linked with both day-to-day weather phenomena like rain, snow, and storms, and with annual variations in the climate at large, like the 'Etesian winds' which

---

[36] On these classifications and my justification for them, see part II, the Catalogue of parapegmata, below.

[37] See Wiegand, 1919– ; Diels and Rehm, 1904.   [38] Rehm, 1949.

[39] See, e.g., Evans, 1998; Rüpke, 2000a; Hannah, 2002.

blow over the eastern Mediterranean at the same time each year. Annual events of agricultural interest are also sometimes noted, as well as seasonal indicators such as solstices and equinoxes, or the return of the swallows in springtime.

Some astrometeorological parapegmata are *attributive*, that is, they link specific predictions to particular astronomers by name. Thus 'the equinox, according to Eudoxus', 'hail, according to Caesar', or 'Aries begins to rise, according to Callippus.' Other parapegmata,[40] on the other hand, simply give the phases or weather with no attributions whatsoever. Ptolemy's *Phaseis* gives his own calculations for risings and settings exclusively, but gives attributions for all of the weather predictions. Occasionally, predictions offer specific locations (or perhaps applications) to which they are supposed to apply, like 'a storm at sea'. Elsewhere, location may be implied in predictions such as 'wind, according to the Egyptians'.

Besides the astrometeorological parapegmata, there is a large class of what I have called 'astrological parapegmata', such as *Thermae Traiani*. In contrast to the astrometeorological parapegmata, the astrological ones offer no predictions. They are instead tools for keeping track of one or more astrological cycles such as the sun's position in the zodiac, the day of the moon (easily converted to its phase, but also astrologically important in its own right), the deified planet astrologically presiding over the day (the 'hebdomadal deity'),[41] and occasionally also calendrical cycles such as the days of the month or the days of the so-called *nundinal week*. This combination of calendrical and astrological information is not unique to the parapegmata. It also shows up in inscriptional dates such as the following:

[IA]NUAR(IVS) D<I> E IOVIS CONS(VLATV) FL(AVII) [L]VN(A) PRIM(A)

In January, Thursday, in the Consulship of Flavius, first day of the moon.[42]

We see here a combination of consular and calendrical dating, with exactly the kinds of information which the astrological parapegmata were used to keep track of: hebdomadal day and day of the moon.

---

[40] Miletus I and the Clodius Tuscus parapegmata, e.g.; for a full description of these and other parapegmata, see the Catalogue, below.

[41] For explanations of the origins and function of the hebdomadal deities, see Bouché-Leclercq, 1899, p. 478f.; Pietri, 1984; Richards, 1998, pp. 271–3.

[42] Dating probably from the fifth century AD. See Worp, 1991. For other examples see Erikkson, 1956, pp. 28–9.

II.ii.   Who wrote parapegmata?

There are a handful of literary parapegmata whose authors we know (Ptolemy, Columella, Pliny, Ovid, and a few others) but most others are anonymous or uncertain. I have already mentioned that some parapegmata are attributive, as is Ptolemy's *Phaseis*, such that sources for the astronomical or meteorological information are cited by name. To take an example from the Geminus parapegma:

ἐν δὲ τῇ θ' Καλλίππῳ τοῦ Ταύρου δύνει κεφαλή· ὑετοί.

ἐν δὲ τῇ ι' Εὐκτήμονι Λύρα ἑῷος ἐπιτέλλει· καὶ ἐπιχειμάζεται ὑετῷ.

ἐν δὲ τῇ ιβ' Εὐδόξῳ Ὡρίων ἀκρόνυχος ἄρχεται ἐπιτέλλειν.

ἐν δὲ τῇ ιγ' Δημοκρίτῳ Λύρα ἐπιτέλλει ἅμα ἡλίῳ ἀνίσχοντι· καὶ ὁ ἀὴρ χειμέριος
    γίνεται ὡς ἐπὶ τὰ πολλά.

On the 9th: According to Callippus the head of Taurus sets in the morning; rainy.
On the 10th: According to Euctemon Lyra rises in the morning, and it is stormy
    with rain.
On the 12th: According to Eudoxus Orion begins to rise acronychally.
On the 13th: According to Democritus Lyra rises at the same time as the sun comes
    up, and the air becomes stormy for the most part.[43]

Here Geminus gives us citations from some very important authors: Democritus (the Atomist); Eudoxus of Cnidus (whose theory of proportions formed the basis for book v of Euclid's *Elements*, whose idea of homocentric nested spheres gets adopted by Aristotle in the *Metaphysics*, and whose *Phaenomena* is usually seen as the main source for Aratus' text of the same name);[44] Callippus (Aristotle's contemporary who came up with a $365\frac{1}{4}$-day year, and whose improvements on Eudoxus' homocentric spheres get incorporated into Aristotle's account); and Euctemon of Athens (the fifth-century Athenian astronomer who is credited with the observation of a solstice in 432 BC). But as it turns out, we have no extant texts from any of these authors, let alone the sources for these particular citations. We do not know for certain what form the sources for the attributions took. Whether, for example, Callippus had composed a work that looked anything like a parapegma is unanswerable. It is certainly possible that the work from which his predictions were excerpted in the later parapegmata was of an entirely different structure. For all we know it may have read more like Hesiod than Ptolemy. For example, it may have been a text on some other subject

---

[43] Geminus, 218.23–220.7.
[44] See, e.g., Kidd, 1997; Gee, 2000. Contrast Sider, 2002, who thinks Aratus had separate sources
    for the two main parts of the poem.

containing scattered references to stellar phases and/or weather, without dating or ordering them systematically, let alone being structured as a parapegma in particular. This fact makes it difficult to know how valuable an exercise it is to follow some modern commentators in excerpting the 'calendrical fragments' of, say, Democritus or Euctemon from the extant parapegmata. Nevertheless, a good deal of the modern literature on parapegmata assumes that the sources of the parapegmatic attributions were themselves parapegmata, and various reconstructions of the 'lost parapegma of Euctemon', for example, have been attempted.[45] There are also serious problems with reconciling the different sources on which the reconstructions are based, such that no two parapegmata agree with any degree of reliability on which phases and predictions can be attributed to Euctemon, for example, and when. This makes it look like the sources of the attributions were either multiple, vague, or both. If there were multiple sources for the different attributions found in the different parapegmata then combining these multiple sources into a single reconstructed 'parapegma of Euctemon' may be little more than anachronistic artifice.[46] If the sources were vague, this means that they were probably not structured as parapegmata are, with a clear timing and sequential ordering of phases. Again, this would make the parapegma of Euctemon a *construction* rather than a *reconstruction*. All this by way of saying that, although it is clear that early Greek astronomers were practising *astrometeorology*, there is insufficient evidence to assert, as has long been assumed by modern scholars, that they were composing *parapegmata* in particular.

But even if we cannot be sure that the authorities for astrometeorological attributions were themselves the authors of parapegmata, we can still be certain not only that they were engaged in some kind of astrometeorology more generally, but also that they were contributing authoritatively to the tradition. They were seen by later authors as not just having practised astrometeorology, but as having actively contributed to the body of knowledge which formed the sources of later parapegmata. That this body of knowledge was central to the practice of early classical astronomy can be seen in the fact that the list of authorities cited in the parapegmatic tradition reads like a veritable *Who's Who* of early Greek astronomy: Meton, Euctemon, Eudoxus, Callippus, Conon, Dositheus, and Hipparchus, among others. Philosophers like Democritus the Atomist, his student Metrodorus,

---

[45] See Hannah, 2002; van der Waerden, 1984a; Rehm, 1913.
[46] For more on the problems with reconstructing parapegmata, see Lehoux, forthcoming, and also my comments in section G of the Catalogue, below.

and Plato's student Philippus of Opus, also get cited. There are vaguer references to 'the Chaldaeans' and 'the Egyptians',[47] and later parapegmata cite Roman sources such as 'Caesar' (either Julius or Germanicus)[48] and Varro, among others. For a complete list and discussion of the authorities cited in the parapegmata, see the *Authorities cited in parapegmata* section in part II of this book.

III.   Towards a history of parapegmata

We cannot with any certainty trace the history of parapegmata themselves back to the lost ur-parapegmata of Euctemon, Meton, and Eudoxus. And as we look to the archaeological and textual evidence over the course of this book, it also becomes clear that the history of the parapegma *qua* instrument is to a certain extent distinct from the history of astrometeorology itself. The two do coincide in some places and at some times, but not always and everywhere. By treating parapegmata as a class of *technological* artefact, that is, as instruments or tools for tracking cycles of one sort or another, we will see that astrometeorological parapegmata are only one kind (though an admittedly very important kind) of such instrument. In short, not all parapegmata are astrometeorological, and much of this book will be dedicated to working out the kinds of relationships that obtain between the different types.

The earliest parapegma extant is an inscriptional fragment from the Ceramicus district of Athens, dating from the fifth century BC.[49] It is simply

---

[47] As we shall see in chapters 5 and 6, there is some foundation to the claims that the parapegmatic tradition is drawing on an indigenous Egyptian astrometeorological tradition, but in the case of Mesopotamia ('the Chaldaeans') the situation is much less secure.

[48] The question of whether the *Caesar* here refers to Julius or Germanicus is an open one. The earliest mention in a parapegma is Ptolemy's *Phaseis* (second century AD). This and later parapegmata give no information beyond the name 'Caesar'. Speculation ultimately rests on a judgment as to the weighting of one of two possibilities: either (a) Julius Caesar, in some kind of connection with his calendar reform, may have left some material that was later incorporated into parapegmata under his name (Pliny seems to hint as much at *HN* XVIII.211); or (b) Germanicus Caesar's translation of Aratus may have included (or been related to) new material later incorporated into the parapegmatic tradition but now lost in the original. On Germanicus' Aratus, see Gain, 1976; Gee, 2000; Maurach, 1978; Possanza, 2004.

[49] Published by Brückner, 1931, pp. 23–4. See also *IG* II² 2782.

a list of ordinal numbers from fifth through to ninth. In its entirety, the fragment reads:

• πέμπτη	•	• fifth	•
• ἕκτη	•	• sixth	•
• ἑβδόμη	•	• seventh	•
• ὀγδόη	•	• eighth	•
• ἐν]ά[τ]η [		• ni]n[t]h[	

We have no idea what cycle it was meant to track. It may have been calendrical, but it may just as well have tracked some other cycle. If comparative evidence is any guide, then it looks to be neither astronomical nor astrometeorological.[50]

But, as we have seen, by the time the Ceramicus parapegma was in use at Athens, there was already a long-established textual (and probably also oral) tradition of astrometeorology dating back to at least Hesiod and being carried forward in some form or another in the works of Euctemon, Eudoxus, and others. Some time after Ceramicus, we begin to see the emergence of inscribed astrometeorological parapegmata, which are a combination of the *content* represented by the literary astrometeorological tradition, and the *technology* represented in the Ceramicus parapegma. The earliest archaeological evidence we have for this coming-together are the inscriptional Miletus parapegma fragments collectively called *Miletus II*, dating from the first century BC. Here the basic tool of an inscribed peg-board with a moveable peg was being used to track astrometeorological phenomena. But the Miletus II parapegma, although it is the earliest *inscriptional* astrometeorological parapegma, is not the earliest astrometeorological parapegma generally. It is pre-dated, by about two centuries, by a *literary* astrometeorological parapegma, P. Hibeh 27,[51] which indexes its astrometeorology to the Egyptian calendar. Aratus' *Phaenomena* is also roughly contemporary with P. Hibeh, and like P. Hibeh, is literary.

By distinguishing form and content in parapegmata, and taking morphological relationships more seriously than has traditionally been the case in modern scholarship on these texts, we can more fully appreciate the historical development of, and the relationships between, the different kinds of

---

[50] Rehm, 1949, though it was meant to track days in a zodiacal calendar (otherwise unattested), but the reasoning that led him to this conclusion is both strained and self-serving (see chapter 3, below). Hannah, 2005, treats it as astrometeorological, but does not give a reason. See my further comments on this inscription in the Catalogue in part II.

[51] Published by Grenfell and Hunt, 1906, p. 138–57. See item A.i in the Catalogue, below.

parapegmata. Miletus II, for example, shares morphology with the Ceramicus parapegma, but does not share function: both are inscribed peg-boards, but the one is astrometeorological, and the other is not. Conversely, Miletus II shares a function with P. Hibeh and Aratus, but not morphology: all are astrometeorological, but Miletus II is an inscribed peg-board, and the other two are literary texts. Because of the fact that Miletus II shares some features with almost all other parapegmata, I take it for comparative purposes as paradigmatic.

Using this model, and taking Miletus II as our starting point, we will see in chapter 2 that the Roman astrological parapegmata like Thermae Traiani are morphologically similar, but functionally quite different from the Greek inscriptional astrometeorological parapegmata. While they are physically comparable – insofar as they have holes for moveable pegs that are used for tracking some cyclic phenomena – they are functionally distinct, insofar as they are not astrometeorological. The literary parapegmata, on the other hand, are functionally similar to Miletus II (they are astrometeorological), but morphologically different, insofar as they are textual, and typically index the astrometeorological information to a calendar rather than to a moveable peg.

But there are other relationships between the Latin Thermae Traiani type of parapegma and the astrometeorological ones. We understand the uses of the content of the Latin inscriptional astrological parapegmata of the Thermae Traiani type because of passages in Roman agricultural texts like Columella, Varro, and Pliny. There we get descriptions of the proper (and propitious) times for all kinds of agricultural work (on which, see chapter 2). In particular, we see that these agricultural texts use the content of the astrological parapegmata alongside of the content of astrometeorological parapegmata. Thus the two main functional traditions come together in these texts for their usefulness in regulating agricultural activity.

This is not, however, to say that Miletus II necessarily represents the earliest type of parapegma. There is no denying that the Ceramicus parapegma is much older, and seems to represent some functionally different type of parapegma, since no other known inscriptional parapegma before the Roman ones indexed a series of sequential numbers to a peg. The type of parapegma represented by Miletus II thus stands as a fusion of two distinct traditions: a textual astrometeorological tradition, and a technological tradition represented by the Ceramicus parapegma. We can date this inscriptional astrometeorological tradition with a *terminus ante quem* of 89/88 BC (Miletus II). On the current archaeological evidence it seems that the Miletus II type came into being as a combination of the function of astrometeorological texts

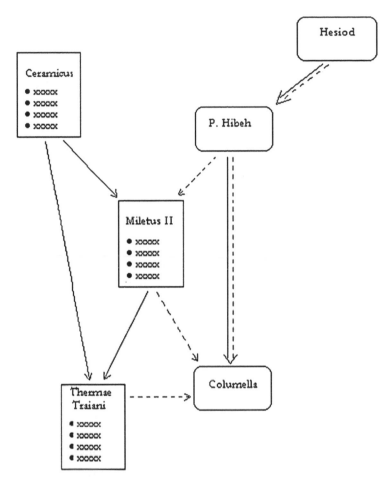

Fig. 1.5 Relationships between the different kinds of parapegmata. Solid lines represent morphological continuity, dashed lines represent functional continuity. The particular parapegmata named are meant to be representative of types.

like P. Hibeh 27, and the morphology of parapegmata of the Ceramicus type. Further functional change took place after this time, which led to the morphologically derived Roman astrological and calendrical parapegmata.

To summarize: the oldest known parapegma is inscriptional, but of unknown use. Apparently unrelated to it is a textual astrometeorological tradition dating as far back as Hesiod. This tradition came to use the morphology of the Ceramicus type of parapegma to track astrometeorological cycles, which resulted in the Miletus II type of parapegma. Alongside of the occasional production of inscriptional astrometeorological parapegmata, literary P. Hibeh type parapegmata were still being produced and used right up into the Middle Ages. From either the Ceramicus or the Miletus type parapegmata, however, another tradition developed in the Roman world which used the basic technology of these inscribed instruments to track a new set of phenomena with a peg: lunar, hebdomadal, and nundinal cycles.[52] Because of the importance of both astrology and astrometeorology in Roman farming, we see the inclusion, in Latin agricultural texts, of material related to the Thermae Traiani type of parapegma alongside textual astrometeorological material.

iv. Questions and problems

Having sketched out the sources, uses, and the historical development of parapegmata, we can now move on to look at the issues, questions, and problems raised by these instruments in more detail. In the following chapters I will explore the relationships between astrometeorology, parapegmata, and the development of various ancient civil and astronomical calendar systems. We will look at the relationship between parapegmata and Greek calendrical cycles, between astronomy, astrometeorology, and Mesopotamian calendars, and also between Mesopotamian calendrical schemes and Greek calendrical cycles. So also, we will examine the question of whether and how the Greco-Latin material is indebted to Mesopotamian astronomy and astrology, and to Egyptian calendrics and astrology. We will see how both astrometeorological parapegmata and astrological considerations informed Roman agricultural practice, and how the development of the Julian calendar relates to this material. From here we will investigate the logic and epistemology of the predictive inferences at the very heart of astrometeorology. The book concludes with a detailed descriptive catalogue of all the parapegmata and related texts and instruments known to me, and the text and translations of all the extant parapegmata and many related texts.

Some of the main threads in this book bear on current debates in the history of astronomy. One of the larger arguments I will flesh out over the course of the book is that classical astronomy and astrology originate in techniques for weather prediction. In Mesopotamia and Egypt, the story is

---

[52] One example of a calendrical adaptation of this type of inscriptional parapegma is extant, the Guidizzolo Fasti (see item D.ii in the Catalogue, below, which uses a peg to track the Roman calendar. From this tradition (or possibly before it?) derived an inscriptional graffiti type of parapegmata, which consisted of the writing out of a cycle on a wall for tracking lunar, hebdomadal, nundinal, and calendrical cycles without the help of pegs. See the Catalogue, below.

slightly different, in that astronomy and astrology don't *originate* there in weather prediction exclusively, but these cultures do use the fixed stars for weather prediction, and in Mesopotamia this shows up relatively early in the astronomical and astral omen texts. This set of arguments gives strong confirmation to a thesis first articulated by Bowen and Goldstein,[53] that early Greek astronomy is concerned with a fundamentally different set of problems than later astronomy. Specifically, early Greek astronomy is more interested in the annual phases of the fixed stars than it is with planetary motion.

This sets my account apart from standard histories of astronomy, which are usually told as stories that begin with the remarkable Mesopotamian techniques for predicting planetary positions. Recently these accounts have begun to be modified by the inclusion of older astral omen texts in the narrative.[54] Moving from Mesopotamia into the classical world, the usual accounts see two plot lines unfolding in tandem: the cosmological one, where we begin with Eudoxus' concentric spheres model and move up to Ptolemy's use of the epicycle-on-deferent constructions for planetary motion, and the mathematical one, where Mesopotamian ideas like the zodiac and certain planetary numerical parameters move into the Greek tradition in the centuries before Ptolemy. Both of these lines see Ptolemy's *Almagest* as the central reference point for defining what ancient astronomy *is*.

But all of this puts an undeserved emphasis on *planetary* phenomena as being the be-all and end-all of ancient astronomy. As we have seen, fixed-star astrometeorology also played a very important role across many genres of ancient literature. Not only is the classical astrometeorological tradition *older* than the mathematical and cosmological planetary ones, it is also remarkably widespread and longer lived. It begins with the earliest classical texts, as we have seen, and forms a continuous thread through to the Middle Ages and beyond. The modern-day *Old Farmer's Almanac* (North America's longest-running periodical) is in many ways the inheritor of this tradition. There is certainly no modern representative of ancient planetary astronomy that is available for purchase as an impulse buy in the check-out line at my local hardware store.

And the reader will probably not be surprised to know that there are forty years of back-issues of the *Old Farmer's Almanac* on Mary McLeod's bookshelf.

---

[53] Bowen and Goldstein, 1988; See also Evans, 1998. Compare Dicks, 1970.
[54] See, e.g., Reiner and Pingree, 1975–98; Rochberg-Halton, 1988; Rochberg, 1999; Reiner, 1999; Hunger and Pingree, 1999.

2 | Spelt and spica

'Now John,' quod Nicholas, 'I wol nat lye,
I have yfounde in myn astrologye
As I have looked in the moone bright,
That now a Monday next, at quarter nyght,
Shal fall a reyn, and that so wilde and wood
That half so greet was nevere Noes flood.'

Chaucer, 'The Miller's Tale'

1. Contexts and margins

It has been taken for granted in virtually all the modern literature on para-pegmata that the Roman inscriptional parapegmata of the Thermae Traiani type (the class I call *astrological parapegmata*) are only marginally related to the Greek *astrometeorological* parapegmata such as we find in Ptolemy and Geminus. While it is true that the astrological parapegmata track different cycles than do the astrometeorological parapegmata, it turns out that the cycles that the astrological parapegmata *do* track are, in the Roman contexts, not unrelated to the cycles tracked by their astrometeorological counter-parts. We know that Roman authors preserved, re-worked, and augmented earlier Greek astrometeorological material. In general, though, the formats of Roman literary parapegmata differ from those of their Greek counter-parts. Greek literary parapegmata are usually independent entities, extant either as texts in their own right (Ptolemy, P. Hibeh), or else as complete units more or less blocked-off from the texts to which they are appended (Geminus, Clodius Tuscus cited in Lydus). Roman literary parapegmata, on the other hand, tend overwhelmingly to be broken up into sections and interwoven with other material, much of it agricultural (Pliny, Columella, Varro), bucolic, or calendrical (Ovid).[1] This interweaving of parapegmata

---

[1] On agriculture in Pliny, see Healy, 1999; Beagon, 1992; Le Bouffle, 1989; Pedersen, 1986. For Varro, see Green, 1997; Flasch, 1996. For Columella, see Noè, 2002. On Roman agricultural

with other material proves invaluable for the modern reader seeking to understand both the fullness of the contents and especially of the contexts of Roman astrometeorology. The interspersing of parapegmatic material with other kinds of agricultural material, for example, allows us to trace the ways in which different temporal cycles – calendrical, astrometeorological, lunar, cultic – interweaved and interacted to produce the particular rhythms of Roman farming and of Roman time in general.

Certain times of the year are favourable for certain activities, where the year is framed by its calendrical or stellar divisions (and after the Julian calendar reform of 45 BC these two sets of divisions became united under one civil calendar that now tracked the seasonal and astrometorological cycles extraordinarily well). But so too, certain phases of the moon are conducive to specific activities, and thus both the stellar *and* lunar cycles needed to be considered in choosing the optimal time for particular tasks. To pick just one example, Cicero tells us to fell trees during a particular season (winter), but also under a particular moon (waning).[2] What we find in Roman agricultural writers are combinations of stellar, seasonal, calendrical, and lunar time that are ultimately *intercyclical*, and this should not in the least be surprising. Even today, my wall calendar contains a host of intercyclical rhythms, all laid out for the year: the 365-day calendar year (with its rhythm of holidays, gardening, birthdays and more) is interwoven with the seven-day week (with its rhythms of the work-week cycle), and the moon even makes an appearance for the determination of one holiday (Easter). And sometimes significance is attached to exactly the ways in which the different cycles interact: Easter is always on a Sunday, Fridays the thirteenth are supposed to be bad luck, and 'blue' moons – defined as the second full moon in a calendrical month – are supposed to be good. Time just is intercyclical, and the Roman parapegmata need to be understood in this light, because the intercyclicality turns out to play an important role in many of the Latin sources. This is true of the textual astrometeorological parapegmata, such as we find in Pliny, Columella, and Ovid, for example, as well as in inscriptional parapegmata, where in almost every example, Latin inscriptional parapegmata can be seen to be tracking, not so much several independent cycles at once, but more the ways in which the different cycles interact.

literature more generally, see White, 1970; Christmann, 2003; Morgan, 1999; Green, 1997; Klingner, 1967. On agricultural poetry, see West, 1978; Farrell, 1991; Nelson, 1996, and esp. 1998; Gibson, 1997; Sharrock, 1997.

[2] Cic. *De div.* II.14.33.

Fig. 2.1. *Menologium rusticum.* Illustration from Gruterus, 1707.[3]

II.   Agriculture and the heavens

Roman agricultural treatises preserve a large amount of detailed practical information about Roman farming. Our other major source of knowledge about Roman farming practices is archaeological, ranging from fragments of farm tools, to evidence of land use patterns, to public inscriptions like the *menologia rustica* (see Fig. 2.1).[4] Tapping these sources, K. D. White, in his monumental work on Roman agricultural practice, discusses the history of

[3] Illustration courtesy of the Thomas Fisher Rare Book Library, University of Toronto.
[4] See Degrassi, 1963, vol. XIII.2, p. 286f.; Broughton, 1936.

almost all aspects of Roman agricultural practice, including estate organization, tools, methods of cultivation, crops, manure, livestock, servants, and so on.[5] Nevertheless he misses an important part of Roman farming practice: the uses of astronomy and astrology, and of methods of time reckoning.[6] In all the Roman agricultural texts, instructions for planting, pruning, manuring, ditch-digging, sheep-shearing, harvesting, and ploughing are given with specific attention to the phases of the fixed stars, as we should expect given the discussion in the previous chapter.[7] The importance of the stars is highlighted, for example, in the very first line of Vergil's first-century BC bucolic poem the *Georgics*:

Quid faciat laetas segetes, quo sidere terram
vertere, Maecenas . . .

What makes the crops happy, and under what star
should we turn the soil, Maecenas . . .[8]

Modern scholarship has tended to treat the relationships between Roman agricultural texts and the Greek parapegmata solely in terms of the Roman uses of fixed-star astrometeorology. While it is true that a good deal of astrometeorological material (much of it related to Greek sources) is attested in Roman agricultural texts, it is also true that more of the heavens is at play in the Latin authors than the fixed stars alone. I have already mentioned that the moon plays a major role in Roman agricultural texts, and it does so in a way that seems largely independent from the Greek astrometeorological traditions. At the same time, unlike Greek civil calendars, the Republican and Julian calendars were not lunar. This meant that lunar phases were not even nominally tied to calendrical dates, and so in order to track these lunar phases extra-calendrical tools – parapegmata – were adopted and adapted.

We saw in passing in the last chapter a kind of parapegma that had nothing to do with the fixed stars. The Roman Thermae Traiani parapegma

---

[5] White, 1970. See also Sirago, 1995; Herz and Waldherr, 2001; Fowler, 2002. On early farming generally, see Grigg, 1974.

[6] Isager and Skydsgaard, 1992 also make almost no mention of stellar phases in their work on Greek agriculture.

[7] Although Wenskus, 1986, has argued that Cato's emphasis on solstices and equinoxes represents an indigenous Roman tradition distinct from the stellar-phase tradition discussed in the previous chapter.

[8] Vergil, *Georg.* I.1–2. Questions around the aims and didacticism of Vergil's *Georgics* have received a lot of attention in the last couple of decades. See, for example Lembke, 2005; Kronenberg, 2000; Battstone, 1997; Thomas, 1995; 1988; 1987; Mynors, 1990; Spurr, 1986; Erren, 1985; Wilkinson, 1982; Putnam, 1979.

(Fig. 1.4, ch. 1) was used to track three distinct cycles: (1) the days of the seven-day week, (2) the days of the moon; and (3) the motion of the sun or moon through the zodiac. Thermae Traiani is far from alone. The Thermae Traiani, or astrological, type of parapegma is the second largest overall class of these instruments, and the best-attested class inscriptionally. Not all parapegmata of this type are limited to tracking just the three kinds of cycle tracked by Thermae Traiani. There are several other kinds of cycle that combine with one or more of those in Thermae Traiani. In the Latium parapegma[9] (Figs. 2.2 and 2.3), for example, we see the days of the hebdomadal week across the top, and the days of the moon tracked by the rosette in the middle. But down the right-hand side we see a new kind of cycle: the nundinal days. The nundinal day is nominally the market-day for a given Italian town, which occurred, from archaic times onward, every ninth day on the Roman reckoning (every eighth day counted as we would do). The local market-day was a holiday from agricultural work,[10] and farmers could thus come to town to exchange wares and produce, as well as to keep up on local affairs, politics, laws, etc.[11] On the Latium parapegma, the local market-day is simply labelled *in vico*, 'here'.

The final cycle we find on the Latium parapegma is that of the seasons. At the four corners of the rosette are the names and dates of the seasons. From left to right and top to bottom, we have:

[*ver ex* XIII *K. Febrar. in* XII *K. Mai.: dies* LXXXXI][12]
aestas ex XI *K. Mai. in* X *K. August.: dies* LXXXXIIII

---

[9] First published by Gruterus, 1707. Later published by Henzen in *CIL* VI.4.2, no. 32505. More fully reconstructed by Degrassi, 1963. My reconstruction here follows Degrassi, with minor emendations (see Lehoux, forthcoming). The fragment is currently in the Museo Archeologico Nazionale di Napoli. Drawings and photograph reproduced with permission.

[10] Macrobius, *Sat.* lists the nundinae as *feriae* (1.16.5) but points out that there was a divergence of opinion in antiquity on the matter (see 1.16.28–31).

[11] Macrobius, *Sat.* 1.16.34. Various *Fasti* (Degrassi, 1963, vol. XIII.2; Michels, 1967, pp. 23f., 187–90) have the days of the month labelled consecutively from *A* through *H* (called the 'nundinal letters'), where one of these days would be the local market-day in a given year (see Rüpke, 2000a; Michels, 1967, pp. 27–8; also, see below Catalogue, D.iii). There are also nundinal lists, as we see in the *Latium Parapegma*, which have the names of eight different towns inscribed, which has usually been read as indicating that the market-day occurred in eight different towns on eight different days, such that the market-day in Rome was followed regularly by that in Capua, then Calatia, etc., and then it would be market-day in Rome again after eight days. I have argued elsewhere, however, that there is evidence that nundinal lists like that in *Latium* are not linked to actual market-days in other towns (see Lehoux, forthcoming). On the nundinae and nundinal lists generally, see MacMullen, 1970; Deman, 1974; Tibiletti, 1976–7; de Ligt, 1993; Andreau, 2000; Marino, 2000.

[12] On the reconstruction here, see Lehoux, forthcoming.

[*autumnus ex* ix *K. August. in* xi *K. Nov.: dies* lxxxxi]
hiemps ex x *K. Nov. in* xiiii *K. Febrar.: die*[*s* lxxx]viiii

Fig. 2.2 The Latium parapegma.[13]

---

[13] Photograph used with permission of the Soprintendenza per i Beni Archeologici delle Province di Napoli e Caserta.

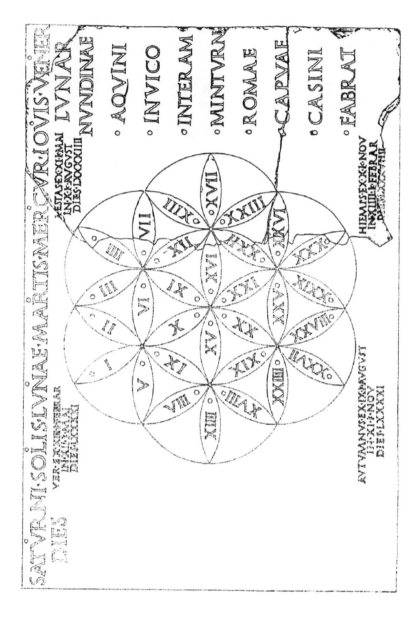

Fig. 2.3. The reconstructed Latium parapegma.[14]

[14] Drawing slightly altered from Degrassi, 1963, with permission.

[Spring (is) from the thirteenth day before the Kalends of February until the twelfth day before the Kalends of May: ninety-one days.] [15]

Summer (is) from the eleventh day before the Kalends of May until the tenth day before the Kalends of August: ninety-four days.

[Autumn (is) from the ninth day before the Kalends of August until the eleventh day before the Kalends of November: ninety-one days.]

Winter (is) from the tenth day before the Kalends of November until the fourteenth day before the Kalends of February: eighty-nine days.

Looking at the different cycles tracked by these parapegmata, we find that some of them are very clearly astrological. As we will see, the Roman seven-day week is (at least at first) a purely astrological contrivance. The days of the moon are also astrologically important in several distinct ways, and even the days of the eight-day nundinal week could be ominous. This particular combination of timekeeping and omens, and timekeeping and astrology more specifically, differs in the Roman contexts from the kinds of relationships tracked by the Greek parapegmata, and will need some looking into. First, though, we should see if we can clarify the categories with which we are framing the discussion before turning to see how these categories apply to Roman agricultural contexts.

III.   Astronomy, astrology, and the calendar

The terms *astronomy* and *astrology* were used fairly frequently in the last chapter, and we would do well to pin them down, both in general, and with respect to astrometeorology and parapegmata in particular. In modern contexts there is a sharp division between the two terms, where astronomy is seen as a perfectly legitimate science (practised in universities, funded by national and international scientific granting agencies), and astrology

---

[15] Where our calendar counts up from the first of the month, the Roman calendar counts *down* to the first day of the *next* month (the Kalends) or to two other dates called the *Nones* and the *Ides*. Thus the eleventh day before the Kalends of February is the *eleventh last day* of January. To convert this to a date in our calendar, though, we must keep in mind the fact that the Romans also counted inclusively, where what we would call the day before the Kalends was thought of by them as the *second day*, which they called the *pridie*. The day before the pridie was called the third day before the Kalends. Thus the eleventh day before the Kalends of February would be 22 January. After the Kalends they then began counting down to the Nones (on either the fifth or seventh of the month as we would reckon it) and then to the Ides (the thirteenth or fifteenth, depending on the month).

is usually portrayed as a kind of quackery, a *pseudo*-science. But such a sharp division has not always been historically the case. As we look at how the categories of astronomy and astrology sometimes diverge and sometimes overlap in the ancient world, a set of interesting questions and problems will begin to emerge. Over the next few chapters we will see how parapegmata stand at the confluence of what are apparently (at least on the face of things) distinct kinds of ancient knowledge about the character and timing of the changes that are happening in the sky, and what these mean for farmers and others. Parapegmata will turn out to have a complex intertwining of astronomical, astrological, and chronometric features.

By *astronomy*, we usually mean the observation of the motions of the heavenly bodies and the mathematical modelling of these motions. In antiquity, the observations were limited to what could be seen with the naked eye, meaning the fixed stars and constellations of the night sky, and the seven moving heavenly bodies (all of which were thought of as 'planets' in Greco-Roman texts): the sun, the moon, Mercury, Venus, Mars, Jupiter, and Saturn. A mathematical table such as we find in Ptolemy's *Almagest*, for example, that lets its user calculate the precise positions of a planet in longitude and latitude for any given date is astronomical in this sense. So too are Ptolemy's geometrical models that underpin these arithmetical tables. Astronomy is thus primarily concerned with observing, modelling, and predicting planetary and stellar motions and positions.

Astrology, on the other hand, is the use of such astronomical data (whether observed, predicted, or retrodicted) to predict future events on earth, events like the fates of individuals or kingdoms and the outcomes of battles, or to determine when the most favourable time to undertake a certain activity might be, and more. The casting of a personal horoscope is a good example of this kind of practice. But the dividing line between astronomy and astrology and how the difference between them is justified, are not always simple matters in ancient sources. If an ancient text says that when the sun enters the sign of Cancer at the summer solstice the weather will be hot, this will not look particularly astrological to us since we readily see that there are physical reasons for the heat: the sun is high overhead and beating directly down on the earth below. But what if we say something analogous about the moon? To take a classical example (we will look at the Near East in later chapters), when the moon enters Sagittarius, the weather will be wet. Since, on our understanding of how the universe is put together, we can conceive of no legitimate *physical* reason for the moisture, we want to class this kind of claim as astrology rather than astronomy or physics. The issues at stake become even clearer when we start to look at claims about how the relative positions of the heavenly bodies with respect to each other and to the signs of the zodiac affect things down here on earth. For example a claim that 'Mars-in-Pisces-and-120°-away-from-the-sun' has a particular effect on the atmosphere looks to be purely astrological. The trouble is that the transition from the first case, the sun in Cancer, to the last case, Mars *trine* the sun in Pisces, may be smoother than a strongly dichotomous reading of the astronomy/astrology distinction can tolerate. One reason for this is that in the ancient texts the distinctions are not nearly as hard and fast as they are for us, so that astronomy and astrology overlap considerably, and this on several levels.

We can, as we saw, come up with a ready physical explanation for the heat of the sun at the solstice. So can the ancients, but their ideas about what counts as *physical* are going to look a little different from ours, since their physics is so very different.[16] And it is here that the borders that *we* see between astronomy and astrology will begin to break down. It turns out that, instead of the geometry of the sun's position, many ancient texts tend to prefer to talk about how Cancer is a 'hot' sign.[17] But this is exactly the same language that they use to describe the astrological effects of Mars (which is hot and dry), or the moisture of the moon in Sagittarius.[18] The fact is that even the most clearly astrological cases are often justified in ancient sources in terms of ancient physics, with its near-standard breakdown of elemental qualities into hot, cold, wet, and dry.[19] This makes some of the distinctions we might like to draw between astronomy and astrology, physics and superstition, or more generally between science and pseudoscience, anachronistic.

But this is not to say that the ancients did not recognize *any* distinction between astronomy and astrology. Sometimes they did, but not always with the same implications as the modern division. Ptolemy, for example, sets out an argument in the introduction to his *Almagest* that holds the certainty attainable in astronomy as being due to the solidity of its foundation, mathematics. By contrast, he tells us that the other sciences – *and he mentions physics here explicitly* – are uncertain, and 'should be called conjecture rather than knowledge . . . because of the instability and inscrutability of

---

[16] See, e.g., Lloyd, 1991b; 1987; 1979; 1973; 1970.

[17] E.g., Ptol. *Tetr.* ii.11; Vettius Valens, *Anth.* i.2.32–3.   [18] See Ptol. *Tetr.* i.4.

[19] Not all ancient physical systems adopted the four qualities. Epicureanism is a notable example, and one that also, not entirely incidentally, rejected astrology. For an overview of the four qualities and their varying places in ancient physics, see Lloyd, 1970; 1973.

matter'.[20] In the *Tetrabiblos* Ptolemy uses very similar language to explain why the results of astrology have less certainty than those of mathematical astronomy.[21] At the same time, though, he goes on to say that astrology provides very useful information that astronomy alone is incapable of furnishing, and he even feels the need to remind his reader that mathematical astronomy is still worthwhile, even if it cannot attain the same useful ends as astrology. Where modern philosophers of science have tended to group astronomy and physics together under one rubric – *science* – and astrology under another – *pseudoscience*[22] – Ptolemy is associating astronomy with mathematics and calling these *knowledge*, while classing astrology and physics together under the rubric of *conjecture*, where their uncertainty is due to the limits of our knowledge about matter on the one hand, and the inherent instability of that matter on the other. Beyond the groupings themselves, there is another salient feature of these categorizations that is worth highlighting: on the modern reading, where astronomy and physics are classed as sciences, astrology is classed in a particular kind of relation to them, expressly as *not*-science, or more precisely, *false* science. The relationships between these modern classes has the effect of highlighting certain features of astrology (its mathematical rigour and its claims to experimental – or at least experiential – verification, for example) as being explicit efforts on the part of its practitioners to make it look science-like, while simultaneously underscoring the importance of the fact that this attempted scientific dress is insufficient, illegitimate, or unfounded. Astrology thus becomes on this classification scheme something that pretends to be what it is not, hence the *pseudo* in pseudoscience. In Ptolemy's classification, on the other hand, we not only see astrology being classed together with physics (notable in itself), even if still in some kind of opposition to astronomy, but we also see that the relationships between the two larger categories, those of *knowledge* and *conjecture*, relate to each other quite differently than the modern categories of *x* and *pseudo-x* do. Rather than having astrology insufficiently pretending to the epistemological status of astronomy, Ptolemy's categories see the science of astrology as having *a degree of* knowledge that is perfectly legitimate and

---

[20] Ptol. *Alm.* VI.12–15.   [21] Ptol. *Tetr.* I.i.

[22] This division is still common. The paradigmatic example is Popper, 1959. See more recently Pingree and Gilbert, 2002. *Contra* this position, see Yates, 1964; Barton, 1994, pp. 5–6. Evans, 1998, does call astrology a 'science', but it merits only two of the 500 pages in his monumental *History and Practice of Ancient Astronomy*. See also the disparaging comments by Neugebauer, 1975, p. 943. In Neugebauer and van Hoesen, 1959, we see horoscopes being used primarily as evidence for ancient methods of astronomical calculation, with little or no interest in their astrological content. Jones, 2000, also mines horoscopes for mathematical astronomy, but he does give some welcome attention to the astrology as well.

useful, if less firmly certain than the science of astronomy. Rather than being *pseudo*-knowledge, astrology is simply *less firmly grounded* knowledge, just as physics, rooted in changeable matter and unreliable experience, is less firmly grounded than axiomatic-deductive mathematics.

Furthermore, it is worth noting that physics and astrology are more closely related than just by their shared classification for Ptolemy and many other ancient authors. As we go on to read more of the *Tetrabiblos*, we see that Ptolemy almost always couches his astrological theories in terms of the physical properties of the signs, stars, and planets.[23] And this is the standard ancient understanding of astrology, one that we can trace all the way up through the Middle Ages and beyond.[24] Thus we find in Columella that there is a physical force, *vis*, emanating from the stars to affect the atmosphere down here on earth.[25] (Of course not all ancient authors agree with the physicalist reading of astrology, and we will look at some of alternate interpretations in the next chapter.)

As we have also seen, the motions of the moon and the fixed stars also have another dimension, where they can play important roles in timekeeping. The question then arises of how genuinely independent those timekeeping functions are from their astrological functions.

iii.i.   Time reckoning

We saw in the last chapter that astronomy plays a large role in methods of time reckoning in classical antiquity. The phases of the fixed stars provide a useful way of tracking the seasons for ancient farmers, sailors, and physicians, and parapegmata provide in turn a useful way of tracking those phases of the fixed stars. Astrometeorological and astronomical parapegmata thus serve important timekeeping functions. By setting out the divisions and subdivisions of the seasons, they help the ancient farmer to know when to perform various activities around the farm. But as I mentioned above in the context of our own calendar, we don't just divide time up into one big 365-day cycle. We subdivide that 365 days into several other cycles. I have a pay cycle at work that runs from the first of one month to the first of the next month. My transit pass and the nursery fees for my children work in the same time frame. At a finer level, we also divide the year up into repeating

---

[23] But causal explanations are only one type of justification for astrology. As we shall see in chapter 3, Sextus Empiricus and Geminus ground astrometeorology in observation rather than causation.

[24] See Barton, 1994.   [25] Columella, *RR* xx.1.32.

seven-day cycles, the weeks, which we use to regulate our work, shopping, recreational, religious, and other activities. Particularly (but not exclusively) in Roman agricultural texts, these smaller cycles play roles in the timing of ancient labour just as they do today, but as we shall see, not always for the same reasons. Trying to determine the reasons the ancients had for timing various activities according to these smaller cycles will run us right back into the astronomy/astrology question.

Thus although the days of the eight-day nundinal week play an important timekeeping function in Roman culture, it is also clear that, like many aspects of the Roman calendar, these had an ominous dimension. We know from Roman sources that it was seen as ominous if the local market-day fell on the first of the year, or on the Nones of any month.[26] Similarly, particular calendar dates could also be seen as good- or bad-luck days as well. After the assassination of Julius Caesar, for example, the Ides of March were often seen as bad luck.[27]

An even better example is that of the seven-day hebdomadal week. This is simply a list of the planets that astrologically preside over each day.[28] The idea is that each planet presides over an hour of the day in turn, and the planet that presides over the first hour of the day is the most powerful that day. The Roman hebdomadal week begins with Saturday, which means that Saturn rules over the first hour of the day and so over the whole day in general. The rest of the planets preside in lesser capacity, hour by hour. The actual ordering of the days of the week is a function of the most common classical conception of the distances of the planets, which are derived from their relative speeds. Saturn, the slowest planet, is the farthest away. Next in is Jupiter, followed by Mars, then the sun, Venus, Mercury, and closest to us is the moon. Thus on Saturday, Saturn rules the first hour, Jupiter the second, Mars the third, and so on (see Fig. 2.4). When we get to the twenty-second hour of the day, we are back again at Saturn, followed by Jupiter, then Mars for the twenty-fourth hour. The first hour of the next day is thus ruled by the planet after Mars, which is the Sun. This is how we arrive at the ordering of the days of even our modern week: Saturday, Sunday, (Moon)day, and then (switching to French) mardi, mercredi, jeudi, and vendredi. It is a historical quirk that English uses Norse equivalents of these last four deities: Tiu, Woden, Thor, and Freya.

---

[26] See, for example, Macrobius, *Sat.* 1.13; Cassius Dio, xlviii.33.4.

[27] On calendar omens generally, see Grafton and Swerdlow, 1988.

[28] See Cassius Dio, xxxvii.18f.; Hübner, 1998–9; Richards, 1998, pp. 268–73; Evans, 1998, pp. 165–6; Bickerman, 1980, p. 61.

Hour	Planet						
1	**Saturn**	**Sun**	**Moon**	**Mars**	**Mercury**	**Jupiter**	**Venus**
2	Jupiter	Venus	Saturn	Sun	Moon	Mars	Mercury
3	Mars	Mercury	Jupiter	Venus	Saturn	Sun	Moon
4	Sun	Moon	Mars	Mercury	Jupiter	Venus	Saturn
5	Venus	Saturn	Sun	Moon	Mars	Mercury	Jupiter
6	Mercury	Jupiter	Venus	Saturn	Sun	Moon	Mars
7	Moon	Mars	Mercury	Jupiter	Venus	Saturn	Sun
8	Saturn	Sun	Moon	Mars	Mercury	Jupiter	Venus
9	Jupiter	Venus	Saturn	Sun	Moon	Mars	Mercury
...	...	...	...	...	...	...	...
22	Saturn	Sun	Moon	Mars	Mercury	Jupiter	Venus
23	Jupiter	Venus	Saturn	Sun	Moon	Mars	Mercury
24	Mars	Mercury	Jupiter	Venus	Saturn	Sun	Moon

Fig. 2.4 Planets ruling the days of the week and the hours of the day.

In addition to the day-to-day astrological influence exerted by the presiding planet, the day of the hebdomadal week on which the Kalends of January fell could have ominous significance for the year as a whole, influencing everything from politics and the social order to health and crop yields.[29]

The relationship between timekeeping and omina is underscored as well by a passage in Petronius' first-century novel, the *Satyricon*, when the character Encolpius describes something that sounds very like an astrological parapegma in the house of the overweeningly proud freedman Trimalchio:

sub eo titulo et lucerna bilychnis de camera pendebat, et duae tabulae in utroque poste defixae, quarum altera, si bene memini, hoc habebat inscriptum: 'III. et pridie Kalendas Ianuarias C. noster foras cenat,' altera lunae cursum stellarumque septem imagines pictas; et qui dies boni quique incommodi essent, distinguente bulla notabantur.

Under this inscription there hung from the ceiling a double lamp, and there were two boards fixed to the two posts, of which the one, if I remember correctly, had this inscribed: 'The third day before and the day before the Kalends of January "our Caius" dines outdoors.' The other [had inscribed] the course of the moon, and painted pictures of the seven stars, and which days were good and which bad were marked by a peg that distinguished them.[30]

---

[29] Lydus, *De mens.* IV.10.

[30] Petronius, *Sat.* 30. The passage is sometimes taken as evidence for the haughty self-image of Trimalchio, since he dares to presume some degree of control over time itself (Walsh, 1975, p. 139, for example, compares Petronius to Augustus here). But the comparative evidence of the Latin inscriptional parapegmata, which range from public inscriptions to common graffiti, show that this is not straightforwardly the case. For a discussion of the interweaving of the ordinary and the overblown in Petronius, see Donahue, 1999. On the satirical importance of the master's dining outdoors and the servant's referring to him by his praenomen ('our Caius'), see Walsh, 1970, p. 119; Smith, 1975, p. 63; Horsfall, 1989. Virtually all modern commentators

The seven stars surely refer to the deities of the hebdomadal week (and if so, this is an early reference to them).[31] The 'course of the moon' was probably simply the numbers I–xxx as in so many astrological parapegmata, and these would seem to have been marked by a separate peg from the weekdays.[32]

Other related Latin inscriptions combine calendrical and astrological functions. Take, for example, a graffito found on a shop's wall in Pompeii (Fig. 2.5).[33]

Here we see the seven days of the astrological week in the leftmost column, followed by the nundinal days in the second. The three middle columns are for tracking any calendar month, and the final three columns are for the astrologically significant days of the moon. The inscription thus combines the calendrical and the astrological quite freely.

III.ii.   The moon

The lunar days columns should be of no surprise to us, since the moon plays such a very important role in virtually every Roman agricultural text, where discussions of, for example, the types of manure to be used on a particular crop, would be followed immediately by a prescription concerning which phase of the moon it was best to spread the manure under.[34] The days of the moon have been mentioned repeatedly as an important astrological indicator. In fact, though, there are three distinct but related lunar influences apparent in Roman agricultural texts. In addition to the days of the moon, there are also the phase of the moon and the position of the moon relative to the horizon. Each of these influences determines the timing of various

---

have missed the fact that this passage describes a parapegma. The one exception I know of is Müller and Ehlers, 1965, p. 455.

[31] Heseltine's belief, echoed by Smith, 1975, p. 63, that the 'seven stars' refers to the 'sun, earth(!), and planets Mercury, Venus, Mars, Saturn, (and) Jupiter' is absurd. See Heseltine, 1913 p. 53, n. 1. The mistake is corrected by Slater, 1990, p. 57 n.

[32] Rehm agrees that the parapegma marked the hebdomadal days and lunar days, but thinks there were many *bullae* which he argues would probably have been of different colours, or painted with letters to mark the two kinds of day (Rehm, 'Parapegma', *RE*, col. 1363). Dölger thinks that a white peg was used to mark lucky days and a black one to mark unlucky days (Dölger, 1950, p. 205). He bases this on Macrobius' comment that the Kalends, Nones, and Ides were 'black' (*atros*) days. See Macrobius, *Sat.* 1.16.21. But I am unconvinced: if particular hebdomadal days were seen in themselves as lucky or unlucky, I see no reason why coloured or marked pegs must have been used. That is, the days (good and bad in themselves) were simply marked with a peg. The peg need not have specified the qualities. Finallly, *distinguente bulla* is in the singular, and so cannot mean 'distinctive knobs,' as Heseltine believes.

[33] For my reconstruction, see Lehoux, forcoming.

[34] It is best under a waning moon. See, e.g., Cato, *Agr.* xxix; Pliny, *HN*, xviii.322; Columella, *RR* ii.5.1, ii.15.9, ii.16.1; Palladius, x.12.

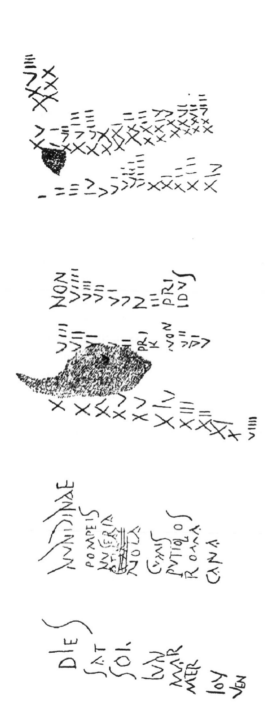

Fig. 2.5. The Pompeii calendar.[35]

[35] Illustration from *CIL*. Used with permission.

farming practices, such that certain activities are best carried out under a waxing moon, others avoided on a particular day of the moon, or still others performed when the moon is above the horizon.

(a) Phase: One should only plant beans when the moon is increasing or full, according to Columella's first-century agricultural treatise the *Rei rusticae*.[36] The idea behind this would seem to be that the increasing moon will bestow an increasing influence on the beans. A waning moon would be seen as destructive. The full moon also bestows beneficial influence. Pliny, however, instructs that vetch and fodder crops be sown at the new moon.[37] As with Columella's full moon, the new moon is here seen as beneficial, though I must confess I am at a loss to explain why. The idea that the new moon is an increasing moon would seem tempting,[38] until one considers that on this logic the full moon would be a waning moon, which would therefore seem to be destructive to planting, contrary to what Columella believes. It is possible that some difference between the uses or growing patterns of beans versus vetch was thought to call for different planting times. We see something analagous in the modern *Farmer's Almanac*, where a distinction is made between above-ground and below-ground crops when maximizing the astrological influence of the moon for planting and harvesting times. Another possibility is that there were several schools of thought on the issue (or at least variant traditions) in antiquity, with different ideas about what influence a particular phase would have.

(b) Day of the moon: Vergil tells us that the seventeenth day of the moon is propitious for planting vines, the ninth lucky for fugitives but unlucky for thieves, and that the fifth day is unlucky for all work.[39] The passage is striking in two respects: first, the seventeenth day would be a waning moon, and second, the unluckiness of the fifth day has nothing to do with the phase of the moon at all: rather it was the lunar day on which, Vergil tells us, 'pale Orcus and the Furies were born',[40] and was therefore seen as unlucky. Vergil is unfortunately silent about what (if anything) happened on the seventeenth to make it fruitful for viniculture, or on the ninth to make it helpful for fugitives. Since the day of the moon is here being used as a way of keeping track of the 'anniversary' (or perhaps more accurately: 'mensiversary') of an event, one may be tempted to see in this a remnant of an ancient lunar calendar. And one would be correct to do so, but the calendar in question is Greek, not Roman: Pliny tells us that Vergil was following Democritus

---

[36] Columella, *RR* ii.10.10, xi.2.85.   [37] Pliny, *HN*, xviii.314.
[38] See Tavenner, 1918, p. 69.   [39] Vergil, *Georg.* i.276–86.   [40] Vergil, *Georg.* i.277–8.

in this dating, and similar passages can be found in Hesiod.[41] I note also that Vergil's admonition differs slightly from the Roman belief that certain calendrical dates (recurring only once a year) were unlucky, since Vergil is branding days of *every* lunar cycle as significant, not just of particular calendar months. Interestingly, however, like many of the ancient ominous calendrical days, a tradition of the occurrence of a particular event on a particular day colours that day in perpetuity for Vergil.[42]

The use of days of the moon to govern certain activities turns up also in Pliny where he reports that:

Varro in fabae utique satu hanc observationem custodiri praecepit. alii plena luna serendam, lentim vero a vicesima quinta ad tricesimam, viciam quoque iisdem lunae diebus: ita demum sine limacibus fore.

Varro has advised us to heed this observation (of the rising of the Pleiades) for sowing beans. Others say they are to be sown at full moon, and lentils between the twenty-fifth and thirtieth, and vetch on the same days of the moon, for this is the only way they shall be free of slugs.[43]

Notice the free transition from stellar phases to lunar phases here. The two are complementary.

(c) Lunar position relative to the horizon: Another kind of lunar influence, one that gets no mention in the modern literature so far as I can see, is found in Pliny and Columella. In the *Natural History*, Pliny tells us that we should 'prepare nurseries when the moon is above the earth' (*seminaria cum luna supra terram sit fieri*),[44] following which he goes to great lengths to explain the slightly tedious business of determining when the moon is above and below the earth *during all its phases*. Pliny demonstrates that the moon is not just 'above the earth' when it is visible, but also when it is invisible, such as at the new moon. During a new moon, the moon is above the earth all day, exactly when the sun is above the earth, and during a full moon it is above the earth from sunset to sunrise, that is, all night. He then gives us a linear interpolation rule for calculating when it will be above the horizon on any subsequent day or night: add ten-and-a-quarter twelfths of an hour to its rising and setting time each day. His detailed explanation shows that Pliny quite literally means 'when the moon is above the earth' rather than simply

---

[41] Pliny, *HN* xviii.321; Hesiod, *Op.* 765f.

[42] For a discussion of these calendrically ominous dates, see Grafton and Swerdlow, 1988. Babylonian parallels can be found in Langdon, 1935, and Egyptian ones in the *Calendar of Lucky and Unlucky Days* (on which, see below, ch. 6).

[43] Pliny, *HN* xviii.228.

[44] Pliny, *HN* xviii.322. Contrast my reading with that of Tavenner, 1918, p. 69.

'when the moon is shining'. Conversely, some operations (felling timber and treading wine must) are best carried out when the moon is below the earth (*cum luna sub terra*), which again does not mean at a new or waning moon, but rather quite literally, when the moon is below the horizon, whatever its phase. So also Columella tells us to plant and store garlic when the moon is below the earth, so that its flavour is not overly sharp and it will not give us bad breath.[45]

The idea that the moon's position above or below the earth matters in agriculture shows up also in the tenth-century agricultural miscellany known as the *Geoponica*, in a chapter attributed to Zoroaster and called ὅτι ἀναγκαῖόν ἐστιν εἰδέναι, πότε ἡ σελήνη γίνεται ὑπὲρ γῆν, πότε δὲ ὑπὸ γῆν, 'That it is necessary to know when the moon is above the earth, and when below the earth'. It is justified on the grounds that 'many of the farmer's tasks are necessarily carried out when the moon is above the earth, or when the moon is below the earth'.[46] Rather than interpolation rules, it simply lists the hours for each day of a thirty-day lunar month during which the moon is above or below the earth, beginning with the new moon.[47]

Thus there are three lunar influences affecting Roman agricultural activity: one, by far the most common, which related beneficence and maleficence to the phases of the moon generally; another, previously confused with the first kind, which believed that the position of the moon above or below the earth, regardless of its phase, would affect certain agricultural operations; and a third, which saw influences of lunar days as affecting farming activity. The position of the moon relative to the horizon could be easily calculated, as Pliny instructs, or looked up in a book, as in the *Geoponica*, provided one knew the lunar day. But the lunar day was also important in its own right, as we have seen. Phase was either simply observed or, when the moon was invisible, likely worked out from lunar days, as we see in a parapegma from Trier,[48] where instead of numbers for the days of the moon, we see illustrations of the changing phases of the moon (see Fig. Cat. 7 in the catalogue, below).

III.ii.a.   The archaic Roman calendar

I have argued that the days of the moon are astrological, and I now want to show that they are astrological *only*. Many treatments of this material see the

---

[45] *RR* xi.3.22.   [46] *Geoponica*, i.7.

[47] Compare also *Enūma Anu Enlil*, tablet xiv, tables A and B, in Al-Rawi and George, 1991.

[48] Currently in the Rheinisches Landesmuseum, Trier. See also Dölger, 1950.

days of the moon as fossilized remnants of an archaic Roman lunar calendar,[49] but as we shall see, we need to be very cautious about the supposition of an original purely lunar calendar. It probably never existed. Instead, the Roman calendar seems to have been luni-solar from its inception.

It is a fact that the earliest *attested* Roman calendar is luni-solar,[50] with a standard year of 355 days divided into twelve schematic months of twenty-eight, twenty-nine, or thirty-one days (roughly equal in length to a lunar month), and with complex rules of intercalation to keep it more-or-less in line with the solar year.[51] Most scholars see this calendar as having developed from an earlier calendar based on actual observation of the phases of the moon, with month lengths of twenty-nine or thirty days alternately, and Flammant argues that the change to the luni-solar calendar from the 'primitive' calendar was both deliberate and sudden.[52] The evidence for this observational lunar calendar centres on the facts that (1) the Kalends, Nones, and Ides of any given month can be seen to correspond roughly (but, I contend, *too* roughly) with the first appearance of the moon, the half moon, and the full moon respectively; (2) Macrobius (fifth century) and Varro (first century BC) report that the Kalends, Nones, and Ides were once feasts timed according to the new moon;[53] and (3) the 355-day 'standard' (i.e., non-intercalated) year is close to a true 354-day lunar year. A fourth and very important pillar of this position is the bald assumption that a lunar calendar is supposed to be somehow inherently more *primitive* that a solar one.

---

[49] See Tavenner, 1918; Rehm, 1941; Gjerstad, 1961. Brind'Amour, 1983, and Hannah, 2005, are less committed to the necessity of an archaic lunar calendar.

[50] The earliest Roman date cited in the ancient sources is a solar eclipse mentioned by Cicero which occurred on the Nones of June in *c.* 401 BC. As Samuel, 1972, pointed out, even if the Roman calendar *had* once been lunar, it certainly was no longer so by 401, since an eclipse can only happen at the conjunction of the sun and moon, i.e., at or before the Kalends rather than on the Nones of a given month. See also Brind'Amour, 1983, p. 227.

[51] For details, see Rüpke, 1995, p. 191f.; Brind'Amour, 1983; Michels, 1967; Samuel, 1972, ch. 5; Gjerstad, 1961.

[52] Flammant, 1984. See also Samuel, 1972, pp. 159, n. 2, 166; Michels, 1967, pp. 16, 119f.; Brind'Amour, 1983, p. 225f.; Nilsson, 1952; Pedroni, 1998; Hannah, 2005. Holleman, 1978, has argued (unconvincingly) for an original 306-day year. An alternate account has been advanced by V. Johnson, 1963, who sees the Roman calendar as derived from an original four-month (120-day) year which was based on the breeding cycles of pigs. The argument rests on facts such as that figs and beans are among the prominent items in the rites of the Caprotine Nones (in March), which products 'would seem more appropriate to pigs than goats', and that *Caprotinus*, the original name for March (!) may have meant 'boar' in primitive times. For the month lengths, see Gjerstad, 1961, p. 197; See also Macrobius, *Sat.* I.13.6–7; Censorinus, *De die nat.*, xx.2–5.

[53] Macrobius, *Sat.* I.15.9–12. Varro, *LL* vi.27.

Of these arguments, none is convincing. The idea that lunar calendars are more primitive and so should precede solar ones is founded on assumptions made explicit by Nilsson's *Primitive Time Reckoning*, cited by virtually everyone writing on the early Roman calendar. There Nilsson simply states, without any argument at all, that 'the moon is indeed the first chronometer' and that 'while the human mind arrives only gradually at the conception of the year, the month is already given by the natural phenomena'.[54] In an attempt to get around the question of what evidence there could possibly be for such a claim, Nilsson appeals to nineteenth-century anthropologists' reports of the 'primitive' temporal conceptions of indigenous South American tribespeople, for example, as indicative of 'primitive' early European calendrics. Apart from both cultures sharing an adjective in Nilsson's mind, however, there is no sense in which such 'comparative evidence' is genuinely comparative.

With respect to the other main argument, that concerning the relationship between the Kalends, the Nones, and the Ides on the one hand and the three corresponding lunar phases on the other, we find that the divergence between the calendar dates and the true lunar phenomena is as much as two days. And if we believe Macrobius, then even in the archaic calendar this strange discrepancy must have existed between the true phases and the idealized dates of the feasts, for on Macrobius' account, only the Kalends was truly observational, the other feasts were schematic:

priscis ergo temporibus, antequam fasti a Cn. Flavio scriba invitis patribus in omnium notitiam proderentur, pontifici minori haec provincia delegabatur ut novae lunae primum observaret aspectum visamque regi sacrificulo nuntiaret. itaque sacrificio a rege et minore pontifice celebrato idem pontifex calata, id est vocata, in Capitolium plebe iuxta curiam Calabram, quae casae Romuli proxima est, quot numero dies a kalendis ad nonas superessent pronuntiabat, et quintanas quidem dicto quinquies verbo καλῶ, septimanas repetito septies praedicabat . . . ideo autem minor pontifex numerum dierum qui ad nonas superesset calando prodebat, quod post novam lunam oportebat nonarum die populares qui in agris essent confluere in urbem, accepturos causas feriarum a rege sacrorum scituros que quid esset eo mense faciendum.

So in ancient times (before the *fasti* was produced by Gn. Flavio the clerk, and made publicly known, against the will of the nation) this duty was assigned to a *Pontifex Minor*: to observe the first (sign) of the new moon, and to note its appearance and

---

[54] See Nilsson, 1920, p. 148f. More recently, Lalonde has used this same method to offer a Marxist interpretation of 'primitive' time reckoning. Lalonde, 1996, p. 247, n. 35, dubiously appeals to 'la structure quasi universelle des "calendriers" primitifs et des calendriers des civilisations historiques'.

report it to the High Priest. And so, a sacrifice being performed by the High Priest, and the *Pontifex*, having called (that is: summoned) the people to the Capitol, beside the *curia* of Calabra, which is near the house of Romulus, the *Pontifex Minor* would declare how many days there were between the Kalends and the Nones, repeating the word καλῶ five times if (the Nones were on) the fifth (day), and seven times if (they were on) the seventh (day) . . . And so the *Pontifex Minor* made known the number of days which remained until the Nones, since after the new moon, on the Nones, the people from the country had to meet in the city to hear from the High Priest the reasons for the festivals and what sacrifices were to be performed that month.[55]

The combination of observation and schematic dating of feasts is disconcerting. Moreover, as Gjerstad noted, the title of *Pontifex Minor* is probably anachronistic here. But these anomalies are easily accounted for by the fact that Macrobius is writing more than 800 years after this supposed lunar calendar must have fallen out of use, and his information was corrupted.[56] Indeed, he has the new moon occurring on the Kalends, but in Varro's earlier version of this same story the new moon is supposed to have occurred on the *Nones*. If Varro is correct, we have other problems: How exactly was the new moon supposed to have been predicted as being either five or seven days after the Kalends? Why not six days? What was observed (if anything) on the Kalends to make this prediction? At least Macrobius has an observation that his pontifex is supposed to be working from, but Varro is silent on the issue. Perhaps sensing some problems with the account he is passing on, Varro hedges his bets by offering a non-lunar alternative to the new-moon (*novus mensis*) reading of the etymology of the word 'Nones': perhaps, he says the word is simply named because it is *nonus*, the 'ninth' day before the Ides. At least, from a modern etymological point of view, this latter alternative is not implausible, whereas *novus* → *Nonae* is at best far-fetched. But if both Macrobius and Varro are confused about what exactly is supposed to have been going on in what was for both of them, we must remember, the remotest past (and we can legitimately ask what kinds of evidence either of them could have had at his disposal), then neither account can be accepted on face value. We need to look around and see if there is any corroborating evidence for their accounts.

Is the 355-day year such corroborating evidence? I would argue not. The reason is that the 355-day year first shows up as the product of the lengths of the twelve schematic months of the luni-solar year, such that four months of

---

[55] Macrobius, *Sat.* 1.15.9–12.

[56] Macrobius' version is based on the same story as Varro's in *LL* vi.27. See also Hannah, 2005.

thirty-one days (not observationally lunar) plus seven of twenty-nine days plus one of twenty-eight days (again, not observationally lunar) equals 355 days. But this year length was seen as unsatisfactory in the earliest records we have of it, such that the Romans irregularly intercalated extra months in order to keep the calendar in line with the solar year, and this intercalation involved shortening February to twenty-three or twenty-four days (certainly not lunar) and adding an intercalary month of twenty-seven or twenty-eight days (yet again: not lunar).[57] Although the earliest known Roman calendar does seem to show some connection to the behaviour of the moon (in dividing the calendrical months on roughly – but importantly inexactly – the same order of magnitude as the lunar month) we cannot ignore the fact that it equally undeniably shows a very strong connection to the behaviour of the sun.

This being the case, we could just as easily argue that the archaic calendar, rather than being a degenerate lunar calendar, was a degenerate solar calendar. Or, what seems most likely based on the evidence available to us for the earliest known Roman calendar, that it was never purely either, but was *from its inception* luni-solar. The evidence of Varro and Macrobius does not weigh in favour of any of these possibilities. Their accounts are nothing more than the repetition of a just-so story based on the rough coincidence of the monthly feasts with idealized lunar phases, and ultimately can be reduced to the identical (and no more convincing) modern argument.

Since we have no really secure evidence for an archaic lunar calendar, the question arises of why the late Republican and early Imperial Romans were so concerned with the moon. The answer is that their interest was, as we have seen, largely astrological.

III.iii.   Stellar phases and the Roman calendar

Lunar phase instructions could apply during any lunar month. But many farming activities are only carried out in certain seasons, or at particular times of the solar year. The Roman agricultural writers give rules for keeping track of these seasons as well. We find two types of dating accompanying the instructions: one is the standard and familiar-looking calendar date, such as 'it is best to plough (sloping ground) between the first and thirteenth of

---

[57] Michels, 1967, has argued that the intercalary month was always twenty-seven days, but most scholars think it was sometimes twenty-seven and sometimes twenty-eight. For the ancient sources and her arguments on the question, see her Appendix 1.

September'.[58] The other is more clearly astronomical: 'between the vernal equinox and the rising of the Pleiades'.[59] These astronomical dates were used in several ways: first, to delimit the seasons generally, as in the season lists, and second, to indicate precisely a particular date in the solar year. As the Roman calendar evolved and established a better connection for itself with the solar year, the use of stellar phases gradually began to be replaced by, or at least intertwined with, Roman calendar dates. This is why we see post-Julian writers using calendar dates where the Greeks and earlier Romans would have used stellar phases as date indicators.

The relationship between the solar year and the Julian calendar is first made explicit in Varro, and we can see the change from stellar to calendar dates being made:

dies primus est veris in aquario, aestatis in tauro, autumni in leone, hiemis in scorpione. cum unius cuiusque horum IIII signorum dies tertius et vicesimus IIII temporum sit primus et efficiat ut ver dies habeat XCI, aestas XCIV, autumnus XCI, hiems XXCIX, quae redacta ad dies civiles nostros, qui nunc sunt, primi verni temporis ex a. d. VII id. Febr., aestivi ex a. d. VII id. Mai., autumnalis ex a. d. III. id. Sextil., hiberni ex a. d. IV id. Nov. . . .

The first day of spring (the sun) is in Aquarius, of summer in Taurus, of autumn in Leo, of winter in Scorpio. The twenty-third day of each of these four signs is the first day of (each of) the four seasons, and this makes it so that spring has ninety-one days, summer ninety-four, autumn ninety-one, and winter eighty-nine. Which (seasons), rendered in our current civil dates (sets) the first (day) of the season of spring on the VII Id. Feb., of summer on the VII Id. May, of autumn on the III Id. Sextilis, of winter on the IV Id. Nov. . . . .[60]

Before the Julian reform, such an attempt at dating would have been very approximate at best, since the seasons and the calendar months had a comparatively sloppy correlation. If we want to be more accurate, though, it is the stars that delimit the seasons of the year. Thus Varro continues:

eaque in partes VIII dividuntur: primum a favonio ad aequinoctium vernum dies XLV, hinc ad vergiliarum exortum dies XLIV, ab hoc ad solstitium dies XLIIX, inde ad caniculae signum dies XXVII, dein ad aequinoctium autumnale dies LXVII, exin ad vergiliarum occasum dies XXXII, ab hoc ad brumam dies LVII, inde ad favonium dies XLV.

(The year is) divided in eight parts: the first from the (beginning of the) west wind to the vernal equinox, forty-five days; from there to the rising of the Pleiades, forty-four days; from this to the solstice, forty-eight days; thence to the sign of the Dog Star,

---

[58] Columella, *RR* II.4.11.   [59] Varro, *Rust.* I.30.   [60] Varro, *Rust.* I.28.

twenty-seven days; from there to the autumn equinox, sixty-seven days; then to the setting of the Pleiades, thirty-two days; from which to the winter solstice fifty-seven days; then to the (beginning of the) west wind, forty-five days.[61]

Note that stellar phases, the solstices and equinoxes, as well as a meteorological phenomenon are all used here. This points to a close connection between certain annual weather patterns and stellar reckoning.

But the links between weather and the stars are the very bread and butter of astrometeorological parapegmata, as we find in Columella's agricultural treatise:

> *XVII cal. Feb. sol in Aquarium transit; Leo mane incipit occidere; Africus, interdum Auster cum pluvia.*
> *XVI cal. Feb. Cancer desinit occidere; hiemat.*

> XVII K. Feb., the sun enters Aquarius, Leo begins to set in the early morning, there is a south-west, or occasionally a south wind with rain.
> xvi K. Feb., Cancer finishes setting, wintry.[62]

Such day-by-day predictions are much more temporally specific than the seasonal indicators, insofar as they apply only to specific days rather than to a span of time. Some of the phases more directly governed agricultural practices; thus the best time for activities such as planting was, as Pliny tells us, *magnaque ex parte rationi siderum conexa*:

> *Vergilius triticum et far a vergiliarum occasu seri iubet, hordeum inter aequinoctium autumni et brumam, viciam vero et passiolos et lentem boote occidente; quo fit ut horum siderum aliorumque exortus et occasus digerendi sint in suos dies.*

Vergil says to plant wheat and spelt after the setting of the Pleiades, barley between the autumn equinox and the winter solstice; vetch, kidney beans and lentils at the setting of Boötes; thus it is that the rising and setting of these and other stars are to be set out according to their days.[63]

So also Ovid tells us that the crops are 'nourished' by the stars,[64] and to his own account, Pliny adds that

> *quidam omissa caelesti subtilitate temporibus definiunt: vere linum et avenam et papaver atque, uti nunc etiam transpadani servant, usque in quinquatrus, fabam, siliginem Novembre mense, far Septembri extremo usque in idus Octobres, alii post hunc diem in Kal. Novembres. ita his nulla naturae cura est, illis nimia, et ideo caeca*

---

[61] Varro, *Rust.* i.28.   [62] Columella, *RR*, xi.2.4.
[63] Pliny, *HN*, xviii.202.   [64] *Fasti* iv.913f.

subtilitas, cum res geratur inter rusticos litterarumque expertes, non modo siderum. et confitendum est caelo maxime constare ea, quippe Vergilio iubente praedisci ventos ante omnia ac siderum mores, neque aliter quam navigantibus servari.

Some, ignoring the exactness of the heavens, designate this calendrically: so flax, oats and poppy (are planted) before the festival of Minerva[65] (as they now do beyond the Po); beans and wheat in the month of November, spelt from the end of September to the Ides of October, (while) others (plant) after this date, until the Kalends of November. Thus these people show no concern for nature, where the previous people show too much,[66] and so their exactness is blind. So the matter is treated by farmers and literary men, not only those who know the stars. But it is acknowledged that it is mostly dependent on the heavens, as Vergil, to be sure, tells us to have foreknowledge of the winds from the habits of the stars above all, just as they are used for navigation.[67]

It was to keep track of these 'habits of the stars' that the astrometeorological and astronomical parapegmata were used, in the setting out of seasons generally, the prediction of day-to-day weather, and in the regulation of farming and navigational activity. This timing function looks on the face of it to be strictly astronomical, but in some sources we can see astrology lurking in the background. Even though he is sceptical of the exactness claimed by 'the Chaldaeans', Columella goes on to acknowledge that the stars do actually *affect* the weather.[68] Varro emphasizes the importance of the moon for everything from crops to haircuts. Neither Varro nor Columella explicitly tells us, though, what the specific causal mechanism is supposed to be between the heavens and the earth. If we look to other sources, though, the mechanism most frequently invoked is that of *sympathy*. Sympathy is usually seen in modern sources as a characteristic doctrine of Stoicism,[69] but it is in fact a widespread belief in antiquity, particularly in astrological sources. And it is always, as I have argued elsewhere, seen as *physical*.[70]

---

[65] Which began on the fifth day after the Ides of March.

[66] *ita his nulla naturae cura est, illis nimia* . . . I think Pliny says 'too much' here not because he thinks that those who use the stars are wrong, but because they do not pay enough attention, in his opinion, to the other causes that can affect the weather. So *HN* xviii: 'Even though all this depends on stationary stars, fixed in the sky, the motion of (other) stars intervene, affecting the hails and clouds to no small extent, as we have shown, and they disturb the expected arrangement.'

[67] Pliny, *HN* xviii.205.   [68] Columella, *RR* xi.1.32.

[69] And Sextus reports that the Stoics saw astrometeorology as evidence for sympathy (*Adv. Math.* viii.79).

[70] On stellar sympathy, see Lehoux, 2006a. See also Lehoux, 2003. The most detailed study of Stoic sympathy is still Reinhardt, 1926.

Conclusion

In a world where the stars and the earth are physically connected, where the qualities, positions, and motions of the stars and planets have specific causal effects on, in the first instance, the earth's atmosphere, it only makes sense that activities that are utterly dependent on rain, heat, and sun – and here I am thinking of farming in particular – should place special importance on the heavens. In the Roman contexts that we have looked at in this chapter we have seen a particular set of developments in agricultural theory and practice that combine the earlier Greek fixed-star astrometeorology with a new set of influences coming from the moon. The lunar influences are complex, with phase, position relative to the horizon, and lunar days all playing different but overlapping roles.

While Roman calendars (particularly after the Julian reform) could do what no Greek civil calendar could do – that is: to track the annual astrometeorological cycle from year to year – what they could not do was to track the lunar motions from month to month. The Romans accordingly adapted the Greek technology of the inscribed parapegma in order to track the temporal cycle marked by the moon, as well as the hebdomadal and nundinal cycles, and occasionally the calendrical cycle itself.

Parapegmata and related texts were used in Roman agriculture to regulate a wide variety of activities with the simultaneous and often overlapping aims of timing agricultural activities and of according the farmer's actions with the important influences of the stars and the moon. These influences, when fleshed out in ancient sources, are most commonly referred to in terms of sympathy, antipathy, and the four qualities. Like hot/cold/wet/dry, sympathy and antipathy are part of a causal chain that links the heavens and the earth, and seem to operate as physical causes. While the stars are seen to exert some power (*vis*, δύναμις) on terrestrial affairs, it is important to note that this causal connection still always underpins another relationship between the stars and the weather: that of *signification*. The stars are *signs* in this instance at least partly because they are causes. But as we shall see in the next chapter, causation was also only one way of justifying the stars as signs.

3 | *De signis*

1. What the predictive sign is and what it is not (or at least not necessarily)[1]

In the astrometeorological parapegmata, the stars function as *signs* for predicting coming weather patterns. But this simple statement hides a multitude of difficulties. We need to be clear in the first place on what we mean by 'sign', and in the second place on how these signs were observed and interpreted. It is taken for granted in almost all of the ancient and modern literature on astrometeorological parapegmata that the *observation* of the stars plays a major role in both the use and the development of parapegmata. But what exactly are the roles of stellar observation? We shall see that, in fact, there are two distinct kinds of observation adduced in two distinct contexts. On the one hand, we have the ancient texts (and some modern ones) claiming that parapegmata were developed through the observation and recording of correlations between the fixed stars and the weather. On the other hand, we find modern texts assuming that the use of parapegmata depended on the observation of the phases of the fixed stars and the subsequent consultation of a parapegma. The ancient is supposed to have looked to the sky to see what significant phases were occurring, and then consulted his parapegma to find the weather corresponding to that phase. We will see, though, that this misrepresents the use of parapegmata. The parapegmata, as predictive texts, instead became a locus of authority that *canonized* the timing and sequence of the stellar phases and weather. These texts were to a large extent normative, rather than descriptive. Nevertheless, the ancient authors themselves emphasized observation as one of the foundations of the parapegmatic tradition, and in this chapter we will explore how and why these claims were made, and to what extent they are justified.

We will see that, in making his predictions, the astronomer/astrologer (in spite of frequent rhetoric to the contrary) can be seen to be primarily working from texts and instruments, rather than from observations in the natural

[1] Much of the content of this chapter appeared in Lehoux, 2004a.

world. This means that the actual *sign* observed in making a prediction is no longer a stellar phenomenon. Instead, the stellar phenomenon functions as the sign-in-theory, but no longer in practice, of astrometeorological prediction. The sign-in-practice is now a text, a table, or an instrument. A parallel case can be made for astrology as a whole.

II. Two sorts of observational claim

Like the proponents of all the other major branches of ancient learning, *the astrologers* get their fair turn as the objects of Sextus Empiricus' Sceptical assault in the *Adversus mathematicos*. Sextus begins his attack by distinguishing between two types of astrologer. On the one side, there are the horoscopic astrologers who, Sextus tells us, work from an inherently uncertain account of stellar causation. They think the stars exert some kind of influence on earthly affairs, and on the births of individuals in particular. These horoscopic astrologers are the main focus of Sextus' criticism. On the other side, there is a group of astrologers that Sextus wants to leave out of his criticism, on the grounds of what he sees as their sound methodology. Instead of working from hypothetical accounts of stellar *causation* as the horoscopic astrologers do, Sextus tells us, these others work strictly from *observation*. They predict atmospheric changes based on the observation and recording of past correlations between events, with no causal substructure supposed. As Sextus says:

περὶ ἀστρολογίας ἢ μαθηματικῆς πρόκειται ζητῆσαι οὔτε τῆς τελείου ἐξ ἀριθμητικῆς καὶ γεωμετρίας συνεστώσης . . . οὔτε τῆς παρὰ τοῖς περὶ Εὔδοξον καὶ Ἵππαρχον καὶ τοὺς ὁμοίους προρρητικῆς δυνάμεως, ἣν δὴ καὶ ἀστρονομίαν τινὲς καλοῦσι (τήρησις γάρ ἐστιν ἐπὶ φαινομένοις ὡς γεωργία καὶ κυβερνητική, ἀφ' ἧς ἔστιν αὐχμούς τε καὶ ἐπομβρίας λοιμούς τε καὶ σεισμοὺς καὶ ἄλλας τοιουτώδεις τοῦ περιέχοντος μεταβολὰς προθεσπίζειν) . . .

It now lies before us to inquire concerning astrology, or the mathematical art, [by which I do] not mean the complete practice of arithmetic and geometry taken together . . . nor the predictive ability of the followers of Eudoxus, Hipparchus, and other such men, which is also called 'astronomy', *for this is the observation of phenomena*, as in agriculture and navigation, from which it is possible to foretell droughts and downpours, plagues and earthquakes, and other such atmospheric changes . . .[2]

---

[2] *Adv. math.* v.1–2, italics mine.

Since the correlations drawn by the astrometeorologist between stellar phases (the annual risings and settings of the fixed stars) and weather are *observational*, rather than theoretical, Sextus – known as *Empiricus*, after all – has no objection to them.[3]

Geminus, writing in the first century BC,[4] makes a similar observation claim on behalf of astrometeorology:

αἱ δὲ γινόμεναι προρρήσεις τῶν ἐπισημασιῶν ἐν τοῖς παραπήγμασιν οὐκ ἀπό τινων παραγγελμάτων ὡρισμένων γίνονται, οὐδὲ τέχνῃ τινὶ μεθοδεύονται κατηναγκασμένον ἔχουσαι τὸ ἀποτέλεσμα, ἀλλ᾽ ἐκ τοῦ ὡς ἐπίπαν γινομένου διὰ τῆς καθ᾽ ἡμέραν παρατηρήσεως τὸ σύμφωνον λαμβάνοντες εἰς τὰ παραπήγματα κατεχώρισαν.

The actual predictions of changes in the weather in the parapegmata do not happen because of some kind of regular rules, nor are they calculated by some craft [as if] the effects of the stars were constrained. Rather the harmony was perceived by daily observation of what generally happens, and written down in parapegmata.[5]

Ptolemy also tells us that the weather predictions in his *Phaseis* are derived from observation, and he even tells where each of his observers did their observing:

. . . καὶ τῶν ἀναγραψάντων αὐτῶν τὰς ἐπισημασίας ἄλλοι κατ᾽ ἄλλας χώρας τυγχάνουσι τετηρηκότες καὶ πολλαχῇ μηδ᾽ ὁμοίας καταστάσεσι περιπεπτωκότες ἤτοι . . .

. . . and (regarding) those (authors) who wrote down the changes in weather, different ones happened to observe in different places, and to be in altogether different climates . . .[6]

And later,

. . . Αἰγύπτιοι ἐτήρησαν παρ᾽ ἡμῖν, Δοσίθεος δ᾽ ἐν Κῷ, Φίλιππος ἐν Πελοποννήσῳ καὶ Λοκρίδι καὶ Φωκίδι, Κάλλιππος ἐν Ἑλλησπόντῳ, Μέτων καὶ Εὐκτήμων Ἀθήνησιν καὶ ταῖς Κυκλάσι καὶ Μακεδονίᾳ καὶ Θρᾴκῃ, Κόνων δὲ καὶ Μητρόδωρος ἐν Ἰταλίᾳ καὶ Σικελίᾳ, Εὔδοξος ἐν Ἀσίᾳ καὶ Σικελίᾳ καὶ Ἰταλίᾳ, Καῖσαρ ἐν Ἰταλίᾳ, Ἵππαρχος ἐν Βιθυνίᾳ, Δημόκριτος ἐν Μακεδονίᾳ καὶ Θρᾴκῃ.

. . . the Egyptians observed here; Dositheus in Cos; Philippus in the Peloponnesus, Locris, and Phocis; Callippus in the Hellespont; Meton and Euctemon in Athens, the Cyclades, Macedonia, and Thrace; Conon and Metrodorus in Italy and Sicily;

---

[3] On the roles of Sextus' medical empiricism and his contribution to the debates on ancient sign theory, see Allen, 2001.
[4] For the date, see Jones, 1999b.   [5] Geminus, XVII.182.6f.   [6] *Phas.* 11.19f.

Eudoxus in Asia, Sicily, and Italy; Caesar in Italy; Hipparchus in Bithynia; and Democritus in Macedonia and Thrace.[7]

In the modern literature on parapegmata we also find common and often casual acceptance of the centrality of observation in the parapegmatic tradition.[8] But we would do well here to distinguish two different senses of *observation* in this context. The observation claim as we find it in the ancient texts (and a few modern ones)[9] works as part of the epistemological *justification* of astrometeorology, by presenting a core of empirical data that the tradition is supposed to be based on. In this sense, observation is central to the original correlation between particular stellar phases and particular weather predictions: this stellar phase and this weather phenomenon were observed to coincide, at some historical point in time, by such-and-such an authority. This is how the ancients understand the attributions '*according to x*' that we find so commonly in parapegmata.

The second sense of observation is confined to the modern literature, and has to do with the actual *use* of parapegmata. Unlike Douglas Adams' famous package of toothpicks, ancient parapegmata do not come with instructions for use. But how they were used seems on the face of it simple and obvious. Modern authors generally suppose that an astronomer, or at least an astronomically aware observer, would go out on a particular night or morning and observe any stellar risings or settings of note. He or she would then turn to a parapegma where the observed stellar phase would be looked up and the weather prediction read off. For example, in arguing that the parapegmata use apparent rather than true phases, Bowen and Goldstein say:

... given that the practical value of a parapegma lies in its treating astronomical events at the horizon as signs or indicators and correlating them with meteorological changes (the *significata*), it would be odd to introduce theoretical and, hence, unobservable events as signs. Moreover, the literary tradition which lies behind and is the context of the invention of the parapegmata is limited to relating the weather to *visible* astral horizon-phenomena.[10]

---

[7] *Phas.* LXVI.23f.

[8] See, e.g., Hellman, 1917; Rehm, 1941; van der Waerden, 1949; van der Waerden, 1985; Hannah, 2002.

[9] Hellmann, 1917 and Leitz, 1995 both depend for much of their readings on the assumption of an original observational correlation between stellar phenomena and weather. See also van der Waerden, 1984a, b, c.

[10] Bowen and Goldstein, 1988, p. 54, italics theirs.

And other examples of explicit or implicit acceptance of the idea that the observation of stars is central to the use of parapegmata are common.[11] We need to keep this *practical* observation claim distinct from the *foundational* observation claim we find in the ancient literature, and we shall look at each of these two claims in turn. As we shall see, the practical observation claim has trouble sustaining itself when we turn to look at how parapegmata were used. This is because the model of prediction assumed by modern historians, in which daily astronomical observations were referenced to astrometeorological texts in order to derive day-to-day weather predictions, turns out to be impractical for parapegmata (although it does work well for texts like Hesiod). There are some interesting problems around the foundational claim as well, in that the foundational observational correlation of phase and weather is not possible without a prior schematization of the stellar phases.

ii.i.   Watering down the foundational observation claim

Let us begin by looking at the foundational observation claim. It turns out that, for the astrometeorological parapegmata it is, at a very basic level, technically impossible. One simply cannot observe the co-incidence (in the literal sense of that word) of the morning rising of Arcturus and a rainstorm. The rain precludes the possibility of making an astronomical observation that day. That being said, a watered-down version of the observational correlation can be maintained if we presume the astronomical cycle to have been at least partially canonized first. Once we have ordained a sequence of stellar phases for the year, with at least a rough idea of the date differences between them, then one rainy morning we can observe the weather and, consulting our scheme for the sequence of stellar phases, associate the weather with the stellar phase that we know from the text should be happening today. We see something of this kind happening in a handful of cases in the astrometeorological sources, such as Ovid:

instinerint nonae, missi tibi nubibus atris
signa dabunt imbres exoriente Lyra.

---

[11]   Hannah, 2001a; Evans, 1998, pp. 6–7, 190–1 (but contrast p. 201); Rehm, 'Parapegma', *RE*; Rehm, 1941. Hannah, 2001b, pp. 62, 74f. and Hannah, 2002 describe parapegmata as tools for ordering 'observations'. On the other hand, Wenskus, 1990, pp. 17 and 27, says she finds it implausible that the ancient farmer could have relied on practical observation of stellar phases on any kind of regular basis.

When it is the Nones [of January, i.e., 5th)], the rains sent to you from dark clouds will give the sign that the Lyre is rising.[12]

So, too, Pliny says that we can tell 'from storms that a star is completing (its phase)' at the equinoxes (*HN* II.108).

This watered-down foundational observation claim is complicated by the evidence from some parapegmata that weather could be associated with a stellar phase a few days before or after. See, for example, Geminus:

πολλάκις δὲ μεθ' ἡμέρας τρεῖς ἢ τέσσαρας ἐπεσήμηνε τῇ ἐπιτολῇ ἢ τῇ δύσει τοῦ ἄστρου, ἔστι δ' ὅτε προέλαβε τὴν ἐπισημασίαν πρὸ ἡμερῶν τεσσάρων.

Often (the parapegmatist) has marked a change in the weather[13] with the rising or setting of a star three or four days too late, and sometimes he has anticipated the change by four days.[14]

In the Aëtius parapegma[15] there is a similar flexibility with regard to the temporal sequence of weather and phase, and Columella tells us that 'the force (*vis*) of a star is sometimes before, sometimes after, and sometimes on the actual day of its rising or setting.'[16] While these passages would admittedly let us make some actual observations of the delayed coincidence of a stellar phase and a weather phenomenon, they are notably not the usual association made in the parapegmata, and I think serve as the exception that proves the rule.

But such examples alert us to one other possibility for the correlation between weather and stellar phases, what we might call *observational interpolation*: a Euctemon or a Callippus may miss the exact day of the morning rising of Arcturus, but when the sky finally clears the next day or the day after, they can see that Arcturus is then too high in the sky to just be rising for the first time that day. So the actual date of the rising of Arcturus could then be interpolated back to a day or two earlier, and a correlation canonized. But again, this is a watered-down version of the kinds of strong observational claims we see in Sextus and Geminus.

Thus a strong foundational observation claim is, strictly speaking, not tenable. Nevertheless, we can argue (I think plausibly) for a watered-down version of such a claim, but this can only work if we presume (a) a prior

---

[12] *Fast.* I.315. Compare also IV.901f.

[13] ἐπισημαίνει is here being used in an extended sense meaning 'to mark an ἐπισημασία'. The grammar of this sentence is odd, but the sense is clear. See Lehoux, 2004c.

[14] Geminus, 188.18f.

[15] Published in Wachsmuth, 1897, pp. 295–9. See also A.xv in the Catalogue, below.

[16] *RR* XI.1.32.

schematization of the annual sequence of phases for a given latitude, and/or (b) an interpolation of phases from different observed positions of stars relative to the horizon several days apart. That (a), (b), or both – and I suspect it is both – must hold can also be seen by the consideration that in no single year will an observer be lucky enough to get a string of uninterrupted observations of phases. Weather must intervene from time to time, forcing the events unobservable this year to be interpolated, or inserted from a different year's observations. But we should keep in mind that by thus watering down the foundational observation claim, we are simultaneously – and to the same extent as we watered down the claim – moving away from any strict definition of the word *observational*.

Ptolemy himself seems to recognize something of the sort in his discussion of the calculation of stellar phases for each of the different latitudes. In both the *Phaseis* and the *Almagest*,[17] he admits that his values for the order and timing of the stellar phases for each latitude are based on calculation. When he lays out the method for this calculation in the *Almagest*, he argues that, although in a perfect world he *would* make or collect observations of each star at each latitude, in practice he can do little better than to make observations from one latitude and trust that the results are in fact generalizable. He next gives the geometrical construction and method by which such generalizations could, in theory, be calculated. But then he admits that even this method is still too cumbersome to be practical, and that he will be satisfied to use the records of his predecessors, and/or a celestial sphere to compute the phases for the different latitudes.

II.ii.   Problems with the practical claim

Now to the practical claim. Let us look again at the structure of a parapegma. In every parapegma astronomical, astrometeorological, or astrological information is *indexed* to some date-marking function. In an inscriptional parapegma, this date marker is the peg itself, which by its very presence acts as a temporal 'you are here' marker for the astrometeorological or other cycle tracked by the parapegma. In a literary parapegma, the cycle is indexed to a calendar. So, depending on the type of parapegma, by either (a) knowing the date, or (b) glancing at the entry beside the peg, the astronomer is able to look up the current astrometeorological situation in a parapegma. To see this, look again at the purely astronomical inscriptional parapegma Miletus I:

---

[17] Ptolemy, *Phas.* p. 3–4; *Alm.* VIII.6.

- The sun is in Aquarius
- [. . . . .] begins setting in the morning and Lyra sets.
  • •
- Cygnus begins to set acronychally.
  • • • • • • • • •
- Andromeda begins rising in the morning.
  • •
- Aquarius is in the middle of rising.
- Pegasus begins to rise in the morning.
  •
- The whole of Centaurus sets in the morning.
- The whole of Hydra sets in the morning.
- Cetus begins to set acronychally.
- Sagitta sets, the season of the west wind accompanying.
  • • • •
- The whole of Cygnus sets acronychally.

If the practical observation claim were correct, we should see the astronomer observing a stellar phase and then consulting this parapegma to – do what? Find the peg? Obviously not. Move the peg? No. The peg was simply moved from one hole to the next each and every day. Instead it is the *peg* that was looked for – we should properly say *observed* here – by the user of the parapegma, and the astronomical situation was then read from beside it. Quite the contrary to the practical observational claim, what the parapegma does is to obviate the need for astronomical observation. What is being observed in practice is instead a peg.

So also in literary parapegmata, the calendar functions in place of the peg. Knowing the date, the user looks up the astrometeorological situation. The very organization of the parapegma, with the stellar phases and weather indexed to the calendar, shows that it was via the calendar that the astrometeorological situation was referenced. To see this, let us imagine a user making an observation of, say, the morning rising of Arcturus (let's assume from a Clima of $14\frac{1}{2}$ hours), and then trying to find that phase in a parapegma like Ptolemy's. We can see just how difficult it would be to find anything. Look at the following eight-day excerpt from the month of Thoth in Ptolemy's *Phaseis*:

κγ΄. ὡρῶν ιδ΄ ∟΄ ὁ καλούμενος Αἲξ ἑσπέριος ἀνατέλλει. ὡρῶν ιε΄ ∟΄ Ἀρκτοῦρος ἑῷος ἀνατέλλει. Αἰγυπτίοις ψακὰς καὶ ἄνεμος, ἐπισημαίνει. Καλλίππῳ καὶ Μητροδώρῳ ὑετία.

κδ. ὡρῶν ιγ´ ∟´ ὁ κοινὸς Ἵππου καὶ Ἀνδρομέδας ἑῷος δύνει.

κε. ὡρῶν ιγ´ ∟´ ὁ λαμπρὸς τῆς νοτίου Χηλῆς κρύπτεται. ὡρῶν ιε´ ὁ λαμπρὸς τοῦ Ὄρνιθος ἑῷος δύνει. Αἰγυπτίοις ζέφυρος ἢ νότος καὶ δι᾽ ἡμέρας
ὄμβρος.

κϛ´. ὡρῶν ιε´ Ἀρκτοῦρος ἑῷος ἀνατέλλει. Εὐδόξῳ ὑετός. Ἱππάρχῳ ζέφυρος ἢ νότος.

κζ´. ὡρῶν ιδ´ ὁ κοινὸς Ἵππου καὶ Ἀνδρομέδας ἑῷος δύνει, καὶ ὁ ἔσχατος τοῦ Ποταμοῦ
ἑῷος δύνει.

κη´. μετοπωρινὴ ἰσημερία. Αἰγυπτίοις καὶ Εὐδόξῳ ἐπισημαίνει.

κθ´. ὡρῶν ιδ´ ὁ καλούμενος Ἀντάρης κρύπτεται. ὡρῶν ιδ´ ∟´ Ἀρκτοῦρος
ἑῷος ἀνατέλλει. Εὐκτήμονι ἐπισημαίνει. Δημοκρίτῳ ὑετὸς καὶ ἀνέμων ἀταξία.

λ´. ὡρῶν ιδ´∟´ ὁ κοινὸς Ἵππου καὶ Ἀνδρομέδας ἑῷος δύνει. Εὐκτήμονι καὶ Φιλίππῳ
καὶ Κόνωνι ἐπισημαίνει

23. $14\frac{1}{2}$ hours: the star called Capella rises in the evening. $15\frac{1}{2}$ hours:
Arcturus rises in the morning. According to the Egyptians drizzle and
wind, there is a change in the weather. According to Callippus and
Metrodorus, rainy.

24. $13\frac{1}{2}$ hours: the star shared by Pegasus and Andromeda sets in the morning.

25. $13\frac{1}{2}$ hours: the bright star in the southern claw disappears. 15 hours:
the bright star in Cygnus sets in the morning. According to the Egyptians west wind or south wind, and thunder storms throughout the
day.

26. 15 hours: Arcturus rises in the morning. According to Eudoxus rain.
According to Hipparchus west wind or south wind.

27. 14 hours: the star shared by Pegasus and Andromeda sets in the morning, and the rearmost star of Eridanus sets in the morning.

28. Autumnal equinox. According to the Egyptians and Eudoxus there is a
change in the weather.

29. 14 hours: the star called Antares disappears. $14\frac{1}{2}$ hours: Arcturus rises in
the morning. According to Euctemon there is a change in the weather.
According to Democritus rain and unsettled winds.

30. $14\frac{1}{2}$ hours: the star shared by Pegasus and Andromeda sets in the morning. According to Euctemon, Philippus, and Conon there is a change in
the weather.

It quickly becomes apparent just how impractical the practical observation
claim is.

Another important clue that it is dates and not astronomical phenomena
that are 'observed' by the user can be seen at, for example, Thoth 7. In its

entirety, it reads: 'According to Metrodorus bad air. According to Callippus, Euctemon, and Philippus bad air and unsettled air. According to Eudoxus rain; thunder; variable winds.' We see that on this day we can expect one or a combination of (a) bad air, (b) unsettled air, (c) rain, (d) thunder, and (e) variable winds. But these are all indexed to the date *only*. There is no astronomical phenomena that they are tied to at all, so any observation we may or may not have made that day is irrelevant to actually finding this entry. And by the time of the *Polemius Silvius Fasti*,[18] (fifth century AD) the stellar phases have dropped out of the parapegma entirely and *all* the meteorological entries are indexed to dates alone.

What the parapegma does is to canonize a temporal cycle (astronomical, astrometeorological, astrological) in its entirety, one event after another, and then to provide a handy means (the peg or the date) of locating ourselves in that cycle. It tells us where we are in the year, for example, and it associates the different temporal locations in that cycle with both stellar phases and weather.[19] But the associations of stellar phases with dates, once thus canonized, are no longer associations referenced by stellar observation, but are instead *normative* statements that particular events, stellar and/or meteorological, happen on particular days or in a particular order, and the instrument itself now serves as the tool for locating ourselves in that cycle.

Now, saying that the day-to-day use of parapegmata does not depend on observation is *not* to say that astronomical observation goes out the window entirely. On the contrary, it does still have some roles to play. For example: observation can confirm or check the content of a parapegma, and observation is importantly used to calibrate parapegmata from time to time (as we saw in chapter 2), and observation can serve (as it does in Hesiod and Aratus, for example) when there is no parapegma around, but observation is basically superfluous in the day-to-day use of a parapegma.

---

[18] Published in Degrassi, 1963, vol. XIII.2, pp. 263–76. Text and translation in part II of this book.

[19] For the most part, these cycles are not civic or religious. The use of civil or other calendars only occurs in place of the peg in literary parapegmata as a handy way of locating the current day. Most astrometeorological parapegmata do not mention religious or civil cycles at all. The one obvious exception to this is P. Hibeh 27 (published in Grenfell and Hunt, 1906, pp. 138–57), which correlates both astrometeorology and Egyptian religious festivals with Egyptian civil dates. On the other hand, Roman inscriptional parapegmata do often include hebdomadal and nundinal cycles (the Roman seven- and eight-day 'weeks', on which, see, e.g., Michels, 1967), and some also track civil calendrical cycles as well (see the Catalogue, below). On the relations between Roman civil and religious cycles generally, see Salzman, 1990.

III.   How signs are observable

Look back now at the passage from Bowen and Goldstein that introduced us to the practical observation claim initially:

. . . it would be odd to introduce theoretical and, hence, unobservable events as signs.[20]

It turns out that their central claim holds after all, although we find that we must redirect their conclusion. It *is* true that unobservable (in the sense of imperceptible) events cannot function as predictive signs, since a sign that is imperceptible-in-principle could offer no way of feeding itself in (*qua* sign) to a predictive calculus.[21] There is simply nothing to draw a conclusion from, if nothing has been perceived. This is an important point, and Bowen and Goldstein hit it square on the head. An invisible sign is no sign at all.

But we cannot conclude from this that the signs used to draw predicted conclusions in parapegmata must have been observed *stellar phases* specifically. We have already seen that predictions are arrived at from parapegmata by observing either the peg or the date, not by observing the stars. And since the peg or the date is what is observed in making the astrometeorological prediction, then it is the peg or the date that, properly speaking, functions as the *sign* in the predictive calculus. And that sign is, after all, observable.

IV.   The move from practical to theoretical sign in astrology

I remarked earlier that the practical observation claim does still seem to hold up when dealing with texts such as Hesiod, where the astronomical phenomena associated with the weather are not indexed to a day or date marker. From both the structure of the poem and the paucity of astronomical phenomena to watch for, it seems that to follow Hesiod's advice is simply to remember a few rules of thumb, and to call them to mind when one knows (through observation or otherwise) that a phase is occurring. But by the time we start to see full-blown parapegmata, something has changed, in that the users no longer work primarily from astronomical observation. The instrument itself, by canonizing the entirety of a very detailed cycle, quietly moves us away from the observational to the instrumental. I say quietly here just because the working of the parapegma is always understood by its users

---

[20] Bowen and Goldstein, 1988, p. 54.
[21] On predictive signs, see Lehoux, 2004d. See also Bobzien, 1998; Allen, 2001; Barnow, 2002.

as relating *actually occurring* stellar phases with weather. The stellar phase does still function as a kind of sign, but now only a sign-in-theory. The fact that the user no longer needs to make observations of those actually occurring phases goes unremarked by the ancient authors. And this shift is not unique in the parapegmatic tradition. Other examples are easily found. Take for example Greek horoscopic astrology, where the stars are seen as conditioning the character of an individual by their positions at the time of her birth. To predict a significant event in a person's life, the astrologer looks at the configuration of the sky at that person's moment of birth, and then furnishes predictions based on that configuration. Just as with astrometeorology, the ancients assume that the configuration they are using to make their predictions is an *actual* configuration. But it is not. It is rather a *calculated* – retrodicted, to be specific – configuration.[22] What is observed by the astrologer is not the stars, nor even old observational reports of the stars, but is instead a set of tables of one sort or another which then determines for the astrologer what the positions of the planets had been at a particular instant in the past.[23]

It may be worth noting that this had not always and everywhere been the case. In the very beginning of the astrological traditions, in the Mesopotamian texts usually referred to as 'astral omens' to distinguish them from horoscopic astrology, the astronomical phenomena were not yet predictable. Look at the following example from tablet 59 of the second-millennium-BC Mesopotamian astral omen collection *Enūma Anu Enlil*:[24]

šumma Dilbat ina Du'ūzi ippuḫ-ma māšu ana pānī-ša izzizū šar akkadî iḫalliq.

If Venus rises in the month of Tammuz and Gemini stands in front of it, the king of Akkad will die.[25]

Here we see an observed sign (Gemini in front of a rising Venus in the month of Tammuz) correlated (possibly via a claim of historical precedence) with a prediction.[26] We presume some trained observer looking at the sky with

---

[22] See Jones, forthcoming.

[23] For a description of the types of table used in the Greek astrological tradition, see Jones, 1999c; Jones, 1999a, pp. 113–9, 175–7, 231; Jones, forthcoming.

[24] For the relationship between Mesopotamian astral omens, astrology, and astronomy on the one hand and Greco-Egyptian astrology and astronomy on the other, see Barton, 1994; Evans, 1998; Neugebauer, 1975; Jones, 1999a, pp. 15–34.

[25] *EAE* 59–60.IV.2, text in Reiner, 1998, p. 118. Transliteration and translation mine.

[26] On precedence in the omen tradition, see Lehoux, 2002. The question of whether observations were systematically collected in Mesopotamia for the development or improvement of the omen tradition tends to be centred on the role of the so-called 'Astronomical Diaries'. See chapter 5, below. See also Sachs and Hunger, 1988; Swerdlow, 1998; Slotsky, 1997; Hunger and Pingree, 1999, pp. 139–40.

an eye open for signs of this kind. Early one morning, our observer sees the signature conjunction of Venus and Gemini at Venus' rising and, knowing his or her way around *Enūma Anu Enlil*, looks the observation up in the text to see what it portends. Schematically, we have the following situation:

(1)   Observed Sign — (rule) → Prediction
       Observed conjunction — (omen text) → Death of the King

But at some point around the fifth century BC (give or take, depending on the phenomenon and method of prediction in question) the sign in the protasis of the astrological omen – and this is generally true only of the astrological omen[27] – became itself the subject of a second-order prediction. Mesopotamian astronomers, and their later Greek counterparts, were now able to predict the conjunction of the rising Venus with Gemini.[28] This adds another layer of complexity to the astrological prediction. We now have a two-step predictive process. In the first step, the astrologer is predicting what used to be the protasis of the omen: the conjunction of Venus and Gemini, and in the second step is then shifting the results of that prediction back into the protasis of an omen to furnish a final apodosis: the death of the king.[29] But how do we get the first of these two predictions, that of the conjunction of the two heavenly bodies? As with astrometeorology, it is through the consultation of texts of one sort or another.[30] Comparing this with the

---

[27] To be sure, there are some other kinds of omen apodoses that themselves serve as protases of other omens, for example some liver omens predicted eclipses (e.g., Manzāzu tablet 3.26 in Koch-Westenholz, 2000), which were in turn ominous events that portended doom for kings and such. And while I insist that such examples are not trivial, it is only with astral omens that a significant number of protases become predictable, and it is only in astrology that this predictability is mathematical.

[28] See Koch-Westenholz, 1995, pp. 51–2. I am deliberately sidestepping the controversial question of the relationship between Mesopotamian astronomy and Mesopotamian astrology. We know surprisingly little for certain about whether or how the well-attested mathematical astronomical methods were used by astrologers and diviners. For a sense of the current state of the question, see Rochberg, 1999; 2004.

[29] Modern scholars generally believe that Mesopotamian astral omens still relied on the actual observation of a predicted eclipse for it to have ominous significance, and that unseen eclipses were not ominous. Nonetheless, there are a few letters and reports (e.g., Parpola, 1993, no. 114; Beaulieu and Britton, 1994) which show that precautionary measures were taken (i.e., the appropriate namburbi ritual was performed, on which, see Caplice, 1974) even when the predicted ominous event was not seen locally. On calculated vs. observed eclipses, see Sachs and Hunger, 1988; Koch-Westenholz, 1995, pp. 51–2; Hunger and Pingree, 1999, pp. 154–6.

[30] For my purposes here, anything from the relatively simple goal-year texts to the complex mathematical ephemerides count as second-order predictive texts (for a good description and samples of the different types of Mesopotamian astronomical text see Hunger, 1999, and Evans, 1998, p. 312f.).

astrometeorological texts we have been looking at in this chapter, we see that the old Greek astrometeorological situation (e.g., Hesiod) is structurally identical with the first-order predictions of the old Mesopotamian astral omens:

(1)   Observed Sign — (rule) → Prediction
       Observed rising of the Pleiades — (Hesiod's rule of thumb) → Good time to harvest

But when the signs themselves become predictable, as in both parapegmata and horoscopic astrology, we see the now-predicted sign assume a new place in this scheme, and a new observed sign take over the initial position:

(2)   Observed Sign → Predicted Sign — (rule) → Prediction
       Observed Table → Predicted Eclipse — (omen text) → Death of King

Or for parapegmata:

(2a)   Observed Sign → Predicted Sign — (foundational 'observational' correlation) → Predicted Weather
       Peg (schematically situated) → Stellar Phase — (according to x) → Predicted Weather, or
       Date (legislated) → Stellar Phase — (according to x) → Predicted Weather

where the Stellar Phase is now a sign only *in theory* – but not in actual practice – of the weather predicted.

How significant a change is this for the practitioners? We might expect that the shift from random to predictable signs would have wide-ranging conceptual ramifications for the cosmologies of the astrologers. And some scholars have made just this claim. Speaking of the impact in Mesopotamia, Koch-Westenholz, for example, has said that 'we have here what may well be the earliest documented instance of a scientific revolution',[31] and she thinks the change in question had the cosmological implication that 'celestial phenomena could no longer be regarded as willed communications from the gods, and the old idea, that "signs" in heaven correlate with events on earth, was abandoned'.[32] There is, however, no real historical evidence for such a claim. The Mesopotamian sources make no comment on this supposedly major cosmological shift, this 'scientific revolution'. This in itself would perhaps not be surprising to those familiar with Mesopotamian astronomical

---

[31] Koch-Westenholz, 1995, p. 52.   [32] Koch-Westenholz, 1995, p. 51.

and divinatory texts, who know how notoriously sparse in cosmological, religious, philosophical, and epistemological commentary these texts are. But in the parallel case of the Greek astrometeorological tradition, which underwent a structurally identical shift, we see not only no contemporary comment on the cosmological significance of the shift, but – what is worse – we find that after the shift the emphasis on observation by authors like Ptolemy and Sextus shows that the fact of the change itself was suppressed. It is not just that they did not remark that a change had taken place (absence of evidence is not evidence of absence), but that they implicitly deny it.[33]

We see, then, how the rhetoric and the theory of ancient astrology are distinguished from its practice after the signs themselves become predictable.[34] Although the theoretical signs associated with predictions are the astronomical phenomena, the practical signs – the things actually looked at by the astrologer in working out his predictions – turn out to be texts, tables, and instruments: pegs, not stars.

---

[33] Whether the insistence on observation was intentionally misleading (perhaps to give astrology a more empirical authority?), or whether it was simply what I have elsewhere called 'sloppy empiricism', I leave as an open question.

[34] As Serafina Cuomo pointed out to me, in some instances it may be tempting to see the split between rhetoric and practice as at least partly attributable to different levels of expertise among practitioners. An astronomically sophisticated astrologer like Ptolemy will have a different (and probably more nuanced) idea about how observation works in astrology than a run-of-the-mill market-place astrologer would. The evidence I have been adducing in this chapter comes mostly from the more theoretically sophisticated practitioners, largely because the simpler practitioners left little more than scraps of horoscope fragments (see Neugebauer and van Hoesen, 1959; Jones, 1999a) or equipment (like astrological boards: see Evans, 2004) with no theoretical commentary.

There is a second level of expertise in the sources I have been using, and that is philosophical. Here we see the continuum run from Sextus at one end (philosophically sophisticated, astronomically (probably) naïve) to Ptolemy (astronomically sophisticated, philosophically less so, but see Taub, 1993) with Geminus – straining my continuum metaphor a little – as fairly conversant with general astronomy, but not particularly philosophically showy. And all three of these sources, interestingly, make essentially identical foundational observation claims.

4 | When are thirty days not a month?

Mason for a while had presum'd it but a matter of confusing Dates, which are Names, with Days, which are real Things.

Thomas Pynchon

1. There are calendars, and then there are calendars

Dutch people take birthdays very seriously. I know of no other culture where you are likely to be congratulated because it is your sister's birthday, for example. In order to keep track of the birthdays of friends and family (and of friends' families) the Dutch employ a marvellous tool called a *verjaardagskalender* (birthday calendar). Every Dutch household I have ever been in has one hanging on the wall. It is a very simple perpetual calendar for keeping track of birthdays. The layout is generally by season, with three calendar months per page (see Fig. 4.1). Each month is listed with its twenty-nine, thirty, or thirty-one days and beside each date is a blank space. There are no days of the week, or anything else, listed. The idea is to write down the names of your loved ones beside their birth date, and then to hang the calendar in a visible place (washrooms are popular) in perpetuity. The verjaardagskalender is usable year in and year out, since it indexes birthdays to calendar dates in a most general sense, but not for any *particular* year. It is not really usable as a calendar in our day-to-day sense of that word since the smaller cycles (weeks) by which we regulate our day-to-day activities are missing. Moveable holidays like Passover or Easter are absent, and 29 February is always there, so one needs to know when to ignore it. What this means is that one must use the verjaardagskalender *in conjunction with* a regular annual wall calendar. Thus, for example, I hang my 2004 wall calendar in the kitchen where I write down meetings and appointments and so on, and I periodically consult my verjaardagskalender to see if there are any birthdays I need to remember coming up. The verjaardagskalender is thus a *calendar* only in a loose sense. It is more like a schedule or index of one kind of annually recurring event, rather than being a tool by which to

Januari	Februari	Maart
1 _____	1 _____	1 _____
2 _____	2 _____	2 _____
3 _____	3 _____	3 _____
4 _____	4 _____	4 _____
5 _____	5 _____	5 _____
6 _____	6 _____	6 _____
7 _____	7 _____	7 _____
8 _____	8 _____	8 _____
9 _____	9 _____	9 _____
10 _____	10 _____	10 _____
11 _____	11 _____	11 _____
12 _____	12 _____	12 _____
13 _____	13 _____	13 _____
14 _____	14 _____	14 _____
15 _____	15 _____	15 _____
16 _____	16 _____	16 _____
17 _____	17 _____	17 _____
18 _____	18 _____	18 _____
19 _____	19 _____	19 _____
20 _____	20 _____	20 _____
21 _____	21 _____	21 _____
22 _____	22 _____	22 _____
23 _____	23 _____	23 _____
24 _____	24 _____	24 _____
25 _____	25 _____	25 _____
26 _____	26 _____	26 _____
27 _____	27 _____	27 _____
28 _____	28 _____	28 _____
29 _____	29 _____	29 _____
30 _____		30 _____
31 _____		31 _____

Fig. 4.1 Page from a Dutch verjaardagskalender.

regulate all the cycles of one's life. The calendar dates it contains are simply used to *index* the birthdays.

This point highlights the fact that we use the word *calendar* in several distinct senses. I have my local theatre's calendar that lists the plays for the season. There is a school calendar that lists important events in the life of the university for this year. We might speak (if we don't have small children) of our busy social calendars. These are all calendars, but in a slightly extended sense. We might call them *schedules* to be a little more precise. To see the distinction I am driving at here, contrast the sense of calendar in *school calendar* with the sense of calendar in a statement about the length of the year in the *Gregorian calendar*. When we talk about the Gregorian or Julian, Muslim, or Hebrew calendars we are talking about what we might call 'calendar systems', that is: systems for dividing up the year (or a series of years) for the purposes of chronology and dating. These usually incorporate smaller cycles like months and weeks into them as well. A third use of the word calendar comes in when I talk about the actual physical calendar I hang on my wall. Here *calendar* refers to an object at the conjunction of the calendar system and the schedule. I use my wall calendar to locate dates in the grand cycle of a particular calendar system (for me the Gregorian), and to organize events in my life in terms of that calendar system and my current temporal place in it. I may even copy birthdays from my verjaardagskalender, or important concerts from my opera calendar, into my wall calendar so that I know where they stand in relation to me on a week-to-week basis.

Turning back to parapegmata, we will need to try and determine which of these senses, if any, might be appropriate for describing the way a parapegma works. The answer, it turns out, is not inconsequential. Parapegmata have been frequently called calendars,[1] and the way in which we understand the sense of calendar here has some rather important implications. If parapegmata are calendars in the wall calendar sense, then the dates by which they index their astrometeorological and other phenomena can be taken to be dates in a *calendar system* as well. To take an example: the Geminus parapegma indexes weather and stellar phases to the motion of the sun through the zodiac, as follows:

Καρκίνον διαπορεύεται ὁ ἥλιος ἐν ἡμέραις λα΄.

α΄ ἡμέρα Καλλίππῳ Καρκίνος ἄρχεται ἀνατέλλειν· τροπαὶ θεριναί· καὶ ἐπισημαίνει.

θ΄ ἡμέρα Εὐδόξῳ νότος πνεῖ.

ια΄ ἡμέρα Εὐδόξῳ Ὡρίων ἑῷος ὅλος ἐπιτέλλει.

ιγ΄ ἡμέρα Εὐκτήμονι Ὡρίων ὅλος ἐπιτέλλει.

ις΄ Δοσιθέῳ Στέφανος ἑῷος ἄρχεται δύνειν.

κγ΄ Δοσιθέῳ ἐν Αἰγύπτῳ Κύων ἐκφανὴς γίνεται.

κε΄ Μέτωνι Κύων ἐπιτέλλει ἑῷος.

κζ΄ Εὐκτήμονι Κύων ἐπιτέλλει. Εὐδόξῳ Κύων ἑῷος ἐπιτέλλει· καὶ τὰς ἑπομένας ἡμέρας νε΄ ἐτησίαι πνέουσιν· αἱ δὲ πέντε αἱ πρῶται πρόδρομοι καλοῦνται. Καλλίππῳ Καρκίνος <λήγει> ἀνατέλλων πνευματώδης.

κη΄ Εὐκτήμονι Ἀετὸς ἑῷος δύνει· χειμὼν κατὰ θάλασσαν ἐπιγίνεται.

λ΄ Καλλίππῳ Λέων ἄρχεται ἀνατέλλειν· νότος πνεῖ· καὶ Κύων ἀνατέλλων φανερὸς γίνεται.

λα΄ Εὐδόξῳ νότος πνεῖ.

---

[1] They are called 'star calendars' (Evans, 1999), 'astronomical calendars' (van der Waerden, 1960; 1984a; 1984b; etc.), 'Steckkalender' (Rüpke, 2000b; Wagner-Roser, 1987; Binsfeld, 1973; Urner-Astholz, 1960; Brückner, 1931; Goessler, 1928), 'Wetterkalender', 'Steinkalender' (Rehm, 1941), or just 'Kalender' (Sadurska, 1979; Rehm, 'Parapegma', *RE*, col. 1295). See also Hannah, 2005.

The sun traverses **Cancer** in thirty-one days.

The 1st day: According to Callippus Cancer begins to rise, summer solstice, and
there is a change in the weather.

9th day: According to Eudoxus the south wind blows.

11th day: According to Eudoxus the whole of Orion rises in the morning.

13th day: According to Euctemon the whole of Orion rises.

16th: According to Dositheus Corona begins to set in the morning.

23rd: According to Dositheus Sirius is visible in Egypt.

25th: According to Meton Sirius rises in the morning.

27th: According to Euctemon Sirius rises. According to Eudoxus Sirius rises in the
morning, and for the next fifty-five days the Etesian winds blow. The first
five (days, the winds) are called the *Prodromoi*. According to Callippus Cancer
<finishes> rising, windy.

28th: According to Euctemon Aquila sets in the morning, a storm at sea
follows.

30th: According to Callippus Leo begins to rise, the south wind blows, and Sirius is
rising visibly.

31st: According to Eudoxus the south wind blows.

If this text is exactly analogous to either our wall calendar or our verjaardagskalender, then we will be forced to conclude that the dates listed
here, the first of Cancer, the second of Cancer, and so on, are dates in
some *calendar system*. This means that a group of people, somewhere, is
using this system to date events and to organize chronology. We might
assume from the structure of the thing that it was developed by and for
astronomers, used for their dating and recording of astronomical observations. And this conclusion has been accepted as true by most modern
commentators.[2]

But if this conclusion is wrong, then a whole ancient calendar system has
been created *ex nihilo*. What is worse, as modern scholars have developed
the implications entailed by the purported existence of such a zodiacal calendar, things have begun to spiral outwards quite quickly, with implications
for our understanding of much of the history of early Greek astronomy, not
just calendrics. Watch what happens: Rehm thought, based on parapegmatic
evidence, that he could attribute the origin of a zodiacal calendar system
to Euctemon in the fifth century BC, and that he could further isolate specific features of this calendar. In particular, Rehm argued that it was divided

[2] Rehm, 1913, p. 8f.; Rehm, 1941, p. 14f.; van der Waerden, 1984a; Bowen and Goldstein, 1988,
p. 53f.; Pingree (in Wenskus, 1990, p. 28); Wenskus, 1990; and Rüpke, 2000b. Toomer, 1974,
sensibly disagrees, as does Hannah, 2001a; 2002.

up into zodiacal *months* with the following lengths in days, beginning with
Cancer: 30, 30, 30, 30, 30, 30, 30, 31, 31, 31, 31, 31. Van der Waerden saw
in these numbers a similarity to a Babylonian System-A type step function,[3] where the sun moves at one speed over one part of the zodiac, and
at another speed over a different part. Van der Waerden then used Rehm's
numbers to derive a Ptolemaic-type eccentric solar model which he ascribed
to Euctemon.[4]

We have moved quite quickly here from a claim that a few astronomers
were passing observation records to each other dated in a zodiacal calendar
system, to a conclusion that pushes the origins of the eccentric solar model
in Greek astronomy back *a full three centuries* before we otherwise knew of it.
If correct, this is a major conclusion indeed, and will force us to rethink our
understanding of pre-Hipparchan[5] Greek astronomy. But there is still more.
In chronology and calendrics the existence of a Greek four-year Julian-type
solar-calendar cycle (one that divides the year up into 365 days, with a leap
day added every fourth year) has been argued for based on the evidence in
Geminus and others,[6] which is a radically different conclusion from what we
would otherwise have come to about the origins of the Julian calendar. A fair
bit has landed on the table at this point, much of it contradicting our other
sources for the histories of astronomy and calendrics, and for chronology. It
would seem timely, then, to probe at the foundations of this edifice and see
if it still stands. A caveat: the argument in what follows may look at first to
be a little more technical than will suit the tastes of some readers, but I have
aimed throughout to keep all the explanations clear, and not to presuppose
any special knowledge on the part of the reader. Much of what follows will
be concerned with dealing with the theoretical edifice built up over the

---

[3] Van der Waerden, 1984a, p. 104; followed by Bowen and Goldstein, 1988, p. 59f. 'System A' is
one of the two ways of modelling solar velocity in Babylonian astronomy. It assumes that the
sun moves at one speed (30° per mean synodic month) over one part of the zodiac (from 13° of
Scorpio to 27° of Pisces) and at a different speed (28° 7′ 30″ per mean synodic month) over the
remainder of the zodiac. See *HAMA*, p. 371f.; *ACT*.

[4] The eccentric solar model that Ptolemy takes from Hipparchus is meant to explain the varying
lengths of the seasons by having the sun move in a circular orbit whose centre is not the earth
(thus it is *ec-centric*). Looking out from the earth the sun seems to move more slowly when it is
farther from us and more quickly when it is closer. For details, see Ptolemy, *Alm.* iii.4.

[5] Hipparchus of Rhodes is a second-century BC astronomer, and the most important of Ptolemy's
predecessors. Ptolemy credits him with an eccentric solar model (*Alm.* iii.4), and he is the
earliest astronomer to be associated with such a model, unless van der Waerden is right about
Euctemon. See Neugebauer, 1975, pp. 57–8; Evans, 1998, p. 210f.

[6] Böckh, 1863; Rehm sees this as the direct predecessor of the Julian calendar (see Rehm, 1927);
Kroll, 1930 traces this calendar back to a Babylonian 'dodecaeteris', as does Boll, 'Dodecaeteris',
RE. But I know of no evidence in Mesopotamia for such a cycle.

lifetime of one of the most indefatigable researchers into parapegmata, Albert Rehm. Rehm did some good work on these texts, but at the same time he had a handful of pet theories that he overzealously thought he had found evidence for in parapegmata. Neugebauer, whose work is one of the major landmarks of modern scholarship on ancient astronomy, dismissed most of Rehm's work with one sentence: 'I have never succeeded in separating facts from mere hypotheses in this vast literature.'[7] I agree that this is the main problem, largely because Rehm himself never disambiguated the two. Nonetheless, some of Rehm's claims have had an unfortunate longevity, partly due to the endorsement of van der Waerden. These claims will accordingly need to be addressed in more detail than Neugebauer was willing to give them.

The basic set of problems that follows revolves to some extent around the efforts by modern scholars to deal with the massive problems posed by Greek civil calendars,[8] which are notoriously slippery. Nominally, classical Greek caledars were lunar, but, certainly by Hellenistic times, and probably earlier,[9] extra days and extra months were intercalated non-systematically. Perhaps non-systematically is even too nice a word. 'With almost reckless abandon' might suit the situation better, at least in some instances. One well-known story in chronology and calendrics circles is that of the Athenians in 270 BC needing to delay a festival performance in honour of Dionysus that was supposed to be held on the tenth of Elaphebolion. Since the performance was wedded to the date (presumably for cultic reasons), the clever Athenians just pushed the tenth of Elaphebolion back a few days by inserting a few extra ninths of Elaphebolion.[10] Not only this, but each city had its own calendar in use, with different month names, and because of the intercalated days and months that each city had inserted independently from time to time, nobody's months could be guaranteed to line up with anybody else's. Knowing that today is Hekatombaion 12 in the Athenian calendar, for example, does not give one any idea what the date is in Argos, Miletus, or Thasos. Although the new moon was nominally the start of the Athenian calendar,

---

[7] *HAMA*, p. 95, n. 17.

[8] On Greek calendars generally, see Samuel, 1972; Pritchett and Neugebauer, 1947; Hannah, 2005; Mikalson, 1996.

[9] Samuel, 1972. Contrast Dunn, 1999; Hannah 2005, p. 47f. Hannah wants to see a more stable fifth-century Athenian calendar than most modern commentators, but the evidence for calendrical stability in general in the fifth century is compromised, *inter alia*, by the fact that we do know that the city of Argos intercalated extra days on at least one occasion for the rather questionable purpose of delaying the beginning of the sacred month of Carneus in order to finish off an invasion of Epidaurus (see Thucydides, v.54).

[10] Evans, 1998, p. 183; Bickerman, 1980, p. 36; Samuel, 1972, p. 58.

the intercalation of days had the effect of moving the civil calendar dates out of synch with the actual behaviour of the moon. Thus Athenian writers sometimes distinguished between the *neomenia* 'according to the archon' (first of the month in the civil calendar) and the *neomenia* 'according to the moon' (true new moon).

Because Greek calendars are so loosely connected to the moon, to the seasons, and to each other, they are not particularly useful for preserving dated astronomical observations, or for figuring out when the best time to plant barley is. We have seen that the barley problem was regularly solved by recourse to stellar phases from Hesiod onwards. But what about the astronomical observations? Scholars have long suspected that early Greek astronomers must have had some more reliable way of passing dated information about observations to each other, both to contemporaries in different cities, and to successors as well. Did the kind of stellar 'dating' (using this word *very* loosely) that we find in parapegmata play any role in attempts to regulate the civil calendars of the various Greek cities? What is the evidence that there even *were* attempts to regulate them? Or else did parapegmata and astrometeorological texts help to develop some kind of alternate calendar system for use by astronomers spread out in time and space? (This latter option would be analogous to the use of the Egyptian calendar by astronomers from Ptolemy through Copernicus and beyond.)[11] Both the regulation and the replacement of civil calendars by ancient astronomers have been proposed by modern scholars for different times and different places in the Greek world. Most of these alternate or regulated calendars are proposed or reconstructed based on evidence in parapegmata and related sources, and the *who's who* of Greek astrometeorology has an almost one-to-one correspondence with the modern understanding of the *who's who* of Greek calendar reform, and this can be no coincidence.

In what follows, I will closely examine the evidence for these supposed calendar reforms. I will argue that in many – but not all – instances, the evidence is at best equivocal (and at worst nonexistent) that there was any *real* calendar reform going on at all. This is not to say that there were not *proposals* for ways to reform the calendar being floated by various authors, but that there is in many cases little or no evidence that such proposals were taken up *in practice* by anyone. So today, there is no shortage of proposals for ways to reform our calendar, but more proposals are made than are used: enter 'calendar reform' in any good search engine to see how many proposals there are – some of them quite possibly sensible – that are currently being

[11] See Evans, 1998, p. 175f.

ignored. We should also distinguish ancient proposals for calendar reform from ancient astronomical correlations of lunar and solar phenomena which may have had nothing to do with any calendars. Stating that a certain number of new moons recurs in a certain number of solar years is astronomically interesting independently of any problems one may have with one's local calendar, and so not all ancient luni-solar cycles can necessarily be taken *prima facie* as proposals for calendar reform.

The following few sections of this chapter will of necessity be a little more technical than many readers may need or want, and wary readers may wish to skip to section IV, below, to pick up the thread of my overall argument there. To summarize the technical sections, I.i–III.iv, in non-technical language is no easy task, but the most important points that I argue are that the parapegmata cannot be said to preserve otherwise lost zodiacal calendar systems, nor do they offer us sufficient evidence to determine whether, when, or how early astrometeorologists began dividing the year according to the sun's motion through the zodiac. Instead, the zodiacal signs we find in some parapegmata were used both as astrometeorological indicators in their own rights, as well as, in some cases, a handy way of allowing users in any locality using any civil calendar to synchronize a parapegma with that local calendar, or else to provide an easy astronomical, rather than a calendrical, means of tracking the astrometeorological cycles. The relationships between parapegmata and civil calendar reforms in antiquity are complex. I try to clarify the issues by distinguishing in the first instance between calendar systems and luni-solar cycles, the latter of which do not necessarily entail the former. We will see that many known astrometeorologists are credited with the development of cycles of one sort or another, but we need to be very careful about the available evidence before we ascribe the development of full-blown calendar systems to some or all of them.

I.i.   On the alleged zodiacal months of the early parapegmatists

I have already mentioned Rehm's argument that Euctemon divided the year up into zodiacal months with lengths of 30, 30, 30, 30, 30, 30, 30, 31, 31, 31, 31, and 31 days, and that this argument was expanded by van der Waerden to revise important parts of the history of early Greek astronomy. But there are good reasons for dismissing Rehm's account, and with it van der Waerden's.

Rehm's argument begins by noting that different season lengths are attributed to Euctemon in the Geminus parapegma and in the Eudoxus papyrus. In Geminus, beginning with summer, Rehm counts the lengths as: <92>, 89, 89, and 95 days. In the Eudoxus papyrus we are told

Euctemon's seasons are 90, 90, 92, and <93> days. Now, Rehm's restored season lengths correspond only with the latter set, so he needs to dismiss the former as incorrect. He does this by arguing (following Böckh)[12] that the Geminus parapegma was using a 'Callippan' zodiacal division (more on this in a minute), and goes on to hypothesize that its author simply copied Euctemon's alleged zodiacal-calendar dates into his own zodiacal calendar, but without preserving Euctemon's season or sign lengths. Thus Euctemon's Cancer 1 or Pisces 5 became Geminus' Cancer 1 or Pisces 5, even if Geminus had a longer Cancer and a shorter Pisces. But Rehm ignores the problem posed by the attribution to Euctemon which turns up in the Geminus parapegma at Taurus 32, which is an impossible Euctemonic date on Rehm's reconstruction, since Euctemon's Taurus is supposed to have had only 31 days. Pritchett and van der Waerden likewise pass over this problem without comment, and their translation of the so-called 'Euctemon parapegma'[13] simply *deletes* the attribution to Euctemon (the acronychal rising of Sagitta) at Geminus' Taurus 30 to make room to move Geminus' Taurus 31 to their Taurus 30, and Geminus' Taurus 32 to their Taurus 31, as follows:[14]

Geminus	*Pritchett and van der Waerden*
Taurus 25: Capella sets in the evening	Taurus 25: Vespertinal setting of Capella
30: Sagitta rises in the evening	30: Vespertinal rising of Aquila
31: Aquila rises in the evening	31: Setting of Arcturus. Sign of weather
32: Arcturus sets, there is a change in the weather	———

In addition to his inability to explain the Euctemon entry at Geminus' Taurus 32, Rehm has failed to prove that Euctemon must have divided his seasons up into the *months* of a zodiacal calendar system in the first place. Even if we were to grant that the season lengths as found in the Eudoxus papyrus are the correct ones, there is no reason to assume that Euctemon used the sun's passage through the zodiacal signs to divide those seasons into twelve solar months. Moreover, the particular month lengths given by Rehm are essentially pulled out of a hat, unattested in *any* ancient source.

---

[12] Böckh, 1863, p. 22f.

[13] For my justification for using scare quotes here, see the entry on the Euctemon parapegma (G.ii) in the catalogue (part II), below.

[14] Pritchett and van der Waerden, 1961.

Likewise, Rehm's claim that Callippus must have used zodiacal months, though generally accepted, is again unfounded. He attributes the following sign lengths to Callippus, beginning with Cancer: 31, 31, 30, 30, 30, 29, 30, 30, 30, 31, 31, and 32 days. This is based on the report in the Eudoxus papyrus that Callippus had season lengths of <94,> 92, 89, and 90 days, combined with Rehm's assumptions that: (a) these seasons should be divided up into zodiacal months; (b) the months should have an integer number of days; and (c) they should show a smooth and symmetrical ascent and descent between a maximum and minimum sign length. But all three of (a) through (c) remain to be proved.

I cannot overemphasize this: There *is no evidence* that either Euctemon or Callippus divided the year up into zodiacal months, or even that either assigned specific lengths in days for the sun's passage through each zodiacal sign. What we *can* say with certainty is that early Greek astronomers show a concern with season lengths, and that the Eudoxus papyrus attests to two different systems. The first is attributed to Euctemon, where the seasons are 90, 90, 92, and <93> days, and the second to Callippus with lengths of <94> 92, 89, and 90 days. By the late second century BC we begin to find sources that further subdivide the seasons by looking at how long the sun takes to traverse each zodiacal sign. Thus the Geminus and Miletus parapegmata give us the number of days it takes the sun to go through individual signs. The sign lengths in Geminus, however, do not agree with the season lengths for either Euctemon or Callippus in the Eudoxus papyrus, but then the Eudoxus papyrus does not tell us how the seasons were supposed to have been divided by either author. We cannot necessarily assume that the Eudoxus papyrus is supposed to be counting seasons from solstice to equinox to solstice as the Geminus parapegma does.

I.ii.   Parapegmata and the origins of the Julian calendar

Rehm's reconstruction of the Callippan zodiacal calendar system is closely tied to his reconstruction of an ancient Roman *Bauernkalender*, which, he argues, is the foundation of the Julian calendar,[15] and which can be traced back to an older Greek original which he credits to Callippus. Thus early Greek astronomers get pretty much the full credit for the Julian calendar reform. Van der Waerden, as usual, followed Rehm but Neugebauer had the good sense to be sceptical.[16] The 'rustic calendar' that all of this rests on was

---

[15] See Rehm, 1941, p. 43f.; Rehm, 'Parapegma', *RE*, col. 1347f.; Rehm, 1927.

[16] *HAMA*, pp. 595–8.

originally a hypothesis offered by Mommsen to account for the appearance of early Greek astronomical conventions in Columella, such as the fixing of the vernal equinox at Aries 8°.[17] Although Mommsen does not claim that his Bauernkalender is supposed to be a calendar system, as it gets fleshed out by Rehm it certainly becomes one. Mommsen's argument assumes an archaic Roman lunar calendar, and tries to account for the stellar phases in the Roman agricultural works and the zodiacal signs in the *menologia rustica* by assuming an archaic 'farmer's calendar' which used the stellar phases as signposts in the solar year, and which ran alongside of the lunar calendar. I agree that stellar phases were probably used as seasonal signposts in the Roman agricultural traditions, but this presupposes only some Hesiodic-type rules of thumb, rather than requiring the existence of a fully fledged self-calibrating calendar system, as Rehm wants. The evidence cited by Rehm from Columella, Varro, Clodius Tuscus, and others, is all considerably later than when his complex Bauernkalender is supposed to have been in use, and, as Wenskus has argued, was probably lifted out of Greek texts by Columella and the others rather than having been preserved on the lips of Roman farmers from the early Republican period on.[18]

1.ii.a   *The Julian calendar and the alleged Greek four-year solar cycle*

Pliny tells us that Eudoxus thought there was a four-year weather cycle, and that the cycle began in an intercalary year (*intercalario anno*) at the rising of Sirius.[19] This statement leads Böckh to conclude that (a) Eudoxus must have composed a four-year parapegma with (b) a leap-day added every fourth year.[20] But this assumes that an 'intercalary year' was, for Eudoxus, one in which an extra day was added to make a 366-day year, as in the much later Julian calendar.[21] This assumption is contradicted by the evidence from the Eudoxus papyrus, which shows that, for Eudoxus, years were counted as 'Egyptian years', each of 365 days (none of 366).[22] I would further counter

---

[17] Columella attributes this equinoctial value to Eudoxus, Meton, and 'the ancients' (Columella, *RR* IX.14.12); See Mommsen, 1859, p. 54f.; see also *HAMA*, p. 596; Wenskus, 1986; Bowen and Goldstein, 1988, p. 61.

[18] See Wenskus, 1998. The chart in Rehm, 1941, pp. 106–8 proves nothing beyond this.

[19] Pliny, *HN*, 11.130.

[20] Böckh, 1863, p. 124f. Böckh's reconstruction of Eudoxus' four-year solar-calendar intercalation cycle is very speculative, and I have serious doubts about it. His assertion that it probably came from a Pharaonic Egyptian tradition is simply wrong (Böckh, 1863, p. 254f.).

[21] Rehm, 1927; Kroll, 1930; Boll, 'Dodecaeteris', *RE*.

[22] See *HAMA*, p. 620.

that Eudoxus is otherwise not known for having composed a solar calendar, but we *do* know that Eudoxus composed an *Octaeteris*, an eight-year luni-solar cycle which intercalated lunar *months* (not days!) and so this would be the more natural interpretation of 'intercalary' in this passage.[23] This reading finds further confirmation in Pliny, *HN* xviii.217 where Pliny again mentions a four-year weather cycle (this time unattributed) as somehow part of an eight-year lunar cycle. He is, however, very short on details. In any case neither passage is sufficient evidence to ascribe a four-year calendar-cycle parapegma complete with leap-day to Eudoxus: one can state that the weather is cyclical without having catalogued the entire cycle in a parapegma, nor does a four-year weather cycle necessarily entail a four-year Julian-type calendrical cycle.

I.iii.   So what *is* the zodiac doing in Geminus and Miletus I?

To recap: there is no evidence for a zodiacal calendar system attributable to Euctemon and Callippus, nor is there evidence for a Greek four-year solar cycle in Callippus and Eudoxus. Now the question must be raised of just what the zodiac is doing in Geminus and Miletus I. Is it evidence for *any* zodiacal calendar system at all? In Geminus, as we saw, the year is divided up into zodiacal signs. Each sign is introduced with the formula 'The sun traverses [sign name] in *x* days.' The astrometeorological predictions for the year are then indexed to day numbers in each sign, for example:

τὴν δὲ Παρθένον διαπορεύεται ὁ ἥλιος ἐν ἡμέραις λ´.
ἐν μὲν οὖν τῇ ε´ ἡμέρᾳ Εὐδόξῳ ἄνεμος μέγας πνεῖ, καὶ ἐπιβροντᾷ Καλλίππῳ δὲ οἱ
    ὦμοι τῆς Παρθένου ἐπιτέλλουσι· καὶ ἐτησίαι λήγουσιν.
ἐν δὲ τῇ ι´ ἡμέρᾳ Εὐκτήμονι Προτρυγητὴρ φαίνεται· ἐπιτέλλει δὲ καὶ Ἀρκτοῦρος,
    καὶ Ὀιστὸς δύεται ὄρθρου· χειμὼν κατὰ θάλασσαν· νότος. Εὐδόξῳ ὑετός,
    βρονταί· ἄνεμος μέγας πνεῖ.
ἐν δὲ τῇ ιζ´ Καλλίππῳ Παρθένος μέση ἐπιτέλλουσα ἐπισημαίνει· καὶ Ἀρκτοῦρος
    ἀνατέλλων φανερός.
ἐν δὲ τῇ ιθ´ Εὐδόξῳ Ἀρκτοῦρος ἑῷος ἐπιτέλλει· καὶ τὰς ἑπομένας ἡμέρας ἑπτὰ ἄνεμοι
    πνέουσιν· εὐδία ὡς ἐπὶ τὰ πολλά· λήγοντος δὲ τοῦ χρόνου ἀπ᾽ ἠοῦς πνεῦμα
    γίνεται.
ἐν δὲ τῇ κ´ Ἀρκτοῦρος Εὐκτήμονι ἐκφανής· μετοπώρου ἀρχή· καὶ Αἲξ ἐπιτέλλει,
    ἀστὴρ μέγας ἐπὶ τοῦ Ἡνιόχου· κᾆπειτα ἐπισημαίνει· χειμὼν κατὰ θάλασσαν.
ἐν δὲ τῇ κδ´ ἡμέρᾳ Καλλίππῳ Στάχυς ἐπιτέλλει τῆς Παρθένου· ὕει.

---

[23] See Geminus, viii.25f.; Blass, 1887.

The sun traverses **Virgo** in 30 days.

On the 5th day: According to Eudoxus a strong wind blows and it thunders. According to Callippus the shoulders of Virgo rise, and the Etesian winds stop.

On the 10th day: According to Euctemon Vindemiatrix appears, Arcturus rises, and Sagitta sets at dawn, storm at sea, south wind. According to Eudoxus rain, thundery, a strong wind blows.

On the 17th: According to Callippus there is a change in the weather while the middle of Virgo is rising, and Arcturus is rising visibly.

On the 19th: According to Eudoxus Arcturus rises in the morning, and winds blow for the next seven days. Generally nice weather. At the end of this time wind is from the east.

On the 20th: Arcturus is visible according to Euctemon. Beginning of autumn, and Capella (the bright star in Auriga) rises, and after that there is a change in the weather, storm at sea.

On the 24th day: According to Callippus Spica in Virgo rises, it rains.

Miletus I works slightly differently. Being an inscriptional parapegma, it uses a peg, rather than a day number, to index its predictions.[24] At the top of each sign, though, we do still get a length, in days, for the sun's passage through that sign, as in the '30' in the following passage:

30 (days)

- The sun is in Aquarius
- [. . . . .] begins setting in the morning and Lyra sets.
  • •
- Cygnus begins to set acronychally.
  • • • • • • • • •

As in Geminus, not every day has an astronomical or meteorological entry, but unlike Geminus, the empty days in Miletus I are indicated with place-holding peg holes.

So is there a zodiacal calendar system lurking behind Miletus I and Geminus? And if so, how is it organized? Certainly we can say that there existed at least one zodiacal calendar in antiquity: the so-called 'Dionysian' calendar, attested in some observation reports in Ptolemy's *Almagest*. The Dionysian calendar divided the year into months whose names were derived from the signs of the zodiac. Böckh shows that this calendar had an epoch

---

[24] The day numbers in the original edition of the fragments were added by Diels and Rehm. See Diels and Rehm, 1904; compare Lehoux, 2005.

at the summer solstice of 285 BC. He reconstructs the calendar itself such that it had twelve months of thirty days each with five epagomenal days at the end, or six every fourth year, as in the Alexandrian calendar.[25] But in order to do this, he needs to emend two of the Dionysian dates (and one of his emendations has since been vindicated by a manuscript not known to him). Other reconstructions are also possible, depending on how the observations reported by Ptolemy are handled.[26] But it does not seem possible to reconcile the month lengths of the Geminus parapegma with any of the proposed reconstructions for the Dionysian calendar. Even more importantly, the Geminus sign names are not the same as the Dionysian month names (e.g., Geminus has Ὑδροχόος where Dionysius has Ὑδρῶν).[27] Miletus I, on the other hand, is too fragmentary to be certain about the lengths of most signs, but again the sign names are different from the Dionysian month names: as in Geminus, we find Ὑδροχόος instead of Ὑδρῶν. And no other attested Greek calendar was zodiacal.

Furthermore, as I mentioned above, the days in Miletus I are not numbered. If we recall how Miletus I was to be used, we see that a peg would have sat beside the current day, and then have been moved each day to the next hole so that each day's astronomical situation could be easily read off. One kind of astronomical information was the sun's entry into the successive zodiacal signs. Thus the entry beside the first peg hole in Aquarius reads: 'The sun is in Aquarius.' Since the days following that entry are not numbered, they were probably not meant to be counted as calendar dates are. If they were to be counted, then what would the date be when the peg was beside the entry that reads 'The whole of Centaurus sets in the morning?' Counting holes is both inefficient and error-prone. If the Aquarius here were chronometric rather than astronomical – were it a month in a calendar system rather than a zodiacal sign the sun moves through – we should have expected dates to have been indicated as well. We can conclude then that Miletus I does not use a zodiacal calendar. Rather, it includes, among the other astronomical phenomena that it was used to keep track of, the sun's entry into each of the twelve signs. Indeed, Ptolemy, in the introduction to the *Phaseis*, tells us that

[25] See Böckh, 1863, pp. 286–340; see also van der Waerden, 1984c; Samuel, 1972, pp. 50–1; *HAMA*, pp. 1066–7.

[26] On the problems with these observations, see Swerdlow, 1989.

[27] Compare Geminus, p. 224, l. 17 with Ptolemy, *Alm.* ix.7.169. Goldstein and Bowen, 1991, p. 123, make the improbable claim that Ὑδρῶν is derived from the constellation Hydra, and Αἴγων from Capella (Αἴξ), rather than from the zodiacal signs.

the sun's entry into the different signs was astrometeorologically significant: it is one of the causes of changes in the weather.[28]

But what about Geminus? It is quite clear that the astrometeorological information is being indexed according to zodiacal days. But does this imply a zodiacal calendar system? Certainly the zodiacal signs and days sit exactly where we should expect calendar dates to be in a literary parapegma. But since there is no attested zodiacal calendar which seems to fit the Geminus parapegma, I am tempted to think that its layout merely implies that its users would need to have a modicum of astronomical knowledge, specifically: knowing what sign the sun is in and how long it has been there. Looking at the regular formula by which the signs are introduced: 'The sun traverses Cancer in thirty-one days. On the first day . . .' seems also to indicate that the signs were seen as astronomical rather than calendrical divisions.

This is a fine differentiation, but the distinction is an important one: if we think of the zodiacal days in Geminus as part of a zodiacal calendar system, then we must presume that someone else besides Geminus (probably some astronomers somewhere, given the nature of the thing), actually used this calendar system as a chronometric tool to mark time and to date events in. But such a system is otherwise unattested, and the evidence from Geminus himself unconvincing. Instead, it seems we should interpret the formulae at the beginnings of the signs literally, and see the zodiacal days simply as 'counters' from one known astronomical event to the next. This known astronomical event can thus be seen to *calibrate* the parapegma, and so to render it independent of the shiftiness of whatever local calendar the reader may have been using – an important function, given the vagaries and variety of Greek civil calendars.

i.iv.   Parapegmata and calendars, generally

Parapegmata track temporal cycles, they tell us when particular events are occurring or will occur, and they situate them in a rich context of other events. In astrometeorological parapegmata we see what stellar phase and what weather phenomena should be happening now, what ones will happen and when, and what ones have happened in the near (or not-so-near)

---

[28] Ptolemy, *Phas.* 11.15f. Curiously, though, Ptolemy does not list the sun's entry into the various signs in the parapegma itself. The other supplemental causes which he mentions (the planets and the quarter and full moons) do not get reported in the parapegma, but this should not be surprising, since they would vary so much from year to year. A lunar eclipse does get mentioned in Clodius Tuscus, however. Ptolemy also talks about the moon and planets as causes of weather in *Tetr.* ii.11.

past. Inscriptional parapegmata indicate this with a peg or pegs, while literary ones generally – but not always – do this by linking the information up to dates in a calendar system of some sort. We have so far been dealing in this chapter only with Greek astrometeorological and astronomical parapegmata, but we should also keep in mind that, as we saw in the last chapter, the Latin astrological parapegmata also track cycles. They show us combinations of hebdomadal days, lunar days, and occasionally calendrical and nundinal days. And this is one of the things that the Roman inscriptional parapegmata excelled at: keeping track of unrelated cycles. Thus at a glance, a Latin astrological parapegma could reveal where the current day stood in the eight-day nundinal cycle, the seven-day hebdomadal cycle, the twenty-nine or thirty-day lunar cycle, and/or the 365-day calendar cycle.

The Greek inscriptional parapegmata, and the Greek and Roman literary ones are by comparison rather pedestrian devices. They simply indicate the current day in, as near as we can tell, a 365-day combined astronomical-and-meteorological cycle, sometimes including calendrical information. We should note that a variety of different calendars are represented in the literary parapegmata: thus Ptolemy uses the Egyptian calendar, and Columella, Pliny, and Polemius Silvius use the Julian calendar, to name a few examples.

To generalize: parapegmata are tools for keeping track of cycles (astrometeorological, hebdomadal, nundinal, lunar). Specifically, they allow us to track our current temporal position in those cycles. One type of parapegma uses a peg or pegs to indicate the current situation, and another generally uses a calendar of some sort to do the same thing. The calendars found in many literary parapegmata simply serve to replace the peg as a way of indexing the current day. But the fact that literary parapegmata often incorporate dates from a calendar system does not mean that all parapegmata, even inscriptional ones, must do so. We have seen that not even all literary parapegmata can be shown to be indexed to calendar dates. Nevertheless, there is a common contention in the modern literature that inscriptional Greek parapegmata had calendars somehow incorporated into them for dating the astrometeorological phenomena.[29] Typically, the claim is that calendrical dates were somehow inscribed on the pegs. But, besides such a system being incredibly unwieldy and (given the sizes the pegs would need to be) probably physically impossible, the assumption of dated pegs is also unnecessary, since a single moveable peg which marks the current day is

---

[29] See Rehm, 'Parapegma', *RE*; *HAMA*, p. 587. For a less specific account of the incorporation of calendars, see Bowen and Goldstein, 1988, p. 52f.

sufficient for finding our position in a current cycle. Calendrical information is superfluous.

It is true that many parapegmata are concerned with astronomical phenomena – the stellar phases – which are closely tied to the annual motion of the sun. And it is exactly this annual motion of the sun which Greek civil calendars were unable to keep good track of. So it is no accident that we find pegged astronomical and astrometeorological parapegmata almost exclusively in Greece.[30] On the other hand, the late Republican and early Imperial Romans, whose calendar kept much better track of the solar year (but was useless for lunar cycles), largely used their pegged parapegmata to keep track of lunar and other cycles. In this light, parapegmata can be seen as *extra-calendrical* devices used for keeping track of non-calendrical cycles.[31] The appearance of calendrical information in literary parapegmata does not contradict this point. The calendar simply takes the place of the peg for referencing the current and future situation relative to the current date.

II.   Calibration of parapegmata

Thinking of the pegs, calendars, and zodiacal days as 'counters' necessitates, as I have said, that there be some astronomical event or events, which were either observed, calculated, or dated, and which were used from time to time to calibrate the parapegma, that is: to synchronize astrometeorological parapegmata with the tropical or sidereal year, or astrological parapegmata with the lunar days. For astrological parapegmata, lunar phenomena may need calibrating once every month or two, for which a simple observation would suffice.[32] But what about for astrometeorological and astronomical parapegmata? Would the users of the Geminus parapegma have needed some external way of knowing when the sun moved into each individual sign? This would require that the external source in question used the same season and

---

[30] The one exception being the Puteoli parapegma, inscribed in Latin, on which, see Lehoux, 2006b.

[31] Rüpke, 2000b, goes so far as to call these non-calendrical cycles 'calendar systems', and sees parapegmata as tools for 'enabling one to track calendar systems that differ from [one's] particular civil calendar'. But this is simply an imprecise use of language. Neither the lunar days (Rüpke mistakenly calls these 'phases') tracked by astrological parapegmata, nor the stellar phases tracked by astronomical and astrometeorological parapegmata, can be called 'calendar systems' in the stricter sense I have been using here.

[32] For the more complex Coligny calendar McCluskey has argued, I think correctly, that it would need observational calibration only about once a year.

sign lengths as Geminus, which is no small qualification considering the multitude of available options. Another, simpler, possibility is that the users of the Geminus parapegma needed only an external way of knowing when the summer solstice, the first day of the parapegma, occurred. From there, a simple count or a correspondence with whatever the reader's local calendar was, would situate the reader in the parapegma for the rest of the year. One synchronization with the solstice would thus calibrate the parapegma for the year.

This means that for users of any calendar at all there would have been a calibration mechanism where each year the Geminus parapegma could be re-aligned with the solstice. But this raises an important question: how *does* a Greek know when the summer solstice occurs? We will return to this momentarily.

II.i. Self-calibration in the four-year Julian calendar cycle

After the calendar reform of Julius Caesar, every third or fourth year contained 366 days.[33] But the post-Julian astrometeorological parapegmata (all of which are literary) have only 365 days. One wonders then what these parapegmatists did in leap-years, and how they moved their stellar phases and weather predictions around to fit them to the intercalary 366-day years. Only the Coligny calendar shows any evidence of keeping track of anything longer than a one-year cycle, and in that case it is a lunar calendar, not a solar one. No parapegma contains an entry for a leap-day.

We have, as I see it, several options to choose from: (1) the post-Julian parapegmatists simply ignored leap-years, and were content to let the leap-day pass without astrometeorological comment; (2) every parapegma was written only for a particular year, and all the ones preserved just happen to be from non-leap-years; (3) there was some rule of thumb for dealing with the leap-years, of which we are not aware; or (4) the parapegma was periodically calibrated to the solar year. I think we can rule out (2), since there is never any indication of what particular year any parapegma was written for.[34] As for (3) and (4), we need to ask whether they would be necessary in the case of the Julian calendar. Certainly, the Julian calendar

---

[33] For the first few years, the admonition to add a day 'every fourth year' was interpreted, through the peculiar Roman counting system ('first, second, third, fourth (which is first), second third, fourth (which is first), second . . .), to mean that there was an intercalary year (I), then two normal years (-), then an intercalary year, as follows: I - - I - - I . . . , rather than I - - - I - - - I . . . This was only sorted out and corrected in 8 BC.

[34] With the possible exception of Clodius Tuscus, on which, see A.xi in the Catalogue, below.

was meant to self-calibrate with the solar year, and so any parapegma linked to it should stay synchronized in perpetuity, without need for calibration. The same is true for the Alexandrian calendar, which Ptolemy used. Thus (3) and (4) are reducible to (1), the ignoring of the leap-day, and since the ignoring of the leap-day should have posed no special problems, we can say that post-Julian parapegmata, all of which were literary, were probably meant to be self-calibrating. Indeed, Ptolemy specifically justifies his choice of the Alexandrian calendar based on the fact that it keeps the phases for the most part on the same dates year after year.[35]

III.   Luni-solar cycles

I said above that some of the parapegmata preserve astrometeorological attributions from Greeks who are also believed to have composed luni-solar cycles of one sort or another. This is the case with Meton, Euctemon, Eudoxus, Philip, Callippus, Dositheus, Hipparchus, and possibly (though doubtfully) Democritus.[36] I will argue that in all of these cases, the lunar cycles are correlated with astrometeorological cycles. This should not be surprising, since any attempt to reconcile the lunar months with the solar year would synchronize the lunar calendar with either (a) the solstices and equinoxes or (b) the phases of the fixed stars, or (c) both sets of phenomena.[37] But since both (a) and (b) were closely connected to meteorology from the time of the earliest Greek records, then any luni-solar cycle would have immediate meteorological consequences for a Greek writer. Moreover, exploring possibility (a) may provide us with a full or partial answer to a question I left open in a previous section: how does a Greek know when the summer solstice occurs?

III.i.   How does a Greek know when the summer solstice occurs?

It has been long assumed that observing a solstice is astronomically a fairly simple matter, and that Meton did just that in 432 BC in Athens using some

---

[35] Ptolemy, *Phas.* x.5f. Like all the other parapegmatists, he seems to ignore the precession of the equinoxes, which would have caused the solstices and equinoxes to slip slowly relative to the stellar phases.

[36] On Censorinus' attribution of a luni-solar cycle to Democritus, see *HAMA*, pp. 619–20.

[37] Before the discovery of precession by Hipparchus in the second century BC (c) would have been seen as an exact correlation. Afterwards (a) would be seen to slip slowly away from (b) at a rate just barely perceptible over a single lifetime.

sort of sundial or gnomon.[38] But Bowen and Goldstein have argued that Meton could simply have defined some date as the summer solstice using an Uruk-type lunar calendar scheme. The Uruk scheme is a Babylonian convention for schematically determining the dates of solstices, equinoxes, and the phases of Sirius in a nineteen-year lunar calendar cycle. It is attested from the early fourth century BC in Babylon.[39] There is evidence of a nineteen-year Babylonian calendar cycle predating the Uruk scheme by about a century, but it is unclear whether it is simply a calendrical intercalation cycle or if it could have included schematic solstice dates, as in the Uruk scheme itself. Bowen and Goldstein have argued for the latter possibility, and called this a 'precursor' to the Uruk scheme. Besides Bowen and Goldstein's argument, there is no solid evidence for the adoption of an Uruk-type scheme in Greece. But the possibility that the Metonic nineteen-year cycle was influenced by contemporary Babylonian calendrics may be tempting, considering the identical number of years in the two cycles, and the prominence of the summer solstice in both the Uruk scheme and the Metonic cycle.[40]

Importantly, Bowen and Goldstein have not fully eliminated the possibility of Meton's actually having observed the solstice. Solstice observation is not particularly difficult.[41] Meton's observation of one may be remarkable since no other solstice is known to have been observed in Greece until 280 BC, but *someone* had to be the first, and if we follow Bowen and Goldstein to discount Meton, then Aristarchus' observation in 280 BC becomes suspect for the same reasons as Meton's, and we get a kind of vicious spiral.

---

[38] See Kubitschek, 1932; Rehm, 'Parapegma', *RE*; Toomer, 1974.

[39] Following Neugebauer and Sachs' dating. See *HAMA*, pp. 358, 362f.; Sachs, 1952, p. 106.

[40] Note that Hunger and Pingree, 1999, pp. 199–200, have argued that 'any observation-based or computation-based method of guaranteeing that, say, the vernal equinox always falls in a synodic month bearing a particular name will necessarily, if successful, involve the intercalation of seven synodic months in nineteen years'. While this is true, the coincidence of the emergence of such a method in Babylon, followed shortly thereafter in Greece (if indeed this is what the Metonic cycle is), points to the possibility that this method, or at least the idea of such a method, was borrowed by Meton. Van der Waerden has argued that Euctemon, Meton's associate, did derive a nineteen-year calendar cycle from the Babylonian Uruk scheme, which would be a rare and important instance of direct transfer of methods and data from Babylonian to early Greek astronomy, if he is right. But van der Waerden's argument (van der Waerden, 1984a, p. 113f.) is unfortunately circular. It is based on the completely unfounded assumption that Euctemon, copying the Uruk scheme, divided months into *tithis* rather than days. The similarities to the Uruk scheme which van der Waerden finds in Euctemon's cycle are then seen as evidence that Euctemon was copying the Uruk scheme.

[41] As A. Jones has pointed out to me, observing shadow lengths or sunrise/sunset points over several days around the solstice will give a few days where there is little or no apparent change. Taking the halfway point of this will give a solstice date accurate to half a day. Looking at both the sunrise *and* the sunset points, will give a date accurate to a quarter day.

But even if we cannot be sure whether or not Meton observed a solstice, we do have other information about what he was doing on the Pnyx on the day of the solstice in 432. Diodorus Siculus tells us that Meton publicly inaugurated his nineteen-year cycle on Skirophorion 13 in 432 BC, the very day Ptolemy has him observing a solstice.[42] If we believe Ptolemy that this date *was* a solstice, whether observed, as the standard view has it, or schematically defined, as Bowen and Goldstein contend, then we can combine this with the Diodorus report to conclude that Meton's nineteen-year cycle began with a summer solstice.

Now, if Bowen and Goldstein are correct that Meton's nineteen-year cycle *was* an Uruk-type scheme for defining the dates of solstices, equinoxes, and possibly some stellar phases, then we can see how it would be a useful tool for calibrating an annual astrometeorological cycle. But whether or not that cycle was written up as a parapegma is, I think, not determinable on the current evidence.

But when we say *Uruk-type scheme*, how flexible are we willing to be? How much does Meton's scheme need to look like the Uruk scheme? The evidence is nowhere near specific enough to allow us to assert that what Meton did was specifically to set up an inscription with a list of dates for solstices in a lunar calendar cycle. This does, nonetheless, seem like at least a possibility. In any case, he did make some sort of nineteen-year cycle public on or about the date of the solstice in 432 BC, and he did do something which saw his name cited in later parapegmata.

While the question of whether Meton (or Euctemon, for that matter) did schematically define solstice dates in particular in a nineteen-year cycle must remain open, we should be careful not to overstate the evidence for the possibility that they did so. At the same time, though, we should not rule out the possibility that solstices were still meant to be observed. In any case, either the schematic dating or the actual observation of a solstice would be sufficient to calibrate a parapegma, by the time there were parapegmata to be calibrated.

III.ii.   The Metonic cycle more generally

Speculating that the nineteen-year Metonic cycle may have been a calendrical system for schematically dating solstices and equinoxes rests on the

---

[42] Diodorus Siculus, XII.36.1f. Ptolemy dates it in the Egyptian calendar, Phamenoth 21 (*Alm.* pp. 205–6), and Miletus fragment 84 (see Lehoux, 2005) equates this date with the Athenian Skirophorion 13 for the year 432.

underlying assumption of an unattested calendar of some sort in which the solstices and equinoxes were so dated. The most we can say with reasonable certainty about the Metonic cycle is that it equated nineteen years with 235 months and 6,940 days, that this relationship would give a year length of 365 $\frac{5}{19}$ d,[43] and that Meton inaugurated this cycle on Skirophorion 13, in the archonship of Apseudes (= 27 June, 432 BC), the same day he is supposed to have observed the summer solstice. Diodorus' claim that Meton started his cycle at the solstice of 432 BC seems to indicate that the Metonic cycle is more than just an astronomical correlation for determining the mean lunar month in days. It certainly seems to be a repeating cycle with a definite start date.[44]

Moreover, the description of the Metonic cycle in Geminus seems to indicate that it was used to determine a sequence of *full* (thirty-day) and *hollow* (twenty-nine-day) months which would repeat after nineteen years.[45] Diodorus indicates that the Metonic cycle was found to be in agreement with the cycle of the motions of the stars and their attendant changes in the weather, and he seems to imply that Meton wrote on this subject. He says:

ἐν δὲ ταῖς Ἀθήναις Μέτων ὁ Παυσανίου μὲν υἱός, δεδοξασμένος δὲ ἐν ἀστρολογίᾳ, ἐξέθηκε τὴν ὀνομαζομένην ἐννεακαιδεκαετηρίδα, τὴν ἀρχὴν ποιησάμενος ἀπὸ μηνὸς ἐν Ἀθήναις σκιροφοριῶνος τρισκαιδεκάτης. ἐν δὲ τοῖς εἰρημένοις ἔτεσι τὰ ἄστρα τὴν ἀποκατάστασιν ποιεῖται καὶ καθάπερ ἐνιαυτοῦ τινος μεγάλου τὸν ἀνακυκλισμὸν λαμβάνει· διὸ καί τινες αὐτὸν Μέτωνος ἐνιαυτόν ὀνομάζουσι. δοκεῖ δὲ ὁ ἀνὴρ οὗτος ἐν τῇ προρρήσει καὶ προγραφῇ ταύτῃ θαυμαστῶς ἐπιτετευχέ-ναι· τὰ γὰρ ἄστρα τήν τε κίνησιν καὶ τὰς ἐπισημασίας ποιεῖται συμφώνως τῇ γραφῇ· διὸ μέχρι τῶν καθ᾽ ἡμᾶς χρόνων οἱ πλεῖστοι τῶν Ἑλλήνων χρώμενοι τῇ ἐννεακαιδεκαετηρίδι οὐ διαψεύδονται τῆς ἀληθείας.

In Athens, Meton the son of Pausanius, renowned for astronomy, set out what is called the nineteen-year cycle, starting it on the thirteenth of the Athenian month

---

[43] This cycle is also attributed to 'Meton and Euctemon' (Ptolemy, *Alm.* III.1; Theodosius, *De diebus*, II.18), and to οἱ περὶ Εὐκτήμονα καὶ Φίλιππον καὶ Κάλλιππον, the followers of Euctemon and Philippus and Callippus. (Geminus, p. 120). It is unclear whether the year length was a by-product of the equation of days to months to years, or vice versa. The same is true of the Callippic modification of this cycle, on which, see below.

[44] Hannah, 2001a, pp. 148–9, has also recently brought forth corroborating evidence for this claim from a scholiast to Aratus. My only hesitation with this evidence is that the inscriptions mentioned by the scholiast are attributed to 'the followers of Meton', and the description there may only be a paraphrasing of Diodorus.

[45] The reliability of the details of Geminus' account, i.e., exactly how the full and hollow months were determined, has been questioned by Neugebauer, in *HAMA*, p. 617; Bowen and Goldstein, 1988, p. 43, n. 27. But Jones, 2000, p. 152f., sees it as less worrisome. Van der Waerden's interpretation of this passage is extremely strained (van der Waerden, 1984b, p. 121f.).

Skirophorion. In that number of years the stars make a return as they complete the cycle of a kind of great year. Because of this some people call it the *year of Meton*. And it seems that this man was wondrously successful in this prediction and forecast, for the stars make both their motion and their changes in the weather in agreement with his writing. Thus the majority of the Greeks down to our own time, who measure time with the nineteen-year cycle, are not misled from the truth.[46]

It is clear from this that the nineteen-year cycle was meant to correlate lunar and astrometeorological cycles in *some* way. Because of this, it has been long assumed that the Metonic cycle had something to do with the setting up of a parapegma,[47] but as I have continually emphasized: astrometeorologist does not necessarily mean parapegmatist.

I have already remarked on the similarities of the Metonic cycle with the Babylonian nineteen-year calendar cycle (the Uruk scheme), and I should perhaps say a few words on the differences. Unlike the Metonic, the Babylonian cycle is *not* concerned with the exact sequence of full and hollow months. In the Babylonian calendar the beginning of the month was determined by the observation of the new moon, and so their nineteen-year cycle always allowed for any particular month to be determined as either hollow or full. The Metonic cycle by contrast, seems to have *prescribed the order* of hollow and full months. The real purpose of the Babylonian cycle was to prescribe when intercalary months would be inserted over the course of nineteen years, in order to keep the Babylonian months in line with the solar year. It had the effect that the summer solstice was almost always in the same month from year to year. Geminus tells us that the Metonic nineteen-year cycle contained seven intercalary months, which happens to be the same number as the Babylonian system, although we do not know how they were to be distributed.[48] Finally, there are no dates preserved which are expressly in a 'Metonic calendar' or counted from a 'Metonic epoch'.[49]

---

[46] Diodorus Siculus, XII.36.2.

[47] See Heath, 1913; Rehm, 1941, p. 7f.; Kubitschek, 1932; van der Waerden, 1960; Toomer, 1974; *HAMA*, p. 622; Bowen and Goldstein, 1988.

[48] Toomer believes that Meton may have used the exact same spacings for his intercalations as the Babylonians (Toomer, 1996).

[49] Toomer, following Fotheringham, believes three eclipse-dates in the *Almagest* to be in the 'astronomical calendar of Meton' (Toomer, 1984, p. 211, n. 63), but the argument is inconclusive. Müller, 1991, argues that the Metonic cycle was used to determine intercalations at Athens from 126/5 BC to AD 198/9, but his argument is based on very little data. For this period of more than 300 years he has no complete cycles attested, and only four cycles with data for more than just one or two years. Only two cycles are more than half complete, and of these, one contains data that contradicts his thesis. The conclusion he reaches is thus extremely underdetermined, *pace* the praise he offers himself in Müller, 1994.

This mention of (a) a specific prescribed number of intercalary months and (b) a sequence of full and hollow months sets the Metonic cycle apart from the irregularly intercalated Athenian calendar,[50] and is, I think, the best evidence that the Metonic cycle was somehow calendric.

III.iii. The octaeteris

The *octaeteris* is an eight-year lunar intercalation scheme sometimes associated with Eudoxus and Dositheus.[51] It prescribes intercalary months in the years 3, 6, and 8 of the cycle, and also prescribes the sequence of full and hollow months. Unfortunately the details of this scheme have been lost.[52] Although the attributions to Eudoxus and Dositheus are sketchy, there is other evidence to connect the octaeteris with astrometeorology more generally: there is a passage in Pliny, where he situates a four-year astrometeorological cycle within an eight-year lunar cycle.[53] There is no evidence that this cycle was ever used as a civil or astronomical calendar, however.

III.iv. The Callippic cycle

The Callippic cycle was a modification of the nineteen-year Metonic cycle which removed one day every fourth cycle (i.e., every seventy-six years). This reduced the year length from $365 \frac{5}{19}$ d to $365 \frac{1}{4}$ d. There were twenty-eight intercalary months over the course of the seventy-six years. The epoch of the Callippic cycle is the summer solstice of 330 BC, and several observations cited in the *Almagest* are dated in years from this epoch, in the following format: 'in the thirty-second year of the third Callippic cycle'. There are, however, some problems with seeing this as a calendar system *per se*. Hipparchus records observations counted in years from the Callippic epoch, but dated in the *Egyptian* calendar, whereas *Almagest* observations attributed to Timocharis are counted in years from the Callippic epoch but dated with *Athenian and Egyptian* month names, and a recently published first-century AD eclipse canon from Oxyrhynchus also uses both types of dates.[54] The Timocharis observations have led scholars to attempt

---

[50] See Pritchett and Neugebauer, 1947.
[51] For a discussion of its authorship, see Neugebauer, in *HAMA*, pp. 620–1.
[52] See Blass, 1887, col. XIII.12f. [53] Pliny, *HN* XVIII.217.
[54] Compare, e.g., Ptolemy, *Alm.* III.1, 204 with VII.3, 25 and with *P. Oxy.* LXI.4137, l. 18f. (in Jones, 1999a). For commentary, see also Jones, 2000, p. 146.

reconstructions of the 'Callippic calendar' as an artificial astronomical calendar system using Athenian month names and dating conventions, but none of the reconstructions is satisfying, and most require emendations of one or another of the Timocharis citations.[55] The Hipparchus dates are even worse, leading Toomer to conclude that Hipparchus used the Egyptian *calendar* for days and months, but the Callippic *cycle* for years, and that he may have used different dating conventions for different observations. Jones has, I think convincingly, refuted this.[56]

The conflicting nature of this evidence leads Neugebauer, Goldstein, and Bowen to be sceptical that the Callippic and Metonic cycles are calendrical.[57] Toomer and Jones argue, on the other hand, that the Metonic and Callippic cycles do represent an astronomical calendar system,[58] and that this would best explain how the Athenian month names came to turn up in Miletus fragment 84, in Timocharis' observations at Alexandria, and in the Oxyrhynchus eclipse canon. As far as I can see, the gamut of possibilities can be enumerated as follows: the Callippic cycle represents (a) a simple correlation of a whole number of days to a whole number of months to a whole number of years; (b) an Uruk-type scheme; (c) a calendar used by astronomers, including Timocharis and Hipparchus, from 330 BC until at least the first century AD.

Point (a) seems not to offer enough to account for the use of a Callippic epoch in some of Ptolemy's sources in the *Almagest*. Neugebauer argues against (c), and seems rather to prefer something like (b) when he says that 'Meton did not attempt to introduce a new lunar calendar but intended to establish a definite starting point in the solar year for the construction of parapegmata.'[59] He offers no argument against Toomer's point about why the Athenian calendar should have shown up in Miletus I or Timocharis, or at Oxyrhynchus, for that matter. On balance, I find – cautiously, mind you – the evidence to weigh in favour of the Callippic cycle being a regulated lunar calendar which probably used the same month names as the Athenian civil calendar. This calendar is attested as being used for the dating of astronomical observations, and also probably served to calibrate lunar and astrometeorological cycles.

[55] For details, see Ginzel, 1914, vol. II, pp. 409–19; Samuel, 1972, pp. 42–9.
[56] See Toomer, 1984, p. 13; Jones, 2000, pp. 148–50.
[57] Neugebauer, 1975; Goldstein and Bowen, 1989.
[58] Toomer, 1974; Jones, 1999a, pp. 89–90; Jones, 2000, pp. 143, 156f. See also Fotheringham, 1924.
[59] *HAMA*, p. 622.

IV.   What sorts of sources could the parapegmatic attributions be taken from?

I have said repeatedly that we cannot be certain that the attributions in the astrometeorological parapegmata were taken from earlier texts that were themselves substantially like parapegmata. They may, for example, have been cobbled together from very different sources of very different types. When it comes to reconstructing or even imagining the original sources that the parapegmatic attributions are taken from, we are faced with some problems. The first and most important is that it has so far proven impossible to reconcile the attributed phases and their *sequentia* in different parapegmata to any satisfying degree.[60] Although we occasionally see what look like identical meteorological quotations on similar dates (either when counted from the summer solstice or reduced to Julian dates),[61] the overwhelming majority of cases show very poor correspondence, if any at all. To take one example, Ptolemy includes attributions to Meton, Euctemon, Callippus, and Dositheus (to name a few) that have no counterpart in Geminus, and vice versa.

Bowen and Goldstein argue that there may have been some problems of translation of star names from the time of the early astrometeorologists to the time of the attributive parapegmata.[62] But this should not present a problem if the original sources of the attributions looked substantially like the later parapegmata. It would, however, be a problem if the original sources lacked calendrical and date-differential information, so that obscure star names could not be figured out from their rising-times relative to known stars.

But even if the astral vocabulary *was* well understood, it is still possible to see how incompatible dates and attributions could have found their way in to the different sources. The attributions may have been taken from a number of different sources. Euctemon, for example, seems to have included stellar phases in his *Phaenomena*, but he could well have mentioned others in some other works, along with a smattering of astrometeorological information here and there. Not all his works need to have been equally detailed or in

---

[60] For more on this problem, see Lehoux, forthcoming.

[61] For example, compare Ptolemy's (P) Phaophi 5 to Geminus (G) Libra 6. Likewise: P Phaophi 27 = G Libra 30; P Mechir 28 = G Pisces 2; P Phamenoth 8 = G Pisces 12; P Epeiph 29 = G Cancer 28.

[62] Bowen and Goldstein, 1988, pp. 56–7.

perfect agreement, and so we can imagine a Geminus or a Ptolemy trying to collect as much of this material as he could, but either incorporating it into his individual schema differently, stressing different bits of it, or even having access to different works or conflicting versions of the same work. This is not even to mention the possibility of mistaken attribution or falsification, as in the *Ars Eudoxica*.[63] All of this by way of saying that we need some way of accounting for the fact that no two extant parapegmata agree on the sequence and timing of more than a handful of attributed phases or weather predictions. The degree of disagreement is so great as to need some accounting for, and scribal error seems an insufficient explanation, particularly if the original texts are supposed to have been formatted as parapegmata, a format that should have to some extent actually discouraged such errors.[64]

We are told by Aelian that Meton set up a stele inscribed with the 'turnings of the sun',[65] which many modern commentators interpret as a full-blown parapegma.[66] But the stele could easily have been something as basic as a list of season lengths, or even something like an Uruk-scheme. This, combined with a simple list of phases and their dates or day-differences (what I call their *order*), would provide enough raw material to translate a set of phases into a later parapegma – and such a list of dated or ordered phases need not even have been inscribed on the stele of 433/2, nor even have all been originally found in one source. Moreover, astrometeorological information need not have been attached to the source for the ordered phases, but may have been culled from other sources, which may have been quite vague, indeed. I offer this simply as an extreme-case possibility to illustrate how cautious we need to be in speculating about the nature of the sources of the attributive parapegmata. I see it as a counter-balance for the common

[63] See *HAMA*, p. 687f.

[64] These issues are most fully discussed in the section on the Codex Vindobonensis parapegma, C.iii in the Catalogue, below.

[65] Aelian, *Var. hist.* x.7.

[66] Rehm, 1941; Pritchett and van der Waerden, 1961. To get a sense of just how detailed Rehm's original hypothesis became, see Wenskus, 1990, p. 28:

Der erste Parapegmatiker, von dem wir wissen, ist der Athener Meton, der sein inschriftliches Parapegma im Jahre 433/32 v. Chr. veröffentlichte. Er scheint nur altbekanntes Phasenmaterial berücksichtigt zu haben: die Sonnenpunkte, den Morgenaufgang des Sirius und den des Orion sowie den Morgenuntergang der Plejaden. Sicher ist, daß sein Parapegma zu den Phasen auch die dazugehörigen Episemasien, also die erwarteten Wetteränderungen, vermerkte: ein Verfahren, das echt griechisch zu sein scheint. Zumindest gibt es in Babylon nichts Vergleichbares.

I wish we *could* be so confident.

contention that the earlier astrometeorologists wrote parapegmata which looked substantially like the much later preserved ones. Not every astrometeorologist is necessarily the author of a parapegma, nor can he be assumed to be the author of a calendar system or of a calendrical cycle (although some certainly are). Conversely, authors of calendar systems are not necessarily the authors of parapegmata. I emphasize this point because a great deal of the modern literature haphazardly associates parapegmata with calendar systems and astrometeorologists with parapegmatists.

Conclusion

The parapegmata are *extra-calendrical* tools for tracking phenomena which are not directly linked to a local calendar. In the case of the Greeks, the calendars in question were lunar, and the parapegmata are overwhelmingly concerned with phenomena tied to the sidereal year. In the case of the Romans, the calendar is fairly well linked with the solar year, and the phenomena tracked in inscriptional parapegmata are frequently non-solar. Some Greek parapegmata treat the sun's entry into zodiacal signs as an important phenomenon, because it was seen as astrometeorologically significant in itself, or as a means to calibrate the parapegma from year to year. The summer solstice, I have argued, was probably used for calibration. On the evidence found in parapegmata and related texts, we find that the Metonic cycle may have been (or included) schematic lists of solstice dates, and the Callippic cycle probably functioned as a regulated astronomical calendar.

5 | Calendars, weather, and stars in Babylon

Never mix your cement when the moon is in Libra.
Kevin McNamee

I. Crossing boundaries

The problems with establishing what relationships there may be between classical Greco-Roman astrometeorology and Mesopotamian astronomy/ astrology may be best highlighted if we consider this sexagesimal number: 29; 31, 50, 8, 20.

It represents 29 days, plus $\frac{31}{60}$ of a day, plus $\frac{50}{3,600}$ (60^2) of a day, plus $\frac{8}{216,000}$ (60^3) of a day, plus $\frac{20}{12,960,000}$ (60^4) of a day. Needless to say, this number is very precise. To put it in decimal notation, it equals 29.53059414 days, a number specific to the eighth decimal place (one part in just under 13,000,000 of a day, or about 6 milliseconds). Just *stating* a number to this many decimal places betrays a real amount of confidence, not to mention skill, in its calculation.

29; 31, 50, 8, 20 days is the value for the mean lunar month in the so-called System B lunar theory of Babylonian astronomy.[1] This number was brought to the light of modern scholarship in 1900 by Francis Xavier Kugler, whose work has been called, probably quite justifiably, 'the finest, the most original, and the most difficult study carried out in the history of science up to the date of its publication', which 'revolutionized . . . the entire history of ancient science'.[2] One of the most important results Kugler published in 1900, one of the findings that was so very revolutionary for our under-standing of ancient science, was the fact that this number – 29; 31, 50, 8, 20 – is also the value for the mean lunar month in Ptolemy's account of the lunar theory of his second-century BC predecessor Hipparchus of

---

[1] The System B lunar tablets come from both Babylon and (especially) Uruk, and they date from the mid-third to the first century BC.

[2] Swerdlow, 1999, p. 7. Kugler's publication is Kugler, 1900.

Rhodes. Among other evidence set out by Kugler,[3] historians saw for the first time that Greek astronomy was indebted to Babylon for some of its most fundamental parameters. 29.53059414 days is not a value that two investigators working independently in different countries and in different centuries could possibly both come up with on their own. It depends on using particular observations of lunar eclipses, with a particular value for their distance apart in days, and (crucially) with small built-in errors due to the inaccuracies of timekeeping at night in antiquity. 29; 31, 50, 8, 20 thus became a kind of fingerprint, identifying the originator of some crucial and celebrated values in Greek astronomy as Babylonian. In the light of the evidence set out by Kugler – and even more material that has been published since – historians have spent the last century re-writing the history of Greek astronomy as one of profound and significant contact with, borrowing from, and development on, a Babylonian original.[4]

But not all sciences preserve fingerprints the way that mathematical astronomy does. In medicine and pharmacology, for example, contact between Greece and Mesopotamia has been much harder to trace. So, too, in mathematics. There is speculation, but the evidence in these other sciences is often of an entirely different nature.[5] This is not to say that contact is impossible to prove, just that the work is different, and so far nothing has come to light to clinch the case one way or the other. But as work in Assyriology progresses, and as we learn more about the Mesopotamian sciences, I am confident that we will find that there was in fact important contact in more areas than just astronomy and astrology.

Turning to astrometeorology, we can ask the question of whether or not Greek and Roman parapegmata follow the pattern of planetary astronomy and astrology in being influenced by or drawing on Babylonian predecessors. The question turns out to be a tricky one to tackle. There are no Babylonian parapegmata, although there *are* texts that show some similarities to some

---

[3] Specifically the numerical values for the relationships between the synodic, draconitic, and anomalistic months. A synodic month is the length of time from full moon to full moon, or new moon to new moon. The draconitic month is the period for the moon's variation in latitude, that is, from the moon's crossing of the ecliptic to its next similar crossing. The anomalistic month is the period of the moon's variation in longitudinal speed, from its moment of maximum or minimum speed to its next maximum or minimum.

[4] For the full version of contact between Greece and Babylon in astronomy, see Kugler, 1900; Aaboe, 1955; Neugebauer, 1975; Toomer, 1988; and a string of publications by Jones: 1991; 1993; 1996; and especially 1999a; also Britton and Jones, 2000. For astrology, see Barton, 1994; Rochberg, 1998; 2004. On Babylonian lunar theory generally, see Neugebauer, 1955; 1975; Brack-Bernsen, 1997.

[5] For medicine, see Nutton, 2004. On mathematics, see Robson, 2003.

aspects of the later Greek and Latin texts, and these are interesting in their own right. The questions to be addressed in this chapter revolve around on the one hand how extensive these similarities are and whether or not they will allow us to assert or speculate on a Babylonian influence on Greek astrometeorology, and on the other how the Mesopotamian texts themselves work and interrelate within their own scientific, historical, and cultural contexts. A caveat: I do not want to be understood here as being engaged in a search for Mesopotamian 'precursors' to Greek practice – *precursors* in that vague and outdated sense that saw Aristarchus of Samos as the 'Ancient Copernicus', for example.[6] Instead, the point is to address a century's worth of arguments by a wide range of scholars that claim to have found some evidence for a direct borrowing of astrometeorological material by Greece from Babylon. In order to give their arguments the fullest airing possible, I have chosen in this chapter to collate and examine as full a range of evidence as possible, including evidence that may not have been available or fully understood when the various claims were first made. This approach also has the side effect that the material in this chapter is treated thematically (meteorology, calendrics, stellar astronomy) rather than strictly historically or geographically. That is: it does not aim to be a history of Mesopotamian astrometeorology and calendrics (a subject that would occupy a much-needed monograph unto itself) but to be an examination of the relationship of the Greco-Latin traditions to the Mesopotamian, which of necessity begins with a collation and examination of the relevant Mesopotamian traditions.

On the question of Greco-Roman indebtedness, modern scholars have generally conjectured that Greco-Roman astrometeorology borrowed from a Babylonian original.[7] Recently, for example, Wenskus has favoured a Babylonian influence on some of the early Greek conceptions of the constellations and astronomical time reckoning, although she admits that the evidence is far from complete.[8] Graßhof, on the other hand, has argued that in at least one parapegma, Ptolemy's, there is evidence that Babylonian methods and/or data were in fact *not* used.[9] This chapter will attempt an exhaustive presentation and a weighing of the current evidence for the relationships between Babylonian and Greek stellar astronomy and astrometeorology. We will see that, although there are scattered omina which show similarities to parts of the different types of parapegmata, none preserves anything more

---

[6] See Heath, 1913.
[7] Rehm, 1941; van der Waerden, 1984a,b,c; See also Bowen and Goldstein, 1988.
[8] Wenskus, 1990, pp. 21–4.   [9] Graßhof, 1993.

than a similarity in kind, indicating less a borrowing than an independant pursuit of similar goals.

Most similar to the classical parapegmata is a text called MUL.APIN, which has a list of schematic heliacal rising dates, and some seasonal meteorological predictions. It is certainly not a parapegma, however, and the similarities between it and classical parapegmata are not close enough to warrant a claim of Mesopotamian influence. Nevertheless, MUL.APIN and other Mesopotamian texts do show that problems with timing annual climactic cycles were sometimes handled in analogous ways in Mesopotamia, Greece, and Rome, and this in itself is interesting as it points to similar sets of solutions to similar kinds of problems being found independently in these different cultures. On the other hand, other texts that superficially show some similarities to parapegmata, such as the so-called *astrolabes*, which associate calendar months and stars, will be seen to be really quite different from classical parapegmata. In the case of the astrolabes, I will show that the associations between the months and the fixed stars were most likely mythological (and possibly ominous), rather than truly representing the *dates* of any stellar phases.

This contrast between one set of Mesopotamian texts that runs roughly parallel (but *only* roughly parallel) to classical sources and another set that proceeds differently and to a different destination will help to highlight the ways in which different cultures both frame and flesh out their approaches to the natural world, and will, I hope, lead to a better understanding of each of the two traditions.[10]

II. Astronomy and the Babylonian calendar

The Mesopotamian calendars were luni-solar.[11] Months were observationally determined, and were counted from first sighting of new moon crescent to first sighting of new moon crescent, although it would seem (from the lack of any attested months longer than thirty days) that some rule was in place to prevent the month from running beyond thirty days, even in cloudy weather.[12] Months were intercalated from time to time, to keep the

---

[10] Such a comparative method has been successfully used In Lloyd's recent work. See, e.g., Lloyd, 1996; 2002; 2004; Lloyd and Sivin, 2002.

[11] With the exception of the Assyrian calendar before the late second millennium, which was lunar.

[12] For a discussion of month length, see Huber, *et al.*, 1982.

year roughly in line with the seasons. There seem to have been, by the late second or early first millennium BC at the latest, rules for when to add and when not to add a thirteenth month to any given year. Determining exactly how early such rules were in place depends crucially on determining which texts did or did not contain intercalation rules. A second problem, by no means trivial, is to then try and determine if a given intercalation rule, once found, was actually used in practice. Finally, the question of whether or to what extent such intercalation rules (or the texts in which they are found more generally) are related to known Greek or Latin luni-solar cycles, such as the Metonic and Callippic cycles, can be asked.

As early as the Sumerian period (third millennium BC), there is some indication that astronomy was associated with the regulation of the agricultural year in Mesopotamia. Koch-Westenholz says:

Two of [the goddess] Nisaba's characteristic traits [are] her lapis-lazuli tablet, originally associated with astronomy, and her management and fair distribution of agricultural products . . . The overall impression given by the Sumerian sources is that Nisaba was mainly concerned with the management of agriculture and the timing of activities that were dependent on the yearly seasons. The knowledge of astronomy . . . attributed to her was used to correct the vagaries of the lunar calendar.[13]

In a note, she adds:

This is a fairly exact rephrasing in our terms of the statement in *Lugal-e* 721, 'together with Suen, she counts the days'.

Just saying that she 'counts the days', though, is less specific than saying that she has a role in timing seasonal activities and correcting the lunar calendar. While we can see that agriculture, astronomy, and dating are somehow combined in Nisaba's functions and attributes, the specifics are still obscure. The best we can say here is that astronomy, dating, and agriculture seem to have represented, to some extent, a set of related practices in Sumerian society.

Although the evidence for the very early periods is scant, we do know that in later times the periodic insertion of an intercalary month was meant to keep the lunar calendar roughly in line with the agricultural year,[14] and that by the fifth century, intercalation was regulated by a nineteen-year cycle (the Uruk scheme),[15] but before then the intercalations were sporadic.

---

[13] Koch-Westenholz, 1995, p. 33.

[14] The practice of intercalation was also taken up in Assyria beginning in the late second millennium. See Weidner, 1935–6, pp. 28–9; Horowitz, 1996, p. 37, n. 5.

[15] On which, see Rochberg-Halton, 1992, p. 810; Slotsky, 1993; and also see chapter 4, above.

II.i. Intercalation and the astrolabes generally

But if the pre-fifth-century calendar was intercalated irregularly, how was it determined, in any given year, whether or not to add an extra month? Horowitz has argued that the heliacal risings of selected fixed stars were used to determine whether or not a particular year would have twelve months or thirteen.[16] More specifically, he argues that the so-called *astrolabes*,[17] which were basically lists of three different stars associated with each month of the year, were used to determine which stars should rise in which lunar month. He bases this in part on a report by one of Aššurbanipal's scribes: 'Let them intercalate a month. All the stars of heaven are late. Let Adar not pass unluckily. Let them intercalate it.'[18] There are also omens pertaining to 'early' and 'late' risings of stars (e.g., in *EAE* 51, on which, see below).

But there is also abundant evidence for a three-year intercalation cycle, as well as for a 360-day stellar year,[19] and it is unclear how or if these were supposed to be used in conjunction with the lunar calendar and the astrolabes, or whether there were regional, professional, or other uses for each cycle. Lastly, Hunger and Reiner, and Cohen after them, raise the (not unlikely) possibility that intercalations were based on observations of agricultural and meteorological phenomena up until the first millennium.[20]

Moreover, there are some problems with seeing the astrolabes as listing ideal heliacal risings. In the first place, the associations of the stars with particular months and the three 'paths'[21] seems to be in part purely mythological,[22] in the second place, some of the stars in the astrolabes are *planets*, with no annual cycles to preserve in a calendar, and in the third place, two

---

[16] Horowitz, 1996. This claim was earlier made by Weidner, 1928– , pp. 72–3.

[17] The name astrolabe is perhaps unfortunate, since these texts are completely unrelated to the more modern instruments known as astrolabes. On Mesopotamian astrolabes, see Hunger and Pingree, 1999, pp. 50–7; Donbaz and Koch, 1995; Cagirgan, 1985; Reiner and Pingree, 1975–98, vol. II; van der Waerden, 1949; Weidner, 1915. Horowitz has promised a new edition of the astrolabes as part of the *AfO* Beiheft Series (see Horowitz, 1996, p. 37, n. 5), which we eagerly await. Such an edition would be very helpful for clarifying what we know and can say about these texts.

[18] Hunger, 1992, no. 98, rev. 8–10; Horowitz, 1996, p. 38.

[19] See Horowitz, 1996, pp. 38–41; On the intercalation cycle, see Britton, 2002; Hunger and Pingree, *MUL.APIN*, II.ii.11–12, on the stellar year, I.iii.35–50.

[20] Hunger and Reiner, 1975, p. 21; Cohen, 1993, p. 5.

[21] The 'paths' or Ea, Anu, and Enlil probably refer to segments of the horizon over which the stars were seen to rise. See Reiner and Pingree, 1975–98, vol. II, p. 17; Brack-Bernsen, 2003;

[22] Reiner and Pingree, 1975–98, vol. II, p. 3; Hunger and Pingree, 1999, p. 63; Langdon, 1935, p. 10f.

of the stars are circumpolar, and it is difficult to see how these might have been used in a calendar.[23]

II.i.a   Intercalation and 'Astrolabe B' specifically

Astrolabe B,[24] unlike the other astrolabes, states *explicitly* that the stars named rise in their month. For example:

Month of Arahsamna: the Mad Dog, the Scorpion, and the King rise.
The Pleiades, the Old Man, and *Anunītu* set.[25]

The text itself was copied at Assur in the late second millennium,[26] and is the oldest of the astrolabes. It is a bilingual Sumerian/Akkadian text.[27] Part A of this text associates each of the twelve months with a constellation, a god, and with mythological events:

Month of Nisannu, Pegasus, seat of Anu[28]
the king is installed, the king is appointed;[29]
a good beginning[30] for Anu
and Enlil; month of Nanna-Suen,
oldest son of Enlil.[31]

Some agricultural activities are associated with months as well.[32]

---

[23] See Hunger and Pingree, 1999, p. 50f. Langdon tries to use the modern composite reconstruction known as 'Pinches' astrolabe' together with the menologies to reconstruct features of (a) an archaic Sumerian calendar, which started with the month of Nisannu at the rising of the Pleiades, and (b) a prehistoric calendar which began with the autumn equinox (Langdon, 1935, pp. 1–2, 21, 67), but he falls prey to the same problems with the astrolabes as Horowitz does.

[24] Casaburi, 2003; Schroeder, 1920, no. 218; Transcribed in Weidner, 1915, pp. 66–7, 76–9, 85–7. A much improved transcription of section A can be found in Reiner and Pingree, 1975–98, vol. II, pp. 81–2. For duplicates of Astrolabe B, see Cagirgan, 1985.

[25] Astrolabe B, part C, ll. 27–8.

[26] During the reign of Tiglath-Pileser I (1113–1075 BC), according to Horowitz, 1996, p. 37, n. 5; van der Waerden, 1949, p. 10. Hunger and Pingree, 1999, p. 51, date it to the reign of Ninurta-apil-Ekur (1191–1179 BC).

[27] Cohen notes that it is unclear whether Akkadian or Sumerian is the original language, and that the translation from the one language to the other is 'inexact'. Cohen, 1993, p. 306, n. 4.

[28] Weidner, 1915, p. 87, adds a verb to this sentence in his translation: 'Zum monat Nisan gehört der kakkab*DIL-GAN* . . .'.

[29] *šarru iššaqan.* Weidner translates this as 'der König wird proklamiert' and Cohen, 1993, p. 306, as 'the king is invested (with authority)'.

[30] *šurrum damqum.* I am unsure why Cohen leaves *šurrum* untranslated: 'a good . . . for Anu and Enlil'.

[31] Astrolabe B, A.I.7–11; For a text closely related to this section, see Reiner and Pingree, 1975–98, vol. II, text X, ll. 24–35 and 37–49.

[32] E.g., at Astrolabe B. A.I.19–25, I.45–50, II.43–6.

Part B of this text[33] is a list of stars in the paths of Ea, Anu, and Enlil, and part C schematically lists three constellations, each in its path, for each month, and then says that those three constellations rise in that month, and that three other constellations set.[34] The constellations which are said to set are specifically those that rise six months later. For example:

Month of Nisannu: Pegasus, *Dilbat*, and the Plough Star rise; Vela, the Scales, and *Entenabarḫum*[35] set.

Month of Ajaru: the Pleiades, the Old Man, and *Anunītu* rise; the Mad Dog, the Scorpion, and the King set.[36]

and six months after Ajaru:

Month of Arahsamna: the Mad Dog, the Scorpion, and the King rise; the Pleiades, the Old Man, and *Anunītu* set.[37]

The primary obstacle to reading this as some kind of intercalation rule is that this schematic six-month difference is astronomically impossible, and so could not have been used to regulate the calendar. Hunger and Pingree, in a discussion of a similar passage in MUL.APIN, point this out most forcefully:

This list is meaningless as an astronomical document (it is basically mythological), as is also the list at the end of *Astrolabe B* where this list is mechanically converted into one in which three constellations rise in a month and three set. The person who composed this totally misunderstood the nature of his source (already in − 1100!) and unfortunately misled several scholars of this century.[38]

Donbaz and Koch have published yet another type of astrolabe, which they argue was used for the determination of intercalary years, but this interpretation has also been called into question by Hunger and Pingree.[39]

On balance, I think the fact that the astrolabes in general prove unworkable as astronomical calendar regulation devices indicates that they were probably not intended to be used for intercalation. Finally, to address a claim made by van der Waerden, I would add that this means that there is no Greek or Latin analogue for the Mesopotamian astrolabes, because the

---

[33] Part B is transcribed in Weidner, 1915, pp. 76–9, and part C at pp. 66–7.

[34] There is a minor departure from this formula in the month of Tešritu, ll. 25–6.

[35] Weidner transcribes this as EN.TE.NA.MAŠ.ŠÍG. *Entenabarḫum* is equated with part of Centaurus in Koch-Westenholz, 1995, p. 207. In Reiner and Pingree's catalogue, MUL.EN.TE.NA.BAR.ḪUM = *ḫabaṣīrānu*, 'Centaurus' (Reiner and Pingree, 1975–98, vol. II, pp. 10–16). Labat, 1988, p. 303, reads MUL.EN.TE.NA.BAR.HUM as *ḫaṣīrānu*.

[36] Astrolabe B, part C, ll. 13–16.   [37] Astrolabe B, part C, ll. 25–8.

[38] Hunger and Pingree, 1999, p. 63.

[39] Donbaz and Koch, 1995; Hunger and Pingree, 1999, p. 57.

astrolabes show no real similarity with anything in classical stellar astronomy or astrometeorology.[40]

II.ii.   Intercalation in MUL.APIN

MUL.APIN ('*The Plough Star*') is an early first-millennium astronomical compilation known from copies at Assur and Nineveh, which includes a star list, a list of heliacal rising-dates, and another of their date differences, a dated list of *ziqpu* stars (stars that culminate at known time intervals), and a list of constellations in the path of the moon.[41] There is some planetary material as well, including periods of planetary visibility and invisibility. In MUL.APIN we also find explicit observational astronomical rules for determining intercalation, according to two distinct schemes:[42] one at II.i.9–24, and the other at II Gap A.1–ii.20. The first uses stellar phases together with the moon's position relative to the Pleiades or to the constellation called the *Hired Man* (our Aries) in order to determine whether a given year is intercalary. The second uses ideal equinox dates and the risings of certain stars, together with the conjunctions of the moon with the Pleiades on particular dates to determine intercalation. We should note the central role of the moon in both of these schemes, which distinguishes them from the strictly stellar schemes that scholars have argued for in the astrolabes. As Hunger and Reiner have shown, evidence from attested dates indicates that the *Pleiaden-Schaltregel* may not have actually been implemented in practice, and different versions of it are contradictory.[43] As with the astrolabes, we again see no similarities with any Greek or Latin texts, at least in the intercalation schemes of MUL.APIN. We shall see in a moment that the astrometeorological parts of the text, on the other hand, look more promising.

III.   Astrometeorology

There are two sources for Mesopotamian astrometeorology: MUL.APIN and *Enūma Anu Enlil*. Both are compilations dating from around the late second

---

[40] Van der Waerden's claim to the contrary (in van der Waerden, 1949) is based on a misunderstanding of how the astrolabes work.

[41] Published by Hunger and Pingree, *MUL.APIN*. See also Koch, 1995–6; Horowitz, 1989–90. On *ziqpu* stars, see Roughton, Steele, and Walker, 2004; Schaumberger, 1952.

[42] *MUL.APIN*, pp. 150–2.

[43] See Hunger and Reiner, 1975, pp. 26–8; Rochberg-Halton, 1992, p. 811.

or early first millennium BC, and both are probably based on much older material. We shall see from a close look at these texts that they do in some instances show similarity to the Greco-Roman material, but not enough to argue positively for an influence. Probably what they represent are distinct traditions that solve some of the problems of agricultural timekeeping and weather prediction in ways that are reminiscent of, but never identical to, the classical parapegmata.

III.i.   The fixed stars and weather omina in *EAE* 50 and 51

The central celestial omen series, *Enūma Anu Enlil (When Anu and Enlil)*, was first published as *L'Astrologie Chaldéenne* by Virolleaud between 1908 and 1912. More tablets have since been discovered, and Virolleaud's original publication has been improved on, although publication is not yet complete. Newer editions and translations of parts but by no means all of this important series have since been published,[44] and this work continues. The series itself consists of sixty-eight or seventy tablets of omina, and dates as a collection from around the beginning of the first millennium BC,[45] although much of it is probably derived from older material. Tablets 1–22/3 contain lunar omina, 23/4–39/40 solar omina, 40–49/50 meteorological omina, and the remainder planetary and stellar omina.[46] A sample of weather apodoses from *EAE* tablet 50 will give the flavour:

I.17: The Star of the sunset (is) for *(ana)* raining.
II.1: The Rainbow (is) for raining [. . .
II.3: The False star (is) for the rising of wind.
II.14b: *Dāpinu*[47] (is) for [. . .] wind.
III.4: The Rainbow (is) for not raining.
III.4a: 'When on a cloudy day when it has rained a rainbow has arched, it will not rain.'
III.5: *Entenabarḫum* (is) for early wind.
III.5c: . . . the early sown cultivated field will be fine, at the end of the year rain will cease.

---

[44] Reiner and Pingree, 1975–98; Rochberg-Halton, 1988; Al-Rawi and George, 1991; van Soldt, 1995; Verderame, 2002. For commentary, see Hunger and Pingree, 1999; D. Brown, 2000.

[45] Hunger and Pingree, 1999, p. 12 give this date, but Koch-Westenholz, 1995, p. 78, dates it to the eleventh century.

[46] Different recensions sometimes have different tablet numbers, and occasionally even combine the omens differently from more than one tablet. For a discussion of the variations in the recension, see Koch-Westenholz, 1995, pp. 75–6; 79–82.

[47] MUL.UD.AL.TAR = *Dāpinu*, a name for Jupiter. Reiner and Pingree translate this literally, as 'the Heroic'.

III.6: *Ṣāriru*[48] (is) for the rising of wind.

III 6b: 'When *Ṣāriru* has been red, the flood will increase.'

III.6c: 'When the Field's stars[49] have been very red, the flood will increase.'

III.7: The False star (is) for the rising of wind.

What we should note here is that only a minority of the omens have clearly stated protases and apodoses ('When the Flashing star is very red, the flood will increase'). For the most part, there are only associations of stars with weather phenomena ('*Entenabarḫum* (is) for early wind'), where the precise nature of this association is not made explicit. A similar formula appears in the 'Great Star List'.[50]

Tablet 51 contains a list of stars and the lunar months of their ideal risings (and occasionally their settings), as well as omens derived from their risings. The months of their risings are 'ideal' in the sense that when the star rises in this month, it is generally taken as a non-ominous event. It is usually seen as ominous when the star rises either before or after its ideal month. For example:

IX.1: The Field rises heliacally in Nisannu. When this star has risen early: [. . .] When this star has risen late and has passed by its month and has risen [. . .

IX.2: The Bristle rises heliacally in Ajaru. When this star has risen early, the gods will give good counsel to the land. When this star is late and has passed by its month and has risen [. . .[51]

There are some agricultural and meteorological omens, such as:

IX.11: In Nisannu, the star of the field [. . .]. When it has risen heliacally at its specified time, the irrigated land and the cultivated lands in the land will prosper. When it has risen heliacally not at its specified time, the irrigated land and the cultivated lands will not prosper. [. . .] will not bring forth [. . .

---

[48] MUL.AN.TA.SUR.RA. Reiner and Pingree translate this as 'the Flashing Star', and comment that it probably refers to a shooting star or meteor. Labat, 1988, p. 301, simply translates it as 'Brillante'.

[49] The Field, MUL.AŠ.GÁN, is α, β, and γ Peg, and α And. See Reiner and Pingree, 1975–98, vol. II, p. 11.

[50] Published and translated by Koch-Westenholz, 1995, Appendix B.

[51] This follows Reiner and Pingree's translation, with minor changes. Throughout this book I translate *šumma* + preterite as 'When *x* has happened' rather than as 'If *x* happens'. My reasons for this are outlined in Lehoux, 2002.

X.23: In Addaru the Fish, Ea, [lord of] mankind [. . .] high water will mount in the
    springs. When [it has risen] not at its specified time, rain and high water will
    be scarce in the springs.[52]

As Reiner and Pingree show, tablet 51 derives from a tradition closely
related to Astrolabe B, and indeed, parts of their text X duplicate parts of
Astrolabe B.[53]

But there are some important differences between the *EAE* omens and
the astrometeorology we have been looking at in the other chapters of this
work. The weather omens in *EAE* do not derive simply from the phases of
the stars. We do not see the familiar Greco-Latin pattern of 'when star *x* rises,
there will be rain'. Rather we find the unspecified relation of a star with a
weather phenomena: 'star *x* is for weather *y*'. And the 'stars' mentioned may
not even all be stars: there are the 'Flashing star' (a meteor?) and the 'False
star', for example. There are also omens derived from the apparent colours of
stars.[54] The most complex of the fixed-star omens involve the relationships
of the phases and the Babylonian lunar calendar. A phase which occurs in
a specified month is seen as non-ominous, and a phase earlier or later than
this is ominous. There is no parallel for these kinds of omens in Egypt,
Greece, or Rome. *EAE* is a different sort of text from classical parapegmata.

I should perhaps also mention a unique weather omen, which seems
to be derived from an intercalation rule, of all things, in the text K.
3923 + 6140 + 83-1-18, 479.[55] It says:

When, in the month of Addaru, on the 25th d[ay, you
    [have observed the Pleiades and the moon and they]
    [had the same longitude, then this year is normal;]
    [when they have fallen down, then it is left]
    [behind . . .][56]
Good winds will come [. . .[57]

It is unclear whether this may be related to the *EAE* omens.

---

[52] Text X is related to, but not a duplicate of Text IX (from which the previous three examples
were taken). For commentary, see Reiner and Pingree, 1975–98, vol. II, pp. 52–4.

[53] Reiner and Pingree, 1975–98, vol. II, pp. 1, 52–4.

[54] Compare also ll. 150–3 of the *Great Star List* in Koch-Westenholz, 1995, pp. 192–5 and
MUL.APIN II.iii.33, II.iv.1–2.

[55] Text A in Hunger and Reiner, 1975, p. 22. My translation follows Hunger and Reiner's, with
minor modifications.

[56] The reconstruction here is based on the formula throughout the rest of the text. It is unclear
whether part of this formula or some extra (now lost) material was meant to be the protasis to
the meteorological apodosis.

[57] Line I A in Hunger and Reiner, 1975, p. 22.

iii.ii.   Meteorology in MUL.APIN

It is in MUL.APIN that we see texts which most resemble the parapegmata, although without the level of precision and detail generally found in the classical texts, and it is MUL.APIN that is most often cited as the possible Babylonian analogue or precursor of classical astrometeorology. MUL.APIN i.ii.36–iii.12 schematically dates the heliacal risings of certain fixed stars. These ideal dates of visibility are paired with the culminations of other constellations (*ziqpu* stars) later in the text.[58] The list is schematic, insofar as all the dates given are either the first, fifth, tenth, fifteenth, twentieth, or twenty-fifth of the month. It is useful as a sequential ordering of risings, but it notably uses only approximate dates.

These heliacal risings are followed by the related lists of the simultaneous risings and settings of fixed stars, and the schematic date differences (again, all multiples of five days) between the heliacal risings, but it is unclear how these lists were used. Much has been made in the German literature of MUL.APIN's list of date differences between stellar phases, and this has been seen as a parallel to a handful of such lists in classical texts.[59] Rehm talks about these and the MUL.APIN list as though they were all representatives of the same genre of text, a genre called ἄστρων διαστήματα, *The Differences between Stars.* Van der Waerden and Wenskus follow him in their discussions of texts of this type.[60] But the genre, as such, is not actually attested in ancient sources, Rehm's artificial coining of the name in classical Greek notwithstanding. The fact that two cultures should both come up with lists of day-differences between phases is not enough to allow us to posit a whole lost genre of text that made its way from Babylon to Greece. If the list of stars and their exact date-differences were similar enough in detail, then a case could be made for a cross-cultural transfer, but there is no Greek or Latin text that preserves anything like MUL.APIN's schematic multiple-of-five day-differences.

Elsewhere in MUL.APIN, at the risings of certain stars (but not of all) we are told to 'observe the wind that blows',[61] and later, certain stars are related to wind directions with the formula 'constellation *x* lies across (*ina*

---

[58] MUL.APIN i.iv.13–I.iv.30

[59] Purported classical parallels include: Varro's list at *RR*, 1.28, the Eudoxus papyrus (see F.ix in the Catalogue, below), a short excerpt from Diocles of Carystus (see F.v in the Catalogue), and the manuscript C. Vindob. Gr. philos. 108, f. 282ᵛ (C.iii in the Catalogue).

[60] Rehm, 'Parapegma,' *RE*, col. 1304f.; Rehm, 1941; van der Waerden, 1984a,b,c; Wenskus, 1990, p. 27f.

[61] MUL.APIN ii.i.25–37; a similar admonition for the observation of Mercury at harvest time occurs at ii.i.58.

ZI) wind y, which has been generally taken to mean simply that it lies in one of the four cardinal directions, rather than being an astrometeorological association.[62]

General seasonal meteorological predictions are, however, derived from the sun's presence in each of the three paths of Anu, Enlil, and Ea, and these are related to their calendar months:

From the first of Addaru until the 30th of Ajaru the sun stands in the path of the
    Anu stars; wind and weather.
From the first of Simanu until the 30th of Abu the sun stands in the path of the Enlil
    stars, harvest and heat.
From the first of Ulūlu until the 30th of Arahsamna the sun stands in the path of
    the Anu stars; wind and weather.
From the first of Kislīmu until the 30th of Šabatu the sun stands in the path of the
    Ea stars; cold.[63]

The omens at the end of MUL.APIN include some meteorological and agricultural predictions, such as 'When the U.RI.RI-star (Mercury?) has been seen, rain and flood';[64] 'When the star of Marduk has been seen at the beginning of the year, in this year the crop will prosper.'[65] I note that all such omens seem to be either (a) derived from what appear to be planets, or (b) are related to the colour or brightness of stars at their rising, rather than to the phases themselves.

On balance, we see that the differences between MUL.APIN and the classical texts seem to outweigh their similarities. What we have is probably independent traditions that sometimes solve similar problems in analogous ways, but which do not share any information at a detailed level.

iv.   The *Astronomical Diaries*

The last main locus of an interrelation between astronomy, weather, and calendars in Mesopotamia is in the so-called *Astronomical Diaries* from Babylon. The *Diaries* are a truly remarkable series of texts that record night-by-night dated observations of astronomical, celestial, and meteorological

---

[62] MUL.APIN ii.i.68–71. I follow Hunger and Pingree's interpretation of *ina* ZI tentatively. See, e.g., Hunger and Pingree, 1999, p. 73. For the normal meanings of ZI in the astronomical texts, see, *ACT*, pp. 496–7.

[63] MUL.APIN ii.Gap A.1–8.

[64] MUL.APIN ii.iii.22. Compare ii.i.57. For the association with Mercury, see *MUL.APIN*, p. 134, commentary on iii.22.

[65] MUL.APIN ii.Gap B.1.

phenomena, as well as such things as commodity prices, river levels, and political events.[66] The observers seem to have been employed by the temple of Marduk.[67] We get the mundane and the monumental together, as in a happily extant dated report from 323 BC, understated to the point of poignancy: 'The 29th: the king died; clouds [. . . .]'. Here the death of Alexander the Great is just one fact among many ('clouds . . .') to be recorded without subjective commentary. But what is most fascinating about these texts is their duration. They seem to have been recorded, uninterruptedly, from the seventh or eighth centuries through to the first century BC, and possibly even into the first century AD, a period of six, seven, or even eight centuries.[68] Even on the short dating, this would make the Babylonian recording of the *Diaries* the longest-running continuous research programme in the history of the sciences.

To get a feel for the *Diaries*, look at the following excerpt:

Night of the 15th, last part of the night, the moon was $2\frac{1}{2}$ cubits in front of [γ] Capricorn. The 15th, gusty north wind.[69] . . .

That month, the equivalent (of 1 shekel of silver was): barley, 1 pān 1 sūt $1\frac{1}{2}$ qa, in the middle of the month, 1 pān 1 sūt 3 qa, at the end of the month, 1 pān 1 sūt $1\frac{1}{2}$ qa; dates, 1 pān 3 qa [. . . .] At that time, Jupiter and Saturn were in Gemini; Venu[s was in Virgo; Ma]rs was in Virgo, at the end of the month, in Libra; Mercury, which had set, was not visible.

That month the river level – remainder $\frac{1}{2}$ cubit and 8 fingers . . .[. . .'[70]

There has been much debate about the purpose of the *Diaries*. Swerdlow argues that they were tied to omen-watching, and Slotsky, Hunger, and Pingree have argued that they were primarily for the determination of astronomical parameters.[71] But two parts of Hunger and Pingree's argument will not stand: (1) they claim that 'the weather is reported [in the *Diaries*] because it affects observations', but this cannot account for the frequent mentions of wind direction or temperature, entries such as 'The 26th, cold

---

[66] See Sachs and Hunger, 1988. Although the *Diaries* are generally referred to as observation texts, some astronomical phenomena are clearly computed. See Sachs, 1948, p. 271. John Steele has pointed out to me how frequent such computations are in the *Diaries*: solstices, equinoxes, and Sirius phenomena are always computed. Computations of eclipses and planetary phenomena that could not be seen either because of cloud cover or because they happened in daytime are also common. I would also add that many of the historical events are second-hand reports rather than the records of the scribe's own observations.

[67] See van der Spek, 1985.   [68] Hunger and Pingree, 1999, p. 139f.

[69] Sachs and Hunger, 1988, −324, B obv. 21.   [70] Sachs and Hunger, 1988, −324, B rev. 9–11.

[71] Swerdlow, 1998; Slotsky, 1997; Hunger and Pingree, 1999, pp. 139–40; but Hunger, 1999, p. 80 is qualified.

north wind',[72] which do not affect observation at all; and (2) they claim that 'the *Diaries* treat periodic phenomena as predictable; this deprives them of their meaning as omens'. If this were true, then there can be no explanation for the frequent use of, for example, eclipses (which were predictable) as ominous portents in official reports and letters. I think here they pin too much on the idea that *randomness* must somehow be inherent in omens. A third point I would raise is that there is no reason to suppose that economic phenomena could be seen as periodic, but that weather was not.

Rochberg has set out the most comprehensive evidence for a close connection between diary texts, mathematical astronomy, astrology, and omens.[73] She has shown that the scribes who were copying and reading mathematical astronomical texts were also copying and reading astrological and omen texts, and in at least one text (*CT* 49, 144) are reported as having been also responsible for the nightly watch associated with the *Astronomical Diaries*.[74] Hers is the strongest argument I have seen for a picture of late Babylonian science as including both mathematical astronomy and divination under the aegis of the same practitioners, who are further operating under the auspices of particular temples at Babylon and Uruk, and who, finally, are also associated with the compilation of the *Diaries*. If this picture is correct, then the *Diaries* become part of a set of related practices that include *both* astronomy and divination, which makes perfect sense given the contents of the texts themselves. The institutional setting, the organization of a long-term continuous research programme, and the particular combination of interests reflected by the *Diaries* have no parallel in Greece or Rome.

v.   Good and bad luck days in the Babylonian calendar

As in Egypt and Rome, the Mesopotamians ascribed good and bad luck to different calendar dates. There are a group of Akkadian texts, called *hemerologies* or *menologies*[75] which tell us that particular lunar calendar dates were in general favourable or unfavourable, and more specifically also offer lists of activities which were prescribed or proscribed for each day. Many of these activities were of a religious nature (e.g., 'an offering to Anu

---

[72] Sachs and Hunger, 1988, −369, rev. 12.   [73] Rochberg, 1993; 2004.

[74] See also van der Spek, 1985.

[75] For a detailed description as well as an edition and translation, see Labat, 1939 and Labat, 1965. Compare the Egyptian *Calendar of Lucky and Unlucky Days*, in chapter 6, below. In this work, I will refer to both the menologies and the hemerologies as 'hemerologies'.

will be accepted'), but some economic, medical, divinatory, and dietary proscriptions are also found. There are some astrological omens, and some stellar divinities (the Pleiades, Orion) are mentioned as being receptive to offerings on particular dates in one of the texts.[76] Some tablets list the months favourable for certain activities, and others derive omina from activities performed in certain months.[77] These texts are more formulaically and topically diverse than their Egyptian and Roman counterparts.

VI.   Other texts

There are two interesting astrometeorological texts from Uruk (*TU* 19 and 20)[78] which use certain planetary phenomena to predict weather. The phenomena of the protases include oppositions of Jupiter and Mars, conjunctions of Mercury and Venus, planetary passings through certain constellations (including the Pleiades and Perseus), and more. *TU* 20, like *TU* 11, has a 'goal-year' type of scheme for predicting the weather.[79]

I should also note, at least in passing, the existence of a late tenth- or early ninth-century BC Hebrew agricultural calendar that reports the activities for each (unnamed) month of the year, beginning with the autumn harvest. It is written in verse on a stone with a single hole in the middle, presumably for mounting it for display. In its entirety it reads:

His two months are (olive) harvest,
　　His two months are planting (grain),
　　　　His two months are late planting;
His month is hoeing up of flax,
　　His month is harvest of barley,
　　　　His month is harvest and feasting;
His two months are vine-tending,
　　His month is summer fruit.[80]

---

[76] See *KAR* 178 (in Labat, 1939, pp. 61–2), on the 18th and 19th of Nisannu, respectively.

[77] E.g., 'The king purifies his garments: Nisannu, Ajaru, Simanu, Abu, and Tašrītu are favourable' (*KAR* 177 obv. 2.25–7, in Labat, 1939, p. 155); 'When, in the month of Nisan, the foundations of a house have been built on the 16th day, the house will not be finished' (Labat, 1965, §2.1).

[78] Hunger, 1976.

[79] *TU* 20, Rs 2–4; *TU* 11, Rs 23. A goal-year text is one which predicts planetary phenomena for the coming year based on a cycle of *x* years for that planet, and the occurrence of identical phenomena *x* years ago. See Sachs, 1948.

[80] Albright's translation, as republished with further commentary in Cohen, 1993, p. 383.

Lastly, there is a third-millennium Sumerian text, sometimes called the *Sumerian Farmer's Almanac*, or the *Georgica Sumerica*,[81] which I mention only for completeness. It consists of a list of farming instructions, but these are listed only sequentially, and with no reference to astronomy or a calendar.[82] The name *Farmer's Almanac*, and indeed also the implied comparison with Vergil's *Georgics*, is therefore unfortunate.

Conclusion

Contrary to what many scholars have argued, none of the Mesopotamian material shows a clear relationship to the parapegma tradition, although we can see from the material surveyed here that there were a number of parallels to various aspects of the parapegmata. Where Roman parapegmata were partly concerned with good and bad luck days, so we find the Babylonian hemerologies with a similar concern, although differently realized. Likewise, we find some Babylonian texts which report the dates of stellar phases, but this is schematic and perhaps not meant to be precise. Lastly, weather omina are numerous but scattered in the Babylonian omen literature.

Although we find no simple cognate of a parapegma in Mesopotamia, we do find diverse texts which here and there show interesting similarities. While there is not enough evidence to argue for a line of descendancy from Babylon to the classical parapegmata, it does show how these two different ancient cultures addressed agricultural timing, weather prediction, and lucky calendar days in sometimes remarkably similar ways. But the organization of astronomy and astrology in Mesopotamia as a set of disciplines under, probably, the auspices of particular temples, combined with the continuity of the tradition and the particular styles of the organization of data, simply have no parallel in the Greco-Latin traditions. In Greece and Rome, we see by contrast what appears to be a handful of scattered practitioners, separated in both place and time, who have varying interests, research objectives, and varying degrees of access to each other's work.[83]

---

[81] Civil, 1994; Bauer, 1998; Salonen, 1968. For commentary, see the *Reallexikon der Assyriologie*, 'Landwirtschaft', §8.

[82] 'Stars' do get mentioned once (at §39), but the import is very unclear.

[83] Netz, 1999, has painted pretty much this picture for Greek mathematics, and it seems plausible that the same is true for astronomy, although the analysis still needs to be done in detail.

6 | Egyptian astrometeorology

Alle Welt von Anfang des menschlichen
Geschlechts hatt erkennt, das die järliche
Witterung durch das Gestirne gubernirt werde.

Johannes Kepler, 1603

1. Fount or fabrication?

We saw in the last chapter that there are a number of practices in Mesopotamia that are loosely related, at least in kind, to the classical parapegmata. But the relationships between the Mesopotamian material and the classical material are never close enough to argue for a direct transmission from east to the west. Indeed, the evidence seems to point to independent development in Mesopotamia on the one hand and Greece and Rome on the other. This stands in contrast to the case of mathematical planetary astronomy, where Greco-Roman astronomy and astrology are often deeply and directly influenced by their Mesopotamian predecessors. In short: astrometeorology does not follow exactly the same trajectories of transmission and influence as planetary astronomy and astrology. It turns out that this general conclusion is also true for Egypt, but here the terms are reversed. Where Egypt has little or no detectable influence on Greco-Roman mathematical astronomy, it turns out that there are Egyptian practices going back to at least the fourth century BC that may be directly related to the Greco-Roman parapegmatic traditions. Before looking at these Egyptian sources, it will be worthwhile to briefly sketch the general picture we have of Egyptian astronomy.

To some extent this picture can be summed up by the small joke Otto Neugebauer allows himself in his monumental 1975 *History of Ancient Mathematical Astronomy*. In what has since become the standard reference book on early mathematical astronomy, Neugebauer includes a short but nearly empty chapter on Egyptian astronomy, whose primary purpose, he says, is to draw attention to the fact that the chapter itself is basically nonexistent:

Egypt has no place in a work on the history of mathematical astronomy. Nevertheless I devote a separate 'Book' on this subject in order to draw the reader's attention to its insignificance which cannot be too strongly emphasized . . . [1]

These are rather harsh words. And Neugebauer does not just criticize astronomy here. He also goes on to criticize Egyptian mathematics, and Egyptian science in general. He uses the word *primitive* a lot.

At the same time, though, Neugebauer himself did do some important work on Egyptian science. He and Parker wrote what is still the standard book on Egyptian astronomy, *Egyptian Astronomical Texts* (1969), and he elsewhere contributed to our understanding of Egyptian mathematics and calendrics as well.[2] Neugebauer's point in the *History of Ancient Mathematical Astronomy* chapter was to draw a distinction between the copious and complex *mathematical* astronomies of Babylon and Greece, and the more qualitative and sparser astronomy of Egypt. Much work has been done since Neugebauer's 1975 book to rehabilitate our impression of Egyptian science, but still no one has convincingly found any complex mathematical astronomy there.[3] This historical picture of Egyptian astronomy stands in contrast to a now-popular view of Egypt as the most ancient seat of culture, knowledge, and wisdom. Unfounded though it may be, the popular conception of an Egypt that harbours a deep and hidden well of profound learning has a long and quite distinguished lineage. The Greeks and Romans themselves often refer to the Egyptians in terms resounding with sheer awe. Herodotus, to pick a familiar example, is charmingly naïve in his hyperbolic praise of Egypt, the country, and the knowledge of its savants. And Herodotus is not alone. In classical philosophical and scientific texts, we often see claims that Egypt was the *Urquell* of one branch of learning or another. Thales is supposed to have gone there. So, too, Pythagoras, Plato, Solon, Democritus, and many more. Egypt was often thought by the Greeks to be the birthplace of geometry, astronomy, astrology, and medicine, among other disciplines.[4]

---

[1] Neugebauer, 1975, p. 559.   [2] See, e.g., Neugebauer, 1969; Neugebauer and Parker, 1969.

[3] See, e.g., Clagett, 1995; Depuydt, 1997; Ritner, 2000; Schaefer, 2000; von Spaeth, 2000; Symons, 2002; Kraus, 2002; Leitz, 1995; Quack, 2003; Høyrup, 1999; Wells, 1996. Depuydt, 1998, p. 6, points out that it is probably unfair to accuse Egypt of having failed to invent a complex mathematical astronomy, since the development of such complex astronomies elsewhere actually post-dates the 'great epochs of Egyptian civilization'. For medicine: Kolta and Schwarzmann-Schafhauser, 2000. For mathematics: Imhausen, 2002; 2003.

[4] There is some evidence that the claims about Egyptian medicine may have had some foundation. See Saunders, 1963; Veith, 1965; Lloyd, 1983; Lloyd, 1991a, pp. 279–98; Marganne, 1993. Von Staden, 1989, is cautious. On mathematics, see e.g., Lloyd, 1991a, pp. 279–98; Neugebauer, 1969.

The disjunction here between what the Greeks thought and what scholars now believe about Egyptian science and mathematics is interesting and informative in a number of ways. The Greeks and Romans, it seems, were quick to invoke the Egyptians and the Chaldaeans (a term that generally referred to the Babylonians) whenever they wanted to add a little longevity, a little temporal *gravitas*, to a discipline.[5] As it turns out, in mathematical astronomy and astrology, some of the references to the advanced state of knowledge among the Chaldaeans may not have been so very far off the mark. On the other hand, the claims about Egypt were, it seems, rather exaggerated.

But were *all* such claims? Consider the following excerpt from Ptolemy's *Phaseis*:

[Θώθ] γ′. ὡρῶν ιγ′ L′ ὁ ἐπὶ τῆς οὐρᾶς τοῦ Λέοντος ἐπιτέλλει. ὡρῶν ιε′ ὁ καλούμενος Αἲξ ἑσπέριος ἀνατέλλει. Αἰγυπτίοις ἐτησίαι παύονται. Εὐδόξῳ ἄνεμοι μεταπίπτοντες. Καίσαρι ἄνεμος, ὑετός, βρονταί. Ἱππάρχῳ ἀπηλιώτης πνεῖ.

[Thoth] 3. [For the latitude where the longest day is] 13 $\frac{1}{2}$ hours: the star on the tail of Leo rises. 15 hours: the star called Capella rises in the evening. According to the Egyptians the Etesian winds stop. According to Eudoxus variable winds. According to Caesar wind, rain, thundery. According to Hipparchus the east wind blows.

In addition to the Greek and Latin sources from which Ptolemy claims to be getting his information, he mentions also 'the Egyptians' as an authority. Unfortunately, it has been unclear whether this refers to Egyptians, to some unnamed Greek astronomers living in Egypt, or whether it is just another instance of the rhetorical invocation of Egypt for the sake of *gravitas*, either by Ptolemy himself or by his sources.

Neugebauer and Rehm dismiss the idea of an indigenous Egyptian tradition, and believe that the references are to Greek authors living in Hellenistic Egypt. Neugebauer dates the references to an unknown source from the second century BC, and Rehm pushes it as far back as the third century.[6] Van der Waerden, on the other hand, has argued at some length that attributions to the Egyptians are actually based on *Greek* observations made in Phoenicia, Cyprus, or Cilicia, and lifted from the – I argue spurious – 'Parapegma of

---

[5] On different classical uses of Egypt, a very good introduction is von Staden, 1992. See also La'da, 2003; Vasunia, 2001; Elsner, 1997; Nesselrath, 1996; McEvoy, 1993; Erbse, 1992; Marganne 1992; Lloyd, 1991a, pp. 279–98; MacCoull, 1991; Cowley, 1990; Fehling, 1989; Hartog, 1988; Lloyd, 1973; Gill, 1979; Davis, 1979; West, 1971; Wilson, 1970. For an informative, if polemical, look at how an Egyptologist sees modern classicsists' perceptions of Hellenistic Egypt, see Ritner, 1992.

[6] See Neugebauer in *HAMA*, p. 562; Rehm, 1941, pp. 101–4.

Dionysius'.[7] Part of the reason for the universal dismissal of an indigenous Egyptian tradition rests on Hellmann's work that argues, based on modern observations, that the weather predictions ascribed to the Egyptians must have been made in northern Greece rather than in Egypt.[8] The central problem with such an argument is that it assumes the astrometeorology in the parapegmata to be based exclusively on observed weather phenomena across a series of years whose weather does not substantially differ from our determination of normal weather in the same locations 2,500 years later. As we saw in previous chapters, observation is never an unproblematic category, and it is certainly not at all clear that we can use parapegmata as evidence for millennia-old weather patterns in Greece or Egypt. Do the parapegmatic attributions to Hipparchus let us reconstruct the climate in Rhodes in the second century BC? The case has certainly never been made, and I'm not sure that it could be. There is the secondary problem that we also cannot simply assume that the weather for any given year 2,500 years ago is going to substantially correlate with modern averages, and this in two ways: (1) a good deal of work since Hellman has shown how global climate has changed over the course of the centuries, but there is still, to my knowledge, no paleoclimatological data on the week-by-week or day-to-day weather patterns for the various regions of Greece and Egypt that would allow us to locate any particular astrometeorologist in one specific location based only on his weather predictions as preserved in parapegmata;[9] and (2) if the weather in the parapegmata is assumed (as it is by van der Waerden and others) to be a record of observations taken over a small sampling of years at most, then there are unaddressed statistical problems with expecting it substantially to conform to *any* average, modern or otherwise.

There is also, as it turns out, *positive* evidence for Egyptian astrometeorology which indicates that there was in fact an indigenous Egyptian tradition of weather prediction linked to stellar phases. The evidence is twofold. A new look at the fourth-century BC autobiography of Harkhebi reveals that some Egyptians, at least, seem to have been interested in predicting weather phenomena in the early Ptolemaic period. Even more interestingly, the naos from Saft el-Henna[10] (dated to the time of Nektanebos I, 381–364 BC), makes

---

[7] Van der Waerden, 1985; on the alleged parapegma of Dionysius, see G.iii in the catalogue, below.

[8] Hellmann, 1916; 1917.

[9] On historical climate change see, e.g., Budyko, 1982; Rampino, *et al.*, 1987; Lamb, 1995; Fleming, 1998; Bradley, 1999; Brown, 2001. Even a treatment as detailed as Le Roy Ladurie's (1967), were one available for our period, would not let us make the kinds of claims Hellmann (and, as we shall see presently, Leitz) want to make.

[10] See Leitz, 1995.

an *explicit* connection between certain Egyptian constellations (decans)[11] and particular weather phenomena, and looks to be astrometeorological. These two texts point to an *indigenous* Egyptian tradition of astrometeorology in existence a full two centuries before Neugebauer's date for the *Greco*-Egyptian tradition he had posited.

In this connection, a further argument has been made by Leitz that there are both meteorological forecasting and stellar observations in the Ramesside (thirteenth-century BC) *Calendar of Lucky and Unlucky Days*, but a close look at the *Calendar*'s contents will show that, although there are some references to weather phenomena in the *Calendar*, there is no reason to suppose any specifically astrometeorological content. Indeed, even the claims to *astronomical* content by Leitz and others turn out to be crucially underdetermined. I will spend a few pages towards the end of the chapter looking at these claims and arguing for more rigorous criteria of proof for purported astronomical content in non-astronomical texts.

II.   Harkhebi, astrometeorologist

Harkhebi was a native Egyptian astronomer/astrologer living in the early Ptolemaic period. The only definite evidence for his life and work is contained in a single autobiographical inscription written in late Egyptian hieroglyphs on a small damaged statue of, probably, Harkhebi himself, which was found in a field near Tell Far'un by a farmer in 1906.[12] The *Autobiography of Harkhebi* was first published by Kamal in 1906. It was dated by Daressy to the early Ptolemaic period, but Yoyotte has more recently argued for the association of Harkhebi with the διοικητής Archibius mentioned in a Tebtunis papyrus (P. Tebt. 1.61b), which would, if he is right, date him to the mid to late second century BC.[13]

In 1916 Daressy republished the text with a French translation, and included a number of editorial changes.[14] Daressy's translation of Harkhebi's autobiography is unsatisfactory, partly because the Egyptian text itself is sometimes obscure and difficult. Neugebauer and Parker include an improved translation in their *Egyptian Astronomical Texts*. Following De

---

[11] The decans are thirty-six constellations situated in a band roughly parallel to the ecliptic, and slightly south of it. By Greco-Roman times they had come to represent 10° sections of the zodiac.

[12] Cairo: JE 38545.

[13] Yoyotte, 1989. Oates, *et al.*, 2001, dates the papyrus itself to March–April 117 BC.

[14] Daressy's text was republished in Clagett, 1995, vol. II, fig. iii.105.

Meulenaere,[15] they occasionally change the readings of many otherwise straightforwardly understandable signs in order to obtain particular meanings, even when there is a much simpler reading available. In one instance of particular relevance, they squeeze the improbable reading of *ḥnm* 'to unite' out of a perfectly normal writing of the wind sign, *ṯ3*. While this is the one really significant sentence in Harkhebi for our present purposes, I nonetheless offer a new translation of the whole astronomical part of the Harkhebi text:

Hereditary prince, noble, sole friend, skilled and wise of heart in divine words, who sees all that is seen in heaven and earth; skilled and wise of heart in observing the starry skies, who does not make mistakes among / / them, who tells the rising and setting[16] in their time(s) and the gods who foretell the future. He purified himself for them in their time when (the decan) Akh rose heliacally[17] / / beside Venus from the earth in order for him to calm the land with his words, being one who sees every star[18] in the sky, who knows the heliacal rising of every (star) in a good year, / / who foretells the heliacal rising of Sirius at the first of the year. He observes her in the first day of her festival, in order to calculate her motion for the times appointed for it (the festival), observing / / everything she does every day. She has foretold everything through him, he being one who knows the northward and southward motions of the sun, telling all of its properties, and what they (i.e., the properties) cause the day to bring.[19] He says what (will) happen / / because of them, coming at their times, (being one who) divides the hours correctly at both times (i.e., night and day), not ever going into error at night, being wise in every thing / / seen in the sky, which is guarded[20] by him on the earth, (being one who) knows their winds[21] and their omens,[22] being entirely complete in setting out (his) opinion, and being exalted[23] because of his reports when / / he discerns the hidden

---

[15] See Neugebauer and Parker, in *EAT*, vol. III, p. 214f.

[16] Following Neugebauer and Parker, I translate *'nḫ ḥtp* as 'rising and setting'.

[17] Following Neugebauer and Parker, I translate the verb *prl* in this context as 'rise heliacally'.

[18] Neugebauer and Parker (and Clagett, following them) have 'the culmination of every star'.

[19] *rdl-sn dl hrw*, following Daressy. Neugebauer and Parker, following De Meulenaere, emend it to *dl r sn sp*. Clagett follows Neugebauer and Parker almost word-for-word in his translation of this passage.

[20] I translate *s3(w)* as 'guard'. Neugebauer and Parker translate it as 'await', but this makes little sense of the text, and it forces them to read Daressy's *r t3* as *s*, following Kamal. Clagett again defers to Neugebauer and Parker's translation.

[21] Neugebauer and Parker propose to read the *ṯ3* sign as standing for *ḥnm*, 'to join', but this would be a unique reading of *ṯ3*.

[22] Neugebauer and Parker emend Daressy's *šm* to *gsgs*, 'to order, regulate', largely, I think, because no likely meaning of *šm* was known to them. For *šm* as 'omen', see Ritner, 1993, p. 36, n. 167.

[23] I read this troublesome passage as *tm* ('be complete' as opposed to Daressy's and Neugebauer and Parker's 'to not be') *prl lb r dr, k3w* (as opposed to De Meulenaere's improbable emendation to *ḥ3p r3*). The meaning of 'be complete' as opposed to 'not be' is necessitated by

language[24] through[25] everything being observed by him, and every end being complete[26] when (he) counsels[27] on account of it, making judgments for the lord of the two lands.[28]

We should note that Harkhebi does not speak here of doing anything resembling horoscopic astrology.[29] His reference to 'knowing the winds and the omens' seems to refer to some kind of meteorological prediction. A clue to one way in which he may have predicted his weather patterns comes from his repeated mention of his knowing the heliacal risings of the fixed stars, which accords nicely, though perhaps only coincidentally, with Ptolemy's report of what certain unnamed Egyptian astrologers were up to.

We should note also the prominence of Sirius in Harkhebi's description of his work. This is not really surprising, given that since the earliest times the rising of Sirius had marked the feast of the beginning of the Egyptian year and the rising of the Nile.[30] While it is true that the long-standing correlation of a stellar phase with a seasonal phenomenon of agricultural and navigational importance means that the Egyptians were, from a very early date, making limited use of the sort of observation that would be needed for the construction of a stellar almanac,[31] there is little evidence for a *detailed* astrometeorology before the fourth century BC, as we shall see in the next sections.

In addition to Sirius, Harkhebi also mentions observing the planet Venus and the northward and southward motions of the sun. How the observations of Venus were used is unclear here. We know that the Babylonians had associated some Venus omina with weather phenomena in *Enūma Anu Enlil*,[32] but there is no evidence that such omina had reached Egypt. On the other hand, there is one Egyptian source, the Demotic text of lunar omina

---

the context, since Harkhebi has been saying all along that he tells what he knows. I see no reason why he should become suddenly reticent about his reports.

[24] *r*, 'language' or 'intent'. It may also be *r-ḫrw*, 'he discerns what is hidden *from below*. Everything is observed by him.'

[25] I see no reason to emend the *ḫr* in the text as *ḥr*, which Neugebauer and Parker need to do in order to make the text mean 'be discreet with' (*ḫ3p r ḥr*).

[26] Again, *tm* as 'be complete'; *r-ꜥ*, 'end, limit'.

[27] I take the writing of *sḫr*, 'to overthrow' as a mistake for *sḫr*, 'plan, counsel'.

[28] From here, Harkhebi goes on to tell us of his skill at charming snakes and scorpions.

[29] *Pace* Daressy.

[30] Katherina Zinn here reminds me that the flooding of the Nile might be thought of as a kind of meteorological phenomenon (or at the very least as *in concert with* meteorological phenomena), given its profound effect on agriculture and navigation.

[31] See Depuydt, 1997, pp. 14–15.   [32] See Reiner and Pingree, 1975–98, vol. III.

published by Parker in 1959,[33] that includes weather predictions such as inundation and wind, derived from observations of the colour and appearance of the moon, which Parker noted was similar to Babylonian methods of forecasting. But Aratus (*Phaen.* 778f.) also uses the colour of the moon to predict weather, and it is possible that all three sources were independent. The northward and southward motions of the sun are directly related to seasonal changes and these too may have been meteorological indicators for Harkhebi, although the text itself is unclear on this point.

Whether or how the Greco-Latin material is related to what we are seeing in Harkhebi is an open question. As we saw in the case of Babylon, what the Greeks have to say about Chaldaean astrometeorology is not borne out by the actual evidence so far found in Mesopotamia, and so the parallel evidence here may not amount to much. But there *is* as it turns out other Egyptian evidence for a tradition of astrometeorology. We shall see from a look at the fourth-century BC Ṣafṭ el-Ḥenna naos that the hypothesis that Harkhebi was doing astrometeorology is strengthened by comparison with an earlier Egyptian tradition.

III.   The Ṣafṭ el-Ḥenna naos

The Ṣafṭ el-Ḥenna naos (sometimes called 'the naos of the decades') is now in several major pieces. The top half has been in the Louvre for 200 years, and most of the lower half was pulled from the Bay of Abukir in the 1940s and is now in Alexandria.[34] More parts were being raised from the ocean as recently as 2001.[35] Most recently published by Leitz, the naos is valuable in that it preserves the earliest known Egyptian correlation between celestial phenomena and weather. The inscription dates from about fifty years before the Ptolemies began to rule Egypt.

The text of the naos contains descriptions of the powers of each of the decans during some phase of their appearance, presumably their heliacal rising, their culmination, or their setting. We have more-or-less complete entries for *decades* (ten-day periods) 3, 6, 9, 11–13, 17, 18–21, and 25–37, with scraps of decades 8 and 24. Entries are schematic: 'The great god in

---

[33] Parker, 1959. This text dates from the late second or early third century AD, but is probably a copy of a sixth-century BC original.

[34] See Leitz, 1995, p. 3. The two main fragments thus far published are: Louvre D 37 and Alexandria JE 25774. The total size of the naos is estimated by Habachi and Habachi, 1952, as 1.78 × 0.88 × 0.80 m.

[35] The excavation has been under the direction of Franck Goddio.

the beginning; he causes such-and-such', where the phenomena caused by the decans range from battles to sickness to rain and wind. I have excerpted here all of the entries which pertain to weather:

3rd decade, l. 2:	He causes heat.[36]
6th decade, l. 3:	He will be hot for five days.
8th decade, l. 3:	. . . [rai]n for the earth.[37]
9th decade, l. 2–3:	He causes rain in heaven.
12th decade, l. 1:	He causes ra[in].[38]
13th decade, l. 3:	. . . in bitterness for twelve days.[39]
20th decade, l. 1:	[He cau]ses the south winds in heaven . . . . [He causes] three days of bitterness.
21st decade, l. 1:	[He causes] the north [wind]s in heaven.[40]
24th decade, l. 2–3:	. . . cold(?) . . . bitterness . . .[41]
25th decade, l. 2–3:	. . . rain in heaven.[42]
26th decade, l. 1:	[He c]auses an evil wind in the night.
27th decade, l. 2–3:	. . . and consuming everything on earth in their bitterness.
28th decade, l. 1–3:	He draws forth the flood-water from its cavern . . . (text unclear) . . . He causes sickness in *Mnty*[43] by bitterness in the house[44] for four days and by a sickness in the belly.
29th decade, l. 2–3:	He causes bitterness of nine days.
32nd decade, l. 2:	. . . bitterness . . .

Here we see a clear connection between decans and weather phenomena. One interesting detail is that the expression of the activity of each decan is

---

[36] *ntf sḫpr šmm*. Leitz thinks this refers to fever, which is possible.

[37] The text is fragmentary, having only the last wave of the water determinative with the pool determinative followed by *r t3*. The two determinatives, however, match the usual writing of 'rain' in this text.

[38] The text breaks off part-way through the word: *ntf sḫpr h[w]y[t]*.

[39] *m dḥrt n hrw 12*. Leitz takes this to refer to sickness, which seems probable, but I include these references to 'bitterness' here since it seems remotely possible that they are references to bad weather. In any case, it is interesting to note that by late antiquity Greek and Latin parapegmata often contained information on medical matters. See Burnett, 1993, p. 28f.

[40] *[ntf sḫpr t3]w mḥwt m pt*. 'Wind' reconstructed based on similarity with previous entry.

[41] Parts of *ḥs* are visible, as is all of *dḥrt*.

[42] *. . . ḥwyt m pt*. Leitz notes that 'ein Substantiv, *ḥwyt*: "Regen" erscheint aber angeseits der Jahreszeit eher als unwahrscheinlich' (Leitz, 1995, p. 24, n. 94). I see no reason to emend away what appears in the text here. For my criticisms of Leitz's method of determining which season this decan would fall in, and the likelihood of rain under any given decan, see below.

[43] Or possibly *mr mn t3*: 'a lasting and painful sickness'.

[44] Contrary to Leitz, who sees the house as superfluous.

highly formalized throughout the text, saying repeatedly that the particular decan 'causes (*špr*) such-and-such to happen'. There are only two exceptions to this in the weather sections,[45] which occur in decans 6 and 28, respectively. These read simply 'He will be hot for five days' (*šmm-f n hrw 5*) and 'He draws forth the flood-water from its cavern' (*ntf šdi ḥ'py m ṭpḥt-f*). Even in this last instance, though, the idea of causation is implied by the active verb in 'He draws forth . . .'. The author of this text seems to be representing the decans as acting upon the earth and bringing about the weather conditions described. This is – coincidentally? – paralleled by a second- or third-century AD Greek papyrus from Egypt, *P. Oxy.* LXV.4473, which says that a particular decan brings (φέρει) the flood.

That the decans exert a causal influence is further strengthened by the introduction to the whole inscription, which talks unambiguously of causation:[46]

'*Bt-nbs*,[47] the door of heaven. When Re shines, his face causes the great mooring-post[48] . . . / / . . . the disks[49] . . . / / . . . great . . . the 36 decanal stars . . . / / Heaven, earth and the underworld[50] are under their counsel.[51] They rise and they set so that the temples of '*Bt-nbs* / / . . . / / . . . [They are the ones who c]ause the flood / / . . . his secrets. They are the ones who cause storms. They are the ones who open the sky and who prevent rain. They bring the day. They bring the night. They rise. They set. They refresh themselves in the northern sea.[52]

And in the formulaic inscription beside each decan, it says:

Water, wind and (fertile) fields are requested from him in his decade in '*Bt-nbs*.
   Month III of *šmw*, day 1 to 10: offerings are made to this god by the king in
   '*Bt-nbs* in order to protect the land from disaster.
He rises[53] in this form in the decade of (his) motion, (as) master of the earth.[54]
   He is the one who causes sickness and death.[55]

[45] Slight variations do occur in the context of some other powers, e.g., decan 18.2 has *ntf di ḫpr* . . .
[46] For the text, see Habachi and Habachi, 1952; Leitz, 1995.   [47] Ṣafṭ el-Ḥenna.
[48] Following the text in Habachi and Habachi: *iw ḥr-f rdi-f n'yt wrt n* . . . Leitz follows Habachi and Habachi, 1952, p. 255, in emending the text to *iw ḥr-f <ḥr> rdit-f <n> n'yt wrt n* . . . , translating it 'Wenn Re aufgeht, wendet sich sein Gesicht <dem> grossen Haus des [. . .] zu [. . .].'
[49] Following Habachi and Habachi, 1952.
[50] The text actually has *niwt*, 'town' rather than *dw3t*, 'netherworld'.
[51] Leitz, 1995, p. 7, n. 22, translates *šḫr* as 'Aufstellung'.
[52] The remainder of the text is mythological.
[53] Leitz argues that '*ḥ*' must mean 'culminate' here.
[54] Leitz translates this phrase as 'wenn auf die Erde ausgesandt wird' (Leitz, 1994, p. 9). The passive of *h3b* is possible, but I think it makes less sense of *ḥry-tp*.
[55] This is a difficult passage. I read *mr mwt*. Leitz reads *mwt dm*, 'den stechenden Tod', seeing the *mr* as a mistake for *mwt*, and the two knives as *dm*. I am treating the knives as determinatives

His living Ba is master of the earth in this influence. Life is requested from him in
his decade in *'Ȝt-nbs*.

It is his image which gives offerings in every temple in the decade with invocation
offerings[56] in his temple.

His living Ba (is) eternal. His body gives offerings in the necropolis. A good burial
is requested of him in his decade in *'Ȝt-nbs*.

The language in these passages makes it clear that some power exerted by the
decans themselves was thought to bring about particular effects, whether
meteorological phenomena, disease, or crop growth.

III.i.   Which phase?

Leitz, in his examination of the weather predictions in the Ṣafṭ el-Ḥenna
naos, takes it for granted that these weather phenomena were established
empirically by the Egyptians. In attempting to determine, for example,
whether we can say which phase of the decans may have been responsible for the wind or rain (whether their rising, their culmination, or their
setting) Leitz argues that rain would be most likely at the time of the ninth
decan's culmination in the twelfth hour of night. He then uses this to argue
that each decade is determined by the midnight culmination of a decan,
rather than by its rising or setting.

Auch hierin liegt wieder ein Argument für die Annahme, dass die in den Beischriften
gennante Wirkung der Dekane tatsächlich in die Dekade mit der Kulmination in
der 12. Nachtstunde fallen. Wäre es der Augenblick ihres heliakischen Aufgangs, so
wären die entsprechenden Daten der 1. und 31. Oktober (greg.), was nicht zu den
meteorologischen Gegebenheiten paßt.[57]

But these meteorological facts are based on averaged rainfall, measured in
recent years, at Cairo and Alexandria, and so subject to the same objections
I raised above about van der Waerden's similar method, with the further
complication that we have no evidence, even if the weather predictions in
this text *were* observationally derived, that such observations were made

---

and the dying man (seen by Leitz as a determinative) as phonetic *mwt*. Another possibility is
that the dying man is a determinative for *mr* (as, e.g., in decade 29) and the two knives are
meant to be phonetic and represent an adjective describing the particular kind of sickness
caused: 'He is the one who causes the *dm*- sickness.' Habachi and Habachi think the knives are
phonetic *dm* = *tm*: 'evil doers'. If this is correct, a preposition is missing.

[56] Following Habachi and Habachi who read *pr-ḫrw* in place of Leitz's (following Brugsch,
1883–91, vol. I, p. 181) *Pr-Spdt*.

[57] Leitz, 1995, p. 18, commenting on the ninth decade.

in Alexandria. Finally, we don't see in the Egyptian text any claim to an observational correlation. As complicated and oversimplified as we have seen even *explicit* observation claims in Greek and Latin sources to be, such claims are not even attempted in the naos. Perhaps there are astrological, physical, or magical correlations between the stars and weather, rather than observational ones. We really can't say, based on this evidence, which phase of the decans was the relevant one for each decade in the naos.

What we can say is that it is at least clear that there is some kind of close relationship between the phases of the stars and the weather in this text, and one that forms a closer analogue to the Greco-Roman material than we saw in the case of Mesopotamia. So now the question arises of where this Egyptian tradition derives from. It has been claimed that the naos is related to some aspects of both time reckoning and meteorology in a much earlier Egyptian text, the thirteenth-century BC *Calendar of Lucky and Unlucky Days*.

iv.   The *Calendar of Lucky and Unlucky Days*

Before we look in detail at the *Calendar of Lucky and Unlucky Days*, I would like to outline a methodological point that is perhaps best illustrated by looking at one small aspect of the work of the nineteenth-century American astronomer Elias Colbert. Colbert was a well respected astronomer, appointed as director of the Dearborn Observatory in 1872, and Professor of Astronomy at the University of Chicago some time later. He was also at one time president of the Chicago Astronomical Society, and he gave the welcoming address at the first meeting of the newly renamed American Astronomical Society in 1914. It is a little unfair to his reputation as an astronomer for me to do this, but I would like to highlight a couple of minor errors he made, as they are illustrative for the present inquiry. In one of his books, Colbert indulged himself in a little nineteenth-century comparative mythology, attempting to find simple explanations in nature for some mythical stories and associations.[58] In particular, he thought he could explain the form and name of the constellation Pisces – a constellation that has relatively few stars – as an allegorical reference to the 'poverty of a fish diet'. Or, more to our present purposes, the biblical story of Jonah and the Whale (where Jonah was swallowed by a whale and regurgitated unharmed three days later) was for Colbert an allegorical reference to the passage of

---

[58] In Colbert, 1869.

the moon through the constellation Cetus, 'The Whale', a passage that takes about three days. Besides its triviality, there are real problems pointed to by Colbert's account that need to be taken seriously. One is that Colbert so deeply divorces a single very small detail of Jonah and the Whale from *everything* we know about the religious and moral contexts and contents of the story itself.[59] Not only is there no reason to think of Jonah in astronomical terms to begin with (certainly the biblical story does not say anything about astronomy, the constellations, or the moon), but Colbert's reading also strips Jonah of his very purpose as a story, which is ethico-religious at its core.[60]

A second problem is that Colbert's interpretation of this story depends on the reader accepting his particular set of associations between literary references and astronomical phenomena. He has a kind of single-use cryptographic key for this story that he is asking us to accept. The problems with cryptographic keys of this sort will be immediately obvious to anyone familiar with the history of attempts to decipher Egyptian or Mayan hieroglyphics.[61] The seventeenth-century Jesuit scholar Athanasius Kircher, to pick the most dramatic example, thought he had a decoding key for Egyptian hieroglyphics that allowed him to read the writings of ancient Egypt, which writings all turned out to be mystico-theological in a particularly 'Kircherian' mode. But Kircher was simply 'reading' the characters as abstract but direct representations of what he thought of as Egyptian ideas (a methodology he borrowed from Plotinus), since he had no conception that the glyphs may have carried mundane phonetic and grammatical information. The symbolism he developed thus served to 'decode' the secret message of the script for him, but for him alone.

I highlight these admittedly extreme examples to make a point: any claims to having deciphered some secret code or hidden meaning underlying a text or object need to be carefully thought through and very rigorously supported. This is not to say that *no* texts have a double meaning or that no texts are meant to carry a submerged message, but only that analytical methods

---

[59] A nice survey of the problems here can be found in chapter 2 of Csapo, 2005.

[60] Kinch Hoekstra has pointed out to me that *if* there were considerable evidence that biblical or mythical stories *were* ethico-religious agglutinations onto an older naturalistic core, then we would need explicit reasons to discount Jonah here. But since we do not have evidence of, for example, subsequent retellings of the same story that become less naturalistic and more religious, all we are left with is Colbert's assertion of the association itself. And a *just-so* story is no evidence for itself.

[61] For the former, see Gardiner, 1957, pp. 11–12. A good account of the latter can be found in Coe, 1992.

used in reading texts need to be rigorous, and to be closely guided by what we know about the historical, cultural, and scientific contexts of those texts themselves. If we have no other corroboration that the ancient Hebrews had visited pre-Columbian North America, then we should be very, very cautious about seeing an ancient Hebrew inscription in the 'Bat Creek Stone', for example. Likewise, if we have no other evidence that biblical moralists were really interested in astronomical phenomena, then reading such observations into literary or other texts is spurious. On their own, cryptographic hypotheses need to be approached with a good deal of caution, particularly when applied to small sample sets. But enough with the warnings; now on to the text at hand.

We know that, possibly as early as the Ramesside period, there existed in Egypt a system of calendrical omina, typified by the *Calendar of Lucky and Unlucky Days*.[62] It is, on the face of it, simply a collection of calendrical omina, not terribly unlike the hemerologies and menologies of Mesopotamia, or in a looser sense, our own non-astronomical calendrical omina such as Friday the thirteenth bringing bad luck, or 'Monday's child' being full of grace. Here the omens are simply indexed to calendar dates, and if we know the calendar date, we can look up the omen. These dates themselves are framed in the 365-day Egyptian year, and are, beyond some strictly nominal associations, not transparently related to astronomical or astrological phenomena.[63]

In the Ṣaft el-Ḥenna naos we have seen, nearly a millennium after the Ramesside *Calendar*, an instance of the integration of astronomical phenomena and day counts. There the astronomical phenomena are clear and obvious in the text. We are told that the particular decan has a particular influence at a particular time. The *Calendar of Lucky and Unlucky Days*, on the other hand, makes no such explicit associations between astronomical phenomena and time. But it does make rather a lot of references to Egyptian mythology, and it is through these mythological references, with the use of associative systems for decoding the meanings lying behind the

---

[62] Published most recently in Leitz, 1994; earlier by Bakir, 1966. Leitz gives the *Calendar* a *terminus ante quem* of the reign of Ramesses II (Leitz, 1994, p. 6), and Bakir thinks such an early date to be possible, though he is not willing to commit himself fully (Bakir, 1966, pp. 5–6).

[63] Remember that the Egyptian calendar of 365 days is not synchronized with any solar or lunar motions, and it accordingly 'slips' over the years such that every four years it falls approximately one day further out of line with the solar year, continually moving away from its original seasons. Thus the calendrical season of 'Inundation', while it more or less agreed with the actual inundation for a time, would, over the course of the centuries, move away from the seasonal inundation and then back again some fourteen centuries later. For a detailed discussion of the Egyptian calendar, see Parker, 1950; Depuydt, 1997.

mythological entries in the calendar, that several recent scholars have argued for astronomical readings (and datings) of the *Calendar*. If they are right – I will argue that they are not – then we would have in the *Calendar* the earliest attested systematic astronomical observation text, not just in Egypt, but in the world.[64] The further fact that some of the omen apodoses contain unambiguous references to weather also raises the possibility, if the astronomical readings of the *Calendar* are correct, of the *Calendar*'s being the oldest extant precursor of the astrometeorological traditions we have been looking at in this book. These are arguments we should look at carefully, then.

iv.i.   *Meteorology in the* Calendar

The *Calendar* lists each day of the Egyptian year, together with its omens, good and bad. Weather omina include:

I *ȝḥt* 4:[65]   . . . The gods go as evil winds.
II *ȝḥt* 24:   . . . Do not go out in any wind until the sunset.
III *prt* 20:   . . . You will not see sunlight.
III *prt* 25:   . . . This day has a great storm.

In addition to weather omina strictly speaking, we also see for some dates proscriptions (such as 'Do not go out in any wind until sunset'). There are a number of weather entries of these two kinds and the references to meteorological phenomena are clearly stated in plain language in the text.

But the *Calendar* also includes many obscure mythological references to different gods or demons coming forth or battling with each other. Leitz sees in these obscurities veiled references to weather patterns, which he interprets with the aid of both modern meteorological statistics, and the so-called Coptic-Arabic almanacs.[66] So, for example, he is able to interpret a sentence like that in IV *prt* 3: 'The great ones fight the *wpyt*-serpent' as

---

[64] This claim depends on the fact that the Babylonian 'Venus tablet of Ammiṣaduqa' dates, as a text, from the first millennium BC, even if some of its contents may be older. See Hunger and Pingree, 1999, pp. 32–4; Huber, 1987; Huber, *et al.*, 1982; Reiner and Pingree, 1975, pp. 21–5.

[65] The Egyptian calendar has three seasons, *ȝḥt, prt,* and *šmw,* each divided into four months of exactly thirty days. I *ȝḥt* 4 refers to the first month of *ȝḥt,* day 4.

[66] For his use of modern meteorology, see, e.g., Leitz, 1994, commentaries on: I *ȝḥt* 4; III *prt* 24 and many others. For the Coptic-Arabic almanacs, see, e.g., Leitz, 1994, commentaries on I *ȝḥt* 20 and 29 where he sees references to the hot season which are nonapparent to me; II *šmw* 19, 20, and 22 where winds are read into the text based on the almanacs, and III *šmw* 12 where he has the clouds vanishing based on nothing I can determine in the text. I have voiced my

really meaning 'The clouds push themselves before the sun.' But the mytho-logical actors here, *the great ones* and the *wpyt*-serpent, have no parallels as meteorological terminology *anywhere* in the corpus of Egyptian litera-ture, nor are they shown to be consistently correlatable with Leitz's version of their referents. If it could be shown that there is a consistent use of, for example, '*wpyt*-serpent' where we should expect 'sun' in this and other texts, then a case could be made for coding. But we would need multiple more or less clear substitutions. We should also want to see proof that 'the great ones' means 'clouds' across at least *some* body of texts. So, too, 'fighting' for 'moving in front of'. Codes only work if they have a one-to-one correlation between signifiers and referents. Cryptographic keys that use semi-arbitrary, speculative, or (as here) uniquely associated referents are ultimately indis-tinguishable from ad hoc-ery, which weakens their force as evidence rather considerably.

To put the evidence in full perspective, I here set out all of the clearly worded entries pertaining to astronomy or weather in the *Calendar*, and there are not really that many:

I *ꜣḫt* 1:   There is purification when[67] the entire land is under the water of the High Nile,[68] going forth as young Nun, as it is said.[69]

I *ꜣḫt* 4:   The gods go out as evil winds.

I *ꜣḫt* 24:   The majesty of this god sails in good breezes and in peace . . . west.

II *ꜣḫt* 24:   Do not go out in any wind until sunset.

III *ꜣḫt*18:   If there is a great wind[70] on this day it will not be in a good way.[71]

III *ꜣḫt*19:   A great[72] storm is born in the sky.[73] Do not sail upstream or downstream upon the river. Do not travel in any boat on this day.

IV *ꜣḫt* 7:   This day, the wind is guarded against in the entire land.[74]

IV *ꜣḫt*12:   Do not go out on any road in the wind.[75]

I *prt* 19:   The winds in heaven on this day are mixed with the annual pestilence[76] and many diseases.

---

scepticism about the usefulness of modern meteorological averages as comparative evidence, and the use of the Coptic-Arabic almanacs is equally contentious, if for different reasons.

[67] *m-ḫt*. Bakir translates this as 'throughout', and Leitz as 'über'.

[68] *ḥꜣt ḥꜥpy*. Both Bakir and Leitz take this to refer to the beginning of the flood.

[69] Following Bakir. Leitz translates this as '"Jugendlicher" wird er gennant'.

[70] *P. Sallier* only. The sign is uncertain, the determinative strange.

[71] *nn ꜥd st*.   [72] *P. Sallier* only.

[73] Bakir translates this as 'The children of the storm of . . .'.   [74] *P. Cairo* only.

[75] *P. Cairo* has *nn-k prı̓ ı̓m-f r wꜣt nb(t) m t̠ꜣw*, whereas *P. Sallier* IV reads *nn-k prı̓ r wꜣt nbt m hrw pn*: 'Do not go out on any road on this day.'

[76] *ı̓ꜣdt*: 'pestilence' can also mean 'pouring rain'.

III *prt* 19:   The children of Nut[77] are in good breezes[78] . . . Do not go out from your house therein.[79] You will not see sunlight.[80]

III *prt* 20:   You will not see sunlight.[81]

III *prt* 25:   This day has a great storm.

IV *prt* 3:   As for any lion who says the name of the decan Orion, he will die immediately.[82]

IV *prt* 6:   The stars go forth, bitterness before [them].[83] If anyone sees small cattle,[84] he will die immediately.

IV *prt* 13:   Do not go out in any wind[85] on this day.

I *šmw* 2:   Do not go out from your house in any wind on this day.

II *šmw* 20:   There are many deaths. They come from evil winds. Do not go out in any wind on this day.

IV *šmw* 26:   The gods [sail][86] in every wind.

On examination, it is apparent that what predictions there are here are of a different nature from those in the Ṣafṭ el-Ḥenna naos and the parapegmata. The omens are hemerological rather than astrological: the various winds are said simply to happen on particular dates. There is no explicit attempt in the calendar to connect these omina with celestial phenomena, and also no evidence that the omina were based on actual *observations* of weather patterns over the course of the year. It is entirely possible that the winds

---

[77] Bakir has 'The birth of Nut . . .' which is also possible.

[78] Leitz translates this as 'Die Kinder der Nut sind in einem gänzlich günstigen Wind', thus reading *mꜣ ꜥꜣꜣ nfr*. I prefer to read the *mꜣꜥw* as 'breezes' with the wind sign as a determinative rather than as a noun modified by *mꜣ* 'true'. Curiously, this is how Leitz himself reads the identical wording at I *ꜣḫt* 24.

[79] P. *Cairo* only. P. *Sallier* has 'Do not go out of your house on any road on this day.'

[80] P. *Cairo* only.   [81] P. *Sallier* has 'You will not be near sunlight.'

[82] Although this entry does not deal with an astronomical phenomenon as such, I include it since it does mention a decan. Note however that the decan is being conceived of as a deity rather than as a stellar object, as is evidenced by the use of the god determinative in place of the star determinative.

[83] *dḥrt ḫft-ḥr*. Leitz translates this as 'mit bittern Gesicht' and Bakir as 'bitterly and openly', although in a note (p. 76, n. 4) he raises the possibility that *dḥrt* is a mistake for *dšrt*, 'red', and so offers the possible interpretation of the phrase *sbꜣ dšrt ḫft-ḥr* as 'the culmination of Mars'. This seems to me to be extremely unlikely.

[84] *ꜥwt*. Leitz, 1994, p. 307, argues that this refers to the culmination in the first hour of night of the decan *ḥry-ib wiꜣ*, since the small cattle are associated with Seth. I think that if this were true we could expect a star determinative after *ꜥwt* rather than the animal determinative, and in any case the chain of reasoning is extremely weak. I prefer to interpret the *ꜥwt* as animals of the ordinary domestic kind, since the connection drawn by Leitz is highly tentative, and the naming of the decan would be unusually cryptic.

[85] P. *Cairo* only. P. *Sallier* has 'Do not go out in it on any road . . .'.

[86] A hole in the papyrus has obliterated all but the determinative.

were predicted according to other criteria, even if we know not what.[87] We should also note that the majority of entries do not even predict the weather so much as give hortatory statements about how one should behave *in the case of* the occurrence of certain weather phenomena on a particular date. These omina do not say that the phenomena *will* invariably occur on particular dates, but only that if they do, one should not do such-and-such.

IV.ii. Astronomy in the *Calendar*

The only direct reference in the entire *Calendar* to an astronomical phenomenon of any kind (apart from the setting of the sun) is at IV *prt* 6: 'The stars go forth, bitterness before [them].' Nonetheless, there is a current in the modern literature on the *Calendar* that takes an astronomical substrate for granted.[88] A debate that has been running for several years now between Leitz and Krauss is not about *whether* there are astronomical phenomena underlying the mythological entries of the calendar, but about exactly *which* astronomical phenomena. I, on the other hand, want to emphasize here the need for raising the questions of *whether* and *when* we may be justified in seeing such substrata at all.

Let us look at Leitz's idea that certain phrases in the *Calendar* are references to the culmination of particular decans. Leitz's commentary on IV *ȝḥt* 26 offers a considerably different reading of the text from that of Bakir. The two manuscripts are highly corrupted at this point, especially the *P. Sallier* version.[89] As it stands, the *P. Cairo* text for IV *ȝḥt* 24 reads: . . . *nfr smn-in Ḏḥwty srw Rʿ ḫntyw Šm*.[90] Bakir translates it as 'Thoth establishes the nobles in an advanced position in Letopolis' and Leitz as 'Daraufhin setze Thoth die "Fürsten" *(srw)* des Horus von Letopolis ein.' While the German

---

[87] For example, associations of calendrical dates with certain historical or mythological events could have furnished omen lore. Although I know of no evidence from the Egyptian tradition which makes a clear case for this possibility, there is evidence, as we have seen, of just such a practice in the Roman agricultural literature (e.g., Vergil, *Georg.* 1.276–86). For commentary, see chapter 2, above.

[88] E.g., Leitz, 1994; Krauss, 2002. See also Krauss, 1999.

[89] Specifically: (a) IV *ȝḥt* 24 is missing entirely from *P. Cairo*; (b) *P. Sallier*'s IV *ȝḥt* 24 is roughly equivalent to *P. Cairo*'s IV *ȝḥt* 26, though highly corrupted; (c) IV *ȝḥt* 25 is missing from *P. Sallier*; (d) IV *ȝḥt* 25 is almost entirely destroyed in *P. Cairo*, such that Bakir does not even attempt a translation, although Leitz offers a reconstruction, for what it is worth; (e) *P. Sallier*'s IV *ȝḥt* 26 has no parallel in *P. Cairo*, so Leitz assumes it is equivalent to *P. Cairo*'s missing IV *ȝḥt* 24.

[90] The name *Rʿ* is, unusually, followed by the Horus determinative instead of the sun. Bakir reads it as *r* and Leitz as a mistake for *Ḥr*. Leitz thus reads the whole phrase *Rʿ ḫntyw* as an epithet of Horus: *Ḫntyw-Ḥr*.

translation is not in itself objectionable, the astronomical decoding of it is. Leitz begins with the assertion that the *'Fürsten' des Horus von Letopolis* refers to the stars of the decan *phwy-ḏȝt*, which 'begins its work on this day', that is, culminates at the end of the first hour of the night. But Leitz's claim depends on us accepting a rather long string of correlations that: (a) jump around in time (from the Ramesside *Calendar* to the millennium-and-a-half later *P. Carlsberg* Ia);[91] (b) create evidence *ex nihilo* (we are asked to believe that there is an important, but completely unattested, 170-day decanal period in Egyptian astronomy);[92] (c) depend on a multi-stage poly-hypothetical correlation (an association between Horus of Letopolis, the decan Knumis, and the content of the *Calendar*);[93] and (d) rely on an unattested and frankly implausible pun.[94] Likewise the loose string of association which allows

[91] Leitz claims that IV *ȝḫt* 26, the date of Leitz's supposed end-of-first-hour decanal culmination in the *Calendar*, is mentioned as the date of a decanal phase in *P. Carlsberg* Ia (*EAT*, pp. 90–1). But the mention of IV *ȝḫt* 26 in *P. Carlsberg* Ia (dating from the second century AD, sixteen centuries later than Leitz's date for the *Calendar*) is a simple statement of the fact that this date is separated from III *prt* 6 by 290 days. It says: 'If it [a star] goes to the netherworld on <III> *prt* <6>, it will rise on IV *ȝḫt* 26 to I *prt* 6.' The text is highly corrupt at this point, such that the dates of the phases are all curiously reversed, and the reconstructed addendum to *P. Carlsberg* Ia which mentions *phwy-ḏȝt*, proves nothing for the Ramesside *Calendar* of a millennium and a half earlier.

[92] The only two instances of the word *sr*, 'noble', in the *Calendar* (at I *ȝḫt* 26 and II *šmw* 16, respectively) happen to occur 170 days apart, and in order to make *sr* correspond with the decan *phwy-ḏȝt*, Leitz needs to find a 170-day astronomical period to account for their positions in the text. He accordingly asserts that the following decanal periods must have been astronomically important to the Ramesside Egyptians: ten days of marking the end of the first hour of night + ninety days in the west + seventy days in the underworld. But Leitz's emphasis on the significance of this first ten-day period is based on nothing in the Egyptian sources. Leitz is simply adding an extra ten days to the decan's ninety days in the west and its seventy days in the underworld (both of which periods *are* elsewhere attested) in order to get a 170-day interval to match the 170 days between the two instances of *sr* in the *Calendar*.

[93] The Horus of Letopolis/Knumis association is made – based on his reading of one text and his reconstruction of another – by Junker, 1917, pp. 42–4. For this to count as a part of the argument about *phwy-ḏȝt* and *sr* we have to further accept Leitz's decoding of the calendar entry at I *prt* 6 as a veiled reference to Knumis.

[94] Leitz thinks that the Middle Egyptian word *srw*, 'nobles', could *conceivably* be a pun on *sbȝw*, 'stars'. His argument is that (1) the word *sriw*, 'rams', in Middle Egyptian (ME) happens to resemble the word *srw*, 'nobles', and (2) that by *Roman* times the word 'ram' (Coptic ϭⲥⲟⲟⲩ) was used as a pun on the word 'star' (Coptic ⲥⲓⲟⲩ) – notice that 'ram' had lost the phoneme *r* by Roman times, and that 'star,' (ME *sbȝw*) had *never* had one. But this does not prove that in the Rammesside period the word 'nobles' resembled the word 'stars' enough to be a recognizable pun. Indeed, 'nobles' kept its strong *r* right into Coptic times (ME *sr* became Coptic ⲥⲓⲟⲩⲣ). To schematize: Leitz's claim is that

$$(1) \text{ (ME) A} \sim \text{(ME) B,}$$
$$(2) \text{ (Coptic) A} \sim \text{(Coptic) C,}$$
therefore (3) (ME) B $\sim$ (ME) C.

But because the equivalence in (2) depends on the dropped *r* between (ME) A and (Coptic) A, I argue that (3) does not follow, such that (ME) B $\neq$ (Coptic *or* ME) C.

Leitz to interpret the entry at II *ȝḥt* 11, 'The front part of the bark of Re is attached on this day. Life and dominion are before him, stability and honour[95] are established behind him. Everything is good on this day', as being a coded reference to the midnight culmination of Rigel, is simply fantastic.

On another front, Krauss has built up a series of papers that argue for an identification of 'the Eye of Horus' in the *Calendar* with the planet Venus, and 'Seth' with Mercury.[96] All entries relating to the Eye are then read as astronomical observation reports of Venus, and mentions of Seth as reports of observations of Mercury. Thus he reads II *prt* 14: 'Do not go out on [this day] at the beginning of dawn. It is the day of seeing the rebel and of the killing of him by Seth in the prow of the great barque' as an observation report of a sighting of Mercury at dawn. Just given the number of 'ifs' and 'presumings' in his paper, we should find his conclusions unconvincing. But if one also maps out the positions Venus should be at for each of the mentions of the Eye of Horus in the calendar and for Krauss's purported date for the observations (1298–1297 BC), one quickly sees that the usual astronomically significant phases (first appearance, last appearance, and greatest elongation) are not convincingly or consistently attested, and no motions or positions of Venus are consistently mapped onto a closed set of signifiers in the Egyptian. The host of mythological mentions of the Eye of Horus, where the Eye is 'being sought', 'being complete', 'being angry', 'raging', 'being at peace', 'being reconciled', are not consistently associated with particular motions, phases, or appearances of planets. Certainly 'appearing', 'disappearing', and 'achieving first station' are not talked about in the *Calendar* in any code that can be mapped from mythology to astronomy, from signifier to supposed referent. If all we can say is that Venus is somewhere doing something on each of the days when the Eye of Horus gets mentioned – but not *always* on those days, and not *only* on those days – then we have a code that is polyvalent to the point of meaninglessness. Krauss's associations are very weak creatures, only barely managing to stand on the bald assumption that Venus *is* the coded referent for the Eye of Horus, which, rather than proving the question, merely begs it before trying to support it with a series of ad hoc justifications for the (astronomically) inconsistent and imprecise language of the *Calendar*. What we come down to is an idiosyncratic decoding that takes its lead from Venus and tries to see if the *Calendar* could conceivably

---

[95] *P. Cairo* has a lacuna here. *P. Sallier* has two *Ded* columns ('stability') followed by a lacuna. Thus Leitz restores the *Ded* columns to *P. Cairo* and the *šps* sign ('honoured') to *P. Sallier*.
[96] Krauss, 2002.

be saying something about the planet, rather than finding a consistent set of referents in this or other texts to clearly defined and distinct phenomena of Venus.

These attempts to connect the *Calendar*'s omina with celestial events simply fail. There are no clearly worded astronomical phenomena of any significant note to be seen in the *Calendar*. There is also nothing resembling astrometeorology of any kind. The most we can say from the standpoint of the history of science, is that the Egyptians had been open to the idea of calendrical omina, which included *some* weather apodoses, as early as the mid-second millennium BC. But the evidence from this early period indicates that weather patterns were associated with calendar dates rather than with decanal phases or any other astronomical phenomena, and that the weather predictions themselves were of a very different form than those found later in the Ṣaft el-Ḥenna text. There is no clear relationship between the Ramesside calendrical omina and the content or context of the naos of Ṣaft el-Ḥenna.

Conclusion

In Harkhebi and the Ṣaft el-Ḥenna naos, we have found evidence for an indigenous astrometeorological tradition in Egypt dating from as early as the beginning of the fourth century BC. This is two centuries earlier than Neugebauer and Rehm had supposed an Egyptian tradition to date from, but considerably later than some of the recent readings of the *Calendar of Lucky and Unlucky Days* would have it. It is as yet unclear, however, whether the tradition represented by the naos from Ṣaft el-Ḥenna is the source of the attributions to 'the Egyptians' that we find in classical parapegmata. Certainly Harkhebi is far too vague to paint anything more than a general picture that points to an Egyptian interest in some kind of astrometeorology. The Ṣaft el-Ḥenna naos, however, gives us a considerably clearer idea about the kind of astrometeorology being practised by Egyptians in the Nile delta in the fourth century. What we see there is the association of each of the thirty-six decans with a ten-day period, and with various kinds of phenomena including, sometimes, weather patterns. The text itself offers an explanation for how the weather is associated with the decans. We are told that each decan, conceived of as a deity, causes the weather in each decade and we see that the decan is invoked as a deity at the beginning of each decade where requests are made for fertile fields, health, and so on. We see some association with temple practice, insofar as the inscription itself decorates a naos, but how or whether the inscription relates to ritual activity or festival calendars is unknown. No other astrometeorological text known to me from Egypt or elsewhere is directly related to temple activity, so the fact of a relationship here is itself interesting, even if the details of its religious import are not particularly transparent.

Although the *Calendar of Lucky and Unlucky Days* is not astronomical or astrometeorological, it does show a family similarity to several traditions in Mesopotamia, Greece, and Rome. The idea that calendar dates may be ominous is common to each of these cultures, manifesting itself in Mesopotamian hemerologies and menologies, as well as in classical beliefs about good- and bad-luck days. The proscriptions that we find in the *Calendar* to avoid various activities on certain dates are also mirrored – in kind, if not in specifics – in the Mesopotamian and Greco-Latin traditions. In the Greco-Latin tradition, we do also see these calendrical proscriptions beginning to be used alongside of astrometeorological material in the Roman agricultural texts (as we saw in chapter 2), and by the time of the eleventh-century AD Arabic parapegma of al-Bīrūnī,[97] we see similar proscriptions being incorporated right into parapegmata themselves.

---

[97] On which see A.xix in the catalogue, below.

7 | Conclusion

The developmental and cross-cultural currents in astrometeorology move differently than those in astrology and astronomy themselves. We do not get the same seminal moments for the transmission of practice and content in astrometeorology as we see in astronomy or astrology. There is no *Hipparchan synthesis* for astronomical weather prediction. There is no *Greco-Egyptian hothouse* for fixed-star meteorology as there is for planetary astrology. What we see instead is what looks to be the independent development in Babylon, Egypt, and Greece of a variety of sometimes similar methods of weather prediction and timekeeping involving the fixed stars. In Rome we have the amalgamation of Greek material with traditional Roman farming methods, but this is from the outset synthetic. It is never quite a wholesale adoption of just the Greek models by themselves.

Perhaps the most important reason for this independence of astrometeorological traditions is that the traditions themselves do not start out at a professionalized level in any of the cultures. Astrometeorology seems to have been born in the hands of farmers and sailors. Its eventual inclusion in different literary traditions is essentially a trickle-up effect from those humble beginnings. Think back to Polybius' report of the terrible maritime disaster of the first Punic war, in 255 BC. As Polybius tells it, we see the commanders of the collected forces refusing to listen to the advice of their lead sailors, the steersmen:

We must lay the blame of this [disaster] not on fortune, so much as on the commanders, for many of the pilots warned them not to sail along the outer coast of Sicily . . . and also warned that a shift in the weather was not yet over, and another one was coming, for they were sailing between the rising of Orion and that of the Dog Star.[1]

Those in charge of military strategy and tactics refuse to cede a decision-making veto to their subordinates, the pilots of the individual ships. Nevertheless, it is the knowledge of the subordinates that turns out to have been

[1] Polybius, 1.37.4–5.

crucial. The scale of the disaster serves to drive the point home dramatically: sailors know when it's safe to go back on the water.

So, too, do ancient farmers know when to sow, when to reap, when to prune, and when to process their produce. This knowledge begins to bubble up into the literary record quite early in Greece, and quite extensively in Roman times. The Mesopotamian picture is, not surprisingly, quite different. True, if we compare the content of the Mesopotamian texts, we see some general similarities between parapegmata and parts of *Enuma Anu Enlil* and MUL.APIN, among other texts, but there is no Mesopotamian equivalent of a classical parapegma, and the similarities always remain at a very general level. The texts and contexts in which we find the Mesopotamian astrometeorological passages also frame this material in ways that make it look really quite independent of the later Greco-Roman texts. With Egypt, on the other hand, we do see one text, the Ṣaft el-Ḥenna naos, that is structured in such a way as to be possibly related to the unknown sources for the parapegmatic attributions to 'the Egyptians'. This text's reliance on the Egyptian decans, however, points to its being part of an indigenous Egyptian tradition rather than being derived from Greek exemplars.

Although astronomical weather prediction is sometimes differently realized in the various cultures, there are also remarkable similarities in some of the signs they used for prediction: Stellar phases show up in Egyptian, Greek, Roman, and Babylonian sources as seasonal indicators generally, and also as more specific weather indicators. These phases were, in all these cultures, closely tied to the regulation of the agricultural year, for which the respective calendars used in the various places were, to a greater or lesser extent, unsuitable.

Towards the end of the time period we have surveyed we begin to see astrometeorological material being combined with a host of other signs and omina such as lunar days and general hemerological predictions in a group of texts known as *ephemerides*.[2] The astronomical ephemerides often combine information about calendars with lunar, solar, and planetary phenomena, and good and bad luck days. Much of what goes into a standard ephemeris, and much of its usefulness, is closely related to the data and uses of the different kinds of parapegmata we have been looking at. Indeed, the ephemerides stand as a kind of synthetic development on the diverse body of parapegmatic material. A description of the making of ephemerides in some manuscripts of Theon's commentary on Ptolemy's *Handy Tables*

---

[2] For a description and catalogue of the extant ephemerides, see Jones, 1999a, p. 40f.

shows that the ephemerides sometimes included fixed-star astrometeorological predictions such as we find in Greek and Roman astrometeorological
parapegmata, alongside lunar and hebdomadal information as we find in
Roman astrological parapegmata:

περὶ τῆς τῶν ἐφημερίδων ἐκθέσεως.

ἄνω καὶ κάτω τόπους μείζονας καταλιμπάνομεν ὅπως ἄνω μὲν τὰς ἐπιγραφὰς
δέξωνται τῶν σελιδίων, κάτω δὲ καθ' ἕκαστον μῆνα ἐποχὴν τῶν (συνόδων) καὶ
(πλησισελήνων), ἐν δὲ τῷ μεταξὺ τούτων ιε΄ χώρας, τὴν κατωτέραν μείζονα
ποιοῦντες, ἵνα αἱ μὲν ἄλλαι ἀνὰ β΄ δέξωνται στίχων, ἡ δὲ τελευταία γ΄ διὰ τὸ τὸν
ῥωμαικὸν μῆνα πολλάκις λα΄ ἔχειν. ἐπὶ δὲ τοῦ Φεβρουαρίου ιδ΄ δεῖ μόνας γράφειν.

τὰ δὲ σελίδια ποιήσωμεν οὕτως. πρότερον πάντων πλατύτερον τὸ τὰς ἐπισημασίας τῶν ἀπλανῶν δεχόμενον· καὶ ἐφεξῆς τὸ τρίτον ἢ τέταρτον στενώτερα
τὰ δεχόμενα τοὺς μῆνας, τὸ μὲν πρῶτον τὴν κατὰ ῥωμαικὸν πρὸ πόσων, τὸ δὲ
δεύτερον τῶν ἀλεξανδρέων τὸν μῆνα, τὸ δὲ τρίτον οἵαν τις προαιρῆται τῆς ἑαυ-
τοῦ χώρας, τὸ δὲ δ΄ τὸν κατὰ (σελήνην). ἐπὶ δὲ τούτων σελιδίων καταγράφομεν γ΄
πλατύτερα καὶ τέταρτον στενώτερον· καὶ ἐπιγράφεται μὲν πᾶσιν ἡ ἐπιγραφὴ α΄
(σελήνης) κίνησις, ἰδικῶς δὲ τὸ μὲν πρῶτον ζωδίων, τὸ δὲ β΄ μοιρῶν καὶ λεπτῶν, τὸ
δὲ γ΄ ὥρας μεταβάσεων, τὸ δὲ δ΄ ἀνέμων. ἐφεξῆς γίνονται σελίδια ζ, (ἡλίου) καὶ τῶν ε΄
πλανωμένων. ἐπὶ τούτοις σελίδιον ποιοῦμεν στενώτερον δεχόμενον τὸν ῥωμαικὸν
μῆνα, ἢ ὃν ἄν τις βουληθῇ κατὰ τάξιν. καὶ ἐπὶ πᾶσι τούτοις πλατύτερον ὀφεῖλον
δέξασθαι τὰς καθολικὰς καταρχάς, ἃς δεῖ ποιεῖν, οὕτω τὰ σχήματα καταγράφειν
τῆς (σελήνης) καὶ ἕκαστον τῶν (ἀστέρων) καὶ εἶθ' οὕτω τὰς καθολικὰς καταρχάς.

On the Making of an Ephemeris:

At the top and bottom, we leave a larger space so that the top can have the
headings of the columns, and the bottom the time of new and full moons for each
month; in between these, fifteen rows, making the bottom one larger, such that
the (rest) will have two lines each, and the (bottom one) three, since the Roman
month often has thirty-one days. For February, it is necessary to mark only fourteen
(columns).

We make the columns thus: before all of them, a wider one for showing the
weather changes of the fixed stars [τὰς ἐπισημασίας τῶν ἀπλανῶν], and the third
or fourth from there narrower ones, to hold the months: the first for counting down
the Roman month, the second for the Alexandrian month, the third, if one wishes,
for the calendar of his own land, and the fourth, for that according to the moon.
After these columns, we draw three wider and the fourth narrower, and the heading
'First Lunar Motion' is written over them all. In particular, the first (column) is
the zodiacal signs, the second the degrees and minutes, the third the times of sign
entry, and the fourth the winds. From here there are six[3] columns (the sun and five

---

[3] The MSS all say 'seven', but I have only ever seen six in an ephemeris, and only six are described
here.

planets). After this we make a narrow column for holding the Roman month, or whatever one wishes, in order, and after all of these, a wider (column) to hold the general predictions [τὰς καθολικὰς καταρχάς] which must be made, thus writing the configurations of the moon, and the stars, and from these, then, the general predictions.[4]

Here the hemerological predictions were derived from a complex interplay of lunar, planetary, and calendrical factors. So also in the eleventh-century Arabic-language al-Bīrūnī parapegma, we find hemerological prediction combined with astrometeorology. This coming together of weather omina and hemerological omina represents the unification of two closely related technological traditions: the Roman astrological and the Greek astrometeorological parapegmata.

At the other end of the time scale, in one of the earliest texts we have looked at, the Babylonian omen series *Enūma Anu Enlil*, there is a complex intertwining of weather omina with all sorts of other omina. This is also the case with the powers of the decans in the Ṣaft el-Ḥenna naos. It is only in the middle period, in the classical astrometeorological parapegmata, that we see the subject of astrometeorology being treated as strictly independent from other types of omina. This is not to say that astrometeorology was seen as separate from other types of astrology or divination, however. Rather, the texts in which we find it were but one type of astrological tool for facilitating a very specific set of predictions based on simple annual cycles. While all these cycles are temporal, not all of them are system-calendrical and this distinction is, as I argued in chapter 4, an important one. Zodiacal signs are not zodiacal months, just as lunar days are not necessarily lunar-calendar dates.

The earliest classical literature (Hesiod in the seventh or eighth century BC) shows that the Greeks were already at that point using stellar phases in the regulation of agricultural and navigational seasons. At about the same time as this the Babylonians were keeping careful dated observations of astronomical and meteorological phenomena in the *Diaries*. But this may have been related to the older tradition of weather and other omina, as found in *Enūma Anu Enlil*, and to the schematic dating of stellar and meteorological events in MUL.APIN. MUL.APIN's intercalation schemes have no parallel in the Greek literature, however. I have argued that the Babylonian astrolabes, and in particular *Astrolabe B*, were not intercalation devices, but were probably used for the determination of omina. And this

---

[4] Published in Curtis and Robbins, 1935, p. 83.

highlights how the Babylonian sources differ most notably from classical ones: the Babylonian material shows a close connection with omina which tend to be topically much more diverse than the types of predictions derived from stellar phases in the classical world. So also in Egypt, we find many types of omens connected with the astronomical decans, and with the calendrical year. This contrast nicely highlights a preference in the classical world for organizing the signs of omina in a particular way, such that astronomical protases were treated, or at least organized and listed, distinctly from other kinds of signs.

The heliacal rising of a single star, Sirius, had for millennia been used by the Egyptians as an annual marker of the beginning of the Nile flood: it heralded the rising of the Nile in the earliest Egyptian texts, an event of supreme agricultural importance in Egypt. So also in the Babylonian *Diaries*, Sirius alone among the fixed stars has its heliacal rising recorded. The rising of Sirius had also been one of the dated events (and the only dated *stellar* event) in the Babylonian 'Uruk scheme' (fifth century BC). This contrasts with Greece and Rome, where a *multitude* of stellar phases was used for timing agricultural activity from the earliest literary records, right through to the Middle Ages and beyond. In particular, the Greeks connected weather and seasonal markers with stellar phases. Hesiod has scattered such rules of thumb throughout his *Works and Days*, and astrometeorological rules and predictions could be found in works by later writers such as Euctemon, Meton, Eudoxus, and Callippus. By the third century BC at the latest (the date of P. Hibeh 27), these rules and predictions were being collected and assembled as parapegmata proper, and some of these parapegmata were explicit in crediting predictions to the earlier authors. Shortly after P. Hibeh, the Geminus parapegma also incorporated zodiacal signs as tracking devices, and these turn up also in the late second-century or early first-century Miletus I. While P. Hibeh, the earliest extant astrometeorological parapegma, is literary, it may have been an adaptation of an earlier inscriptional type of parapegma. Nevertheless, the earliest *extant* inscriptional parapegma, the Ceramicus parapegma, was almost certainly neither astrometeorological nor astronomical. Whatever its purpose was, the idea of marking cyclical events with a movable peg was adapted to astronomy and astrometeorology, if not before the P. Hibeh parapegma, then soon after it, as attested by the early first-century BC Miletus fragments. This view contrasts with the generally accepted picture of the development of parapegmata, which sees full-blown parapegmata being erected as early as the fifth century BC.

As for the classification of parapegmata, in the Greek tradition we find both inscriptional and literary astrometeorological parapegmata, but in

Rome there is only one attested inscriptional astrometeorological para-pegma (the Puteoli parapegma). Other Latin inscriptional parapegmata were used for tracking hebdomadal, lunar, and nundinal cycles. The heb-domadal and lunar cycles were used for the determination of good- and bad-luck days, and the lunar cycle also for regulation of agricultural and other activities. The nundinal column may have been used to track the local market-day. The fact that these astrological parapegmata turn up in graffiti leads me to believe that they were fairly common in late Republican and early Imperial Rome. I argued in chapter 2 that in the Roman context this type of parapegma is not so far removed from the astrometeorological type as has been generally supposed.

The core of the parapegma tradition is centred on a need to regulate seasonal activities such as agriculture, navigation, and perhaps also war-fare. These activities are so universal that it comes as no surprise that we find apparently independent traditions in Mesopotamia, Egypt, and Greece. The Roman tradition is sometimes difficult to disentangle from the Greek, but there are astronomical and astrological elements of Roman agricultural practice that are indigenous,[5] in addition to the many derivative elements.

By careful attention to the workings of astronomical signs and their rela-tionships to both texts and observations, I have argued that ancient systems of astronomical weather prediction relate to calendrical systems in fairly complex and diverse ways. While we cannot assume that all systematiza-tions of astrometeorological cycles also reflect parallel systematizations of calendrical cycles, it is apparent that ancient stellar, meteorological, and cli-mactic cycles did sometimes interact with calendrical systems of one sort or another. We see inscriptional parapegmata acting as extra-calendrical tracking devices for astronomical and astrometeorological cycles (among others) in the Greek tradition, and for lunar, hebdomadal, and nundinal cycles (among others) in the Latin tradition. We also see Roman-era literary parapegmata indexing astrometeorological cycles to calendars (Julian and Alexandrian), calendars whose organization was clearly related to attempts at synchronizing civic, astronomical, and astrometeorological cycles.

The cyclicality of these kinds of time is a recurrent theme in this book, and the different types of parapegmata feature as tools for tracking this cyclicality. Different types function in different ways, and interact with different aspects of day-to-day life, but all of them help to define particular rhythms of ancient life.

---

[5] As argued in chapter 2, above. See also Wenskus, 1998.

PART II

———

Sources

Catalogue of parapegmata

What follows is a descriptive catalogue of all known extant parapegmata, classified as follows:

(A) Astrometeorological parapegmata
(B) Astrological parapegmata
(C) Astronomical parapegmata
(D) Other parapegmata
(E) Reports of parapegmata
(F) Related texts and instruments
(G) Dubia

(A) *Astrometeorological parapegmata* are those which relate astronomical phenomena with weather. (B) *Astrological parapegmata* are those which were used to keep track of astrological cycles such as the days of the moon, the hebdomadal deities, and the zodiacal sign of the sun or moon. Some of these include civil calendrical information and nundinal days as well. (C) *Astronomical parapegmata* are those which provided a means to keep track of the phases of the fixed stars, with, so far as we can tell, little or no accompanying meteorological or astrological information. (D) *Other parapegmata* are those which are either too fragmentary to determine their use, or which do not fit my other classes. (E) *Reports of parapegmata* are ancient accounts or descriptions of parapegmata. (F) *Related texts and instruments* are those which show sufficient similarity to parapegmata to warrant some discussion. Lastly (G) *Dubia* is where I list any calendars, inscriptions, etc. which seem not to be parapegmata, but have been claimed to be so by previous authors. I give my reasons in each case for not counting them as parapegmata.

It will be noticed that my classification differs greatly from Rehm's (see Rehm, 'Parapegma', *RE*). In the first place, he distinguishes primarily between inscriptional and literary parapegmata, whereas I have chosen to class them according to their use, rather than their morphology. Secondly, and more importantly, Rehm lists many of the parapegmata under the heading of '*False (uneigentliche) Parapegmata*' (his class III). His reason for this seems to be that he was working under the assumption that the literary parapegmata, most of which are astrometeorological, were paradigmatic, and therefore only those inscriptional parapegmata which were astrometeorological were counted as genuine. The Ceramicus parapegma was counted on Rehm's assumption that it was used to count days in a zodiacal month, and was therefore seen as related to the Geminus parapegma.[1]

As outlined in chapters 1 and 2, I have chosen to treat all inscriptional parapegmata as genuine, and I class the literary ones relative to these. I classify related texts and inscriptions according to their similar uses. Each text is discussed more or less briefly in the catalogue, although the length of the discussion here is not necessarily indicative of the relative importance of a parapegma. The most important parapegmata have been covered in earlier chapters, and so they get only summary treatments here. For particularly problematic parapegmata I give extended treatments in this catalogue. Parapegmata are undated unless otherwise indicated.

A: Astrometeorological parapegmata

A.i	P. Hibeh 27
A.ii	Miletus II
A.iii.	Geminus parapegma
A.iv	Puteoli parapegma
A.v	Ovid's *Fasti*
A.vi	Columella's parapegma
A.vii	Pliny's parapegma
A.viii	Ptolemy, *Phaseis*
A.ix	Polemius Silvius, *Fasti*
A.x	Antiochus parapegma
A.xi	Clodius Tuscus parapegma
A.xii	Madrid parapegma
A.xiii	Johannes Lydus, *De mensibus*
A.xiv	Oxford parapegma
A.xv	Aëtius parapegma
A.xvi	Quintilius parapegma
A.xvii	Paris parapegma
A.xviii	*Iudicia* parapegma
A.xix	al-Bīrūnī parapegma
A.xx	Codex Marcianus 335 parapegma

---

[1] Against this position, see chapter 4, above.

B: Astrological parapegmata

B.i   Thermae Traiani parapegma
B.ii   Dura-Europus parapegma
B.iii   Latium parapegma
B.iv   Veleia inscription
B.v   Neapolitan Museum 4072
B.vi   Ostia inscription
B.vii   Pompeii calendar
B.viii   Pausilipum parapegma
B.ix   Trier hebdomadal parapegma
B.x   Trier parapegmatic mould
B.xi   Soulosse hedomadal parapegma
B.xii   Arlon hedomadal parapegma
B.xiii   Rottweil parapegma
B.xiv   Bad Rappenau hebdomadal parapegma

C: Astronomical parapegmata

C.i   P. Rylands 589
C.ii   Miletus I
C.iii   Codex Vindobonensis 108
C.iv   Antikythera mechanism

D: Other parapegmata

D.i   Ceramicus parapegma
D.ii   Guidizzolo *Fasti*
D.iii   Calendar of 354
D.iv   Capua *Fasti*
D.v   Coligny calendar
D.vi   Ariminum nundinal parpegma
D.vii   Suessula nundinal list
D.viii   Allifae nundinal lists

E: Reports of parapegmata

E.i   Cicero
E.ii   Petronius
E.iii   Diodorus Siculus

E.iv	Vitruvius
E.v	Diogenes Laertius
E.vi	Proclus
E.vii	Suda

F: Related texts and instruments

F.i	*Enūma Anu Enlil*
F.ii	MUL.APIN
F.iii	Hesiod
F.iv	Ṣafṭ el-Ḥenna naos
F.v	Diocles of Carystus, Letter of
F.vi	[Hippocrates], *Peri hebdomadon*
F.vii	[Hippocrates], *On regimen*
F.viii	Aratus, *Phaenomena*
F.ix	Eudoxus (Leptines) papyrus
F.x	Venusia *Fasti*
F.xi	*Menologia rustica*
F.xii	Varro's season list
F.xiii	Hyginus, *De apibus*
F.xiv	Ara Pacis meridian
F.xv	Sundials
F.xvi	Galen, *Commentary on Epidemics I*
F.xvii	Astronomical Ephemerides
F.xviii	Alésia disk
F.xix	Dijon disk
F.xx	Florentinus season list
F.xxi	*Geoponica* phase list
F.xxii	Byzantine season list

G: Dubia

G.i	*Calendar of Lucky and Unlucky Days*
G.ii	'Euctemon parapegma'
G.iii	Dionysius parapegma
G.iv	Pompeii calendar medallions
G.v	Ostia hebdomadal deities
G.vi	Hoffman *menologium*
G.vii	Hebdomadal deities
G.viii	*Dusari Sacrum* monuments

Index of parapegmata by name:

Esdud inscription	G.ix
'Euctemon parapegma'	G.ii
Eudoxus (Leptines) papyrus	F.ix
Florentinus season list	F.xx
Galen, *Commentary on Epidemics I*	F.xvi
Geminus	A.iii
Geoponica phase list	F.xxi
Guidizzolo *Fasti*	D.ii
Hebdomadal deities	G.vii
Hesiod	F.iii
[Hippocrates], *Peri Hebdomadon*	F.vi
[Hippocrates], *On Regimen*	F.vii
Hoffman *menologium*	G.vi
Hyginus, *De apibus*	F.xiii
Iudicia	A.xviii
Latium	B.iii
Leptines (Eudoxus) papyrus	F.ix
Lydus, *De mensibus*	A.xiii
Madrid	A.xii
Menologia rustica	F.xi
Miletus I	C.ii
Miletus II	A.ii
MUL.APIN	F.ii
Neapolitan Museum 4072	B.v
Ostia hebdomadal deities	G.v
Ostia inscription	B.vi
Ovid	A.v
Oxford	A.xiv
P. Hibeh 27	A.i
P. Rylands 589	C.i
Paris	A.xvii
Pausilipum	B.viii
Petronius	E.ii
Pliny	A.vii
Polemius Silvius	A.ix
Pompeii calendar	B.vii
Pompeii calendar medallions	G.iv
Proclus	E.vi
Ptolemy, *Phaseis*	A.viii
Puteoli	A.iv

Quintilius	A.xvi
Rottweil	B.xiii
Ṣafṭ el-Ḥenna naos	F.iv
Soulosse hebdomadal parapegma	B.xi
Suda	E.vii
Suessela nundinal list	D.vii
Sundials	F.xv
Thermae Traiani	B.i
Trier hebdomadal parapegma	B.ix
Trier parapegmatic mould	B.x
Varro's season list	F.xii
Veleia inscription	B.iv
Venusia *Fasti*	F.x
Vitruvius	E.iv

A: Astrometeorological parapegmata

A.i. P. Hibeh 27[2] is a literary Greek parapegma from the Saïte Nome of Egypt and indexed to the Egyptian calendar. It probably dates from the reign of Ptolemy Euergetes (early third century BC). Smyly argues, based on equinoctial dates and comparison with the Eudoxan phases in Geminus, that it was probably the work of one of Eudoxus' followers.[3] But as Grenfell and Hunt note, the text is 'much disfigured . . . by frequent blunders', and so one wonders how flexible we need to be in our understanding of 'follower' or what value the text has for Eudoxan scholarship generally. Neugebauer is very sceptical of a Eudoxan influence.[4] Some of the stellar phases show some similarity to material in C. Vind. 108 (see C.iii, below).

The text itself consists of a brief letter from the compiler to a student, followed by the parapegma. Part of this letter is word-for-word the same as part of the Eudoxus papyrus (see F.ix, below). The parapegma is not attributive,[5] but simply lists risings and settings, the lengths of days and nights, some weather and, uniquely, Nile-depth forecasts. It is also one of only two known Greek-language parapegmata systematically to contain

---

[2] Published by Grenfell and Hunt, 1906, pp. 138–57. Currently in the collection of Trinity College, Dublin (see Coles, 1974).

[3] In Grenfell and Hunt, 1906. [4] See Grenfell and Hunt, 1906, p. 143; *HAMA*, p. 687f.

[5] *Attributive* parapegmata are those that list either stellar phases or weather predictions as 'according to' particular authors. See chapter 1 for fuller discussion.

cultic information of any sort (the dates of Egyptian religious festivals).[6] Roman *Fasti*, in contrast, frequently contain festival dates.[7]

A.ii. Miletus II[8] is a fragmentary inscriptional Greek parapegma, with a series of holes drilled for a single moveable peg to track stellar phases and weather.[9] It was found, like Miletus I (C.ii, below), at the theatre in Miletus in the winter of 1902/3 by a crew working under the direction of Theodor Wiegand (see Figs. Cat. 1–3). Unlike Miletus I, it is attributive for both stellar phases and weather predictions. It is the only known source which mentions the 'Indian Callaneus', to whom several predictions are attributed.[10] In his *RE* article, Rehm links a fragment of introductory material (inv. no. 456C) to what he thought were the fragments for this parapegma (inv. no. 456A, 456D and 'N'), although in his original publication the introductory fragment 456C had been linked with Miletus I (no. 456B, C.ii, below). He later changed his mind, correctly, for 'epigraphical' reasons: 456C is clearly distinct from 456B in terms of epigraphy, content, and the size of the peg holes (so much so that the same peg could not have been used in both fragments).[11]

---

[6] The other is the Oxford parapegma, A.xiv, below. I note that the Antiochus parapegma does also have one festival listed (for the flooding of the Nile, no less). See A.x, below.

[7] On which, see Rüpke, 1995; Salzman, 1990.

[8] Currently in the Pergamonmuseum, Berlin, inv. no. SK 1606. First published in Diels and Rehm, 1904; see also Rehm, 1904; Wiegand, 1904, p. 82. Most recent publication in Lehoux, 2005.

[9] Rehm thought that a multitude of pegs would be used to correlate the parapegma with the local civil or religious calendar(s). Greek civil calendars were lunar, which means that the twelve-month 'year' was 354 days long. When needed, a thirteenth month would be added (a 'leap-month') to keep the lunar calendar roughly in line with the solar year. These differences in year length meant that the lunar calendar date on which a particular stellar phase occurred could vary by a half month or more from year to year. The stone parapegma could thus be used, Rehm thought, to determine what date in a given lunar year would see what stellar phases or what corresponding weather patterns. This would require that the pegs be labelled somehow to indicate calendrical dates. But no complete pegs have ever been found, and in the Thermae Traiani and Pausilipum parapegmata only the remains of a *single* peg were found in each. Rehm's theory seems to require a host of pegs for each and every parapegma (if Rehm's theory is right, then there would have to be at least thirty pegs per astrometeorological parapegma, and possibly as many as 384). Contrast this with the fact that in the parapegmata described by Petronius and Cicero, a single peg seems to have been used to keep track of each cycle. Moreover, on Rehm's theory, there would be no need for the peg holes (such as we find in Miletus I, for example) marking eventless days. It seems clear, then, that there was only one peg which was moved each day from hole to hole, thus indicating only the current date and the current astronomical or astrometeorological situation. Finally, the small size and high density of the peg holes would often make it impossible to fashion pegs with sufficient room for any information to be written on them.

[10] Diels and Rehm associate him with the Callanus mentioned by Cicero, *De div.* 1.47, in connection with Alexander the Great. Diels and Rehm, 1904, p. 108, n. 1.

[11] See Lehoux, 2005.

Fig. Cat. 1. Miletus 456A.[12]

12 Photographs courtesy of the Pergamonmuseum, Berlin. Reproduced with permission.

Fig. Cat. 2. Miletus 456C.

Fig. Cat. 3. Miletus 456D.

456C has the name [Ep]icrates Pylo[rou] written across the top, who was an ephor in 89/8 BC.

A.iii. The **Geminus parapegma**[13] is sometimes also called the *Pseudo-Geminus Parapegma* since it is often thought not to have been written by Geminus,[14] although it appears appended to the end of all MSS of his *Isagoge*. Although the *Isagoge* dates from the first century BC,[15] the parapegma does not cite any authors after Dositheus (late third century). Specifically, Hipparchus, whose work shows up in the *Isagoge* itself, is notably absent from the parapegma. The absence of Hipparchus indicates to me that the parapegma probably predates the late second century BC, but I do not think this implies that it must have been appended to the *Isagoge* by someone other than Geminus. Böckh points out that the season lengths of the parapegma do

---

[13] Editions: Aujac, 1975; Manitius, 1898. Extensive commentary in Evans and Berggren, 2006.
[14] For the original arguments against its authenticity, see Böckh, 1863, p. 22f. Aujac, 1975, p. 157, treats it as genuine.
[15] For this date, see Jones, 1999b.

not agree with those of the *Isagoge*, and claims that Geminus argues against astrometeorology in book xvii, but I note the following: (1) if Geminus *did* append an older parapegma to his book, there is no reason to expect him to have corrected the season lengths; he may have been simply letting the text stand as he found it; and (2) his argument in book xvii is not against astrometeorology *per se* (*pace* just about everyone who has written on the subject),[16] but against a stellar *influence* on the earth. He nowhere says that parapegmata do not work. What he does say is that the associations of stars and weather can only be based on repeated observations, rather than on accounts of stellar causation. This is *not* an argument against astrometeorology, but an argument against stellar *causation*. Sextus Empiricus makes the same point at *Adv. math.* v.1–2.

The Geminus parapegma is complete for the whole of a year beginning with the summer solstice. It is attributive, both for phases and weather. It correlates phases and weather with zodiacal days, that is, 'the first day of Cancer', 'the second day of Cancer', and so on. On these zodiacal days, see chapter 4, above.

A.iv.   The **Puteoli parapegma**[17] is a marble fragment 14.5 cm high × 8.5 cm wide, which turned up, rather by accident, it seems, in a storage room at the Puteoli amphitheatre some time before 1928 (see Fig. Cat. 4). The fragment contains a numeral, XII, and a partly destroyed Latin inscription correlating the evening setting of Delphinus with a storm: *Delphin[us] occid[it ves]peri, t[empes]tas . . .* Both the number and the weather prediction have peg holes. Rehm thinks that the XII must be a Greek civil calendar date, which would be a unique post-Julian example of a Greek lunar calendar, in Italy, in Latin. But as I have argued elsewhere, comparative evidence compels us to see the XII as representing a lunar day, as in the astrological parapegmata of class B, below,[18] which finds confirmation in the fact that the peg holes for the two entries are of markedly different sizes The hole for the XII is 2.5 mm in diameter, and that for Delphinus is 4 mm. The use of two

---

[16] The one exception is Alan Bowen, who in a recent paper given at the *History of Science Society* rightly saw that Geminus was not arguing against astrometeorology itself, but against a particular account of stellar causation.

[17] Most recently published in Lehoux, 2006b. Originally published by Mingazzini, 1928, pp. 202–4; See also Degrassi, 1963, vol. xiii.2, p. 310. Currently in the Museo Archeologico Nazionale di Napoli, inv. 144808. Note that Degrassi published an incorrect inventory number, as a result of which the inscription itself was rather hard to track down. I finally found it in storage in Naples with much help from Annamaria Ciarallo, Ciro Esposito, and Sergio Venanzoni, whose memory is nothing short of miraculous.

[18] See Lehoux, 2006b.

Fig. Cat. 4 The Puteoli parapegma.[19]

different sizes of pegs for the two different series is a kind of foolproofing for preventing the peg from accidentally migrating into the wrong column. We can see similar differences in peg-hole size in the Rottweil parapegma, the Neapolitan Museum 4072 parapegma, and in the Trier parapegmatic mould. Nevertheless, the particular combination of cycles tracked by the

[19] Photograph reproduced with permission of the Soprintendenza per i Beni Archeologici delle province di Napoli e Caserta.

Puteoli parapegma is unique. No other inscribed parapegma tracks both an astrometeorological and a lunar cycle, and this makes this text an important hybrid of Greek and Latin types. I note, however, that Pliny combines lunar days with astrometeorology in his *literary* parapegma, A.vii, below.

A second important feature of the Puteoli inscription pertains to the ruled lines marking off the different entries. Close examination of the fragment reveals that the vertical line to the left of *Delphinus occidit vesperi, tempestas* ends quite deliberately just below the *tempestas*, rather than being a product of damage to the inscription. The peculiar ruling makes it seem unlikely that there were more astrometeorological entries under the one for Delphinus here. The inscription seems instead to have been markedly horizontal.

A.v.   Ovid's *Fasti*[20] (early first century AD) includes the dates of some stellar phases, often with weather predictions. One entry, though, has a weather prediction with *no* stellar phase accompanying:

XVIII Kal. Mai.:
Luce secutura tutos pete, navita, portus:
ventus ab occasu grandine mixtus erit. (IV.625–6)

XVIII K. May
The next morning take refuge in a safe harbour, sailor:
there will be a wind from the west, mixed with hail.

The days of the nundinal week (*A* though *H*) are included throughout, as are the designations of *dies fasti, nefasti,* or *comitales.*

A.vi.   Columella's parapegma[21] is included as part of his treatise on agriculture, dating to the mid-first century AD. It is a long literary parapegma, annotated and interspersed with agricultural instructions. It correlates dates in the Julian calendar with stellar phases, weather, and seasonal indicators such as the migratory patterns of birds. It begins at the Ides of January and

---

[20] Edition: Wormell, Alton, and Courtney, 1997. Commentary on individual books: Fantham, 1998; Green, 2004. On the place of the astronomy in Ovid's poem, see Gee, 2002; Herbert-Brown, 2002b; Gee, 2000; Reydellet, 1999; Newlands, 1995; Santini, 1975; Ideler, 1825.

[21] The only complete recent edition is the Loeb, by Ash, Forster, and Heffner, 1948–55, but it leaves something to be desired, particularly as it makes no attempt to examine all the MSS for any of the books, nor any MS at all for those books previously edited by Lundström. Before the Loeb, the previous complete edition is Schneider, 1794. Budé is slowly issuing an edition but book XI, which contains the parapegma, has not yet been published. The parapegma is at *RR* XI.2.4f. For Commentary, see Wenskus, 1986; 1998. I do not follow Wenskus' (1986) identification of no less than four distinct calendars and two distinct parapegmata in this text. On Columella's text generally, see Noé, 2002.

runs through the year and back to the Ides again. For a fuller discussion, see chapter 2, above.

A.vii. Pliny's parapegma,[22] in book XVIII of his *Natural History* (first century AD) shows up first as a short excerpt of seasonal indicators, combined with stellar phases, Julian dates, weather, and agricultural instructions (many attributed to previous authors), as well as lunar days and other information. It is a complex intertwining of many different sources, as we should expect from Pliny. For a fuller discussion, see chapter 2, above.

A.viii. Ptolemy's *Phaseis*[23] is the longest and most detailed of the surviving parapegmata, and dates from the second century AD. It indexes stellar phases to the Alexandrian calendar. It is the only known parapegma which systematically offers different rising and setting dates for various latitudes, although Pliny mentions stellar phases happening for different regions from time to time. Uniquely, all stellar phases are calculated by Ptolemy himself,[24] and the weather predictions are attributed to a host of astrometeorologists including Euctemon, Eudoxus, Callippus, Meton, Caesar, and Hipparchus. How the weather predictions are timed is not yet clear. Ptolemy probably either linked particular predictions with particular phases according to the *Climata* in which the various authorities were observing, or else he arranged things by day counts from some temporal marker like the equinoxes or solstices.

A.ix.   The Polemius Silvius *Fasti*[25] is a Latin codex-calendar dating from the fifth century AD and organized according to the Julian calendar. It is extant in one twelfth-century manuscript. It includes meteorological predictions for the year, which were largely taken from Columella,[26] as well as Christian and some pagan festivals. The only astronomical entries are the solstices and equinoxes, which do not have weather associated with them. All meteorological entries are tied solely to calendar dates, but the fact that these weather predictions seem to have been excerpted from Columella, who does

---

[22] Edition: Mayhoff, 1897–1906. The parapegma is at *HN* XVIII.202f. Rehm lists Pliny's as two distinct parapegmata, his B6 and B20. I see no reason to do so. On Pliny generally, see Healy, 1999; Maxwell-Stewart, 1995; French, 1994; Beagon, 1992; French and Greenaway, 1986.

[23] Edition: Heiberg, 1907.

[24] On one possible method of calculation, see Graßhoff, 1993.

[25] MS in the Bibliothèque Royale de Belgique, nos. 10615–10729. Editions: *CIL* I[1], pp. 254–79; Degrassi, 1963, vol. XIII.2, pp. 263–76. For a discussion, see Salzman, 1990, pp. 242–6.

[26] See Degrassi, 1963, vol. XIII.2, p. 263.

correlate stellar phases with these weather phenomena on the same dates as Polemius Silvius, leads me to class this parapegma as astrometeorological. As I argued in chapter 3, above, the suppression of the stellar phases in this *Fasti* is a product of the usage of parapegmata, in which the primary indexing criterion is the date, rather than the stellar phase.

A.x.   The **Antiochus parapegma**[27] is a short Greek parapegma that correlates stellar phases with changes in the weather and occasionally with causal statements such as 'July 14: The whole of Orion rises at the same time as the sun; it causes (ποιεῖ) rain and wind.' All dates are in what I call the modified Julian calendar (i.e., dates are given as 1 July, 2 July, etc. rather than by the traditional method of counting down to the Kalends, Nones and Ides), which system seems to have begun to be used in the fourth century AD, rather than the sixth, as Mommsen thought.[28] Unique features of this parapegma are its mention of the 'ὕψωμα of the sun' on 10 April, and the duration of a change in the weather 'for seven days' on 23 May and 5 November, 'nine days' on 5 October, and 'fifteen days' on 6 November. It mentions a religious festival to celebrate the Nile flood on 22 October. Only one stellar phase is attributive (19 July: 'Rising of Sirius, according to the Egyptians'), and it also has 'birth of the sun, light increases' on 25 December.

A.xi.   The **Clodius Tuscus parapegma**,[29] is quoted in its entirety in Lydus' sixth-century *De ostentiis*, and referred to there as *Ephemeris for the whole year, by Clodius Tuscus, in translation, according to his text*. It is a long and detailed parapegma that correlates calendar dates with stellar phases, weather, the sun's entry into the zodiacal signs, the beginnings of the seasons,[30] and some bird behaviours. In many ways it is a puzzling text, with

---

[27] Extant in six MSS, of which the earliest is fourteenth-century, and the latest is seventeenth. Edition: Boll, 1910a.

[28] For this argument, see Ferrua, 1985.

[29] Preserved in Lydus, *De ost.* p. 117f., of which the edition is in Wachsmuth, 1897; Prior to Hase's 1823 publication of Lydus, versions of this parapegma were published independently of Lydus' text and credited to Ptolemy. For details, see Bianchi, 1914. For publication of further MSS of this parapegma, one of which is attributed to Hermes Trismegistus, see Bianchi, 1914, pp. 3–48. I find one of Bianchi's MSS (*Parisinus gr. 2419*) to be sufficiently different from the others to warrant counting it as a distinct parapegma (catalogued as A.xvii, below).

[30] The seasons are sometimes associated with zodiacal signs and sometimes not. Compare, for example, 17 August, ἀρχὴ φθινοπώρου, with 16 September, τὸ δωδεκατημόριον ἄρχεται τοῦ μετοπώρου. I am unclear, however, on why this latter entry should fall three days before ὁ ἥλιος ἐν ζυγῷ (19 Sept.), which is two days before ἡ μετοπωρινὴ ἰσημερία (21 Sept.). Compare also Nov. 9 and 10. I am likewise puzzled by the seven-day difference between 'the sun is in Sagittarius' (18 Nov.) and 'the sun is at the first degree of Sagittarius' (25 Nov.).

entries not shared by other parapegmata. The entry for 1 April tells us that on that date, 'the sun moves one degree'. On 27 July we are told that 'grapes are beginning to ripen', and on the 30th, 'fruits begin to ripen'. Another peculiarity of this parapegma is some entries that are tempting to interpret as observational reports: in addition to frequent entries which read ἄστρον κρυπτόν or κρύφιον ἄστρον, 'the stars are hidden',[31] we find τὸν δὲ ὀιστόν φασι δύεσθαι, 'They say that Sagitta sets' (19 Feb.). Even more unusual, however, are the entries on 10 August and 24 September: ἔκλειψις σεληνιακή, 'a lunar eclipse'. Each of these entries could only apply to one particular year, and the two together could not happen in the same year. I have been unable to find any years in close succession between 50 BC and AD 560 that fit the bill either. This is the only parapegma that refers to an eclipse. There are two entries referring to planetary 'houses' (8 Jan. and 5 Sept.). I am also puzzled by entries such as 29 January: ὁ δελφὶν δύεσθαι μελετᾷ, 'Delphinus tends to set', and 26 April: δύεται καθόλου ἡ ὑάς, 'the Hyades generally set'. The only other source that talks this way are the attributions to Democritus in Geminus.

There are also explicit descriptions of the stellar phases as causing particular weather phenomena, as at 1 January, 6 February, 13 September, 8 and 11 October and 24 November.

A.xii. The **Madrid parapegma**[32] is a Greek parapegma extant in one fifteenth-century manuscript and correlating some stellar phases with weather predictions and dates in the modified Julian calendar. It also predicts the behaviour of birds, such as χελιδόνες παίζουσι, 'the swallows are playful'. Bianchi dates it to 300 or 400 years after Clodius Tuscus, although the preserved MS dates from the late fifteenth century. It is related to the Aëtius parapegma, below, but is much more complete. The year begins with the month of March, and entries are in the reformed Julian calendar.

A.xiii. **Johannes Lydus, *De mensibus***[33] (sixth-century), contains some dated astronomical and astrometeorological predictions interspersed in a rich miscellany of civic, cultic, mythological, astronomical, medical, and other information organized by months and dates. The astrometeorological data have been excerpted from the larger text by Wachsmuth and put into

---

[31] E.g., 14 Jan., 1 Feb., 9 Feb., 14 Apr., 2 July, 15 Aug., 11 Dec., and 18 Dec.

[32] Madrid, Biblioteca Nacional, *Matritensis gr. CX*, fo. 160f. Published by Bianchi, 1914, p. 49f.

[33] Complete text: Wünsch, 1898. Astrometeorological parts excerpted in Wachsmuth, 1897, pp. 295–9.

parapegmatic form. Sources such as Democritus, Caesar, and Eudoxus are cited for both stellar phases and weather predictions.

A.xiv.   The **Oxford parapegma**[34] is extant in one manuscript in the Bodleian Library (C. Baroccianus 131, fos. 423–423v). Weinstock dates it to the early first century BC based on its date for the 'Egyptian New Year' on 20 August. If this is in fact the Julian equivalent in a particular year for the Egyptian Thoth 1, then, knowing that the Egyptian 365-day calendar slips one day every four years relative to the Julian, and knowing that Thoth 1 = 19 July in AD 139, we can conclude that this entry was composed around AD 15. This date is not inconsistent with other material in the parapegma dealing with festivals. Four festivals are mentioned: a feast of Ares on 1 March, a feast of the launching of the boat of Isis on 9 March, the Egyptian New Year already mentioned, the 'New Year' on 23 September, which, being celebrated on the birthday of Augustus links this text with the cultic calendar of Roman Asia Minor.[35] As mentioned in discussion of P. Hibeh (A.i, above), this is one of only two or three Greek parapegmata containing cultic information.

The astrometeorological contents of this parapegma show close connections with the Clodius Tuscus/Aëtius/Quintilius/Paris/Madrid group of parapegmata, including some verbatim passages shared with Aëtius.

A.xv.   The **Aëtius parapegma**[36] appears as a chapter titled περὶ ἐπισημασιῶν, *On Weather Changes*, in Aëtius of Amida's sixth-century medical work, the *Tetrabiblos*. The parapegma includes stellar phases, weather predictions, and some medical information directly associated with the weather. Aëtius explains the usefulness of astrometeorology for physicians as follows:

τῶν ὑγιαινόντων τὰ σώματα καὶ πολλῷ μᾶλλον τῶν νοσούντων ἀλλοιοῦται πρὸς τὴν τοῦ ἀέρος κατάστασιν.[37]

The bodies of healthy people, and especially those of sick people, change with the condition of the air.

This general idea can have some quite specific implications, as in the following:

---

[34] Edition: *CCAG* ix.1, pp. 128–37.

[35] Weinstock, 1948, argues for two more festivals: a ἡλιοδύσια on 22 November, and a ὕψωμα ἡλίον on 12 April. Neither of these is attested as a festival name in extant sources, however, and I am inclined to see them as being remarks of astrological significance rather than cultic.

[36] Edition: Olivieri, 1935 (*Corpus medicorum graecorum* viii.1). The parapegma is in *Tetrabiblos* iii.164.

[37] Aëtius of Amida, *Tetr.* iii.164.

μηνὶ Σεπτεμβρίῳ κε΄ ἰσημερία φθινοπωρινή· καὶ ἐστι μεγίστη ταραχὴ τοῦ ἀέρος πρὸ τριῶν ἡμερῶν· διὸ παραφυλάττεσθαι χρὴ μηδὲ φλεβοτομεῖν μηδὲ καθαίρειν μηδ᾽ ἄλλως τὸ σῶμα κινεῖν σφοδρᾷ κινήσει ἀπὸ ιε΄ τοῦ Σεπτεμβρίου μέχρι κδ΄.[38]

On the 25th of the same month (September): Autumnal equinox. There is the greatest disturbance in the air for three days previous. Thus it is necessary to be careful neither to phlebotomize, nor purge, nor otherwise to change the body violently from the 15th of September through the 24th.

This entry, like many others in this parapegma, associates the phase of a star with the weather several days before the occurrence of the phase.[39] The dates are listed in a schematic but incomplete combination of the Macedonian calendar and the Julian calendar, where three of the Macedonian months are straightforwardly equated with Roman ones, both beginning on the same day, rather than with the proper lag (contrast this with the Quintilius parapegma (A.xvi, below). The year begins in March. Roman dates are listed in the modified manner, as, for example, *26 February*, rather than *IV Kal. Mar.*

A.xvi. The **Quintilius parapegma**[40] shows strong similarities to that of Aëtius. It begins with a passage lifted from Aëtius (the whole introduction to his *On the Weather Changes of the Fixed Stars*, followed by the heading ἀστέρων ἐπιτολαὶ καὶ δύσεις κατὰ Κυιντίλλιον, *The Risings and Settings of the Stars, According to Quintilius*. The dates are listed in a curious combination of the Macedonian and Roman calendars, with the months organized according to the Macedonian calendar (but beginning in Xanthicos rather than Dios), and the individual dates given in the Roman calendar. For the latter, it mostly uses the traditional Roman dating system, although frequent modified dates such as 3 July or 7 September do creep in. The impossible IX Nones October also occurs. Quintilius has a better correspondence between the Macedonian and Roman calendars than Aëtius does.[41]

A.xvii. The **Paris parapegma**, extant in the fifteenth-century MS Par. gr. 2419, was published by Bianchi, thinking it to be a particularly corrupt MS of the Clodius Tuscus parapegma,[42] but it is sufficiently different, I think, to list it separately here. While a comparison with the Clodius Tuscus parapegma does reveal many similarities, particularly in the dating and wording of

---

[38] Aëtius of Amida, *Tetr.* III.164.
[39] Compare also the Madrid, Paris, and Quintilius parapegmata.
[40] Edition: Boll, 1910b.   [41] See Boll, 1910b, p. 24f.
[42] In the Bibliothèque Nationale de France. Edition: Bianchi, 1914, pp. 22–48.

meteorological predictions, the stellar phases are often quite different from Clodius Tuscus. In many places where there are differences between this text and Clodius Tuscus we find similarities in wording with the Aëtius parapegma, which indicates that this text borrows from or shares sources with Aëtius, Madrid, and Clodius Tuscus, even when Aëtius/Madrid and Clodius Tuscus do not share sources or borrow from each other. One notable instance of this is the inclusion of Aëtius' proscription of phlebotomy on 25 September. Entries are indexed to the reformed Julian calendar, beginning with the month of September. Entries become noticeably sparser toward the end of the year.

A.xviii.   The *Iudicia* **parapegma**[43] appears, incomplete, appended to one twelfth-century manuscript[44] of Pseudo-Ptolemy's *Iudicia*, and then turns up again, complete, in the P. Liechtenstein 1509 printing of the *Iudicia*. It seems, as Burnett has argued, to be largely (or even wholly) based on ancient parapegmata, quite probably sharing a source or sources with the Aëtius parapegma (though lacking its medical information). It may also derive some of its information from Pliny, either directly or by way of astronomical Scholia. The text is very corrupt, and information shared with Aëtius is translated fairly roughly. For example, where Aëtius has

μηνὶ τῷ αὐτῷ (Σεπτεμβρίῳ) ιδ΄ ἀρκτοῦρος ἐπιτέλλει· καὶ ἀλλοιοῖ τῇ ἑξῆς ἡμέρᾳ <τὸν ἀέρα>

On the fourteenth of the same month (September) Arcturus rises, and it changes [the air] for six days.

the *Iudicia* has

Quarta decima die eiusdem (sc. mensis septembris) Arcturus – id est Septemtrion – apparet cum Solis ortu. Mutatur aer in crastinum.

On the fourteenth day of the same (month, September) Arcturus – that is, Septemtrion – appears at sunrise. The air is changed the next day.

This parapegma is unique in that a few weather predictions specify the duration of the weather phenomena *in hours*:[45] *Prima die mensis septembris, Icarus custos plaustri apparet cum Solis ortu, et mutatur aer in .vii. horis, hoc fit inter diem et noctem*, 'On the first day of the month of September, Boötes,

---

[43] Edition with commentary: Burnett, 1993. The MS is in the British Library, Harley 5402.

[44] Burnett, following Krchnàk, 1963, p. 176–7, dates it to before 1160, although Lehmann, 1930, dates it to between the thirteenth and fifteenth centuries.

[45] But compare the Antiochus parapegma, which specifies a change happening 'for seven days'.

guardian of the Wain, appears at sunrise, and the air is changed for seven hours. This happens between the day and the night.' Or: *Vicesima septa die (Aprilis) Orion vespertinus ponet, et mutatur aer usque in 9 die<i> horas,*[46] 'On the twenty-seventh day (of April), Orion sets in the evening, and the air is changed continuously for nine hours of the day.' And lastly: *Die 9 (Iunii) vespertinus apparet Delphinus et mutatur aer in 10 hora<m> diei,* 'Day 9 (of June) Delphinus appears in the evening and the air is changed until the tenth hour of the day.' Like Aëtius, the dates are in the modified Roman calendar.

A.xix.   The **al-Bīrūnī parapegma,**[47] called *On the Days of the Greek Calendar* is an eleventh-century summary of Sinān ibn Thābit's tenth-century work *Kītāb al-anwā',* a now-lost work on meteorology. It is a compendium of the Greek parapegmata known to Sinān (Ptolemy, Geminus, and Johannes Lydus), with the addition of Arabic weather traditions and information from other Greek sources such as the Hippocratic corpus. It includes some dietary and medical advice for certain dates, the dates of festivals among different peoples, and the risings of the Euphrates and the Nile. There are also ritual prescriptions, such as the admonition that on the sixteenth of *Hazīrān* one is best 'to rise in the morning from sleep on the left side, and to fumigate with saffron before speaking'.

It mentions but a few astronomical phenomena, and those only in passing. This stems from Sinān's argument that the weather phenomena are tied to particular dates rather than to stellar phases:

Regarding the cause of these *anwā',*[48] scholars do not agree among each other. Some derive them from the rising and the setting of the fixed stars, among them the Arabs . . . Others again from the days themselves, maintaining that they are peculiarities of them, that such is their nature, at least, on an average, and that besides they are increased or diminished by other causes.[49]

---

[46] Burnett does not explain the switch to Arabic numerals in the later part of his edition of the calendar (beginning on the 19 April). It may possibly reflect the usage of the 1509 edition, versus the earlier incomplete MS version, although Burnett does mention 'oriental forms of Arabic numerals beginning on fol. 69r' in the MS.

[47] Azkāyi, 2001. Translation and commentary: Sachau, 1879. See also Neugebauer, 1971.

[48] *anwā'* is the plural of the noun *nau',* which is used in the singular throughout al-Bīrūnī to translate the Greek verb (!) ἐπισημαίνει. I think *anwā'* is being used here in the same general sense as ἐπισημασία is used in Ptolemy's introduction and conclusion to the *Phaseis*: i.e., to mean 'weather changes'. I note that *anwā'* cannot here mean 'influence of a lunar mansion' (as it usually does in astrological contexts), since al-Bīrūnī tells us that *anwā'* as effects of the heavens are explicitly ruled out by the 'other authors', who we are later told include Sinān.

[49] al-Bīrūnī, *Chronology,* p. 231.15–25, translation Sachau's. For further arguments on the issue of the date vs. the phase, see al-Bīrūnī, *Chronology,* p. 261.33f.

The exact relation between this text and other Arabic astrometeorological texts is still unclear.[50]

A.xx.   The **Codex Marcianus 335 parapegma** is preserved in a fifteenth-century Byzantine codex and published in *CCAG*.[51] Dagron dates the text to the tenth century. It is unusual in discussing mostly storms *at sea*. I include it here for only the sake of completeness, because Rehm lists it as a parapegma.

B: Astrological parapegmata

B.i.   The **Thermae Traiani parapegma**[52] was a graffito in wall plaster '$13\frac{1}{2}$ inches high by $16\frac{1}{2}$ inches wide',[53] found in the early nineteenth century in a house at the baths of Trajan on the Esquiline hill at Rome. De Romanis dates the inscription to the fourth century AD based on the 'crudity' of the red squares painted on the wall, into one of which the parapegma was carved, but this date is little better than a guess (*quadrilaterographical* dating is not, I am afraid, an exact science). Left exposed to the elements, the inscription disappeared shortly after de Romanis drew it in 1822. As we saw in chapter 1, Fig. 4, It consists of pictures of the gods presiding over the days of the week inscribed horizontally across the top, with their corresponding holes bored just beneath them. The days of the moon from *I–XV* run vertically down the left side, and from *XVI–XXX* down the right. A hole appears in de Romanis' illustration just above and to the right of the hole for *XXX* but it is not, as has been speculated, for an unnumbered day *XXXI*.[54] Erikkson argued that this hole was used to mark either full or hollow months, but his argument is uncompelling. This in two respects: first, comparison with other, similar parapegmata (see especially Dura-Europus,

---

[50] See, e.g., Dozy, 1961; Pellat, 1986.

[51] MS in the Biblioteca Nazionale di San Marco, Venice. Edition: *CCAG*, II.214. See also Dagron, 1990.

[52] Now destroyed. Publications in Marulli, 1813; de Romanis, 1822; Stern, 1953, p. 177, pl. XXXII; Degrassi, 1963, vol. XIII.2, pp. 308–9. There is a terracotta copy made from the original by someone named Ruspi, which is now in the Kunstgeschichtliches Museum der Universität Würzburg, although this copy has been 'improved' by the restoration of Saturn and Jupiter (see Stern, 1953, p. 177, n. 3; contrast Goessler, 1928, who wrongly thinks that the Ruspi copy may have been made before the original was damaged – Piale's excavation report (in Guattani, 1817, pp. 160–2) says that the graffito was already damaged when it was unearthed). A plaster cast of this copy can also be found in the Museo della Civiltà Romana, Rome. See Manicoli, 1981.

[53] According to Guattani, 1817. De Romanis says that his illustration is one third the size of the original, which would make the original about 25 cm high by 30 cm wide.

[54] See Rehm, 'Parapegma', *RE*, col. 1364; Erikkson, 1956. For my argument, see Lehoux, forthcoming.

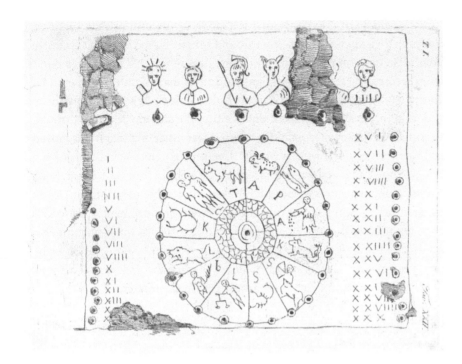

Fig. Cat. 5. Piale's 1816 illustration of Thermae Traiani.[55]

Latium, Ostia, and Pompeii, below) shows that a thirty-first hole is never part of a parapegma of this sort. Furthermore, and even more compellingly, the thirty-first hole *only* shows up in de Romanis' 1822 illustration (and, curiously, is less prominent in de Romanis himself than in later copies from him). In Piale's excellent illustration (see Fig. Cat. 5) of six years earlier, there is no hole to be seen.[56] So also in the copy made by Ruspi in around 1813 (and now in the Kunstgeschichtliches Museum der Universität Würzburg), there are only thirty holes. The Piale illustration of 1816 does also include a text description of the graffito, and in the description no thirty-first hole is mentioned, although the author there speculates that there may have been an extra hole *outside the line* and between the twenty-ninth and thirtieth hole (damage is indicated in Piale's drawing) to act as a stopper for twenty-nine-day months. But Piale's wording makes it look like this was a guess rather than an observation.[57]

---

[55] Reproduced from Guattani, 1817.

[56] The illustration is by Stefano Piale, published in Guattani, 1817, pp. 160–2, pl. xxii. This illustration is reproduced in Stern, 1953, pl. xxxii.5.

[57] Piale, in Guattani, 1817, p. 161.

In the middle of the parapegma is a circle divided into twelve sections, with each division bearing the picture and initial (Latin) letter of successive zodiacal signs running counter-clockwise. There are two holes drilled per sign, representing the beginning and middle of each sign. A fragment of a bone peg was found in one of Gemini's holes. This circle seems to have been used to keep track of the movement of either the sun or the moon through the zodiac. Erikkson speculates that the peg was lunar and that it was moved each day for six days, resting on the seventh, but this would still leave a deficit of one day every three lunar months, or about four days a year. Such a snowballing inaccuracy would have been intolerable over the long term, and so either (a) some more complicated system of movement was used, or (b) the motion of the peg, whether representing solar or lunar phenomena, was dictated or at least calibrated from time to time by observation or by some external text. Unfortunately neither other parapegmata nor testimony about the use of parapegmata shed any light on the matter.

B.ii.   The **Dura-Europus parapegma**[58] is a graffito found scratched into a plastered wall in a Syrian house which served as a military barracks during the Roman occupation (see Fig. Cat. 6). The inscription therefore dates from between AD 165 and 257.[59] Much of the inscription was destroyed during excavation, but the remainder is now at Yale. This parapegma is similar in many respects to Thermae Traiani in that it has the gods of the hebdomadal days represented across the top, apparently with holes drilled into the plaster above them for a moveable peg. The word *LUNA* appears just above the head of that goddess. The numbers from *I* to *XXX* are inscribed vertically down the two sides of the parapegma (beneath the heading *LUNA*) but these do not seem to have had peg holes beside them. Just to the left of the middle is a column that seems to read ~~NUNDINE~~, ~~VIII, VII~~, *VI*, ~~V~~, *IIII, III, PRI[DIE]*, again, without peg holes. Snyder argues that this column should be read as counting down (in the manner peculiar to the Roman calendar) to the nundinal day, which may have had some ceremonial or military significance for the local Roman population.[60] We should be generally cautious, however, about this parapegma, since the published drawing and description were both executed *after* the parapegma itself had been badly damaged in

[58] The drawing is based on a partial tracing, a partial photograph, and a 'careful description', originally published in Rostovtzeff, Bellinger, *et al.*, 1936, p. 42; also in Snyder, 1936, p. 15. The remaining fragments are supposed to be at Yale, but curators there have not so far succeeded in locating them for me.

[59] For these dates, see Drower, Gray, *et al.*, 1996, pp. 574–5.

[60] For my reservations about Snyder's claim, see Lehoux, forthcoming.

the excavation. The drawing was made largely from memory aided by a partial photograph.

Fig. Cat. 6. The Dura Europus parapegma.[61]

B.iii.   The **Latium parapegma**[62] is a marble fragment 53.5 cm high, 33 cm wide, and 3 cm thick, comprising probably the right-hand two fifths of the

---

[61] Drawing from Rostovtzeff, Bellinger, *et al.*, 1936. Reproduced with permission of Yale University Press.

[62] Formerly part of the Farnese collection (hence the inventory number inscribed in the lower right-hand corner). Currently in the Museo Archeologico Nazionale di Napoli, inv. 2635.

original. It has the names of the days of the week (with their corresponding holes) written horizontally across the top, the nundinal days (with holes) written vertically down the right-hand side, and the remains of the numbers *I–XXX* (i.e., days of the lunar month) with their peg holes arranged in a unique floral design in the middle. The inscription *LUNAR* appears in the upper right corner, and Degrassi has speculated that the parapegma originally had *[DIES] LUNAR[ES]*. The preserved fragment also includes the dates and lengths for two seasons as follows: *[A]estas ex XI k. Mai. in X k. August. Dies LXXXXIIII*, 'Summer is from XI K. May to X K. Aug.: ninety-four days' and *Hiemps ex X k. Nov. in XIIII k. Febrar. Die[s LXX]XVIIII*, 'Winter is from X K. Nov. to XIV K. Feb.: eighty-nine days.' Based on the preserved fragment, we cannot be certain whether or not this parapegma originally contained either a column for the zodiacal signs, or even possibly for weather predictions. Degrassi's speculation that there was another column of nundinal days down the left side is based only on an assumption of symmetry, and I think it unlikely. What purpose could a second column of nundinae possibly have? A more likely candidate would, following Thermae Traiani, seem to be zodiacal signs.

B.iv. The **Veleia inscription**,[63] found in 1762 in Veleia in northern Italy, is a group of marble fragments from the upper left corner of a larger inscription, with two smaller bits from elsewhere in the inscription found later. It has drawings of a few five- and six-pointed stars and a crescent moon on the top part. The numbers *III–X[VII]* and *[XV]III–XX[I]* are extant, written horizontally beneath the pictures. These presumably represent lunar days. The number *XV* is written twice; the second time it is circled (perhaps as a deletion mark?). Below the row of numbers are more pictures, with at least a small lunar crescent visible. These may be parts of images for the hebdomadal days of Luna and perhaps Mars, as in the Dura-Europus and Thermae Traiani parapegmata. No peg holes are mentioned by the editors, although there are small holes, probably just decorative indentations, apparent in the centre of each star in Degrassi's photograph.

Published in *CIL* VI. 32505 (see also *CIL* I¹, p. 218; Mommsen, 1852, no. 6747; Gruterus, 1707, p. 136 – note that the upper right-hand corner seems to have been still intact when Gruterus saw the inscription); Degrassi, 1963, vol. XIII.2, pp. 300–1. For my detailed arguments about the reconstruction, see Lehoux, forthcoming. For photograph see Fig. 2.2, above.

[63] Currently in the Museo Archeologico Nazionale in Parma. Inventory number unknown. Fragment sizes: approx. 20 × 24 cm, 11 × 13 cm, and 6 × 7 cm. Published in: *CIL* XI, no. 1194; Degrassi, 1963, vol. XIII.2, p. 313. Earlier publications of individual fragments include: Costa, 1762, p. 48, p. 251, tab. 81; de Lama, 1818, p. 81f.

B.v.   Neapolitan Museum 4072[64] is a marble fragment of uncertain origin measuring 41 cm high, 41 cm wide, and 2.5 cm thick. It contains a partial list of days of the moon (*XVI–X[IX]*) with the word *DIES* partially preserved above them, and the days of the hebdomadal week (*MERCU]RI, IOVIS, VENERIS [...]*) below. The named days and the numbers have corresponding peg holes drilled above them, with the peg holes for the hebdomadal days being noticeably larger than those for the days of the moon.

B.vi.   The Ostia inscription[65] is a partially preserved graffito found in a third-century private house in the so-called *Via dei dipinti* in Ostia (reg. I, ins. IV). It has the word *LVNE [....]* over a row of numbers from *[...] II* through *XXX [...]* Lanciani sees a partial *V* at the end of the third line, after the *XXX*. He does not mention holes near the numbers. The inscription may be for the purpose of correlating lunar days with hebdomadal days.

B.vii.   The Pompeii calendar[66] was found in 1927 carved in the plaster of a shop (reg. III, ins. IV, n. I) in Pompeii. It consists of hebdomadal days, nundinal days, the dates of a calendar month, and the lunar days *I–XXX* written in successive vertical columns. There were apparently no holes for pegs. It reads:

DIES	NVNDINAE	X[VIIII]	VIII	NON	I	XV	XXVIIII
SAT	POMPEIS	X[VIII]	VII	VIIII	II	[X]VI	XXX
SOL	NVCERIA	X[VII]	VI	VIII	III	XVII	
LUN	ATILLA	XV[I]	[V]	VII	IV	XVIII	
MAR	NOLA	XV	[I]V	VI	V	XVIIII	
MERC	CVMIS	XIV	[I]II	V	VI	XX	
IOV	PVTIOLOS	XIII	PRI	IV	VII	XXI	
VEN	ROMA	XII	K	III	VIII	XXII	
	CAPVA	XI	VIII	PRI	VIIII	XXIII	
		X	VII	IDUS	X	XXIV	
		VIIII	VI		XI	XXV	
			V		XII	XXVI	
			[IV]		XIII	XXVII	
			[III]		XIV	XXVIII	
			[PRI]				

---

[64] In the Museo archeologico Nazionale di Napoli. Publication: *CIL* X.1605; Degrassi, 1963, vol. XIII.2, p. 307; Originally published by Fiorelli, 1867–8.

[65] Now lost, so far as I have been able to determine. Size unknown. Editions: *CIL* XIV.2037; Lanciani, 1878, p. 67; Degrassi, 1963 vol. XIII.2, p. 312.

[66] Now apparently lost. Size unknown. Editions: Della Corte, 1927, p. 98; *CIL* IV.8863; Degrassi, 1963, vol. XIII.2, p. 305; Kubitschek, 1928, p. 232. See also Fig. 2.5, above.

In the drawing reproduced in the *CIL*, there is damage to the wall at the top of the inscription between columns 3 and 4, and in a small part of column 7. Curiously, no obvious damage is indicated below the *V* in column 4, but some dates are nonetheless not visible. They may have been erased in antiquity, or else the reproduction is faulty. The *VIII* in column 4, just below the *K* (for *kalendae*) is read as a month name by Zangemeister, Della Corte, and Degrassi: *NOV* by Zangemeister and Della Corte, and *IAN* by Degrassi. But I think neither of these suggestions is plausible. Degrassi sees the entries beneath this, the *VII, VI, V*, as mistakes for *IV, III, PRI*, but I doubt this very much.[67] Instead, it is fair to infer from comparison to other calendrical and astrological parapegmata that this inscription was meant as a kind of perpetual calendar, and that the dates are meant to be used for *any* month, not just one particular month. Moreover, *VIII* follows the pattern throughout the rest of this particular calendar nicely, where the dates begin counting down immediately after the feast day, and the next entry is clearly *VII*, followed by *VI* and *V*. This is an idealized month.

But even so, this calendar has a peculiarity not pointed out by either Degrassi or Della Corte: it seems to contain impossible dates. In no month of the Roman calendar is there an *ante diem VIII* or *VII Nonas*, nor an *ante diem VIIII Idus*. I can offer no suggestion to account for this feature, except to note that the Quintilius parapegma also has an impossible date.

B.viii.   The **Pausilipum parapegma**[68] is a broken marble fragment measuring 26 cm high and 59 cm long, which lists the names of the days of the week, with their corresponding holes, and the nundinal days below these. It reads:

•	•	•	• [...
SATVR	*SOLIS*	*LVNAE*	*MARTIS* [...
•	•	•	• [...
ROMAE	*CAPVAE*	*CALATIAE*	*BENEV [ENTI*...

Maaß dates it to the first century BC.[69] For a photograph, see Fig. 1 in chapter 1, above.

---

[67] For my full argument, see Lehoux, forthcoming.

[68] Excavated in 1891 at Pausilipum, currently in the Johns Hopkins University Archaeological Collection, inv. 5384 a/b. Publication: Fulvio, 1891, p. 238; Degrassi, 1963, p. 304. See also *CIL* I¹, p. 218.

[69] Maaß, 1902, p. 265.

Fig. Cat. 7. The Trier parapegmatic mould (photo courtesy of the Rheinisches Landesmuseum, Trier)

Fig. Cat. 8 Modern pressing from the Trier parapegmatic mould (photo courtesy of the Rheinisches Landesmuseum, Trier).

B.ix. The **Trier hebdomadal parapegma**[70] is a clay fragment found in Trier in 1930, consisting of the images of the first five hebdomadal deities, from left to right beginning with Saturn, each above its own peg hole.

B.x. The **Trier parapegmatic mould**[71] is a unique find from a pottery site in Germany. It is a mould meant to impress on wet clay the image of a hebdomadal parapegma, with the personified images of the four seasons below, and the thirty holes for the days of the moon down the two sides (fifteen per side). The lunar holes are set inside a series of graphic representations of lunar phases. It seems to indicate some scale of production for parapegmata, and therefore a corresponding local demand. (See Figs. Cat. 7 and 8 for the mould itself and a modern pressing taken from it.) Michael Wright has helpfully pointed out to me that ancient pressings would probably have put thirty small sticks into the lunar-day holes in the mould in order to produce a series of small holes in the finished parapegma, rather than the small bumps shown in Fig. Cat. 8. Note also that, as in the Puteoli

[70] Currently in the Rheinisches Landesmuseum Trier, inv. ST12014. Size: 8 × 15 cm. Publication: Dölger, 1950; Binsfeld, 1973; Sadurska, 1979.

[71] Currently in the Rheinisches Landesmuseum Trier, inv. ST14726. Size: 13 × 22.5 cm. Publication: Dölger, 1950, pl. 5; Binsfeld, 1973; Sadurska, 1979.

Fig. Cat. 9. The Arlon hebdomadal parapegma.[72]

and Rottweil parapegmata, the hole sizes are dramatically different for the two cycles tracked by this parapegma.

B.xi.   The **Soulosse hebdomadal parapegma**[73] is a fragmentary limestone relief, 16 cm high by 27 cm long, showing images of four of the hebdomadal deities, from left to right: Mars, Mercury, Jupiter, and Venus. Under each is a peg hole.

B.xii.   The **Arlon hebdomadal parapegma**[74] (see Fig. Cat. 8) is a stone fragment of unknown provenance measuring 20 cm by 19 cm, and 12 cm thick. It seems to be the leftmost part of a bas-relief. On the left is an image I take to be Saturn,[75] and to his right, Sol. There are peg holes beneath both of the images. The stone breaks off to the right of the sun.

---

[72] Photo courtesy of the Institut Archéologique du Luxembourg (Arlon). Used with permission.

[73] Currently in the Musée Départemental des Vosges in Épinal. Inventory number unknown. Publication: Espérandieu, 1907, vol. vii, no. 4857.

[74] Currently in the Musée Luxembourgeois in Arlon, inv. GR/C 63. Publication: Espérandieu, 1907, vol. v, no. 4016.

[75] The image is described by Espérandieu as a veiled goddess which he presumes to be the moon, (although notably lacking her usual lunar crescent). He agrees that the relief depicts the

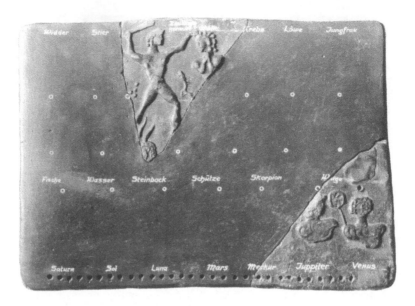

Fig. Cat. 10 The Rottweil parapegma.[76]

B.xiii.   The **Rottweil parapegma**[77] is clay parapegma, apparently pressed from a mould, consisting of two small fragments, one 12 by 12 cm, and the other 9.7 by 11.6 cm. The larger fragment shows Jupiter and Venus with a peg

hebdomadal deities, but thinks it is meant, oddly, to be read from right to left, and that it must have begun, again strangely, with Mars rather than Saturn. Duval, 1953, p. 287, notes that among the representations of hebdomadal deities, this one would be unique in both respects. The relief is very poorly preserved, and Espérandieu's 'veiled goddess' is neither clearly veiled nor clearly a goddess. It very well may be Saturn, who has a headdress in the Dura-Europus and Trier parapegmata, as well as the Alésia disk. Indeed Saturn, lacking obvious characteristics such as the Sun's rays, Luna's crescent, or Mars' helmet, must be deduced by Dölger, 1950, p. 202, as the first god in the Trier fragment. The cowl typically worn by Saturn in Roman iconography can be quite subtle. Moreover, the Moon is usually accompanied in pictures with a lunar crescent sitting prominently either on her head or on her shoulders, a feature which is conspicuous by its absence in this image. I do note, however, that in the earliest representation of the hebdomadal deities, the Pompeii calendar medallions, she has an almost full moon behind her head, looking rather more like a halo than a moon (see Long, 1992). But given the weight of the comparative evidence of other images of the deities and their usual ordering, I think it safe to say that the Arlon relief begins with Saturn on the left, with the Sun to its right, followed (in the lost part) by the Moon, etc. as would be expected.

[76] Photo courtesy of the Württembergisches Landesmuseum, Stuttgart. Used with permission.

[77] Currently in the Württembergisches Landesmuseum, Stuttgart, inv. R 171,55. Thanks to Nina Willburger for her help with examining the fragment. The parapegma itself has been rather unfortunately 'restored' by a well-meaning curator who has embedded the fragments in some kind of plastic matrix, and labelled everything in white paint. Publications: Goessler, 1928; Duval, 1953, p. 287.

hole (4.5 mm diameter) above Venus. There is an incomplete series of holes beneath these (each 3.5 mm diameter), but it is unclear what information they were used to keep track of. Given the different peg size in this row, it is clear that it is an independent row from the hebdomadal one. From comparison with other parapegmata, I do not think it unlikely that these were for counting lunar days. Another fragment shows what may be part of the image of Capricorn (see Fig. Cat. 10).

B.xiv. The **Bad Rappenau hebdomadal parapegma**[78] is a fragmentary sandstone relief excavated in two pieces (9 × 11 cm and 5 × 5 cm) in 1980 and 1982. It contains the busts of the last three hebdomadal deities, Mercury, Jupiter, and Venus, with peg holes extant under the last two. Wagner-Roser dates it to either the late second or the third century AD.

C: Astronomical parapegmata

C.i. **P. Rylands 589,**[79] a second-century BC papyrus from Egypt, originally included what the text itself describes as a parapegma, but apart from an introductory section, most of the text has been lost. The introduction says:

. . . παράπ[ηγμα τ]ῶν κατ[ὰ σ]ελήνην νουμηνιῶ[ν ὡς εἰσι κ]ατὰ [τὰς ἡ]μέρας τῶ[ν] κατ' Αἰγυπ[τίους δωδε]καμή[νω]ν τεταγμέναι, οὗ ἐστὶν ἡ π[ερ]ίοδ[ος ἔ]τη μὲν εἴκ[ο]σι πέντε, μῆνες δ[ὲ σ]ὺν ἐμβ[ο]λίμ[ο]ις τριακόσιοι ἐννέα, ἡμέρ[αι] δὲ ἐν[ακι]σχ[ίλ]ιαι ἑκατὸν εἴκ[ο]σι πέντε. [σ]ημαί[νε]ι δὲ καὶ τοὺς κατὰ σελήνην μῆνας καὶ τούτων τίνες ἠσὶ πλήρη[ς] κ[αὶ] τίνες κοῖλ[οι κ]αὶ π[ο]ῖο[ι] αὐτῶν ἐμβόλιμοι κα[ὶ ἐ]ν τίνι ζω[ιδί]ωι ἥλι[ο]ς καθ' ἕκαστον μῆν[α στή]σεται. ὅ[ταν] διέλθει τὰ εἴκοσι πέ[ν]τε ἔ[τη] πάλιν ἐπ[ὶ τὴ]ν αὐτὴν ἀρχὴν ἥξει καὶ τὸ[ν α]ὐτὸν τρόπ[ο]ν ἀλ[λάξ]εται. ἔστιν δὲ πρῶ[το]ν ἔτος τῆς περιόδου [τ]ὸ αὐτὸ τῶι πρώτω[ι] ὡς βασίλισσα Κλεοπάτ[ρ]α καὶ βασιλεὺ[ς Π]τολεμαῖος ὁ υ[ἱ]ὸς θεοὶ Ἐπιφ[αν]εῖς ἄγουσιν[80] ἐν [ὧι] καὶ τὴν βασιλείαν [π]αρ[ελ]άβοσαν. ὁ δὲ ἥλ[ιος] καθ[έστ]η . . .

A parap[egma o]f the beginning[s of l]unar months (as they are ordered according to [the d]ays of the Egyp[tian ye]ar) of which the cy[cl]e is twe[n]ty-five years, (having) 309 months (including inter[ca]lary months), (or) 9,125 day[s]. It sh[o]ws the months according to the moon, and which of these will be [f]ull, which hollo[w],

---

[78] Published by Wagner-Roser, 1987. Current location unknown.

[79] Currently in the John Rylands Collection, University of Manchester. Publication: Turner and Neugebauer, 1949.

[80] Turner and Neugebauer note that this is 'a formal and technical expression', which is true enough, though its exact meaning is opaque to me.

and w[h]i[c]h intercalary, an[d i]n which zo[dia]cal sign the sun [wi]ll be for each
mont[h]. W[hen] 25 ye[ars] have passed, it will go t[o th]e same beginning, and it
will change in th[e s]ame way. The fir[s]t year of the cycle is [t]he same as the firs[t]
(year) that Queen Cleopat[r]a and Kin[g P]tolemy the y[ou]nger were taken as gods
manifest and in [which] they [r]e[ce]ived the kingdom. The su[n] s[tand]s [...

Of the parapegma itself, only bits remain, including a table equating Egyptian months with the sun's position in the zodiacal signs for one year, the
Egyptian dates of νουμηνίαι (lit: 'new moons', but often referring to the
first days of schematic lunar months) for four months of one year and nine
months of another, and some mention of what may be religious festivals.

It is interesting to note that this text seems to have been used only to fix
the motions of the sun and moon in terms of Egyptian calendar dates in a
twenty-five-year cycle. It is unclear from the text if this would have primarily
had astrological or calendrical significance. Another possibility is that the
cycle was used to correlate parapegmata (which we know from Geminus
and Miletus I were sometimes being indexed to the sun's motion through
the zodiac by the late second century BC) with the Egyptian calendar. The
νουμηνίαι dates may have been used to count lunar days, as in the astrological
parapegmata, above, or may have reflected a local civil or cult calendar.

C.ii.   Miletus I[81] is a fragmentary inscriptional Greek parapegma excavated
with Miletus II (A.ii, above) by Wiegand in 1902/3 and published by Diels,
Rehm, and Dessau in 1904. There is an introduction on a separate fragment,
linked to the parapegma by Rehm[82] based on 'epigraphic similarity', which,
if correct, would date the parapegma to the year 110/9 BC. Unfortunately, I
have not been able to find the introductory inscription (Rehm says only that
it was 'left in Miletus'), and so have been unable to verify whether Rehm's
link is plausible. He certainly made some glaring epigraphical errors in his
original publication of the Miletus fragments, and I am disinclined to trust
him here without verification.[83]

The parapegma fragment itself has stellar phases listed, and holes for
a moveable peg around the phases. The phases are organized into groups
according to the sun's position in the zodiac. There is almost no weather

---

[81] Currently in the Pergamonmuseum, Berlin, inv. no: SK1606; Excavation inv. no. 456B.
Publication: Diels and Rehm, 1904; Dessau, 1904; a new edition has been published in Lehoux,
2005.

[82] Rehm, 'Parapegma', *RE*, cols. 1299–1300. He had originally linked a different introduction to it.
See Diels and Rehm, 1904, p. 102.

[83] See Lehoux, 2005.

information in the preserved fragment. The only meteorological entry is: 'Cygnus sets. The season of the continuous west wind.'

Rehm's theory that Miletus I was meant to be used alongside something like Miletus II assumes that a strictly astronomical parapegma would have been unthinkable.[84] I do not agree. If we assume that the people using the parapegma know certain rules of thumb, then the correlations of phases and weather do not need to be explicitly stated in the parapegma: to take a modern example, I do not need to buy a special calendar that tells me to plant my annuals (in Southern Ontario) on the Victoria Day long weekend in May. I only need to know which weekend is the holiday and I trust my memory for the rest. For the complete argument on this point, see chapters 1 and 4, above.

C.iii.   The **Codex Vindobonensis 108**[85] parapegma exists in one fifteenth-century manuscript originally from Constantinople and now in Vienna. It is a short list of the day-differences between various stellar phases. The Etesian winds get a mention in association with the appearance of Sirius, but other than that there is no astrometeorological content. Based on apparent similarities between this phase list and the attributions to Euctemon in the parapegmata of Geminus and Ptolemy, Rehm and Hannah have argued that this text is attributable to Euctemon in some form or other. Rehm thinks that the day differences in C. Vind. were excerpted from an original Euctemon parapegma (on which, see G.ii, below) that was dated in a zodiacal calendar. Hannah has argued against an original zodiacal organization of Euctemon's data, and instead thinks that Euctemon originally organized his parapegma as a list of day-differences, just as we find in C. Vind.[86] On Rehm's theory, then, C. Vind. would be a modification of Euctemon's original, and on Hannah's version it is something like a fragment of Euctemon.

In order to determine whether either of these accounts is true, there are three separate questions at issue here that I would like to distinguish before proceeding: (1) Is the data in Geminus and C. Vind. unified enough to warrant a claim of derivation from any single source? (2) Can we reconstruct the original form of that source?, and (3) Can we attribute that source to Euctemon? Both Hannah and Rehm are agreed on (1) and (3) that C. Vind. and Geminus' Euctemon attributions derive from a single source, and that

---

[84] Rehm, 'Parapegma', *RE*, col. 1300.

[85] *C. Vind. phil. gr.* 108, fos. 282v–283r, currently in the Österreichische Nationalbibliothek. Described in Hunger, 1961– ; *CCAG* VI, p. 13. Published in an exceptionally unhelpful format by Rehm, 1913.

[86] Hannah, 2002; 2001a.

the source can be attributed to Euctemon himself. They also agree at a basic level on (2), that we can reconstruct the form of the source from the extant sources, but they disagree on *what form in particular* the actual reconstruction should take. Hannah is in favour of day counts, Rehm of a zodiacal calendar. I have argued in chapter 4, above, against reading too strong a notion of 'calendar' into the Geminus data, but I did not there rule out the possibility that some of Geminus' sources may have structured their data according to the sun's motion through the zodiac. Certainly we can see that this is what Geminus himself is doing, but of course there is no reason to suppose that all of Geminus' sources did the same thing. It is probable that some or all of his sources did things differently, that is, in terms of day counts, civil calendars, what I have called 'schedules' of some sort, or else in terms of holes and a moving peg. Any one of these systems can be easily converted into any other, and such conversions won't, unfortunately, leave any tell-tale errors or approximations to show that such a conversion has taken place at all. Given only data in one format, we certainly can't say that the original was in the same format, a point Hannah makes quite forcefully in his argument against Rehm. But Hannah's argument that the original source for C. Vind. and Geminus' Euctemon must instead have used day counts relies on (a) the fact that C. Vind. used day counts, and (b) the fact that Hesiod used day counts. Point (a) is subject to the same objection as Hannah levelled against Rehm, but (b) calls in a piece of comparative evidence. Of course the vast majority of texts dealing with stellar phases do *not* use day counts, and other than the fact that most of them post-date Euctemon whereas Hesiod pre-dates him, we don't have much to go on. Someone obviously made the switch from Hesiod-style data organization to each of the other styles. And Euctemon could well have been one such innovator. The possibility of his using day counts is not ruled out, of course, but neither is it confirmable. On the current state of the evidence, we have to answer question (2), about whether we can reconstruct the original form of the data, in the negative. There are certainly several possibilities, and one guess is as good as another.

So what about questions (1) and (3)? The answer to (3), whether we can attribute C. Vind. to Euctemon, will depend on the answer to (1), whether we can attribute C. Vind. to any one single source. If there are multiple sources at play in C. Vind. and Geminus, then it becomes difficult to pin any C. Vind. material that is independent of Geminus (and therefore uncited) on Euctemon. I am willing enough to grant that the attributions in Geminus are probably authentic insofar as they are derived from some source(s) attributed to Euctemon in Geminus' day, and Euctemon as the genuine author of that source is certainly plausible given the amount of testimony that he was involved in astrometeorology.[87]

But the evidence available to attempt an answer to question (1), whether C. Vind. and Geminus' Euctemon share a single source, is very complex. When comparing C. Vind. to Geminus only – and only Geminus is attributive in a way that would be helpful[88] – we see that there are a handful of entries that combine several stellar phases on the same day in C. Vind. that agree pretty well with Euctemon entries in Geminus. The apparent connection is further strengthened by the fact that there are four such combinations in succession that agree not just in which stars are mentioned, but in the day-differences between them.

(i) *C. Vind.*: Setting of Lyra and rising of Pegasus.

   Geminus' Euctemon: (17 Leo) Lyra sets and it rains, and the Etesian winds stop, and Pegasus rises.

(ii) twenty-four days later in Geminus, and (probably)[89] twenty-five days later in *C. Vind.*, we find

   C. Vind.: Appearance of Vindemiatrix and rising of Arcturus and setting of Sagitta.

   Geminus' Euctemon: (10 Virgo) Vindemiatrix is visible, and also Arcturus rises and Sagitta (corr. to Pegasus by Manitius) sets in the morning; stormy at sea, wind.

(iii) ten days later in both sources, we find

   C. Vind.: Visibility of Arcturus and rising of Capella.

   Geminus' Euctemon: (20 Virgo) Arcturus is visible, beginning of autumn, and Capella rises.

(iv) and then after eleven days for both sources, the following group begins

---

[87] I make this point because my scepticism about question (2) is often confused for scepticism about the existence of any astrometeorology *at all* in Euctemon, Eudoxus, Callippus, and company. My saying that Euctemon is not necessarily a *parapegmatist* is not the same thing as my saying that he is not an *astrometeorologist*, which latter claim I would certainly not want to be on record as having maintained.

[88] Rehm, 1913, and Hannah, 2002, both adduce evidence from Ptolemy's *Phaseis* as well, but there all we have are weather predictions rather than stellar phases attributed, and these agree so poorly with Geminus' Euctemon as to prove nothing at all for the relationship between Geminus' Euctemon and C. Vind.

[89] The C. Vind. entry for the visibility of Arcturus and the rising of Capella (entry (iii) in this list) can be calculated based on the 124-day difference between it and the rising of the Pleiades as indicated in the introduction to the C. Vind. text. From there we are told by C. Vind. that Arcturus/Capella is ten days after Vindemiatrix/Arcturus/Sagitta, so we can count back from Arcturus/Capella to Vindemiatrix/Arcturus/Sagitta to get the day count between it and the previous entry (Lyra/Pegasus) as twenty-five. This method is not infallible.

C. Vind.: Autumnal equinox. (Two days later) the rising of the Kids. (Two days later) the evening setting of the Pleiades. And the rising of Corona some time after that.

Geminus' Euctemon: (1 Libra) Autumnal equinox, and there is a change in the weather. (Two days later) the Kids rise in the evening, stormy. (Two days later) the Pleiades appear in the evening, because of which there is a change in the weather towards dawn. (Two days later) Corona rises, stormy.

The strengths here that argue for a single source behind these texts are two: we have what appear to be characteristic couplings of particular stars on the same day, and we have similar or identical day-differences between the phases. On the other hand, we see that C. Vind. systematically avoids the weather predictions in these entries, even though it preserves some elsewhere (the Etesian winds are associated with the rising of Sirius), and even though there is evidence that many stellar phases were robustly wedded to their weather predictions in early sources (similarities between Geminus' Euctemon and Columella, for example, show that the connections between the Euctemonic stellar phases and weather predictions were already tightly fused in the sources that the two parapegmata were using).[90] Whatever the sources for these, it was – unlike C. Vind – *not* a list of stellar phases only.

There are other problems as well. The fact that C. Vind. has the Pleiades setting in the evening when Geminus' Euctemon has them rising gets explained away by Rehm as a copyist's error in C. Vind. This is plausible enough if we assume that the two sources should correspond, but of course that is exactly the question at issue. The correspondence of day differences, although very good in (i)–(iv), above, and in a few other entries, is not always so close between the two sources. Jut a few days above this string of entries, for example, C. Vind. has six days between the appearance of Orion and the setting of Aquila, but Geminus' Euctemon has fifteen. C. Vind. has seventeen days from the setting of Aquila to the rising of Lyra and Pegasus. Geminus' Euctemon has twenty days here. Geminus' *Eudoxus*, however, has exactly seventeen days between Aquila and Lyra.

We should also look at how these same entries stack up against other parapegmata. If it is the case that multiple sources deal with particular groupings of stars in similar ways, then there may be nothing uniquely

---

[90] Compare, e.g., Geminus' 5 Scorpio with Columella's IV Kal. Nov.; Geminus' 10 Scorpio with Columella's III non. Nov.; Geminus' 7 Capricorn with Columella's IV Kal. Jan.; Geminus' 3 Aquarius with Columella's XI Kal. Feb.; Geminus' 32 Taurus with Columella's XI and X Kal. Jun., and with Miletus fragment 456A.

Euctemon-ish about the associations and datings themselves. Alternately and perplexingly, though, a multiplicity of correspondences could be taken as evidence that all of them *were* looking at Euctemon, since all of the sources we have post-date Euctemon. The difficulty here will be in deciding whether multiple correspondences count for or against single-sourcing the material and attributing it to Euctemon in particular. Before making that particular decision, let us compare these entries in other parapegmata and see where it gets us. We find that in other sources there is a passable correspondence in the series of entries (i)–(iv), above, with both Columella and Pliny, each of whom has all four entries in sequence as follows:

C. Vind.: (i) → (ii) probably 25 days; (ii) → (iii) 10 days; (iii) → beginning of (iv) 11 days. Total: 46 days.

Geminus' Euctemon: (i) → (ii) 24 days; (ii) → (iii) 10 days; (iii) → beginning of (iv) 11 days. Total: 45 days.

Columella: (i) → (ii) 14 days; (ii) → (iii) 10 or 22 days; (iii) → beginning of (iv) either 8 or 20 days. Total: 43 days

Pliny: (i) → (ii) 24 or 25 days; (ii) → (iii) 3 to 7 days; (iii) → beginning of (iv) 12 days. Total: 44 days.

Pliny breaks up the setting of Lyra and the rising of Pegasus onto two consecutive days. This would indicate that his source differs from that for Geminus' Euctemon and from that for C. Vind. (Scribal error seems unlikely to account for turning an *and* into a *the following day*.) I also note that the two entries get a nonsimple sourcing in Pliny, such that the setting of Lyra on 11 August is *according to Caesar*, where the rising of Pegasus is *for Attica*, but as *reckoned by* Caesar. Pliny also seems to indicate multiple sources when he explains the Vindemiatrix/Arcturus/Sagitta entry (ii) such that 'Vindemiatrix rises *for Egypt*, and Arcturus rises in the morning and Sagitta sets at dawn *for Attica*.' Pliny also breaks up the rising of Capella and the appearance of Arcturus onto two separate dates, following Caesar. The string of phases grouped together under number (iv) are fairly common. They show up in C. Vind., Geminus' Euctemon, Pliny, Columella, and (with one omission) P. Hibeh. But if this is to count as evidence for all of these texts working from the same source, we will need to explain, for example, why there are so many entries in P. Hibeh that are completely lacking in C. Vind. and vice versa.

I called the correspondence between Pliny and Columella and C. Vind. 'passable' above, and the reader will no doubt have noted that, although passable, it is definitely inferior to the correspondence between C. Vind. and Geminus' Euctemon. This is true for these particular entries, but if we are to count this looser correspondence *against* Pliny and Columella, then we will be faced with a problem in other parts of C. Vind., where Pliny and Columella are *better* matches for C. Vind. than Geminus' Euctemon is. The entries for the rising of Corona and the rising of the Hyades are one example, where Pliny is one day off from C. Vind., but Geminus' Euctemon omits the phase entirely. Geminus' *Eudoxus*, on the other hand, shows good correspondence with C. Vind. here, setting the rising of the Hyades only one day farther from the rising of the Pleiades than C. Vind. seems to do. P. Hibeh sets the rising of the Pleiades and the rising of the Hyades apart by the same interval as C. Vind. P. Hibeh also gives identical values for the interval from the rising of the Hyades to the setting of Arcturus (sixteen days), and Geminus' Eudoxus has a close fourteen days, but Geminus' Euctemon is again silent. So also C. Vind. has the kite appearing at the vernal equinox, whereas Geminus' Euctemon separates these two events by nine days. Geminus' *Callippus*, on the other hand, has them on subsequent days. Geminus' Callippus also trumps Geminus' Euctemon by not omitting entries for the summer solstice or the rising of Taurus. Geminus' Euctemon omits all or part of C. Vind.'s combination of the setting of Sagitta and the rising of Pegasus. C. Vind. omits the setting of the Hyades attributed by Geminus to Euctemon at 27 Scorpio. C. Vind.'s (acronychal) setting of Orion is not attested in Geminus' Euctemon, but it *is* attested in Geminus' Eudoxus. Finally, neither the rising of Aquila nor the setting of Capricorn in C. Vind. are attested in Geminus' Euctemon. All told, of the thirty-eight entries for stellar phases in C. Vind., ten have no counterpart in Geminus' Euctemon. For date-differences of stellar phases, we find that of the thirty-eight intervals listed in C. Vind., eleven are exactly the same in Geminus' Euctemon, and four more are plus or minus one or two days. Fourteen disagree to a greater extent than this or are for phases unattested in the other text, and nine are unknown.[91]

As far as the *combinations* of stars on single days goes, although it does look fairly good between Geminus' Euctemon and C. Vind. in the string of entries (i)–(iv) above, we find that there are important differences in other entries. C. Vind.'s rising of the Pleiades is separated from the setting of Orion by three or more days, but Geminus' Euctemon has them on the same day (15 Scorpio). C. Vind.'s setting of Sirius and rising of Lyra on the same day get broken up by Geminus' Euctemon (2 Taurus f.), but are united by Pliny. Pliny also hints that he is working from two different sources when he says the two phases happen together 'for Attica *and* Boeotia'. C. Vind.'s setting

---

[91] Rehm's edition tends to make the correspondences look a little better than this as he emends some of the disagreements.

of Vindemiatrix and Arcturus and Pegasus on the same day gets split up by Geminus' Euctemon onto the 12th and 14th of Pisces.

On balance, it looks like there is more than one source being used by each of C. Vind., Geminus, Columella, and Pliny. Some of the discrepancies above can best be explained by multiple sources, some of which C. Vind., Geminus, and the others share, and some of which they do not. This explains how they privilege different information, as well as how each of them has a good deal of unique data. It also explains how some of the stellar phases and day counts in C. Vind. find better correspondence with other astrometeorologists than they do with Geminus' Euctemon. Where C. Vind. agrees nontrivially with Geminus' Euctemon, we can be reasonably certain that they gathered their data from the same source *for those entries*, but where they disagree nontrivially we are led to hypothesizing multiple sources. If we can take the Geminus attributions at face value, then we can ascribe the sources shared by C. Vind. to Euctemon, but where C. Vind. has unique data, we cannot do so necessarily. There is no reason to assume that the author of C. Vind. was working only from sources written by Euctemon, and some data seems to point to other authors, including Callippus and Eudoxus in particular.

A final point I would like to make is that to a certain extent C. Vind. has suffered from its treatment primarily as a source for Euctemon, and has not been studied for its own sake yet as a text. It is one of only a few classical astronomical texts that are organized as lists of date-differences and this in itself is interesting, whether or not it is partly, primarily, or even exclusively the work of Euctemon. Work on the text is also severely hampered by the very real need for a new edition.

C.iv. The **Antikythera mechanism**[92] is a truly remarkable geared astronomical computer pulled from a first-century BC shipwreck by sponge divers (Fig. Cat. 11). It seems to have been used to show the relative motions of the sun and the moon according to a nineteen-year cycle and the Egyptian calendar. It includes an astronomical parapegma. One of the geared dials on the front of this unique calendrical/astronomical instrument is divided into graduated zodiacal signs. Part of Virgo, all of Libra, and the very beginning of Scorpio are preserved. There are thirty graduated divisions in Libra, which Price reasonably interprets as degrees. Above the first degree of Libra the letter A is clearly marked. Price also thinks (but is not certain) that he can see a B above Libra 11, a Γ above Libra 14, a Δ above Libra 16, an E above

---

[92] Currently in the National Archaeological Museum, Athens. Inventory number X 15087. Published by Price, 1974; compare also the sixth-century AD device published by Field and Wright, 1985.

Scorpio 1, and, most of the way around the dial, an ω above Virgo 18. Price proposes (I think plausibly) that these letters are meant to be keyed to the 'parapegma inscription' just below the dial. The largest preserved fragment of this inscription reads:[93]

Z	Ι[...
Η	ΟΡ[...
Θ [...	
Ι [...	
[Κ	ἑ]σπέρ[ιος
[Λ]	ἑ[σ]π[έ]ρια[
[Μ]	ἐπ]ιτέλλ[ει
[Ν]	]ι ἑ[σ]πέρια[
[Ξ	]ΑΣ ἐπιτέλλει ἑῶια[
Ο	ὑάς ἐπιτέλλει ἑῶια[
Π	δίδυμοι ἄρχονται ἐπιτ[έλλειν
Ρ	ἀετὸς ἐπιτέλλει ἐσπέ[ριος
Σ	ἀρκτοῦρος δύνει Ε[

Z	Ι[...
Η	ΟΡ[...
Θ [...	
Ι [...	
[Κ]	e]veni[ng
[Λ]	e]v[e]ni[ng[94]
[Μ]	r]is[e.
[Ν]	]e[v]ening.
[Ξ	The Pleiad]es ri[s]e in the morning.
Ο	The Hyades rise in the morning.
Π	Gemini begins to ri[se.
Ρ	Aquila rises in the even[ing.
Σ	Arcturus sets in the [...

---

[93] The remaining fragments of the parapegma inscription are all insignificant: ΚΑΡΚ[...]; [ἐπι]τέλλουσιν[...]; [ἐπιτ]έλλε[ι]; [ἐπιτέ]λλει ἑῶια[...]

[94] Price, 1974, p. 46, remarks that he has seen Rehm's notes on the parapegma inscription. In Rehm's notes, the entries for lines Λ, Μ, Ν, and Ξ are more complete than the text on the instrument itself was when Price saw it. Price believes that this was due to damage to the instrument between Rehm's viewing and his own. Price does not say whether there is any possibility that the extra text was reconstructed by Rehm, or how carefully Rehm's notes distinguished between text and reconstruction in general. I am here erring on the side of caution. Rehm records these entries as follows (with no significant lacunae): Λ *The Hyades set in the* evening; Μ *Taurus begins to* rise; Ν *Lyra rises in the* evening; Ξ *The Pleiades* rise in the morning.

Fig. Cat. 11 The Antikythera mechanism parapegma face.[95]

It appears that the Greek letters (in the left column of the inscription) were meant to correspond to the letters inscribed over particular degrees of the zodiac on the parapegma dial itself, and the stellar phase corresponding to each letter on the dial could thus be read off from the inscription.

Much of the inscription following the letters Λ through N has now been unfortunately destroyed (Price hints that some degree of damage was done by Rehm's efforts to study and preserve the instrument). Price tentatively reconstructs the remainder of the inscription by comparing it with Geminus, but this is misguided, since even the preserved fragment cannot be made to fit well with Geminus. By counting day-differences from the autumnal equinox (A on the Antikythera dial) to the other letters (if indeed they are secure) and comparing this to the day-differences in other parapegmata, there is closer correspondence to stellar phases in the following parapegmata (the equinox is marked with an asterisk):

(a) Columella: 11, 24*, Sept.; 4, 8 (one day late), 10, 13 (one day late), and 28 Oct.;
(b) Ptolemy, for the clima where the day is 14 hours long: (no entry), Thoth 28*, Phaophi 7 (one day early), 11, 12 (one day early), and 26 (two days early);

[95] Photograph courtesy of the Hellenic Republic, Ministry of Culture. Used with permission.

(c) Ptolemy, for the clima where the day is 14 ½ hours long: Thoth 17 (two days late), 28*, Phaophi 8, 10 (one day early), (no entry), (no entry), 27 (one day early);

(d) Paris parapegma: 6, 19*, 29 Sept.; 3 (one day late), 5 (one day late), 20 (one day late) Oct.

Since so many different possibilities are attested, and since most of the letters on the dial are insecure, I think any reconstruction of the entries for A through E, and ω would be doubtful. Moreover, there is no guarantee that all of the letters on the dial represented purely astronomical phenomena. They may have included important seasonal markers or weather predictions independent of stellar phases.

D: Other parapegmata

D.i.   The **Ceramicus parapegma**,[96] the oldest known parapegma, is represented by a small fragment found in 1929 in the Ceramicus district of Athens, dated by Brückner to the fifth century BC, but by Höpfner to the third. Höpfner thinks it originally belonged to the interior equipment of the Pompeion.[97] It simply has the ordinal numbers 'fifth' through 'ninth' inscribed in Greek beside peg holes:

] • πέμπτη •[	]• fifth •[
]ᴺ • ἕκτη •[	]• sixth •[
] • ἑβδόμη •[	]• seventh •[
•]ὀγδόη •[	•]eighth •[
•]ἐν]ά[τ]η[	•]ni]n[t]h[

Rehm's theory that this parapegma was meant to correlate civil lunar dates inscribed on the pegs, with zodiacal dates (i.e., 'fifth [day of Virgo . . .]') inscribed on the stone itself, goes against much of what I have argued in this book.[98]

Since this fragment is so small and sparse, we cannot rule out the possibility that the Ceramicus parapegma was solely calendrical, or even that

---

[96] Published by Brückner, 1931, pp. 23–4, with photograph. Size: 6 × 7.5 cm. Currently in the Ceramicus collection. Inventory number I 22. The column of holes after the numbers is difficult to make out in the photograph published by Brückner, but shows up clearly in the drawing published in *IG* II² 2782 as well as in the squeeze in the Epigraphical Library of the Institute for Advanced Study, Princeton.

[97] Höpfner, 1976, p. 122.   [98] Rehm, 'Parapegma', *RE*, col. 1301.

it served some other function, completely different from any known later parapegma (can we even assume that this parapegmata must track *temporal* cycles?). Hannah has associated it with astronomical and astrometeorological parapegmata, but it very clearly is neither:[99] the only parapegmata that count numbered sequences of any kind are for tracking either calendrical or lunar phenomena, not stellar.

D.ii. The **Guidizzolo *Fasti***[100] is a crude fragmentary stone parapegma measuring 16.5 cm by 16 cm by 4 cm thick, dating from some time after 8 BC, and excavated in 1891 just south of the ancient town of Brixia. It is a unique parapegma insofar as it seems to be strictly calendrical, with only dates and feasts listed. Holes are drilled beside each entry, and the preserved fragment reads as follows:

[•] *XIII*	• *X[V]*	• *[II]I ID[V]S I[VLIAS*
[•] *XII*	• *XIIII*	*APOLLI[NARIA*
[•] *XI*	• *XIII*	• *X K AVG[VSTIS*
[• *X]*	• *XII*	*NEPTVN[ALIA*
• *VIIII*	• *XI*	• *IDIBVS AV[GVSTIS*
• *VII*	• *X*	*DIANA(E)*
• *VI*	• *VIIII*	• *X K SEPTEM[BRES*
• *V*	• *VIII*	*VOLKANALIA*
• *IIII*	• *VII*	• *III IDVS*
• *III*	• *VI*	*DECEM[BRES*
• *PR[IDIE]*	• *V*	*SEPTIMONTIV[M*
	• *IIII*	• *XVI K IANVAR(IAS)*
	• *III*	*SATVRNALIA*
	• *PRID(IE)*	• *XV K IA[N]VA[RIAS*
		EPON(A)E

The first two columns may well represent November and December, as Degrassi supposed, assuming the festivals were written at the end of the whole calendar sections. It seems clear that this parapegma was meant to keep track of the calendar year and its various festivals, such that a single peg would have been moved to track the day of the calendar month. A separate peg seems to have been used from time to time to mark the particular *Fasti*,

---

[99] Hannah 2002.

[100] *CIL* I¹, p. 253; Degrassi, 1963, vol. XIII.2, pp. 234–5. See also Rüpke, 1995, p. 160f. Now in the Museo Civico, Brescia. Inventory number undetermined.

but this kind of intermittent column (i.e., lacking holes for the *dies nefasti* between the *dies fasti*) is unique, and I am not sure what to make of it. There may have been a single hole marked *DIES NEFASTVS*, so that a feast-marking peg was either there or in one of the fixed (or moveable?) *fasti* on any given day. It is also, I suppose, possible (though probably prohibitively awkward) that the peg was moved from the calendar section over to the feast section for the feast days only, and then moved back to mark normal days.

D.iii.   The **Calendar of 354**, also called the *Fasti Furii Filocali*, is a codex calendar extant in ten manuscripts, the earliest of which date from the ninth century.[101] Many of the MSS are richly illustrated. Besides the standard run of calendar dates with their hebdomadal letters (*A* through *G*) and the nundinal letters (*A* through *H*) that are typical of such *Fasti*,[102] the Calendar of 354 also dates the sun's entry into each zodiacal sign, and includes a dated entry for the *solstitium* on VIII K. July, which is nine days after the sun's entry into Cancer on XVII K. July. A very interesting feature of this text is the (to my knowledge) unique set of *lunar letters* running through the calendar. The series of ten letters from *A* through *K* runs alongside the nundinal and hebdomadal series, but with entries only every three days. Every other month the gap in the exact middle of the cycle, between *E* and *F*, is shortened to two days, as around the Ides of February as in Fig. Cat. 12.

   To understand the function of the lunar letters, we should perhaps first look in a little detail at the nundinal and hebdomadal letters. *Fasti* typically begin the year with the Kalends of January marked as nundinal day *A*. There is general agreement among modern scholars that this was meant to be read as follows: let us suppose that for this year, the first nundinal day of the year fell on the day after the Kalends of January, here labelled *B* in the nundinal cycle. Since the nundinae will recur every eight days, then the next nundinal day will fall on the next day marked *B* in the nundinal column on the calendar, the *IIII pr. id. Ian.*, here. And, of course, each subsequent nundinal day will fall on each subsequent *B*. If on the other hand the first nundinal day of the year fell on a *G*, then for the rest of the year all the nundinae will fall always and only on days labelled *G*. So, too with the hebdomadal week: if the first Saturday of the year fell on the *IIII pr. non.*, labelled *B* here, then

[101] Publication: Stern, 1953; Salzman, 1990; Degrassi, 1963, vol. xiii.2, pp. 237–62; Rüpke, 1995.
[102] See Michels, 1967; Rüpke, 1995; Radke, 1990.

January				February				March			
A	**A**	**A**	**kal. Ian.**		**D**	**H**	**kal. Feb.**	**A**	**D**	**D**	**kal. Mar.**
	B	B	IIII non.		E	A	IIII non.		E	E	VI non.
	C	C	III	B	F	B	III		F	F	V
B	D	D	pr.		G	C	pr.	B	G	G	IIII
	E	E	non.		A	D	non.		A	H	III
	F	F	VIII idus	C	B	E	VIII idus		B	A	pr.
C	G	G	VII		C	F	VII	C	C	B	non.
	A	H	VI		D	G	VI		D	C	VIII idus
	B	A	V	D	E	H	V		E	D	VII
D	C	B	IIII		F	A	IIII	D	F	E	VI
	D	C	III		G	B	III		G	F	V
	E	D	pr.	E	A	C	pr.		A	G	IIII
E	F	E	idib.		B	D	idib.	E	B	H	III
	G	F	XIX kal. Feb.	F	C	E	XVI kal. Mar.		C	A	pr
	A	G	XVIII		D	F	XV		D	B	idib.
F	B	H	XVII		E	G	XIIII	F	E	C	XVII kal. Apr.
	C	A	XVI	G	F	H	XIII		F	D	XVI *sol Ariete*
	D	B	XV		G	A	XII		G	E	XV
G	E	C	XIIII		A	B	XI	G	A	F	XIIII
	F	D	XIII	H	B	C	X *sol Piscibus*		B	G	XIII
	G	E	XII		C	D	IX		C	H	XII
H	A	F	XI		D	E	VIII	H	D	A	XI
	B	G	X *sol Aquario*	I	E	F	VII		E	B	X
	C	H	IX		F	G	VI		F	C	IX
I	D	A	VIII		G	H	V	I	G	D	VIII
	E	B	VII	K	A	A	IIII		A	E	VII
	F	C	VI		B	B	III		B	F	VI
K	G	D	V		C	C	prid(ie)	K	C	G	V
	A	E	IIII						D	H	IIII
	B	F	III						E	A	III
A	C	G	prid.					A	F	B	prid.

Fig. Cat. 12. Lunar, hebdomadal, and nundinal letters in Calendar of 354

all subsequent Saturdays for the year will be on hebdomadal days labelled *B*. The system is quite elegant in its simplicity, and enables the Romans to publish calendars that include nundinal and hebdomadal cycles, but which are not tied to any one particular year. Where we are spoiled by the low costs of paper and printing and so can luxuriate in disposable wall calendars, the Romans opt instead for *inscribing* what are essentially perpetual calendars and remembering a few rules for their use. The only difficulties with the Roman system are that users must know the nundinal and hebdomadal letters for a given year, and users must be careful just at the transition from one year to the next. What exactly was done in leap years is not known.[103]

If the lunar letters are working analogously to the nundinal and hebdomadal letters, then users of the calendar could keep reasonable track of the phases of the moon for the year by simply noting the lunar letter closest to the day of the first full moon (or the first new moon) of the year. Each recurrence of that letter in the lunar column would mark the recurrence of the same phase. The correspondence, while not perfect, would certainly be within a reasonable degree of approximation for most purposes over the course of a given year.

The combination of zodiacal-sign entry, lunar phenomena, and nundinal and hebdomadal weeks tracked in this *Fasti* is unique.

D.iv.   The **Capua *Fasti* parapegma**,[104] first published in 1984, consists of two fragments of white marble, found by D'Istanto in the collection of the Antiquarium of Santa Maria Capua Vetere. D'Istanto dates it epigraphically to the Augustan era. The fragments were labelled as having been found in the amphitheatre's sewer. The original inscription seems to have been a large one, with letters 3.3–4.1 cm high. The fragments themselves are 23.5 by 19.5 cm and 25.3 by 23 cm. One fragment has only three nundinal letters with two of their peg holes preserved,[105] and the other has three nundinal letters, a festival and a *dies comitalis* marked. This *Fasti* is unique in having peg holes indexing both nundinal *letters* and *fasti/nefasti/comitales* days.[106] Like other *Fasti* it seems to have had festivals listed (in these fragments, the *Lar[entalia]* is labelled).

D.v.   The **Coligny calendar**[107] is a long inscription (1.48 m by 90 cm high, Fig. Cat. 14) on bronze of five years of a lunar calendar (probably

---

[103] See Michels, 1967, p. 26f. for a discussion of the problems and one possible solution.

[104] Currently in the Antiquarium in Santa Maria Capua Vetere, inv. no. 163713. Published by D'Isanto, 1984, pp. 144–5.

[105] Both D'Isanto, 1984, and Rüpke, 1995, presume that all the letters in both fragments are nundinal.

[106] See Rüpke, 1995, p. 107.

[107] Currently in the Musée Gallo-Romain in Lyon, inv. no. BR 1. First published in *Comptes rendus des séances de l'année de l'Académie des inscriptions et belles-lettres* (1897), p. 703, pls. I–VI; and (1898) p. 299f.; see also de Ricci, 1898; 1900. The most complete and sober publication is in Duval and Pinault, 1986, including excellent photographs and commentary. See also McCluskey, 1998, pp. 54–60. More speculative reconstructions,

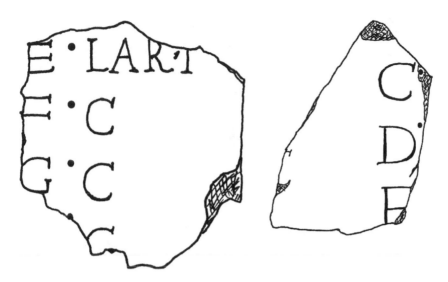

Fig. Cat. 13 The Capua *Fasti* parapegma.

a five-year calendar cycle). It is written in what is believed to be an ancient Celtic tongue, using the Latin script. The days of the month are each accompanied by a peg hole, and they are written vertically, beginning with the name of the month, followed by the numbers *I–XV*, then the word *ATENOVX* and then the numbers *I–XIIII* or *I–XV*, depending on the month.

The year has seven months of thirty days and five of twenty-nine (for a total of 355). A thirty-day intercalary month was added twice every five years, bringing the average year length to 367 days, although it has been argued that one of the thirty-day months, *Equos*, occasionally had twenty-eight days to bring the average year length down. The symbols, letters, and words beside the dates are still largely unexplained.[108] A sample from the calendar[109] should give the flavour:

such as those of Hitz, 1991, and Olmstead, 1992, have been attempted, but these are best approached with caution. More recently, Gaspani and Cernuti, 1997, have tried to put the Coligny calendar into a broader astronomical context, but again, much of their treatment is speculative.

[108] Not to say that attempts have not been made: For a summation of the issues, see Ginzel, 1914, vol. III, p. 80f.; Kubitschek, 1928; and especially Duval and Pinault, 1986, p. 421f. Generally *D* and *N* are thought to refer to *day* and *night* respectively, but I do not see what this could mean in the context of the calendar.

[109] De Ricci, 1898, pl. II.7.

Fig. Cat. 14. The Coligny calendar.

• XIIII M	D	IVOS	• XIIII |||D	IVODOBCANT
• XV	D	IVOS	DIVERTOMV	

MEQVOSANM   MSAMONMAT

• I	D		• I	[........]MANIVOS
• II PRINI LACIVOS			• II |[...........]	
• III M	D	SIMI IVOS	• III † |	[... ...] IVOS
• IIII	D	IVOS	• IIII [...]	D [............]VMIVO
• V	D	AMB	• V	AMB
• VI M	D	SIMIVISO	• VI M	D
• VII	D	ELEMBI	• VII	PRINI LOVD N
• VIII	D	ELEMBI	• VIII	D DVM
• VIIII	D	ELEMBI	• VIIII	D
• X	D		• X M	D
[•] XI	D	AMB	• XI	D
[• X]II	D		• XII M	D
[•] XIII	D	SEMIVIS	• XIII † ||	D
[•] XIIII	D	SEMIV*I*S	• XIIII |†|	D
[•] XV	D	SEMICANO	• XV || †	D

ATENOUX   ATENOUX

• I M	D	SEMIVIS	• I	D DVMAN
• II M	D	SEMIVIS	• II ||⊢ M	D TRINVXSAMO
• III	D	AMBSIMIV	• III	D AMB
• IIII	D		• IIII ⊣||	D
• V	D	AMB	• V |†|	D AMB
• VI |†|	D	S*I*MISO	• VI | |⊢ M D	

Equos and *Samon* are taken to be the names of months. The column here usually labelled *D* is sometimes elsewhere labeled *N*. Twenty-nine-day months end with the word *DIVERTOMV*, as above *MSAMON* here. All month names are preceded by the letter *M* as are *EQVOS* and *SAMON* in this fragment.

The abbreviations beside the month names, *ANM* and *MAT* (sometimes *MATV*) seem, based on Celtic cognates, to mean 'ungood' (*ANMATV*-) and 'good' (*MATV*-, cognate with the Irish *maith?*) respectively. *M* beside the dates may be an abbreviation for *MATV*. All twenty-nine-day months are marked *ANM* and all thirty-day months are marked *MAT* except the month of *EQVOS* which seems to have fluctuated between thirty days (in years I, III and V) and twenty-eight days (in years II and IV).[110] MacNeill argued that *IVOS* referred to a feast of some kind, but Duval and Pinault believe it to be uncertain. Parisot's theory that it refers to possible solar eclipses is

---

[110] See MacNeill, 1928, pp. 27–30; Duval and Pinault, 1986, pp. 411–15.

implausible, as is his idea that *DIVERTOMV* is an extra day occasionally inserted to correct for the lateness of the significant phase of the moon. If *DIVERTOMV* were a day occurring after the twenty-ninth, then it should have a peg hole beside it, which it never does.

The calendar represents the oldest known document in a Celtic tongue, possibly dating from the second century AD[111] and as such is of great importance, but its highly abbreviated character and lack of parallels (apart from the tiny fragments of the Villards d'Héria calendar)[112] make it difficult to interpret. There may be references to either feasts or solstices and equinoxes, or both.[113]

D.vi.   The **Ariminum nundinal parapegma**[114] is a puzzling fragment bearing an inscription read by Degrassi as possibly pertaining to a nundinal list, with two columns of peg holes preserved. If this is a nundinal list, the day names are otherwise unknown to me. They do not seem to be place names, as in other nundinal lists.

- *[. . .*
- *[. . .*
- *[. . .*
- *Q   [. . .*
- *SISTINI   [. . .*
- *ARQUT   [. . .*
- *] PROMO* • *[. . .*
- *]Q   • [. . .*
- *]RANASI* • *Q [. . .*

D.vii.   The **Suessula nundinal list**[115] is a small stone inscription (16 cm by 16 cm), lacking peg holes, with a partial list of nundinal days written vertically:

---

[111] According to Duval and Pinault, 1986; MacNeill, 1928 dates it to the late first century BC or early first century AD. See also Duval and Boucher, 1976.

[112] Published in Duval and Pinault, 1986, p. 251f. The Villards d'Héria calendar seems not to be a parapegma, as near as can be determined from the few remaining fragments. It does seem to be written in the same language, and to use some of the same abbreviations as the Coligny calendar, however.

[113] See McCluskey, 1998, pp. 59–60.

[114] Currently in the Museo della Città, Rimini, inv. no. unknown. Publication: *CIL* XI², 6709, 30; Degrassi, 1963, vol. XIII.2, p. 306.

[115] Current location undetermined. Publication: *CIL* I², p. 218; Degrassi, 1963, vol. XIII.2, p. 303.

[T]IANESI[BVS]
CAMPAN[IS]
ATELLANI[S]
SVESSELA[NIS]
NOLANIS
CVMANI[S]
CALINI[S].

D.viii. The **Allifae nundinal lists**[116] consist of two fragments each apparently listing eight different nundinal days vertically, with no peg holes preserved. A third fragment was lost some time after 1750. Unlike Degrassi, I see them as essentially three different lists. They read:

	(fr. 1)	(fr. 2)
...]M	*[ALTINATIBUS]*	*...]EN[...*[117]
	[INTERA]MN[ATIBUS]	*NUCERI[NIS]*
	[TELESIN]IS	*[L]UCERIN[IS] APVLIS*
	[SAEPI]NATIBVS	*[S]VESSANIS*
	PVTEOLANIS	*[CA]LENIS*
	ATELLANIS	*[SVES]VLANIS*
	CUMANIS	*[SIN]VESANIS*
	NOLANIS	*[CALA]TINIS*

Fragment 1 is 2 cm thick (the same as the *Fasti Allifani*), and fragment 2 is 2.5 cm thick, which leads me to believe they may not have been originally hung on the same wall. Fragment 1 has also been damaged since Degrassi published his photographs of it, and most of the top four lines are now missing, although I reproduce his text here. The third and long-lost fragment reportedly had only *ALLIFANIS, CEREATIS.*

It is possible that, assuming the three lists *were* originally part of the same inscription, they may have been meant as a kind of conversion calculator for figuring out where one was in a different nundinal cycle than one's own.[118]

---

[116] *CIL* IX, 2318; Degrassi, 1963, vol. XIII.2, p. 302. The two extant fragments are in the Museo Archeologico Nazionale di Napoli, inv. 3006. The fragments were found with the *Fasti Allifani* (inv. 3005).

[117] Probably either Beneventum or Venafrum.

[118] I would like to thank James Ker for suggesting this.

E: Reports of parapegmata

E.i. Cicero, in a letter to Atticus[119] announcing that tomorrow will be the beginning of his year of 'exile' in Laodicea, says (following all modern editions): *Ex ea die, si me amas,* παράπηγμα ἐνιαύσιον *commoveto.* 'From this day, if you love me, move the *yearly parapegma.*' The next day he sends another letter saying *Ex hoc die clavum anni movebis.* 'From this day, you should move the nail of the year.' Tyrell and Purser's 1890 edition offers the following commentary:

παράπηγμα] The very same meaning is conveyed by *clavum anni movebis* in the next letter. The phrase is said by the old commentators to take its rise 'from an old custom which came from Etruria to Rome, whereby the *Pontifex Maximus,* on the Ides of September, stuck a nail into the right wall of the temple of Jupiter Optimus Maximus, to keep count of the years.' *Commoveto,* like *movebis,* in the next letter, is used as a Latin equivalent for κινεῖν in the sense of 'to take in hand,' e.g. ἐκίνουν θύρσον ἐς βακχεύματα. Eur. Bacch. 724.[120]

A few years after this was written, the Miletus and Puteoli parapegmata were discovered, offering an alternative interpretation noted by Constans and Bayet in the Budé edition. In a note on this same passage, they say:

C'est-à-dire: compte de ce jour mon année de charge. Il s'agit d'un calendrier à fiches mobiles. On a trouvé récemment à Pouzzoles [Puteoli] un fragment de calendrier latin de ce genre.[121]

But they refer to the older interpretation in their note on Cicero's next letter, commenting on *clavum anni movebis*:

Clauom (sic) *anni mouebis:* littéralement 'tu déplaceras <sur ton calendrier mobile> la fiche marquant l'année'. Cf. *Att.,* V, 14, I: παράπηγμα ἐνιαύσιον *commoueto.* – Mais l'expression semble être une métaphore assez usée, remontant à l'ancien rite de la 'plantation (annuelle) du clou', par le *praetor maximus* ou un dictateur, dans le mur du temple de Jupiter au Capitole.[122]

Shackleton-Bailey, in his edition of the *Letters to Atticus* follows Constans by mentioning the Puteoli parapegma as an example of the kind of timekeeping device referred to by Cicero, but he does not repeat the nail-in-the-temple-wall story.

---

[119] Cic., *Ad Att.* v.14.
[120] Tyrell and Purser, 1890, vol. III, p. 52, n.   [121] Constans and Bayet, 1969, vol. III, p. 242, n. 2.
[122] Constans and Bayet, 1969, vol. IV, p. 246, n.

While I think that Constans and Bayet's *double entendre* is not impossible, I would argue that Cicero primarily has in mind a calendrical parapegma in both passages: in the first, because he says so explicitly, and in the second, because he refers to 'moving' the nail, a practice nowhere attributed to the *Pontifex Maximus*, who was supposed to have added a new nail for each year.[123] I note also that a *clavus* does not correspond directly to our word 'nail', but can refer also to a metal, wooden or bone peg, such as was used in the parapegmata.

Looking at the extant Latin parapegmata, we can, I think, be fairly certain that Cicero was referring *not* to the kind of parapegma found at Puteoli, as Constans and Shackleton-Bailey supposed, but to one more like the Guidizzolo or Capua *Fasti* (D.ii and D.iv, above). Unlike the *astrometeorological* fragment from Puteoli, the Guidizzolo and Capua *Fasti* seem to be strictly calendrical, with the peg moved daily to keep track solely of calendar dates through the course of the civil year.

It is worth pointing out that the text of Cicero is corrupt at just this point, insofar as none of the MSS reads παράπηγμα in this passage. Instead we find παράγγελμα, ΠΑΓΓΕΓΜΑ, or ΠΑΤΤΕΤΜΑ, obviously none of which are acceptable.[124] Curiously, Tunstall's emendation of this to παράπηγμα, made in 1741, was based on no knowledge of inscriptional parapegmata, which were only recognized as such in the early part of the twentieth century.

E.ii. Petronius in the *Satyricon*[125] has Encolpius mention seeing something that sounds very like an astrological parapegma in the house of Trimalchion:

sub eo titulo et lucerna bilychnis de camera pendebat, et duae tabulae in utroque poste defixae, quarum altera, si bene memini, hoc habebat inscriptum: 'III. et pridie Kalendas Ianuarias C. noster foras cenat,' altera lunae cursum stellarumque septem imagines pictas; et qui dies boni quique incommodi essent, distinguente bulla notabantur.

Under this inscription there hung from the ceiling a double lamp, and there were two boards fixed to the two posts, of which the one, if I remember correctly, had this inscribed: '*III.* and *pr. K. Jan.,* our C. dines outdoors.' The other [had inscribed] the course of the moon, and painted pictures of the seven stars, and which days were good and which bad were marked by a peg that distinguished them.

For details, see the discussion in chapter 2, above.

---

[123] The veracity of this ancient story does not concern me here. All that matters is what Cicero may or may not have believed. For an ancient account of driving the nail into the temple, see e.g., Livy, vii.iii.4f. Specifically note that Cincius reports having seen *clavos* (plural) marking the number of years.

[124] See Ernesti, 1810, vol. v, p. 566, n. 93.   [125] Petronius, *Sat.* 30.

E.iii.   Diodorus Siculus[126] paraphrases Hecataeus of Abdera's third-century BC report of an astrometeorological parapegma in the tomb of Ramses II. According to the story told to Hecataeus, the parapegma, made of gold, was plundered by Cambyses in the sixth century BC. It is reported to have had entries for 365 days, listing the stellar risings and settings and the changes in the weather for each day 'according to the Egyptian astrologers'.

If Hecataeus' source is correct, then this would be a very early astrometeorological text. But of course the more-than-third-hand story, told to one Greek by another, who heard it from some unnamed person (perhaps a local 'guide', as trustworthy then as now) carries no conviction on its own. For my doubts on an Egyptian astrometeorology in the Ramesside period, see chapter 6.

E.iv.   Vitruvius lists a number of astrometeorologists known from attributive parapegmata and credits them with discovering the associations between stellar phases and weather. He specifically says that this was done *ex astrologia parapegmatorum disciplinis*. If he is supposed to mean by this that these astrometeorologists were themselves the authors of parapegmata then his wording is odd at best. It looks as though his phrasing is fairly loose here.

> . . . *siderum ortus et occasus tempestatumque significatus Eudoxus, Euctemon, Callippus, Meto, Philippus, Hipparchus, Aratus ceterique ex astrologia parapegmatorum disciplinis invenerunt et eas posteris explicatas reliquerunt. quorum scientiae sunt hominibus suspiciendae, quod tanta cura fuerunt, ut etiam videantur divina mente tempestatium significatus post futuros ante pronuntiare.*[127]

Eudoxus, Euctemon, Callippus, Meton, Philippus, Hipparchus, Aratus, and others discovered, from astronomy and by the science of parapegmata, the risings and settings of stars, as well as the signs of storms. They left this system behind for their successors. Their knowledge should be highly regarded by people, since they took such care that they are seen to declare, as though with divine knowledge, the indications of storms before they come to pass.

E.v.   Diogenes Laertius ascribes a book called *The Great Year, or Astronomy: A Parapegma* to Democritus.[128] Sider has argued, I think convincingly, that the title of the book was originally just *The Great Year, or Astronomy*, which some later editor explained (probably not from first-hand evidence) as 'a parapegma'.[129]

---

[126] Diod. Sic., I.49.5.   [127] Vitr., *De arch.* IX.6.3.   [128] DL IX.48.   [129] Sider, 2002, p. 299.

E.vi.   Proclus mentions the making of parapegmata in his commentary on Plato's *Republic*.[130] In a discussion of how astronomers model irregular stellar phenomena using a combination of several regular mathematical operations, he says:

καὶ γὰρ οἱ τὰ παραπήγματα κατασκευάζοντες λογισμοῖς χρώμενοι μιμοῦνται τὴν πρὸ λογισμῶν καὶ ἐπινοιῶν ἐκεῖνα δημιουργήσασαν φύσιν.

So also those who make parapegmata imitate, with the usual arithmetic, nature, which created those things before arithmetic and contemplation.

E.vii.   The Suda defines 'parapegma' thus:

παράπηγμα· κανών. καὶ εἶδός τι ὀργάνου ἀστρονομικοῦ.

Parapegma: a general rule. Also a kind of astronomical instrument.

F: Related texts and instruments

F.i.   The Babylonian omen series ***Enūma Anu Enlil***[131] includes some weather predictions from fixed stars. These are not ordered or dated, and the exact relation is left unclear: '*Entenabarḫum* is for early wind,' where the relation 'is for' is simply expressed with the preposition *ana*. For a more detailed discussion of this text, see chapter 5.

F.ii.   The Babylonian astronomical compendium **MUL.APIN**[132] associates the annual motion of the sun with seasonal weather changes, as well as giving schematic lunar calendar dates for stellar phases, and a list of stellar-phase date-differences. Evans has referred to it as a parapegma,[133] but it works sufficiently differently from a Greek parapegma, especially in the relationship between the lunar calendar and the stellar phases, that I class it separately. For more on this, see chapter 5.

F.iii.   The astronomical content of **Hesiod's** *Works and Days* has already been discussed, particularly in chapter 1. Hesiod's fragmentary *Astronomia*[134] also included phases of the fixed stars. There is also a lost

---

[130] Proclus, *In R.* ii.234.
[131] There is no satisfactory publication of the whole series. The closest to a complete publication is the 1905–12 edition of Virolleaud, now outdated. Parts of the series have been more recently published by Reiner and Pingree, 1975–98; Rochberg-Halton, 1988; al-Rawi and George, 1991; and van Soldt, 1995. For commentary, see Hunger and Pingree, 1999; Koch-Westenholz, 1995.
[132] Publication: Hunger and Pingree, 1989.   [133] Evans, 1998, pp. 6–7
[134] Hesiod, frs. 288–93.

early text called ναυτικὴ ἀστρολογία, *Nautical Astronomy*, attributed variously to Thales and Phocus of Samos,[135] which may be related, given the
title.

F.iv.   The **Ṣafṭ el-Ḥenna naos**[136] is the only extant Egyptian example of an
astrometeorological text. It dates from the early fourth century BC. For a
detailed description and discussion, see chapter 6.

F.v.   Paul of Aegina preserves a letter attributed to **Diocles of Carystus**
(fourth century BC.) and addressed to King Antigonus of Macedonia,[137]
which ends with a list of the beginnings and ends of the seasons relative to
the solstices, equinoxes, and the phases of certain stars. Diocles correlates
these with the prominent illnesses for each season. He also tells us how long
each season is, in days.

F.vi.   [**Hippocrates**], *Peri hebdomadon*[138] contains a short chapter which
lists the divisions of the seasons and their significance for the physician.
For each season it lists the maladies which are common at that time of
year. It equates the rising of the Pleiades with the beginning of summer,
and marks out the seasons for different diseases from there. It offers no
dates or day-differences. The stellar phases mentioned are: 'the rising of the
Pleiades . . . from there to the summer solstice . . . from there to the rising
of Sirius and then to the rising of the Arcturus . . . from the rising of Sirius
until that of Arcturus . . . after the rising of Arcturus . . . after this, the setting
of Cepheus . . . and afterwards, from the setting of the Pleiades'.

Rehm thought this and *On Regimen* (the following text in this catalogue) were derived from a lost 'parapegma of Eudoxus',[139] which conclusion he based only on both texts mentioning the rising and the setting of the
Pleiades – thin evidence indeed. Because of the supposed connection with
Eudoxus, these two texts were singled out as a pair of related parapegmata
in Rehm's *RE* catalogue, and he treated them more-or-less separately from
other Hippocratic texts that mention stellar phases (*Epidemics I* and *III; Airs,
Waters, Places*).

---

[135] See D-K Thales fr. B.1–2; Phocus.   [136] Edition: Leitz, 1995.

[137] Publication: Paulus Aeginata, *Epitomae medicae*, 1.100.1–6 (*CMG* IX.1, vol. 1), Diocles of
Carystus, fr. 183a in van der Eijk, 2000. For an excellent summary of arguments for and
against the letter's authenticity, see van der Eijk, 2000, vol. II, p. 352f.

[138] Publication: Hippocrates, vol. VIII, pp. 616–73, ch. 23.

[139] See Rehm, 1941, p. 38f.; see also Mansfeld, 1971, p. 25 n.

F.vii. **[Hippocrates]**, *On Regimen*[140] has a short passage that tells the reader what food and activity are proper to each season:

τὸν μὲν ἐνιαυτὸν ἐς τέσσαρα μέρεα διαιρέουσιν, ἅπερ μάλιστα γινώσκουσιν οἱ πολλοί, χειμῶνα, ἦρ, θέρος, φθινόπωρον· καὶ χειμῶνα μὲν ἀπὸ Πλειάδων δύσιος ἄχρι ἰσημερίης ἠαρινῆς, ἦρ δὲ ἀπὸ ἰσημερίης μέχρι Πλειάδων ἐπιτολῆς, θέρος δὲ ἀπὸ Πλειάδων μέχρι Ἀρκτούρου ἐπιτολῆς, φθινόπωρον δὲ ἀπὸ Ἀρκτούρου μέχρι Πλειάδων δύσιος.

I divide the year up into four parts, just as most people know it: winter, spring, summer, and autumn. Winter is from the setting of the Pleiades to the vernal equinox, spring from the equinox until the rising of the Pleiades, summer from the Pleiades until the rising of Arcturus, autumn from Arcturus until the setting of the Pleiades.

A little later in the same chapter it mentions the west wind and the return of the swallows, but these are attached to seasons generally, rather than to specific dates or stellar phases. There are some mentions of day-differences:

χρὴ οὖν τὴν ὥρην ταύτην οὕτω διαιτῆσθαι, ἀπὸ Πλειάδων δύσιος μέχρις ἡλίου τροπῶν ἡμέρας τεσσαράκοντα τέσσαρας· περὶ δὲ τὴν τροπὴν ἐν φυλακῇ ὅτι μάλιστα εἶναι, καὶ ἀπὸ τροπῆς ἡλίου ἄλλας τοσαύτας ἡμέρας τῇ αὐτῇ διαίτῃ χρέεσθαι. μετὰ δὲ ταῦτα ὥρη ἤδη ζέφυρον πνέειν, καὶ μαλακωτέρη ἡ ὥρη· χρὴ δὴ καὶ τῇ διαίτῃ μετὰ τῆς ὥρης ἐφέπεσθαι ἡμέρας πεντεκαίδεκα. εἶτα δὲ Ἀρκτούρου ἐπιτολή, καὶ χελιδόνα ὥρη ἤδη φαίνεσθαι, τὸν ἐχόμενον δὲ χρόνον ποικιλώτερον ἤδη διάγειν μέχρις ἰσημερίης ἡμέρας τριήκοντα δύο . . . χρὴ οὖν, ὡς μὴ ἐξαπίνης τὴν δίαιταν μεταβάλλειν, διελεῖν τὸν χρόνον ἐς μέρεα ἓξ κατὰ ὀκτὼ ἡμέρας . . . μέχρι Πλειάδων ἐπιτολῆς. ἐν τούτῳ θέρος, καὶ τὴν δίαιταν ἤδη χρὴ πρὸς τοῦτο ποιέεσθαι . . . ταύτῃ δὲ τῇ διαίτῃ προσανεχέτω μέχρις ἡλίου τροπέων . . . τὸν ἐχόμενον δὲ χρόνον . . . μέχρις Ἀρκτούρου ἐπιτολῆς καὶ ἰσημερίης ἡμέρας ἐνενήκοντα τρεῖς. ἀπὸ δὲ ἰσημερίης ὧδε χρὴ διαιτῆσθαι, . . . ἐν ἡμέραις δυοῖν δεούσαιν πεντήκοντα μέχρι Πλειάδων δύσιος ἀπὸ ἰσημερίης.

It is necessary to follow this regimen during this season, for forty-four days from the setting of the Pleiades to the solstice, to be most careful around the solstice, and to have the same regimen for the same number of days after the solstice. After that is the season when the west wind blows, and the season is milder, and it is necessary to follow the season with regard to regimen for fifteen days. And then it is the rising of Arcturus, and the season for the swallows to appear, and the time that follows is more variable until the equinox, for thirty-two days . . . It is necessary, in order not to change the diet suddenly, to divide the time into six parts of eight days . . . until the rising of the Pleiades. At this time, it is summer, and around this time you should keep the following regimen . . . Devote yourself to [such-and-such a] regimen until

140 Publication: Hippocrates, vol. vi, p. 594f., book iii. 68.7–13.

the solstice . . . In the time that follows [keep the following regimen . . .] until the rising of Arcturus and the equinox, (which is) ninety-three days . . . [Then do some other things] for fifty-less-two days until the setting of the Pleiades after the equinox.

As with *Peri hebdomadon*, above, Rehm sees the influence of Eudoxus here, but at least in this instance there is a little more to go on than in *Peri hebdomadon*. The similarities he finds in day-differences are not perfect, and they most likely point back to common Greek systems for dividing the year, rather than to a particular line of authorial transmission.[141]

F.viii.   Aratus' *Phaenomena*[142] is a third-century Greek poem, very popular in antiquity, which describes the constellations and various weather signs (*Phaen.* 733–1154), including atmospheric signs, such as the appearance of the moon or sun, and animal signs (as in Theophrastus' *De signis*). It is usually seen as a verse rendering of a now lost work, the *Phaenomena*, by Eudoxus.[143] In the astronomical parts of the poem, Aratus includes some astrometeorological information, such as that the Etesian winds begin just after the rising of Sirius, that Capella heralds storms, or that the north wind is associated with Pisces. None of this is timed with dates or day counts, however. Aratus was translated into Latin by Germanicus Caesar,[144] Cicero, and Ovid, among others. Germanicus' translation may be the source of the parapegmatic attributions to 'Caesar'.

F.ix.   The Eudoxus papyrus (perhaps better called the **Leptines papyrus**, as this is the author's name signed at the end) was published by Friedrich Blass in 1887.[145] It is a Greek text dating from the second century BC and seems to be partly related to material in P. Hibeh 27, though much garbled. Apart from the acrostic, which reads Εὐδόξου τέχνη, *The Craft of Eudoxus*, the text has little obvious connection to what we know of Eudoxus' astronomy. The text is essentially a wide-ranging miscellany of astronomical information. The only content related to parapegmata is a poorly preserved, somewhat

---

[141] See Rehm, 1941, p. 39.   [142] Publication: Kidd, 1997.

[143] See, for example, Gee, 2000; Kidd, 1997. Contrast Sider, 2002, who thinks Aratus had separate sources for the two main parts of the poem.

[144] Gain, 1976, raises some questions about Germanicus' authorship of the translation.

[145] Currently in the Louvre, Department of Egyptian Antiquities, inv. N. 2325. Reprinted by *Zeitschrift für Papyrologie und Epigraphik* in 1997. For the date of the text itself, see Blass, 1887, p. 79, Grenfell and Hunt, 1906, p. 143. The text is explicitly credited to Leptines in the colophon. For a good discussion of the MS itself, see Thompson, 1988, p. 252f. See also Evans, 2004, p. 34.

repetitive, and not particularly well-organized list of the number of days intervening between certain stellar phases, similar to the one found in Varro and C. Vind. (F.xii and C.iii).[146] For example: 'From the (setting of) the Pleiades to the setting of Orion, 22 days; from (the setting of) Orion to the setting of Sirius, 2 days; from (the setting of) Sirius to the solstice, 24 days . . .'.[147] This is followed by a list of season-lengths according to Eudoxus, Democritus, Euctemon, and Callippus. No calendar dates are given other than the report that 'according to Eudoxus and Democritus, the winter solstice (happens) on either the 19th or 20th of Athyr'.[148] There is nothing in the way of weather predictions.

F.x.   The **Venusia *Fasti***[149] is a fragmentary Latin calendar dating from the first century BC. It is complete from the Kalends of May through to XII K. July. Apart from listing the nundinal letters and the *dies fasti, nefasti,* and *comitales,* it has four astronomical entries:

G NON[AE] (sc. *maiae*) *F(ASTUS). VERGILI(AE) EXORI(VNTUR)*
B XV (sc. *K. Iuniae*) *C(OMITALIS). SOL IN GEMIN(IS)*
B XIII (sc. *K. Iuliae*) *C(OMITALIS). SOL IN CANCRO*
A VI (sc. *K. Iuliae*) *C(OMITALIS). SOLSTITIVM CONFECT(VM)*

Unlike the *Menologia rustica* (see the next entry) the Venusia *Fasti* contains the date of a stellar phase, and the dates of the sun's entry into the signs of Gemini and Cancer are given precisely. In the *Menologia,* by contrast, the sun's position in the zodiac is listed only roughly, such that each sign is simply schematically or symbolically associated with a particular month.

---

[146] See *HAMA,* p. 687f. There is some similarity, too, with Columella, *RR* IX.14, where Columella says that there are forty-eight days of spring, which he defines as the time between the equinox 'which occurs on about (*circa*) the VIII K. April, in the eighth degree of Aries' and the rising of the Pleiades. He also mentions that there are roughly thirty days from the (summer) solstice to the rising of Sirius, and roughly fifty from Sirius to the rising of Arcturus. But the ambiguity and incompleteness of his numbers precludes this from being an ἄστρων διαστήμα as Rehm claims (Rehm, 'Parapegma', *RE,* col. 1309).

[147] Blass, 1887, col. XXII.   [148] Blass, 1887, col. XXII.

[149] Publication: *CIL* I, pp. 300–1; *CIL* IX.421; Degrassi, 1963, vol. XIII.2, pp. 55–62. Degrassi's description of the long history of this inscription is worth reading. The inscription itself was dug up in 1470 and lost shortly thereafter. Fortunately Iucundus saw the inscription at Naples in about 1489 and copied it. Until 1854, there was a fair amount of confusion about where the inscription actually came from, which Degrassi clarifies.

F.xi. The *Menologia rustica*[150] are inscriptions that divide the agricultural year up into Roman calendar months. For each month they list such information as: the number of days, the lengths of day and night, the zodiacal sign of the sun, important religious festivals, and some agricultural information. Other than being organized by calendar months, though, no actual dates are typically given.

F.xii. Varro's season list:[151] in his first-century BC work *On Agriculture*, Varro gives a short description of the seasons in terms of the number of days intervening between certain stellar phases and the solstices and equinoxes. In its entirety, it reads as follows:

suptilius descriptus temporibus observanda quaedam sunt, eaque in partes VIII *dividuntur: primum a favonio ad aequinoctium vernum dies* XLV, *hinc ad vergiliarum exortum dies* XLIV, *ab hoc ad solstitium dies* XLIIX, *inde ad caniculae signum dies* XXVII, *dein ad aequinoctium autumnale dies* LXVII, *exin ad vergiliarum occasum dies* XXXII, *ab hoc ad brumam dies* LVII, *inde ad favonium dies* XLV.

In more accurate divisions of the seasons there are some things to be noted, and (the seasons) are reckoned in eight parts: the first from the west wind to the vernal equinox, 45 days; from there to the rising of the Pleiades, 44 days; from this to the solstice, 48 days; then to the rising of Sirius, 27 days; next to the autumnal equinox, 67 days; from that to the setting of the Pleiades, 32 days; from this to the winter solstice, 57 days; then to the west wind, 45 days.

F.xiii. The citation from **Hyginus, *De apibus*** in Columella[152] is a description of the seasons relative to the solstices, equinoxes, and the phases of the fixed stars, with some discussion of the behaviour of bees at each season.

F.xiv. The **Ara Pacis meridian**[153] in Rome used an Egyptian obelisk as a huge gnomon whose noon shadow would indicate the sun's position in the zodiac. The inscriptions are in Greek. Pliny reports that it was designed by a certain Nov(i)us Facundus and erected under the patronage of Augustus.[154] In the preserved part of the meridian, the beginning of summer and the end of the Etesian winds are marked.

---

[150] See Degrassi, 1963, vol. XIII.2, p. 286f.; Broughton, 1936; Salzman, 1990. See also Fig. 1, chapter 2. There is a good exemplar in the Museo Nazionale Archeologico di Napoli.
[151] Varro, *Rust.* I.28.
[152] Columella, *RR* IX.14.   [153] Published by Buchner, 1982.   [154] Pliny, *HN* XXXVI.72f.

F.xv.   Other ancient **Sundials**[155] indicated zodiacal signs, calendrical months, solstices and equinoxes, and the phases of fixed stars. Occasionally the winds are inscribed around sundials as directional indicators,[156] but none other than that in the *Ara Pacis* seem to have been used to actually predict a wind.

F.xvi.   In **Galen,** *Commentary on Epidemics I,*[157] there is a discussion of the divisions of the seasons, marked by the solstices, equinoxes and stellar phases, in the context of Galen's discussion of seasonal weather patterns.

F.xvii.   The **Astronomical Ephemerides**[158] seem to have performed some of the functions of the astrological and astronomical parapegmata. Specifically, the Ephemeris of 140,[159] complete for most of August, correlates dates in the Roman calendar with Alexandrian calendar dates, days of the week (Saturdays are marked, the rest presumably interpolated), and the dates of planetary entries into zodiacal signs.

The Ephemeris of 467[160] correlates days of the week (every seventh day, probably Saturday, is marked with a number), Roman dates, Alexandrian dates, lunar days, the moon's zodiacal sign, longitude, the time of the moon's entry into each sign, the sun's longitude, the daily positions of the other planets, as well as a column which informs us whether the particular day is good, bad, or indifferent. Two later ephemerides, *P. Vind. G.* 29370b and 29370 (from 471 and 489, respectively) also include days of the week.[161]

A description of the making of ephemerides in some MSS of Theon's commentary on Ptolemy's *Handy Tables* shows that the ephemerides could sometimes have also included fixed-star astrometeorological predictions like the parapegmata, as well as lunar and hebdomadal information. The whole of the description reads as follows:

---

[155] The standard collection of sundials is by Gibbs, 1976.

[156] Zodiacal signs: Gibbs, 1976, nos. 1068G, 4007, 4010, 7002G. Calendrical months: e.g., Gibbs, 1976, nos. 1068G, 5021. Solstices and equinoxes: Gibbs, 1976, nos. 1072G, 1074, 1075, 3047, 3050G, 3058G, 3060G, 4001G, 4007, 4008G, 4009, 4010, 5021. Stellar phases: Gibbs, 1976, nos. 1001, 1073, 7001G. Winds: Gibbs, 1976, nos. 4002G, 4008G, 4009, 4010, 5001.

[157] In Galen, vol. xvii.1, pp. 1–83.

[158] For a general description of ephemerides, see Jones, 1999a, pp. 40–2; see also Neugebauer, 1975, p. 1055f.

[159] Jones, 1994a.   [160] Curtis and Robbins, 1935.

[161] See Jones, 1999a, p. 41. For the dating of these texts, see Jones, 1994b who convincingly argues against Gerstinger and Neugebauer's dating (Gerstinger and Neugebauer, 1962).

περὶ τῆς τῶν ἐφημερίδων ἐκθέσεως.

ἄνω καὶ κάτω τόπους μείζονας καταλιμπάνομεν ὅπως ἄνω μὲν τὰς ἐπιγραφὰς
δέξωνται τῶν σελιδίων, κάτω δὲ καθ' ἕκαστον μῆνα ἐποχὴν τῶν (συνόδων) καὶ
(πλησισελήνων), ἐν δὲ τῷ μεταξὺ τούτων ιε΄ χώρας, τὴν κατωτέραν μείζονα
ποιοῦντες, ἵνα αἱ μὲν ἄλλαι ἀνὰ β΄ δέξωνται στίχων, ἡ δὲ τελευταία γ΄ διὰ τὸ τὸν
ῥωμαικὸν μῆνα πολλάκις λα΄ ἔχειν. ἐπὶ δὲ τοῦ Φεβρουαρίου ιδ΄ δεῖ μόνας γράφειν.

τὰ δὲ σελίδια ποιήσωμεν οὕτως. πρότερον πάντων πλατύτερον τὸ τὰς ἐπισημασίας τῶν ἀπλανῶν δεχόμενον· καὶ ἐφεξῆς τὸ τρίτον ἢ τέταρτον στενώτερα
τὰ δεχόμενα τοὺς μῆνας, τὸ μὲν πρῶτον τὴν κατὰ ῥωμαικὸν πρὸ πόσων, τὸ δὲ
δεύτερον τῶν ἀλεξανδρέων τὸν μῆνα, τὸ δὲ τρίτον οἵαν τις προαιρῆται τῆς ἑαυ-
τοῦ χώρας, τὸ δὲ δ΄ τὸν κατὰ (σελήνην). ἐπὶ δὲ τούτων σελιδίων καταγράφομεν γ΄
πλατύτερα καὶ τέταρτον στενώτερον· καὶ ἐπιγράφεται μὲν πᾶσιν ἡ ἐπιγραφὴ α΄
(σελήνης) κίνησις, ἰδικῶς δὲ τὸ μὲν πρῶτον ζωδίων, τὸ δὲ β΄ μοιρῶν καὶ λεπτῶν, τὸ
δὲ γ΄ ὥρας μεταβάσεων, τὸ δὲ δ΄ ἀνέμων. ἐφεξῆς γίνονται σελίδια ζ΄, (ἡλίου) καὶ τῶν ε΄
πλανωμένων. ἐπὶ τούτοις σελίδιον ποιοῦμεν στενώτερον δεχόμενον τὸν ῥωμαικὸν
μῆνα, ἢ ὃν ἄν τις βουληθῆ κατὰ τάξιν. καὶ ἐπὶ πᾶσι τούτοις πλατύτερον ὀφεῖλον
δέξασθαι τὰς καθολικὰς καταρχάς, ἃς δεῖ ποιεῖν, οὕτω τὰ σχήματα καταγράφειν
τῆς (σελήνης) καὶ ἕκαστον τῶν (ἀστέρων) καὶ εἶθ' οὕτω τὰς καθολικὰς καταρχάς.

On the Making of an Ephemeris:

At the top and bottom, we leave a larger space so that the top can have the
headings of the columns, and the bottom the time of new and full moons for each
month; in between these, fifteen rows, making the bottom one larger, such that the
(rest) will have two lines each, and the (bottom one) three, since the Roman month
often has 31 days. For February, it is necessary to mark only fourteen (columns).

We make the columns thus: before all of them, a wider one for showing the
weather changes of the fixed stars [τὰς ἐπισημασίας τῶν ἀπλανῶν], and the third
or fourth from there narrower ones, to hold the months: the first for counting down
the Roman month, the second for the Alexandrian month, the third, if one wishes,
for the calendar of his own land, and the fourth, for that according to the moon.
After these columns, we draw three wider and the fourth narrower, and the heading
'First Lunar Motion' is written over them all. In particular, the first (column) is for
zodiacal signs, the second for degrees and minutes, the third for the times of sign
entry, and the fourth for the winds. From here there are six[162] columns for the sun
and five planets. After this we make a narrow column for holding the Roman month,
or whatever one wishes, in order, and after all of these, a wider (column) to hold
the general predictions that must be made, thus writing the configurations of the
moon and the stars and from these, then, the general predictions.[163]

We see here that an ephemeris could contain both stellar astrometeorology
and the days of the moon (ὁ μὴν κατὰ σελήνην). Unfortunately, I know

---

[162] The MSS all say 'seven', but I have only ever seen six in an ephemeris, and only six are
described here.

[163] Published in Curtis and Robbins, 1935, p. 83.

of no extant ephemeris which does so. If such ephemerides were made, and I think this description good enough reason to suppose they were, then those ephemerides would have combined the functions of *both* the astrometeorological and astrological parapegmata in one document. They thus can be seen as bringing together two distinct parapegmatic traditions, as well as incorporating catarchic astrological information and predictions.

F.xviii.   The **Alésia disk**[164] is a flat bronze disk, 11 cm in diameter, found by Espérandieu in 1933. It is not a parapegma, but seems to be a related type of instrument for keeping track of the hebdomadal days. There is a single hole punched through the centre. It shows the images of the hebdomadal deities around the face of the disk, with the tops of their heads pointing outwards. A curious feature is described by Espérandieu as follows: 'Les bustes sont limités par un cercle et bordés extérieurement de triangles, alternativement vides et couverts de traits, d'où se détachent quatre ou cinq petites lignes dont il est malaisé de comprendre la destination.'[165] As can be seen from his too-crude drawing, there are four or five lines between each deity (a total of thirty-six), but his description seems to indicate four or five more for each one of these. The thirty-six lines may be meant to represent sections of the zodiacal circle, but even so, they are probably only decorative. The disk was probably used, as Espérandieu thought, by simply rotating it one seventh of a turn counterclockwise each day, so that the deity shown at the top was the one presiding over that particular day. Espérandieu also mentions a similar disk in the British Museum, but he fails to give any details about it, and I have been unable to locate it.

F.xix.   The **Dijon disk**[166] is a bronze disk with the names of the hebdomadal deities inscribed in a circle around it. It was probably used in a similar way to the Alésia disk.

F.xx.   The **Florentinus season list**[167] is the first chapter of the *Geoponica* which correlates the beginnings of the prominent winds and seasons with dates in the Roman calendar, and the sun's position in the zodiac (sometimes in degrees). It also dates the rising and setting of the Pleiades to IV Ides January and IV Nones November, respectively.

---

[164] Publication: Espérandieu, 1933. Jim Evans managed to track this object down in Avignon. It had been in Espérandieu's private collection, and is now in the Palais du Roure in Avignon.

[165] Espérandieu, 1933, p. 384.

[166] *CIL*, vol. xiii, no. 2869. Currently the Musée Archéologique de Dijon, inv. no. 263. Size: 7.5 cm.

[167] In *Geoponica*, i.1; also published in Wachsmuth, 1897, pp. 320–1.

F.xxi. The *Geoponica* phase list,[168] attributed to Quintilius, simply lists the rising and setting dates of certain prominent fixed stars and constellations (Arcturus, the Pleiades, Orion, etc.), but it also includes a mention of the Etesian winds. Most of it seems to have been excerpted from the Quintilius parapegma, although there is one entry missing from Quintilius: τῇ ιγ' τοῦ Ἰουλίου, Προκύων ἑῷος ἐπιτέλλει, 'on the 13th of July, Procyon rises in the morning', which occurs in the *Geoponica* phase list at a point corresponding with a corruption in Quintilius.[169] Dates in this phase list are sometimes displaced by one or more days from those in Quintilius, although sometimes they do not differ.

F.xxii. A **Byzantine season list**, of unknown date, is found in a sixteenth-century codex (*C. Berol.* 170) and published in the *CCAG*.[170] It lists the day-differences between various stellar phases, the solstices, equinoxes, and the coming of the west wind. It also gives absolute values for the lengths of the seasons: summer, 94 days; autumn, 90 days; winter, 91 days; and spring, 90 days, for a total of 365.

G: Dubia

G.i. The **Egyptian** *Calendar of Lucky and Unlucky Days* has been claimed by Leitz to be astrometeorological,[171] but I disagree. For my full argument, see chapter 6.

G.ii. The **'Euctemon parapegma'**[172] is a modern reconstruction by Rehm of the presumed fourth-century BC parapegma from which later parapegma-tists are supposed to have excerpted their Euctemon citations. Such reconstructions can be useful, insofar as they allow us to isolate and compare particular groups of observations and predictions, as van der Waerden and Hannah have done.[173] But we cannot be sure that the text from which the citations were taken looked anything like the reconstruction, nor (and this is a very important point) that it was written in the same calendar, as

---

[168] *Geoponica*, i.9.   [169] See from IV Nones July to IX Kal. August in Boll, 1910b, p. 7.

[170] *CCAG*, VII.162–3.   [171] Publication: Leitz, 1994.

[172] Rehm, 1913; See also Pritchett and van der Waerden, 1961; Wenskus, 1990, p. 29f.; Hannah, 2002.

[173] Van der Waerden, 1984a; Hannah, 2002.

Rehm and van der Waerden believe.[174] Moreover, if we are to include the 'Euctemon parapegma' as genuine, then we should, to be methodologically consistent, do the same with Democritus, Meton, Callaneus the Indian, etc.

Rehm's attribution to Euctemon of a list of the date-differences (an ἄστρων διαστήματα, in Rehm's terminology) between various stellar phases (specifically, the very one found in C. Vind., C.iii, above) supposes (a) that the C. Vind. list can be attributed to Euctemon; and (b) that it was written as a list of date-differences, rather than having been derived by some later author from some source such as a parapegma, or even partly from the Geminus parapegma itself.[175] But the correspondence between C. Vind. and the attributions to Euctemon in Geminus and Ptolemy, even were they perfect – and they are not – would not prove that Euctemon wrote out a list of date-differences (see the discussion in C.iii, above). Such a list could quite simply have been excerpted from something he did write. But as we saw in the discussion of C. Vind., above, I do not think that the correspondence *is* close enough to warrant a claim of exclusive derivation from Euctemon.

Hannah has speculated that Euctemon 'probably' also erected a pegged inscriptional parapegma,[176] but this again cannot be shown. It is certainly one possibility, but only one possibility. It is clear that Euctemon wrote at least one astrometeorological text of some sort that was excerpted into later parapegmata, but the form of that text or texts remains open in the absence of more evidence.

G.iii.   Van der Waerden argues that there was a **Parapegma of Dionysius** which was the source for Ptolemy's attributions to 'the Egyptians'.[177] His claim rests on the very dubious argument that (a) the attributions in Ptolemy must have been taken from a source which used a $365\frac{1}{4}$ day year; (b) the calendar of Dionysius, which did so,[178] was 'easier to handle' than that of Callippus; but (c) the Egyptian calendar was 'more convenient' than that of Dionysius for the dating of observations; so (d) the calendar of Dionysius was really for dating solstices, equinoxes, and stellar phases and therefore

---

[174] For my detailed arguments on this, see chapter 4. Hannah, 2002, has recently also argued (albeit for quite different reasons than mine) that Euctemon did not use a zodiacal calendar.

[175] Rehm, 1913; followed by van der Waerden, 1984a.

[176] Hannah, 2002. See also Hannah 2001a, p. 142f.

[177] See van der Waerden, 1985, pp. 99–101. For 'the Egyptians' in Ptolemy, see chapter 6, above.

[178] For this argument, see Böckh, 1863, p. 286f.; van der Waerden, 1984c. For a description of the calendar of Dionysius and its relation to the parapegma tradition, see chapter 4.

Dionysius was a parapegmatist. I should point out that point (a) is just wrong, (b) and (c) unfounded (we know next to nothing about the calendar of Dionysius), and (d) is a non-sequitur.

G.iv.   The **Pompeii calendar medallions**[179] are images of the hebdomadal deities, the months, and the seasons from a room in Pompeii. C. R. Long has argued that there were probably originally also images for the signs of the zodiac, and hence that the room is comparable to the Thermae Traiani parapegma. At a most general level, perhaps, but there are no peg holes, lunar days, or calendrical dates discernible, and the scale and the decorative features of these images are such that they seem to serve a different function than the small, crude Thermae Traiani parapegma.

G.v.   The **Ostia hebdomadal deities**[180] are busts drawn on stone, five of which are extant: the Sun, Moon, and Mars, and on another fragment, Jupiter and Venus. Meyboom and Sadurska have listed this as a parapegma, since holes appear near the heads of the deities.[181] But, as Becatti had earlier pointed out, the holes are rather oddly placed: one below the right ear of the Sun, the next to the right of the Moon's head, one between Jupiter and Venus and a fourth to the right of Venus. Becatti thought they were drilled at some later date, possibly to hold hooks for hanging something on. There are also holes in the bottom of the stone, presumably for mounting the stone on iron posts. I think the strange placement of the holes, 'senza riguardo alle figurazioni', as Becatti says, makes this inscription very doubtful as a parapegma.

G.vi.   The **Hoffmann** *menologium*[182] is a Latin calendar fragment inscribed clockwise on a bronze disc of approximately one Roman foot in diameter. The fragment is complete from the Ides of July through to just before the Ides of December. *VIII K. OCT* is labelled as *AEQVINOCT[IVM]*. Although the drawing in the *CIL* does not reproduce them accurately, Noë says that there are holes punched around the outside of the disk, one for every other day, which leads me to believe that they are not meant to hold a peg or a pebble as a day marker. Noë's suggestion, quoted in *CIL*, that the menologium was meant to represent the length of the days throughout

---

[179]   Publication: Long, 1992.

[180]   Currently in the Museo Ostiense. Published in Becatti, 1954, pp. 116–17, pl. xxxviii.3.

[181]   Meyboom, 1978, p. 785; Sadurska, 1979, p. 74.

[182]   *CIL*, vol. xiii.2.1, no. 5955. See also Froehner, 1886, no. 653. It is listed in the *CIL* as part of the Hoffmann collection, Paris, but I have been unable to verify this.

the year does not seem unreasonable. Degrassi thinks this may be a para-pegma,[183] but it seems to me to be a different sort of instrument.

G.vii. I have chosen to exclude from my list of parapegmata the large number of **images of the hebdomadal deities** which can be found on late antique reliefs and inscriptions, on the grounds that these do not seem to have been used as instruments but rather as decoration. In many cases the publication of these inscriptions leaves much to be desired: details about them are often unclear, and drawings or photographs lacking.[184] This being the case, it would not surprise me if some of them should turn out to be parapegmata, but I shall for the moment err on the side of caution and include as genuine only those of which I am certain.

G.viii The *Dusari sacrum* **monuments**[185] are three apparently related rectangular objects, all bearing the inscription *DVSARI SACRVM*, which were found separately at Puteoli. Dusares was the Latinization of an appar-ently astral deity (Dushara) of the Nabataeans, who seems to have had a cult near Puteoli.[186] One of the monuments has seven rectangular holes in the top, and the other two have three holes each, into which flat tombstone-shaped tablets were fit.[187] Four of these tablets were found with the largest of the three monuments in the port of Pozzuoli in 1965. Meyboom has argued that the seven-holed monument is a hebdomadal parapegma, and that the three-holed ones constitute half of a zodiacal or monthly para-pegma (originally comprising four monuments with three holes each). He thinks the tablets would have functioned in place of pegs. If this is correct, however, we should probably have expected to find only one tablet and not four. If, as Meyboom also suggests, the tablets originally had images of the hebdomadal deities painted or carved on them, then the whole monument would be simply a representation of those gods without being a parapegma, unless only one tablet at a time was inserted, replaced by the next one on the following day, which seems unlikely.

G.ix. The **Esdud inscription**[188] is a puzzling little fragment from Pales-tine, which includes four rectangular holes, 12 cm by 3 cm, each labelled with successive Greek letters. The rectangular holes are reminiscent of the

---

[183] Degrassi, 1963, vol. xiii.2, p. 299.   [184] See, e.g., de Witte, 1877.

[185] Currently in the Museo Nazionale Archeologico di Napoli, inv. nos. 3249, 3250. Publication: Meyboom, 1978; Sadurska, 1979, p. 76.

[186] See Lacerenza, 1988.   [187] See Meyboom, 1978, pl. clvii.1.

[188] Germer-Durand, 1901, p. 74. Listed as a calendar by Sadurska, 1979, p. 75. Current location unknown.

Dusari sacrum monuments, but they are aligned differently, and have letters inscribed above them. Like Germer-Durand, I am at a loss to explain this puzzling inscription. Clermont-Ganneau speculates that it may have been a gaming board of some sort.[189]

G.x.   The **Eschenz object**[190] is a clay artefact from Switzerland with seven holes bored into the top of it, conceivably to receive a peg. There is no inscription of any kind. Urner-Astholz's theory that it is a hebdomadal parapegma seems possible,[191] but is unconfirmable.

G.xi. The **Augst object**[192] is virtually identical to the Eschenz object. Listed as a calendar by Sadurska.

---

[189] Clermont-Ganneau, 1906, p. 208f. Listed as a calendar by Sadurska, 1979, p. 75.
[190] Currently in the Heimatmuseum, Steckborn. Published by Urner-Astholz, 1942. See also Sadurska, 1979, p. 75.
[191] Urner-Astholz, 1960.
[192] Currently in the Römermuseum, Augst. Published by Sadurska, 1979, p. 76.

Extant parapegmata[1]

A.i. P. Hibeh 27[2]

Col. iv.

[Χοιὰκ α΄ . . . ἡ] νὺξ ὡρῶν ιγ΄ ιβ΄ με, ἡ δ᾿ ἡμέρα ι΄ β΄ ε΄ λ΄ ϙ΄. [ι]ϛ΄ Ἀρκτοῦρος ἀκρώνυχος ἐπιτέλλει, [ἡ] νὺξ ὡρῶν ιβ΄ β΄ ιε΄ με, ἡ δ᾿ ἡμέρα ια΄ θ΄ ι΄ λ΄. [κ]ϛ΄ Στέ-φανος ἀκρώνυχος ἐπιτέλλει [κ]αὶ βορέαι πνείουσιν ὀρνιθίαι, ἡ νὺξ [ὡρ]ῶν ιβ΄ ι΄ λ΄, ἡ δ᾿ ἡμέρα ια΄ γ΄ ι΄ λ΄. Ὄσιρις [π]εριπλεῖ καὶ χρυσοῦν πλοῖον ἐξά[γε]ται. Τῦβι [. . .] ἐν τῶι Κριῶι. κ΄ ἰσημερία [ἐα]ρινή, [ἡ] νὺξ ὡρῶν ιβ΄ καὶ ἡμέρα ιβ΄, [κ]αὶ ἑορ[τ]ὴ Φιτωρώιος. κζ΄ Πλειάδες [ἀκ]ρώνυχ[οι] δύνου[σ]ιν, ἡ νὺξ ὡρῶν ια΄ β΄ ϛ΄ ϙ΄, [ἡ] δ᾿ ἡμέρα [ι]β΄ ι΄ λ΄ με. Μεχεὶρ ϛ΄ ἐν τῶι [Τ]αύρωι. Ὑάδες ἀκρώνυχοι δύνουσιν, [ἡ] νὺξ ὡρῶν ια΄ ∠΄ ι΄ λ΄ ε΄

Col. v.

ἡ δ᾿ ἡμέρα ιβ΄ γ΄ με, καὶ Ἥρα κάει. καὶ ἐπ[ι]σημαίνει καὶ νότος π[νεῖ,] ἐὰν δὲ πολὺς γένηται τὰ ἐκ τῆς γῆς κατακάει. ιθ΄ Λύρα ἀκρώνυχος ἐπ[ι]τέλλει, ἡ νὺξ ὡρῶν ια΄ γ΄ ιε΄ με, ἡ δ᾿ ἡμέρα ιβ΄ ∠΄ ιε΄ ο΄, καὶ πανήγυρις ἐν Σάι τῆ[ς] Ἀθηνᾶς, καὶ νότο[ς πνεῖ], ἐὰν δὲ πολὺς γέν[ηται] τὰ [ἐκ τῆς] γῆς κατακάει. [. . . ἀκρώ]νυχος ἐπιτέλλει, [ἡ νὺξ ὡρῶν ια΄ . . .], ἡ δ᾿ ἡμέρα ιβ΄ β΄[. . .] ἄγουσιν κα[. . .]

Col. vi.

κζ΄ Λύρα ἀκρώνυχος δύνει, ἡ νὺξ ὡρῶν ια΄ ϛ΄ ϙ΄, ἡ δ᾿ ἡμέρα ιβ΄ β΄ <ι΄ λ΄> με, Προμηθέως ἑορτὴ ὃν καλοῦσιν Ἰφθῖμιν, καὶ νότος πνεῖ, ἐὰν δὲ πολὺς γένηται τὰ ἐκ τῆς γῆς κατακάει. Φαμενὼτ δ᾿ ἐν τοῖς Διδύμοις. [. . .]

---

[1] I have included the Greek and Latin for most of the translated parapegmata but not for all. My criterion has essentially been to ask whether the Latin or Greek of a specific text will be on the bookshelf of the average classicist or of a moderately good university library. If so, then I have omitted the original here.

[2] Greek text published by Grenfell and Hunt, 1906, pp. 138–57, reproduced here with changes as noted.

ἀνατέλλει,[3] ἡ νὺξ ὡρῶν ιαʹ εʹ μʹ, ἡ δʹ ἡμέρα ιβʹ βʹ δʹ κʹ ϙʹ. εʹ Σκορπίος ἑῶιος [ἄρχ]εται δύνειν, ἡ νὺξ ὡρῶν ιγʹ, [ἡ δʹ ἡμέ]ρα ιγʹ. θʹ παρὰ τοῖς Αἰ[γυπτίοις . . .]εδυ ἑορτή. ιβʹ Σκορπίος [. . .[4]] δύνει, ἡ νὺξ ὡρῶν ιʹ βʹ ϛʹ ϙʹ, [ἡ δʹ ἡμέρα ι]γʹ ιʹ λʹ λʹ μέ. ιγʹ Πλειάδες [. . . ἐπιτέλ]λουσιν.[5]

Col. vii.

[ten lines almost completely lost]
Φαρμοῦ[θι ἐ]ν τῶι Κ[α]ρκίνωι. γʹ Ἀετὸς ἀκρώνυχος ἐπιτέλλει, ἡ νὺξ ὡρῶν ιʹ γʹ λʹ ϙʹ, ἡ δʹ ἡμέρα ιγʹ ιʹ ϙʹ μέ.

Col. viii.

ιαʹ Δελφὶς ἀκ[ρών]υχος ἐπιτέλλει, ἡ νὺξ ὡρῶν [ιʹ εʹ, ἡ δʹ] ἡμέρα ιγʹ βʹ ιαʹ λʹ, [κ]αὶ τῆς Ἥρας [. . .]ιχεια. [ι]ζʹ Ὠρίων ἑῶ[ιος ἐπιτέ]λλει, ἡ νὺξ ὡρῶν ιʹ ιέ, ἡ δʹ [ἡμέρ]α ιγʹ βʹ δʹ ξʹ. κʹ ἡ νὺξ ὡρῶν ιʹ, ἡ δʹ ἡμέρα ιδʹ, καὶ ἐκ τοῦ αὐτοῦ ἀνατέλλει ὁ ἥλιος ἡμέρας γʹ. <καʹ> ἡ νὺξ ὡρῶν ιʹ, ἡ δʹ ἡμέρα ιδʹ. κβʹ ἡ νὺξ ὡρῶν ιʹ, ἡ δʹ ἡμέρα ιδʹ. κγʹ ἡ νὺξ ὡρῶν ιʹ, ἡ δʹ ἡμέρα ιδʹ. κδʹ ἡλίου τροπαὶ εἰς θέρος καὶ ἡ νὺξ μείζω<ν> γίνεται τῆς ἡμέρας ὥρας δωδεκατημόρου μέ,

Col. ix.

καὶ γίνεται ἡ νὺξ ὡρῶν ιʹ εʹ μʹ, ἡ δʹ ἡμέρα ιγʹ βʹ δʹ κʹ ϙʹ. κεʹ ἐτησίαι ἄρχονται πνεῖν καὶ ὁ ποταμὸς ἄρχετ[α]ι ἀναβαίνειν, ἡ νὺξ ὡρῶν ιʹ λʹ ϙʹ, ἡ δʹ ἡμέρα ιγʹ βʹ δʹ λʹ ρπʹ. **Παχὼνς** ϛʹ ἐν τῶι Λέοντι. Προτρυγητὴς ἀνατέλλει, ἡ νὺξ ὡρ[ῶ]ν ιʹ δʹ λʹ ρπʹ, ἡ δʹ ἡμέρα [ιγʹ βʹ λ]ϙʹ. θʹ Ὠρίων ὅλος ἀνατ[έλλει,][6] ἡ νὺξ ὡρῶν ιʹ γʹ μ[έ, ἡ δʹ] ἡμέρα ιγʹ ιʹ ιʹ λʹ ϙʹ. ιηʹ Κύων ἀν[ατέ]λλει, ἡ νὺξ ὡρῶν [ιʹ ιʹ γʹ μέ,]

Col. x.

[ἡ δʹ ἡμέρα ιγʹ γʹ] δʹ. **Παῦνι** [δʹ ἐν τῆι Παρθέ]νω[ι. Ἀ]ετὸς ἠῶιο[ς . . .[7] ἡ νὺξ] ὡρῶ[ν] ιʹ βʹ εʹ λʹ ϙʹ, [ἡ δʹ ἡμέρα ι]γʹ ιέ μ[έ. ιϛʹ . . .] ἑῶιο[ς . . .[8] ἡ ν]ὺξ [ὡρ]ῶν ιαʹ ϛʹ ϙʹ, [ἡ δʹ] ἡμέρα ιβʹ βʹ ιʹ [λʹ μέ,] Β[ου]βάστιος ἑο[ρτή.] κ[.]ʹ Δελφὶς ἑῶιος δύνει, ἡ νὺξ ὡρῶν ιαʹ γʹ [. . .], ἡ δʹ ἡμέρα ιβʹ ιʹ [. . .],

---

[3] Grenfell and Hunt insert <Αἲξ ἑῶια> before ἀνατέλλει. Though there is no lacuna, a subject for the verb has clearly been omitted by the copyist.

[4] Grenfell and Hunt restore [ἑῶιος ὅλος] here.

[5] Grenfell and Hunt restore Πλειάδες [ἑῶιαι ἐπιτέλ]λουσιν.

[6] Grenfell and Hunt insert <ἑῶιος> before ἀνατ[έλλει,].

[7] Grenfell and Hunt restore [δύνει] here.

[8] Grenfell and Hunt restore [ιϛʹ Στέφανος] ἑῶιο[ς δύνει] here.

Col. xi.

[... ἑ]ορτή. [κζ΄...]⁹ δύνει, [ἡ νὺξ ὡρῶν] ια΄ γ΄ ι΄ με΄ ε΄, [ἡ δ᾽ ἡμέρα] ιβ΄ ∠΄ ιε΄ ϙ΄,
[...]ς ἑορτή. [...]α μεγάλα [... ἐπιση]μαίνει, [ἡ νὺξ ὡρῶ]ν ια΄ γ΄ ι΄ λ΄ με΄, [ἡ δ᾽
ἡμέρα ιβ΄] ∠΄ ϙ΄. [᾽Επεὶφ. ἐν τ]αῖς [Χηλαῖς τοῦ σ]κορπίου, [... ἑῶ]ιος [...]¹⁰.

Col. xii.

ἡ [δὲ] νὺξ ὡρῶν ι[α΄...], ἡ δ᾽ ἡμέρα ιβ΄ [...], καὶ ἐν Σάι πανήγ[υρις] Ἀθηνᾶς
καὶ λύχνους κάουσι κατὰ τὴν χώραν, καὶ ὁ ποταμὸς ἐπισημαίνει πρὸς τὴν
ἀνάβασιν. κγ΄ ἰσημερία φθινοπωρινή, ἡ νὺξ ὡρῶν ιβ΄, ἡ δ᾽ ἡμέρα ιβ΄, τοῦ
Ἀνούβιος ἑορτή, καὶ ὁ ποταμὸς ἐπισημαίνει πρὸς τὴν ἀνάβασιν.

Col. xiii.

[... (two lines lost) κζ΄... ἡ νὺξ ὡρῶν ιβ΄ ιε΄ με],¹¹ ἡ δ᾽ ἡμέρα ια΄ β΄ ε΄ λ΄ ϙ΄.
Μεσορεὶ β΄ ἐν τῶι Σκορπίωι. Πλειάδες ἀκρώνυχοι ἐπιτέλλουσιν, ἡ νὺξ ὡρῶν
ιβ΄ ε΄, ἡ δ᾽ ἡμέρα ια΄ β΄ ι΄ λ΄, Ἀπόλλωνος ἑορτή. δ΄ Στέφανος ἑῶιος ἐπιτέλλει, ἡ
νὺξ ὡρῶν ιβ΄ ε΄ λ΄ ϙ΄, ἡ δ᾽ ἡμέρα ια΄ β΄ ιε΄ με΄. θ΄ Σκορπίος ἀκρώνυχος

Col. xiv.

[... (three lines lost) ...]¹² ιδ΄ Σκορπίος ὅλος δύνει, ἡ νὺξ ὡρῶν ιβ΄ γ΄ ι΄ λ΄, ἡ
δ᾽ ἡμέρα ια΄ ∠΄ λ΄. ιζ΄ Ὑάδες ἀκρώνυχ[ο]ι ἐπιτέλλουσιν, ἡ νὺξ ὡρῶν ιβ΄ ∠΄ λ΄. ἐν
ταῖς έ ἡμέραις ταῖς <ἐπ>αγομέναις, δ΄ Ἀρκτοῦρος ἀκρώνυχος δύνει, ἡ νὺξ
ὡρῶν ιβ΄ β΄ ε΄ λ΄ ϙ΄, ἡ δ᾽ ἡμέρα ια΄ ιε΄ με΄, καὶ τῆς Ἴσιος γενέθλια ἔχει [...]

[Choiak]

[1st:]¹³ the night is $13\frac{1}{15} + \frac{1}{45}$ hours,¹⁴ the day $10\frac{2}{3} + \frac{1}{5} + \frac{1}{30} + \frac{1}{90}$.

16th: Arcturus rises acronychally;¹⁵ the night is $12\frac{2}{3} + \frac{1}{15} + \frac{1}{45}$, the day $11\frac{1}{9}$
$+ \frac{1}{10} + \frac{1}{30}$.

---

⁹ Grenfell and Hunt restore [κζ Λύρα ἑώια].

¹⁰ Grenfell and Hunt restore this as [... Ἀρκτοῦρος ἑῶ]ιος [ἐπιτέλλει ...]

¹¹ Grenfell and Hunt restore [κζ Αἲξ ἀκρώνυχος ἐπιτέλλει, ... ἡ νὺξ ὡρῶν ιβ΄ ιε΄ με] in this lacuna.

¹² Grenfell and Hunt restore these lines as [ἄρχεται δύνειν, ἡ νὺξ ὡρῶν ιβ΄ γ΄ με΄, ἡ δ᾽ ἡμέρα ια΄ ∠΄ ι΄ λ΄ ϙ΄].

¹³ Here and elsewhere, the dates are restored by calculation from the length of daylight, assuming a constant increment of $\frac{1}{45}$ hour per day (see Pharmouthi 24).

¹⁴ The text has $13\frac{1}{12} + \frac{1}{45}$ hours. Here as elsewhere the value has been corrected both by comparison with the length of day for the same date, and/or by extrapolating back or forward from other dates, again assuming a constant increment of $\frac{1}{45}$ hour per day.

¹⁵ Meaning in the evening? See Grenfell and Hunt, 1906, pp. 153, n. 56.

26th: Corona rises acronychally, and the bird-bearing north winds blow;[16] the night is $12\frac{1}{2} + \frac{1}{30}$ hours, the day $11\frac{1}{3} + \frac{1}{10} + \frac{1}{30}$. Osiris sails, and the golden boat is brought out.

Tybi

[...]: (The sun is) in Aries.[17]

20th: Vernal equinox; the night is 12 hours, the day 12. The feast of Phi-toroius.

27th: The Pleiades set acronychally; the night is $11\frac{2}{3} + \frac{1}{6} + \frac{1}{90}$ hours, the day $12\frac{1}{10} + \frac{1}{30} + \frac{1}{45}$.

Mecheir

6th: (The sun is) in Taurus. The Hyades set acronychally; the night is $11\frac{1}{2} + \frac{1}{10} + \frac{1}{30} + \frac{1}{90}$ hours,[18] // the day has $12\frac{1}{3} + \frac{1}{45}$. Hera burns, and there is a change in the weather, and the south wind blows. If it is strong, it burns the crops.

19th: Lyra rises acronychally; the night is $11\frac{1}{3} + \frac{1}{15} + \frac{1}{45}$, the day $12\frac{1}{2} + \frac{1}{15} + \frac{1}{90}$.[19] An assembly in Sais for Athena, and the south wind blows. If it is strong, it burns the crops.

[...]th: [...] rises acronychally; [the night is 11 ... hours] the day $12\frac{2}{3}$ [...] they lead [...] //

27th: Lyra sets acronychally; the night is $11\frac{1}{6} + \frac{1}{90}$ hours, the day $12\frac{2}{3} + \frac{1}{10} + \frac{1}{30} + \frac{1}{45}$. Feast of Prometheus, who is called Iphthimis, and the south wind blows. If it is strong, it burns the crops.

Phamenoth

4th: (The sun is) in Gemini. [...] rises; the night is $11\frac{1}{45}$,[20] the day $12\frac{2}{3} + \frac{1}{4} + \frac{1}{20} + \frac{1}{90}$.

5th: Scorpio begins to set in the morning; the night is 11 hours,[21] [the da]y 13.

---

[16] βορέαι ὀρνιθίαι: cf. Geminus, Pisces 2, 14, 21, and 22.
[17] Grenfell and Hunt restore the missing date here as Tybi 5.
[18] The text has $11\frac{1}{2} + \frac{1}{10} + \frac{1}{30} + \frac{1}{5}$.   [19] The text has $12\frac{1}{2} + \frac{1}{15} + \frac{1}{70}$.
[20] The text has $11\frac{1}{5} + \frac{1}{40}$.   [21] The text has 13.

9th: The feast of [. . .] among the E[gyptians].

12th: Scorpio [. . .] sets; the night is $10\frac{2}{3} + \frac{1}{6} + \frac{1}{90}$ hours, [the day 1]$3\frac{1}{10} + \frac{1}{30} + \frac{1}{45}$.[22]

13th: The Pleiades rise [. . .]. //

[. . . (ten lines almost completely lost) . . .]

Pharmou[thi]

[. . .]: (The sun is) in Cancer.[23]

3rd: Aquila rises acronychally; the night is $10\frac{1}{3} + \frac{1}{30} + \frac{1}{90}$ hours, the day $13\frac{1}{2} + \frac{1}{10} + \frac{1}{45}$.[24] //

11th: Delphinus rises ac[ron]ychally; the night is [$10\frac{1}{5}$] hours, [the] day $13\frac{2}{3} + \frac{1}{10} + \frac{1}{30}$,[25] [a]nd Hera's [. . .]

[1]7th: Orion [ri]ses in the mor[ning]; the night is $10\frac{1}{15}$ hours, the [da]y $13\frac{2}{3} + \frac{1}{4} + \frac{1}{60}$.

20th: The night is 10 hours, the day 14, and the sun rises in the same place for three days.

<21st:> The night is 10 hours, the day 14.

22nd: The night is 10 hours, the day 14.

23rd: The night is 10 hours, the day 14.

24th: Summer solstice, and the night becomes greater with respect to the day by $\frac{1}{45}$th of an hour, (where an hour is) the twelfth part (of a day), // so the night is $10\frac{1}{45}$ hours,[26] the day $13\frac{2}{3} + \frac{1}{4} + \frac{1}{20} + \frac{1}{90}$.

25th: The Etesian winds begin to blow, and the river beg[i]ns to rise. The night is $10\frac{1}{30} + \frac{1}{90}$ hours, the day $13\frac{2}{3} + \frac{1}{4} + \frac{1}{30} + \frac{1}{180}$.

Pachon

6th: (The sun is) in Leo. Vindemiatrix rises;[27] the night is $10\frac{1}{4} + \frac{1}{30} + \frac{1}{180}$, the day $\left[13\frac{2}{3} + \frac{1}{30}\right] + \frac{1}{90}$.

9th: The whole of Orion rises; the night is $10\frac{1}{3} + \frac{1}{4[5]}$ [hours, the] day $13\frac{1}{2} + \frac{1}{10} + \frac{1}{30} + \frac{1}{90}$.

---

[22] The text has [1]$3\frac{1}{10} + \frac{1}{30} + \frac{1}{30} + \frac{1}{45}$

[23] Grenfell and Hunt read this line as part of the entry for the third, although it seems to be written separately.

[24] The text has $13\frac{1}{2} + \frac{1}{90} + \frac{1}{45}$.   [25] The text has $13\frac{2}{3} + \frac{1}{11} + \frac{1}{30}$.

[26] The text has $10\frac{1}{5} + \frac{1}{40}$.

[27] Here we have a nice example of why reconstructions are dangerous. Vindemiatrix does not agree with any known parapegma, and yet we find it here, where we would have expected some other star.

18th: Sirius rises; the night is $[10\frac{1}{2} + \frac{1}{30} + \frac{1}{45}]$ hours, // [the day $13 + \frac{1}{3}$] + $\frac{1}{9}$.[28]

Payni

[4th: (The sun is) in Vir]g[o; A]quila [. . .] in the mornin[g; the night is] $10\frac{2}{3} + \frac{1}{5} + \frac{1}{30} + \frac{1}{90}$ hour[s], [the day 1]$3\frac{1}{15} + \frac{1}{4[5]}$.

[16th: . . .] in the morning; [the ni]ght is $11\frac{1}{6} + \frac{1}{90}$ [hour]s, [the] day $12\frac{2}{3} + \frac{1}{10}$ [$+ \frac{1}{30} + \frac{1}{45}$.] The fe[ast] of B[u]bastis.

2[. . .]th: Delphinus sets in the morning; the night is $11\frac{1}{3} + [. . .]$ hours, the day $12\frac{1}{2}$ [$+ . . .$] [. . . f]east [. . .]

[27th: . . .] sets; [the night is] $11\frac{1}{3} + \frac{1}{15} + \frac{1}{45}$ [hour]s,[29] [the day] $12\frac{1}{2} + \frac{1}{15} + \frac{1}{90}$. Feast [. . .]

[. . .] great[30]

[30th: . . . There is a change in the] weather; [the night is] $11\frac{1}{3} + \frac{1}{10} + \frac{1}{30} + \frac{1}{45}$ [hour]s, [the day 12]$\frac{1}{2} + \frac{1}{90}$.

Epiphi

[. . .]

[The sun is in the claws of S]corpio[31]

[. . . mor]ning [. . .] //

the night is 1[1 . . .] hours, the day 12 [. . .], and an assembly of Athena in Sais, and they burn candles throughout the land, and the river changes with respect to its rising.[32]

23rd: Autumnal equinox; the night is 12 hours, the day 12. Feast of Anubis, and the river changes with respect to its rising. //

[(two lines destroyed)]

[. . . 27th: . . . the night is $12\frac{1}{15} + \frac{1}{45}$ hours,] the day $11\frac{2}{3} + \frac{1}{5} + \frac{1}{30} + \frac{1}{90}$.

Mesore

2nd: (The sun is) in Scorpio. The Pleiades rise acronychally; the night is $12\frac{1}{5}$ hours, the day $11\frac{2}{3} + \frac{1}{10} + \frac{1}{30}$. Feast of Apollo.

---

[28] The text has [. . .]$\frac{1}{4}$.

[29] The text has $11\frac{1}{3} + \frac{1}{10} + \frac{1}{45} + \frac{1}{5}$.

[30] It is unclear whether this line is part of the entry for the 27th or the 30th.   [31] I.e., in Libra.

[32] ὁ ποταμὸς ἐπισημαίνει πρὸς τὴν ἀνάβασιν. For an explanation of my translation, see Lehoux, 2004c.

4th: Corona rises in the morning; the night is $12\frac{1}{5} + \frac{1}{30} + \frac{1}{90}$, the day $11\frac{2}{3} + \frac{1}{15} + \frac{1}{45}$.

9th: Scorpio [. . .] acronychally // [three lines lost]

14th: The whole of Scorpio sets; the night is $12\frac{1}{3} + \frac{1}{10} + \frac{1}{30}$ hours, the day $11\frac{1}{2} + \frac{1}{30}$.

17th: The Hyades rise acronychal[l]y; the night is $12\frac{1}{2} + \frac{1}{30}$ hours.[33]

In the five \<ep\>agomenal days

4th: Arcturus sets acronychally; the night is $12\frac{2}{3} + \frac{1}{5} + \frac{1}{30} + \frac{1}{90}$ hours, the day $11\frac{1}{15} + \frac{1}{45}$, and it is the feast of the birth of Isis [. . .]

A.ii. Miletus II[34]

Inv. no. 456 C

```
1        . . . Ε Π Ι] Κ Ρ Α Τ Η Σ Π Υ Λ Ω [Ν Ο Σ . . .                    1
2              . . .]ἡλιακὴν          . . .]όμενον τὸν δ' ἐπιόντα παρα[π]αγῆ      Μ[   2
          . . .]Σ[. . . . .]ΣΤΙΝΑ      ναι τὰς δ' ἡμέρας [ὅτα]ν ὁ μεὶς ΔΙΕ[. . . με-   Ν[
          . . . ζ]ωιδίου [. .]ΔΕ κυκλισ-   τατεθῆνα[ι ε]ὶς [. . . . . . .]ΡΑΦ[.]Ν τῶν   • [. . .
5    μός . . .              . . . ἐκ]άστηι ἀψῖ-   ἡ[μ]ερῶ[ν . . .                   [. . . 5
     δι?. .]ος φερόμενος ΤΑΣ[. . . .]Ν[. . .]   [. . .]Ν[. . . . . . . .]Ν[. . .      [. . .
7      . . .]Υ[. . . . .]ωΝ[. . .          ΜΗΝ[. . .                          • [. . . 7
                                                                              [. . .
                                                                            • [. . . 9
```

Inv. no. 456 D

```
1                                       •                              1
            . . .]Α              ΚΑ[. . .
            . . .]                    [. . .
            . . .]•              • Ὠρίω[ν . . .
5    • . . . . ἀκρώνυ]χος δύνει      κατὰ [. . .                       5
     κατὰ. . . .] καὶ Αἰγυπτίους. •   • Ὑάδε[ς . . .
     • . . . .] νότος πνεῖ κατ' Εὔδοξον   κατὰ [. . .    . . . καὶ
```

[33] Note that no figure for the length of day is given here. Compare the first entry for Pharmouthi. This contrasts with the four days at the summer solstice when only day lengths are given. It should caution us against assuming that both lengths and phases are given in every entry.

[34] Edition in Lehoux, 2005. For photographs, see the catalogue, above.

καὶ Αἰγ]υπτίους, κατὰ δὲ Ἰνδῶν Καλ-
λανέα] Σκο[ρπ]ίος δύνει μετὰ βρον-
10   τῆ]ς καὶ ἀ[ν]έ[μ]ου
      •   •
    • .....]ΕΣ ἀκρώνυχοι ἐπι[τέ]λλουσιν
   .. κατ᾽ Εὔ]δοξον κα[ὶ] Α[ἰγυπτί]ους.
   ....ἑσπέ]ριαι ἐπ[ιτέλ]λουσιν
15   [......................]

Λύρα Ε[...
κατὰ [...
• Ὑάδε[ς...   10
σφόδ[ρα...
• χειμ[αίνει...
• Ὑάδ[ες...
χειμ[αίνει...
• [...   15

Inv. 456 A

1
          ...]ἑσ[πέρ]ας[...   1
     [κα]τ᾽ Εὐκτήμονα. •
     • Αἴξ ἀκρώνυχος δύνει κα[τὰ
     καὶ Φίλιππον καὶ Αἰγυπτί[ους.
5   ... ο]υσιν κατ᾽ ΕΥ-   • Αἴξ ἑσπερία δύνει κατὰ Ἰνδ[ῶν   5
  ... κατὰ Ἰ]νδῶν Καλλανέα   Καλλανέα. •
  ... ἑσπ]έριαι δύνουσιν   • Ἀετὸς ἐπιτέλλει ἑσπέρα[ς
  ... ἐπισ]ημαίνει, χαλάζηι   κατ᾽ Εὐκτήμονα.
  κατὰ ...]• • •   • Ἀρκτοῦρος δύεται ἔωθεν καὶ ἐ[πι-
10 • ...]ΑΣ κρύπτεται ἑσπέρας, χάλαζαι   σημαίνει κατ᾽ Εὐκτήμονα. τῆιδ Ἀ[ε- 10
  ...]οντα[ι] καὶ ζέφυρος ἐπιπνεῖ   τὸς ἐπιτέλλει ἑσπέρας καὶ κ[ατὰ
  ...]ΜΟ[..] κατὰ δὲ Ἰνδῶν   Φίλιππον.
  Καλλανέα ...]

Inv. 456 N

1
         ...]ΑΙ ἐπιση-
  μαίνει ...]Ι κατ᾽ Εὐκτήμονα. τῆι δ᾽ΑΥ   •
  ... κατὰ Φ]ίλιππον. Ἀρκτοῦρος δύεται ...]Ν καὶ ἐπισημαίνει
5   ...]ἐπιτέλλουσιν ἔωθε[ν...
    ...]ΝΕΙ αὐταῖς κατὰ Φίλιππ[ον...
   ...κ]ατ᾽ Εὔδοξον Πλειά[δες
  ἐπιτέ]λλουσιν
    ...]ΙΑΙ ἐπιτέλ[λουσιν...
10   κατ᾽ Ἰνδῶ]ν Καλλα[νέα...

Inv. no. 456 C

Epi]crates, [Son of] Pylo[n

. . .] solar	. . .]ing the following (day?) to peg	M[
. . .]Σ[. . . . .]ΣΤΙΝΑ	the days, [wheneve]r the month[. . .] to	N[
. . . of the z]odiac [. .] and the circular	move [. . . . .] of	• [. . .
motion . . . . . . for e]ach *apsis*	the days [. . . .	[. . .
. . . . . .] bringing the [. . .	[. . .]N[. . . . .]N[. . .	[. . .
. . .]Υ[. . . . .]ωN[. . .	MHN[. . .	• [. . .
		[. . .
		• [. . .

Inv. no. 456 D

	•
. . .]A	KA[. . .
. . .]	[. . .
. . .]•	• Orio[n . . .
• . . . ,]sets [acrony]chally	according to [. . .
according to. . . .] and the Egyptians. •	• The Hyade[s . . .
• . . . .] the south wind blows according to	according to [. . . . . . and
Eudoxus [and the Eg]yptians; and	Lyra E[. . .
according to the Indian Cal[laneus,]	according to [. . .
Scorpio sets with thunder and wind.	• The Hyade[s . . .
• •	very mu[ch . . .
• . . . . .]es r[i]se acronychally	• It is stor[my . . .
. . according to Eu]doxus an[d] the	• The Hyade[s . . .
E[gypti]ans.	It is stor[my . . .
. . . . r[is]es in the [eve]ning.	• [. . .
[. . . . .]	

Inv. no. 456 A

	. . .]ev[en]ing[. . .
	[accordi]ng to Euctemon. •
. . . .]s according to Eu-	• Capella sets acronychally ac[cording
. . . . according to the I]ndian Callaneus	to both Philippus and the Egypti[ans.
. . . .] sets in the [eve]ning	• Capella sets in the evening according to
. . . . there is a ch]ange in the weather,	the Ind[ian] Callaneus. •
with hail, [according to . . .] • • •	• Aquila rises in the evenin[g
• . . .]ας disappears in the evening. It hails	according to Euctemon.
. . . .]οντα[ι] and Zephyrus blows.	• Arcturus sets in the morning and there is
. . . .]MO[. .], and according to the	a cha[nge in the] weather according to
Indian Callaneus . . .	Euctemon. On this day [Aqu-
	ila rises in the evening also, ac[cording to
	Philippus.

Inv. no. 456 N

> . . .]AI there is a change
in the weather . . .]I according to Euctemon, and at the AϒΥ　●
. . . according to Ph]ilippus. And Arcturus se-
ts . . .]N and there is a change in the weather.
(pl.) . . .]rise in the morni[ng . . .
　　. . .]NEI for the same ones, according to Philipp[us . . .
　　. . . acc]ording to Eudoxus. The Pleiades
ri]se
　　. . .]IAI ris[e . . .
According to] Calla[neus] of the India[ns]

A.iii. Geminus parapegma[35]

χρόνοι τῶν ζῳδίων, ἐν οἷς ἕκαστον αὐτῶν ὁ ἥλιος διαπορεύεται, καὶ αἱ
καθ' ἕκαστον ζῴδιον γινόμεναι ἐπισημασίαι, αἵ ὑπογεγραμμέναι εἰσίν.
ἀρξόμεθα δὲ ἀπὸ θερινῆς τροπῆς.

Καρκίνον διαπορεύεται ὁ ἥλιος ἐν ἡμέραις λα´.

α´ ἡμέρᾳ[36] Καλλίππῳ Καρκίνος ἄρχεται ἀνατέλλειν· τροπαὶ θεριναί· καὶ
　　ἐπισημαίνει.

θ´ ἡμέρᾳ Εὐδόξῳ νότος πνεῖ.

ια´ ἡμέρᾳ Εὐδόξῳ Ὠρίων ἑῷος ὅλος ἐπιτέλλει.

ιγ´ ἡμέρᾳ Εὐκτήμονι Ὠρίων ὅλος ἐπιτέλλει.

ις´ Δοσιθέῳ Στέφανος ἑῷος ἄρχεται δύνειν.

κγ´ Δοσιθέῳ ἐν Αἰγύπτῳ Κύων ἐκφανὴς γίνεται.

κε´ Μέτωνι Κύων ἐπιτέλλει ἑῷος.

κζ´ Εὐκτήμονι Κύων ἐπιτέλλει. Εὐδόξῳ Κύων ἑῷος ἐπιτέλλει· καὶ τὰς ἐπομέ-
　　νας ἡμέρας νε´ ἐτησίαι πνέουσιν· αἱ δὲ πέντε αἱ πρῶται πρόδρομοι
　　καλοῦνται. Καλλίππῳ Καρκίνος <λήγει> ἀνατέλλων πνευματώδης.

κη´ Εὐκτήμονι Ἀετὸς ἑῷος δύνει·[37] χειμὼν κατὰ θάλασσαν ἐπιγίνεται.

λ´ Καλλίππῳ Λέων ἄρχεται ἀνατέλλειν· νότος πνεῖ· καὶ Κύων ἀνατέλλων
　　φανερὸς γίνεται.

λα´ Εὐδόξῳ νότος πνεῖ.

---

[35]　Greek text based on Manitius, 1898, with divergences as noted. See also Aujac, 1975;
　　Wachsmuth, 1897.

[36]　Manitius, Wachsmuth, and Aujac emend all the date entries to be of the uniform format 'On
　　the xth day'. I choose to follow the inconsistencies of the MSS.

[37]　Manitius doubts ἑῷος.

τὸν δὲ **Λέοντα** διαπορεύεται ὁ ἥλιος ἐν ἡμέραις λα΄.

ἐν μὲν οὖν τῇ α΄ ἡμέρᾳ Εὐκτήμονι Κύων μὲν ἐκφανής, πνῖγος δὲ ἐπιγίνεται· ἐπισημαίνει.

ἐν δὲ τῇ ε΄ Εὐδόξῳ Ἀετὸς ἑῷος δύνει.

ἐν δὲ τῇ ι΄ ἡμέρᾳ Εὐδόξῳ Στέφανος δύνει.[38]

ἐν δὲ τῇ ιβ΄ Καλλίππῳ Λέων μέσος ἀνατέλλων πνίγη μάλιστα ποιεῖ.

ἐν δὲ τῇ ιδ΄ Εὐκτήμονι πνίγη μάλιστα γίνεται.

ἐν δὲ τῇ ιϛ΄ ἡμέρᾳ Εὐδόξῳ ἐπισημαίνει.

ἐν δὲ τῇ ιζ΄ Εὐκτήμονι Λύρα δύεται· καὶ ἐφύει· καὶ ἐτησίαι παύονται· καὶ Ἵππος ἐπιτέλλει.[39]

ἐν δὲ τῇ ιη΄ Εὐδόξῳ Δελφὶς ἑῷος δύνει. Δοσιθέῳ Προτρυγητὴρ ἀκρόνυχος ἐπιτέλλει.[40]

ἐν δὲ τῇ κβ΄ Εὐδόξῳ Λύρα ἑῷος δύνει· καὶ ἐπισημαίνει.

ἐν δὲ τῇ κθ΄ Εὐδόξῳ ἐπισημαίνει. Καλλίππῳ Παρθένος ἐπιτέλλει· ἐπισημαίνει.

τὴν δὲ **Παρθένον** διαπορεύεται ὁ ἥλιος ἐν ἡμέραις λ΄.

ἐν μὲν οὖν τῇ ε΄ ἡμέρᾳ Εὐδόξῳ ἄνεμος μέγας πνεῖ, καὶ ἐπιβροντᾷ. Καλλίππῳ δὲ οἱ ὦμοι τῆς Παρθένου ἐπιτέλλουσι· καὶ ἐτησίαι λήγουσιν.

ἐν δὲ τῇ ι΄ ἡμέρᾳ Εὐκτήμονι Προτρυγητὴρ φαίνεται· ἐπιτέλλει δὲ καὶ Ἀρκτοῦρος, καὶ Ὀϊστὸς[41] δύεται ὄρθρου·[42] χειμὼν κατὰ θάλασσαν· νότος. Εὐδόξῳ ὑετός, βρονταί· ἄνεμος μέγας πνεῖ.

ἐν δὲ τῇ ιζ΄ Καλλίππῳ Παρθένος μέση ἐπιτέλλουσα ἐπισημαίνει· καὶ Ἀρκτοῦρος ἀνατέλλων φανερός.

ἐν δὲ τῇ ιθ΄ Εὐδόξῳ Ἀρκτοῦρος ἑῷος ἐπιτέλλει· καὶ τὰς ἑπομένας ἡμέρας ἑπτὰ ἄνεμοι πνέουσιν· εὐδία ὡς ἐπὶ τὰ πολλά· λήγοντος δὲ τοῦ χρόνου ἀπ᾽ ἠοῦς πνεῦμα γίνεται.

ἐν δὲ τῇ κ΄ Ἀρκτοῦρος Εὐκτήμονι ἐκφανής· μετοπώρου ἀρχή· καὶ Αἴξ ἐπιτέλλει, ἀστὴρ μέγας ἐπὶ τοῦ Ἡνιόχου·[43] κἄπειτα ἐπισημαίνει· χειμὼν κατὰ θάλασσαν.

ἐν δὲ τῇ κδ΄ ἡμέρᾳ Καλλίππῳ Στάχυς ἐπιτέλλει τῆς Παρθένου· ὕει.

Τὸν δὲ **Ζυγὸν** διαπορεύεται ὁ ἥλιος ἐν ἡμέραις λ΄.

ἐν μὲν οὖν τῇ α΄ ἡμέρᾳ Εὐκτήμονι ἰσημερία μετοπωρινή· καὶ ἐπισημαίνει. Καλλίππῳ ὁ Κριὸς ἄρχεται δύνειν· ἰσημερία μετοπωρινή.

---

[38] Manitius adds <ἑῷος>.   [39] Wachsmuth, following Manitius, adds <ἑσπέριος>.

[40] Following the MSS. Wachsmuth and Manitius correct this to δύνει.

[41] Corrected to 'Pegasus' by Manitius.   [42] ὄρθρου doubted by Manitius.

[43] Following Aujac and the MSS. Wachsmuth, following Manitius, adds <ἑσπέριος> and doubts ἀστὴρ μέγας ἐπὶ τοῦ Ἡνιόχου.

ἐν δὲ τῇ γ´ Εὐκτήμονι Ἔριφοι ἐπιτέλλουσιν ἑσπέριοι· χειμαίνει.

ἐν δὲ τῇ δ´ Εὐδόξῳ Αἴξ ἀκρόνυχος ἐπιτέλλει.

ἐν δὲ τῇ ε´ Εὐκτήμονι Πλειάδες ἑσπέριαι φαίνονται ἐκ τοῦ πρὸς ἕω·[44] ἐπισημαίνει. Καλλίππῳ Παρθένος λήγει ἀνατέλλουσα.

ἐν δὲ τῇ ζ´ ἡμέρᾳ Εὐκτήμονι Στέφανος ἀνατέλλει·[45] χειμαίνει.

ἐν δὲ τῇ η´ Εὐδόξῳ Πλειάδες ἐπιτέλλουσιν.[46]

ἐν δὲ τῇ ι´ Εὐδόξῳ [. . .] ἐῷος ἐπιτέλλει.[47]

ἐν δὲ τῇ ιβ´ ἡμέρᾳ Εὐδόξῳ Σκορπίος ἀκρόνυχος ἄρχεται δύνειν· καὶ χειμὼν
   ἐπιγίνεται, καὶ ἄνεμος μέγας πνεῖ.

ἐν δὲ τῇ ιζ´ Εὐδόξῳ Σκορπίος ἀκρόνυχος ὅλος δύνει. Καλλίππῳ Χηλαὶ
   ἄρχονται ἀνατέλλειν· ἐπισημαίνει.

ἐν δὲ τῇ ιθ´ Εὐδόξῳ βορέαι καὶ νότοι πνέουσιν.

ἐν δὲ τῇ κβ´ Εὐδόξῳ Ὑάδες ἀκρόνυχοι ἐπιτέλλουσιν.

ἐν δὲ τῇ κη´ Καλλίππῳ τοῦ Ταύρου κέρκος δύνει· ἐπισημαίνει.

ἐν δὲ τῇ κθ´ Εὐδόξῳ βορέας καὶ νότος πνέουσιν.

ἐν δὲ τῇ λ´ Εὐκτήμονι χειμὼν κατὰ θάλασσαν πολύς.

τὸν δὲ **Σκορπίον** ὁ ἥλιος διαπορεύεται ἐν ἡμέραις λ´.

ἐν μὲν οὖν τῇ γ´ Δοσιθέῳ χειμαίνει.

ἐν δὲ τῇ δ´ ἡμέρᾳ Δημοκρίτῳ Πλειάδες δύνουσιν ἅμα ᾐοῖ·[48] ἄνεμοι χειμέριοι
   ὡς τὰ πολλά, καὶ ψύχη, ἤδη καὶ πάχνη· ἐπιπνεῖν φιλεῖ· φυλλορροεῖν
   ἄρχεται τὰ δένδρα μάλιστα. Καλλίππῳ τοῦ Σκορπίου τὸ μέτωπον
   ἐπιτέλλει πνευματῶδες.

ἐν δὲ τῇ ε´ Εὐκτήμονι Ἀρκτοῦρος ἑσπέριος δύεται· καὶ ἄνεμοι μεγάλοι
   πνέουσιν.

ἐν δὲ τῇ η´ Εὐδόξῳ Ἀρκτοῦρος ἀκρόνυχος δύνει· καὶ ἐπισημαίνει· καὶ ἄνεμος
   πνεῖ.

ἐν δὲ τῇ θ´ Καλλίππῳ τοῦ Ταύρου δύνει κεφαλή· ὑετοί.

ἐν δὲ τῇ ι´ Εὐκτήμονι Λύρα ἐῷος ἐπιτέλλει·[49] καὶ ἐπιχειμάζεται ὑετῷ.

ἐν δὲ τῇ ιβ´ Εὐδόξῳ Ὠρίων ἀκρόνυχος ἄρχεται ἐπιτέλλειν.

ἐν δὲ τῇ ιγ´ Δημοκρίτῳ Λύρα ἐπιτέλλει ἅμα ἡλίῳ ἀνίσχοντι· καὶ ὁ ἀὴρ
   χειμέριος γίνεται ὡς ἐπὶ τὰ πολλά.

ἐν δὲ τῇ ιδ´ Εὐδόξῳ ὑετία.

---

[44] Manitius excises ἐκ τοῦ πρὸς ἕω.   [45] Manitius has ἐπιτέλλει.

[46] Manitius inserts <ἀκρόνυχοι>

[47] Manitius and Aujac, following Pontedera, insert <Στέφανος> as subject.

[48] ἅμα ᾐοῖ, following the MSS. Manitius, following Unger, emends it to ἅμα ἡλίῳ <ἀνίσχοντι>
   'at the same time as the sun <rises>'. Compare Scorpio 13.

[49] Manitius doubts ἐῷος.

ἐν δὲ τῇ ιε´ Εὐκτήμονι Πλειάδες δύονται· καὶ ἐπισημαίνει· καὶ Ὠρίων ἄρχεται
　　　<δύεσθαι>·[50] καὶ μεσοῦντι καὶ λήγοντι ἐπιχειμάζει.

ἐν δὲ τῇ ιϛ´ Καλλίππῳ <ὁ> ἐν τῷ Σκορπίῳ λαμπρὸς ἀστὴρ ἀνατέλλει·
　　　ἐπισημαίνει· καὶ Πλειάδες δύνουσι φανεραί.

ἐν δὲ τῇ ιη´ Εὐδόξῳ Σκορπίος ἄρχεται ἐπιτέλλειν ἑῷος.

ἐν δὲ τῇ ιθ´ Εὐδόξῳ Πλειάδες ἑῷαι δύνουσι, καὶ Ὠρίων ἄρχεται δύνειν·[51] καὶ
　　　χειμάζει.

ἐν δὲ τῇ κα´ Εὐδόξῳ Λύρα ἑῷος ἐπιτέλλει.

ἐν δὲ τῇ κζ´ Εὐκτήμονι Ὑάδες δύονται· καὶ ἐφύει.

ἐν δὲ τῇ κη´ Καλλίππῳ τοῦ Ταύρου τὰ κέρατα δύεται· ὑετία.

ἐν δὲ τῇ κθ´ Εὐδόξῳ Ὑάδες δύνουσι·[52] καὶ χειμαίνει σφόδρα.

Τὸν δὲ **Τοξότην** ὁ ἥλιος διαπορεύεται ἐν ἡμέραις κθ´.

ἐν μὲν οὖν τῇ ζ´ Εὐκτήμονι Κύων δύεται· καὶ ἐπιχειμάζει. Καλλίππῳ
　　　Τοξότης ἄρχεται ἀνατέλλειν, καὶ Ὠρίων δύνει φανερός· χειμαίνει.

ἐν τῇ η´ Εὐδόξῳ Ὠρίων ἑῷος[53] δύνει.

ἐν δὲ τῇ ι´ Εὐκτήμονι τοῦ Σκορπίου τὸ κέντρον ἐπιτέλλει.

ἐν δὲ τῇ ιβ´ Εὐδόξῳ Κύων ἑῷος δύνει· χειμαίνει.

ἐν δὲ τῇ ιδ´ Εὐδόξῳ ὑετός.

ἐν δὲ τῇ ιε´ Εὐκτήμονι Ἀετὸς ἐπιτέλλει· νότος πνεῖ.

ἐν δὲ τῇ ιϛ´ Δημοκρίτῳ Ἀετὸς ἐπιτέλλει ἅμα ἡλίῳ· καὶ ἐπισημαίνειν φιλεῖ
　　　βροντῇ καὶ ἀστραπῇ καὶ ὕδατι ἢ ἀνέμῳ ἢ ἀμφότερα ὡς ἐπὶ τὰ
　　　πολλά. Εὐδόξῳ Κύων ἀκρόνυχος ἐπιτέλλει· νότια.[54] Καλλίππῳ Δίδυ-
　　　μοι μετίασι[55] δυόμενοι· νότια.

ἐν δὲ τῇ ιθ´ Εὐδόξῳ [. . .] δύνει.[56]

ἐν δὲ τῇ κα´ Εὐδόξῳ Σκορπίος ἑῷος[57] ἐπιτέλλει· καὶ χειμαίνει.

ἐν δὲ τῇ κγ´ Εὐδόξῳ Αἲξ ἑῴα δύνει.

ἐν δὲ τῇ κϛ´ Εὐδόξῳ Ἀετὸς ἑῷος ἐπιτέλλει.

Τὸν δὲ **Αἰγόκερων** ὁ ἥλιος διαπορεύεται ἐν ἡμέραις κθ´.

ἐν μὲν οὖν τῇ α´ ἡμέρα Εὐκτήμονι τροπαὶ χειμεριναί· ἐπισημαίνει. Καλλίπ-
　　　πῳ Τοξότης λήγει ἀνατέλλων· τροπαὶ χειμεριναί· χειμαίνει.

ἐν δὲ τῇ β´ Εὐκτήμονι Δελφὶς ἐπιτέλλει· χειμαίνει.

ἐν δὲ τῇ δ´ Εὐδόξῳ τροπαὶ χειμεριναί· χειμαίνει.

---

[50] <δύεσθαι· καὶ ἀρχομένῳ> added by Wachsmuth, Manitius, and Aujac, following Böckh.

[51] Manitius adds <ἑῷος>.　　　[52] Manitius adds <ἑῷαι>.　　　[53] Manitius adds <ὅλος>.

[54] νότια· Καλίππῳ . . . , following Aujac. Manitius emends νότια to ὑετία. Wachsmuth, following
　　the MSS, has ὅτι Καλίππῳ.

[55] Manitius emends this to μεσοῦσι.

[56] Manitius and Aujac, following Pontedera, emend this to Εὐκτήμονι Αἲξ δύεται.

[57] Manitius and Wachsmuth, following Pontedera, add <ὅλος>.

ἐν δὲ τῇ ζ΄ Εὐκτήμονι Ἀετὸς ἑσπέριος δύεται· καὶ χειμαίνει.

ἐν δὲ τῇ θ΄ Εὐδόξῳ Στέφανος ἀκρόνυχος δύνει.

ἐν δὲ τῇ ιβ΄ Δημοκρίτῳ νότος πνεῖ ὡς <ἐπὶ τὰ πολλά . . . > ἐπιτέλλει.[58]

ἐν δὲ τῇ ιδ΄ Εὐκτήμονι μέσος χειμών· νότος πολὺς ἐπιπνεῖ χειμερινὸς κατὰ θάλασσαν.

ἐν δὲ τῇ ιε΄ Καλλίππῳ Αἰγόκερως ἄρχεται ἀνατέλλειν· νότος.

ἐν δὲ τῇ ις΄ Εὐκτήμονι νότος χειμέριος κατὰ θάλασσαν.

ἐν δὲ τῇ ιη΄ ἀκρόνυχος ἐπιδύνει [. . .]·[59] καὶ νότος πνεῖ.

ἐν δὲ τῇ κζ΄ Εὐκτήμονι Δελφὶς ἑσπέριος δύνει. Καλλίππῳ Καρκίνος λήγει δύνων· χειμαίνει.

Τὸν δὲ Ὑδροχόον διαπορεύεται ὁ ἥλιος ἐν ἡμέραις λ΄.

ἐν μὲν οὖν τῇ β΄ Καλλίππῳ Λέων ἄρχεται δύνειν· ὑετία.

ἐν δὲ τῇ γ΄ Εὐκτήμονι Λύρα ἑσπέριος δύνει· ὑετία. Δημοκρίτῳ χειμών.

ἐν δὲ τῇ δ΄ ἡμέρᾳ Εὐδόξῳ Δελφὶς ἀκρόνυχος δύνει.

ἐν δὲ τῇ ια΄ Εὐδόξῳ Λύρα ἀκρόνυχος δύνει· ὑετός.

ἐν δὲ τῇ ιδ΄ Εὐδόξῳ εὐδία· ἐνίοτε καὶ ζέφυρος πνεῖ.

ἐν δὲ τῇ ις΄ Δημοκρίτῳ ζέφυρος πνεῖν ἄρχεται καὶ παραμένει[60] ἡμέραις γ΄ καὶ μ΄ ἀπὸ τροπῶν.

ἐν δὲ τῇ ιζ΄ Εὐκτήμονι ζέφυρον ὥρα πνεῖν. Καλλίππῳ Ὑδροχόος μέσος ἀνατέλλει· ζέφυρος πνεῖ.

ἐν δὲ τῇ κε΄ Εὐκτήμονι [. . .][61] ἑσπέριος δύνει· καὶ σφόδρα ἐπιχειμάζει.

Τοὺς δὲ Ἰχθύας ὁ ἥλιος διαπορεύεται ἐν ἡμέραις λ΄.

ἐν μὲν οὖν τῇ β΄ χελιδόνα ὥρα φαίνεσθαι·[62] καὶ ὀρνιθίαι πνέουσι. Καλλίππῳ δὲ Λέων δύνων λήγει· καὶ χελιδὼν φαίνεται· ἐπισημαίνει.

ἐν δὲ τῇ δ΄ Δημοκρίτῳ ποικίλαι ἡμέραι γίνονται, ἀλκυονίδες καλούμεναι. Εὐδόξῳ δὲ Ἀρκτοῦρος ἀκρόνυχος ἐπιτέλλει· καὶ ὑετὸς γίνεται· καὶ χελιδὼν φαίνεται· καὶ τὰς ἑπομένας ἡμέρας λ΄ βορέαι πνέουσι, καὶ μάλιστα οἱ προορνιθίαι καλούμενοι.

---

[58] Most MSS have ὡς ἐπιτέλλει, which Manitius and Aujac, following Wachsmuth, emend to ὡς <ἐπὶ τὰ πολλά· Εὐδόξῳ Δελφὶς ἑῷος> ἐπιτέλλει. I agree that something seems to be missing, but I do not share their confidence about what it might be. For the problems with reconstructing parapegmata, see Lehoux, forthcoming. Halma deletes ὡς ἐπιτέλλει altogether.

[59] The MSS have ἐπιδύνει ὁ περσεὺς, with no attribution. Aujac follows Manitius in deleting Perseus and restoring '<According to Eudoxus, Aquila> sets acronychally.' Wachsmuth seems to indicate a lacuna after the date.

[60] Aujac, following Manitius, doubts καὶ παραμένει.

[61] The MSS have no star name here, but Aujac, following Wachsmuth, has proposed Sagitta, and Manitius Pegasus.

[62] Wachsmuth adds 'according to Meton'. Manitius, following Böckh, adds 'according to Euctemon'.

ἐν δὲ τῇ ιβ΄ Εὐκτήμονι Ἀρκτοῦρος ἑσπέριος ἐπιτέλλει, καὶ Προτρυγητὴρ
    ἐκφανής· ἐπιπνεῖ βορέας ψυχρός.

ἐν δὲ τῇ ιδ΄ Δημοκρίτῳ ἄνεμοι πνέουσι ψυχροί, οἱ ὀρνιθίαι καλούμενοι,
    ἡμέρας μάλιστα ἐννέα. Εὐκτήμονι δὲ Ἵππος ἑῷος δύνει·[63] ἐπιπνεῖ
    βορέας ψυχρός.

ἐν δὲ τῇ ιζ΄ Εὐδόξῳ χειμαίνει· καὶ ἰκτῖνος φαίνεται. Καλλίππῳ τῶν Ἰχθύων
    ὁ νότιος ἐπιτέλλει· λήγει βορέας.[64]

ἐν δὲ τῇ κα΄ Εὐδόξῳ Στέφανος ἀκρόνυχος ἐπιτέλλει. ἄρχονται ὀρνιθίαι
    πνέοντες.

ἐν δὲ τῇ κβ΄ Εὐκτήμονι ἰκτῖνος φαίνεται· ὀρνιθίαι πνέουσι μέχρις ἰσημερίας.

ἐν δὲ τῇ κθ΄ Εὐκτήμονι τοῦ Σκορπίου οἱ πρῶτοι ἀστέρες δύνουσιν· ἐπιπνεῖ
    βορέας ψυχρός.

ἐν δὲ τῇ λ΄ Καλλίππῳ τῶν Ἰχθύων ὁ βόρειος[65] ἐπιτέλλων λήγει· ἰκτῖνος
    φαίνεται· βορέας πνεῖ.

Τὸν δὲ **Κριὸν** διαπορεύεται ὁ ἥλιος ἐν ἡμέραις λα΄.

ἐν μὲν τῇ α΄ Καλλίππῳ σύνδεσμος τῶν Ἰχθύων ἀνατέλλει· ἰσημερία ἐαρινή.
    Εὐκτήμονι ἰσημερία, καὶ ψεκὰς λεπτή· χειμαίνει σφόδρα· ἐπισημαίνει.

ἐν δὲ τῇ γ΄ Καλλίππῳ Κριὸς ἄρχεται ἐπιτέλλειν· ὑετὸς ἢ νιφετός.

ἐν δὲ τῇ ϛ΄ Εὐδόξῳ ἰσημερία· ὑετὸς γίνεται.

ἐν δὲ τῇ ι΄ Εὐκτήμονι Πλειάδες ἑσπέριοι κρύπτονται.

ἐν δὲ τῇ ιγ΄ Εὐδόξῳ Πλειάδες ἀκρόνυχοι δύνουσι, καὶ Ὠρίων ἄρχεται δύνειν
    ἀπὸ ἀκρονύχου· ὑετὸς γίνεται.[66] Δημοκρίτῳ Πλειάδες κρύπτονται
    ἅμα ἡλίῳ ἀνίσχοντι[67] καὶ ἀφανεῖς γίνονται νύκτας μ΄.

ἐν δὲ τῇ κα΄ Εὐδόξῳ Ὑάδες ἀκρόνυχοι δύνουσιν.

ἐν δὲ τῇ κγ΄ Εὐκτήμονι Ὑάδες κρύπτονται· καὶ χάλαζα ἐπιγίνεται, καὶ
    ζέφυρος πνεῖ. Καλλίππῳ Χηλαὶ ἄρχονται δύνειν· πολλαχῇ δὲ καὶ
    χάλαζα.[68]

ἐν δὲ τῇ κζ΄ Εὐδόξῳ Λύρα ἀκρόνυχος ἐπιτέλλει.

Τὸν δὲ **Ταῦρον** διαπορεύεται ὁ ἥλιος ἐν ἡμέραις λβ΄.

ἐν μὲν οὖν τῇ α΄ ἡμέρα Εὐδόξῳ Ὠρίων ἀκρόνυχος δύνει·[69] ὑετία. Καλλίππῳ
    ὁ Κριὸς λήγει ἐπιτέλλων· ὑετία, πολλαχῇ δὲ καὶ χάλαζα.

---

[63] Following the MSS. Manitius and Aujac, following Pontedera emend ἑῷος δύνει to ἐπιτέλλει.
    Pfaff reads ἑσπέριος δύνει.
[64] Wachsmuth reads τῶν ἰχθύων ὁ νότιος ἐπιτέλλων λήγει, βορέας.
[65] Emended to νότιος by Manitius.
[66] Wachsmuth punctuates it thus: καὶ ὠρίων ἄρχεται δύνειν· ἀπὸ ἀκρονύχου ὑετὸς γίνεται.
[67] Following the MSS. Manitius and Aujac, following Wachsmuth, emend it to ἅμα ἡλίῳ δύνοντι.
[68] Manitius, following Böckh, adds <ὑετία> πολλαχῇ δὲ καὶ χάλαζα.
[69] Manitius adds ὠρίων ἀκρόνυχος <ὅλος> δύνει.

ἐν δὲ τῇ β΄ [70] Εὐκτήμονι Κύων κρύπτεται· καὶ χάλαζα γίνεται.

τῇ δ΄ <Εὐκτήμονι>[71] Λύρα ἐπιτέλλει.[72] Εὐδόξῳ Κύων ἀκρόνυχος δύνει· καὶ ὑετὸς γίνεται. Καλλίππῳ τοῦ Ταύρου ἡ κέρκος ἐπιτέλλει· νότια.

ἐν δὲ τῇ ζ΄ Εὐδόξῳ ὑετὸς γίνεται.

ἐν δὲ τῇ η΄ Εὐκτήμονι Αἲξ ἑῴα[73] ἐπιτέλλει· εὐδία· ὕει νοτίῳ ὕδατι.

ἐν δὲ τῇ θ΄ Εὐδόξῳ Αἲξ ἑῴα ἐπιτέλλει.

ἐν δὲ τῇ ια΄ Εὐδόξῳ Σκορπίος ἑῷος δύνειν ἄρχεται· καὶ ὑετὸς γίνεται.

ἐν δὲ τῇ ιγ΄ Εὐκτήμονι Πλειὰς ἐπιτέλλει· θέρους ἀρχή· καὶ ἐπισημαίνει. Καλλίππῳ ἡ τοῦ Ταύρου κεφαλὴ ἐπιτέλλει· ἐπισημαίνει.

ἐν δὲ τῇ κα΄ Εὐδόξῳ Σκορπίος ἑῷος ὅλος δύνει.

ἐν δὲ τῇ κβ΄ Εὐδόξῳ Πλειάδες ἐπιτέλλουσι· καὶ ἐπισημαίνει.

ἐν δὲ τῇ κε΄ Εὐκτήμονι Ἀετὸς[74] ἑσπέριος δύνει.

ἐν δὲ τῇ λ΄ Εὐκτήμονι [. . .] ἑσπέριος ἐπιτέλλει.[75]

ἐν δὲ τῇ λα΄ Εὐκτήμονι Ἀετὸς ἑσπέριος ἐπιτέλλει.

ἐν δὲ τῇ λβ΄ Εὐκτήμονι Ἀρκτοῦρος ἑῷος δύνει·[76] ἐπισημαίνει. Καλλίππῳ ὁ ταῦρος λήγει ἀνατέλλων. Εὐκτήμονι Ὑάδες ἑῷαι ἐπιτέλλουσιν·[77] ἐπισημαίνει.

Τοὺς δὲ **Διδύμους** ὁ ἥλιος διαπορεύεται ἐν ἡμέραις λβ΄.

ἐν μὲν οὖν τῇ β΄ Καλλίππῳ οἱ Δίδυμοι ἄρχονται ἐπιτέλλειν· νότια.

ἐν δὲ τῇ ε΄ Εὐδόξῳ Ὑάδες ἑῷαι ἐπιτέλλουσιν.

ἐν δὲ τῇ ζ΄ Εὐδόξῳ Ἀετὸς ἀκρόνυχος ἐπιτέλλει.

ἐν δὲ τῇ ι΄ Δημοκρίτῳ ὕδωρ γίνεται.

ἐν δὲ τῇ ιγ΄ Εὐδόξῳ Ἀρκτοῦρος ἑῷος δύνει.

ἐν δὲ τῇ ιη΄ Εὐδόξῳ Δελφὶς ἀκρόνυχος ἐπιτέλλει.

ἐν δὲ τῇ κδ΄ Εὐκτήμονι Ὠρίωνος ὦμος ἐπιτέλλει. Εὐδόξῳ Ὠρίων ἄρχεται ἐπιτέλλειν.[78]

ἐν δὲ τῇ κθ΄ Δημοκρίτῳ ἄρχεται Ὠρίων ἐπιτέλλειν, καὶ φιλεῖ ἐπισημαίνειν ἐπ᾽ αὐτῷ.

---

[70] Following Aujac and the MSS. Manitius and Wachsmuth follow Unger's correction of the date to the 4th.

[71] Aujac and the MSS have τῇ δ᾽ αὐτῇ, 'On the same (day)', treating this entry as part of the previous one. Manitius, following Wachsmuth, emends this to τῷ δ᾽ αὐτῷ, 'according to him (i.e., Euctemon) also'. I follow Böckh tentatively here.

[72] Manitius adds <ἑσπέριος>.   [73] Manitius doubts ἑῴα.

[74] Corrected by Wachsmuth and Manitius to 'Capella'. It does seem unlikely that, according to Euctemon, Aquila should be setting here and rising only six days later. Aujac deletes the entire entries for the 25th and 30th.

[75] Manitius adds <Ὀιστὸς> as subject.   [76] Manitius doubts the ἑῷος.

[77] Manitius doubts the ἑῷαι.   [78] Manitius adds <ἑῷος>.

The times of the signs of the zodiac, in which the sun traverses each of them, and the changes in the weather occurring in each sign, are here appended. We begin from the summer solstice.

The sun traverses **Cancer** in thirty-one days.

The 1st day: According to Callippus Cancer begins to rise, summer solstice, and there is a change in the weather.
9th day: According to Eudoxus the south wind blows.
11th day: According to Eudoxus the whole of Orion rises in the morning.
13th day: According to Euctemon the whole of Orion rises.
16th: According to Dositheus Corona begins to set in the morning.
23rd: According to Dositheus Sirius is visible in Egypt.
25th: According to Meton Sirius rises in the morning.
27th: According to Euctemon Sirius rises. According to Eudoxus Sirius rises in the morning, and for the next fifty-five days the Etesian winds blow. The first five (days, the winds) are called the *Prodromoi*. According to Callippus Cancer <finishes> rising, windy.
28th: According to Euctemon Aquila sets in the morning, a storm at sea follows.
30th: According to Callippus Leo begins to rise, the south wind blows, and Sirius is rising visibly.[79]
31st: According to Eudoxus the south wind blows.

The sun traverses **Leo** in thirty-one days.

On the 1st day: According to Euctemon Sirius is visible, stifling heat follows, there is a change in the weather.
On the 5th: According to Eudoxus Aquila sets in the morning.
On the 10th day: According to Eudoxus Corona sets.
On the 12th: According to Callippus the rising of the middle of Leo causes really stifling heat.
On the 14th: According to Euctemon there is really stifling heat.
On the 16th day: According to Eudoxus there is a change in the weather.
On the 17th: According to Euctemon Lyra sets, and rain falls, and the Etesian winds stop, and Pegasus rises.
On the 18th: According to Eudoxus Delphinus sets in the morning. According to Dositheus Vindemiatrix rises acronychally.
On the 22nd: According to Eudoxus Lyra sets in the morning, and there is a change in the weather.

---

[79] Meaning it has its apparent rising (as opposed to its 'true' rising).

On the 29th: According to Eudoxus there is a change in the weather. According to Callippus Virgo rises, there is a change in the weather.

The sun traverses **Virgo** in thirty days.

On the 5th day: According to Eudoxus a strong wind blows and it thunders. According to Callippus the shoulders of Virgo rise, and the Etesian winds stop.

On the 10th day: According to Euctemon Vindemiatrix appears, Arcturus rises, and Sagitta sets at dawn, storm at sea, south wind. According to Eudoxus rain, thundery, a strong wind blows.

On the 17th: According to Callippus the rising of the middle of Virgo changes the weather, and Arcturus is rising visibly.

On the 19th: According to Eudoxus Arcturus rises in the morning, and winds blow for the next seven days. Generally nice weather. At the end of this time wind is from the east.

On the 20th: Arcturus is visible according to Euctemon. Beginning of autumn, and Capella (the bright star in Auriga) rises, and after that there is a change in the weather, storm at sea.

On the 24th day: According to Callippus Spica in Virgo rises, it rains.

The sun traverses **Libra** in thirty days.

On the 1st day: According to Euctemon the autumnal equinox, and there is a change in the weather. According to Callippus Aries begins to set, autumnal equinox.

On the 3rd: According to Euctemon the Haedi (the kids) rise in the evening, it is stormy.

On the 4th: According to Eudoxus Capella rises acronychally.

On the 5th: According to Euctemon the Pleiades appear in the evening; from then until dawn there is a change in the weather. According to Callippus Virgo finishes rising.

On the 7th day: According to Euctemon Corona rises, it is stormy.

On the 8th: According to Eudoxus the Pleiades rise.

On the 10th: According to Eudoxus [. . .] rises in the morning.

On the 12th day: According to Eudoxus Scorpio begins to set acronychally, and a storm follows, and a strong wind blows.

On the 17th: According to Eudoxus the whole of Scorpio sets acronychally. According to Callippus the claws (of Scorpio)[80] begin to rise, there is a change in the weather.

---

[80] Meaning Libra.

On the 19th: According to Eudoxus the north winds and the south winds blow.

On the 22nd: According to Eudoxus the Hyades rise acronychally.

On the 28th: According to Callippus the tail of Taurus sets, there is a change in the weather.

On the 29th: According to Eudoxus the north wind and the south wind blow.

On the 30th: According to Euctemon a great storm at sea.

The sun traverses **Scorpio** in thirty days.

On the 3rd: According to Dositheus it is stormy.

On the 4th day: According to Democritus the Pleiades set at the same time as daybreak, wintry winds for the most part, and cold, there is now frost, it tends to be windy, trees really begin to lose their leaves. According to Callippus the forehead of Scorpio rises, blustery.

On the 5th: According to Euctemon Arcturus sets in the evening, and strong winds blow.

On the 8th: According to Eudoxus Arcturus sets acronychally, and there is a change in the weather, and the wind blows.

On the 9th: According to Callippus the head of Taurus sets, rainy.

On the 10th: According to Euctemon Lyra rises in the morning, and it is very stormy with rain.

On the 12th: According to Eudoxus Orion begins to rise acronychally.

On the 13th: According to Democritus Lyra rises at the same time as the sun comes up, and the air becomes stormy for the most part.

On the 14th: According to Eudoxus rainy.

On the 15th: According to Euctemon the Pleiades set, and there is a change in the weather, and Orion begins <to set> and in the middle and end it is stormy.

On the 16th: According to Callippus the bright star in Scorpio rises, there is a change in the weather, and the Pleiades set visibly.

On the 18th: According to Eudoxus Scorpio begins to rise in the morning.

On the 19th: According to Eudoxus the Pleiades set in the morning and Orion begins to set, and it is stormy.

On the 21st: According to Eudoxus Lyra rises in the morning.

On the 27th: According to Euctemon the Hyades set, and rain falls.

On the 28th: According to Callippus the horns of Taurus set, rain.

On the 29th: According to Eudoxus the Hyades set, and it is very stormy.

The sun traverses **Sagittarius** in twenty-nine days.

On the 7th: According to Euctemon Sirius sets, and it is very stormy. According to Callippus Sagittarius begins to rise and Orion sets visibly, it is stormy.

On the 8th: According to Eudoxus Orion sets in the morning.

On the 10th: According to Euctemon the centre of Scorpio rises.

On the 12th: According to Eudoxus Sirius sets in the morning, it is stormy.

On the 14th: Rain according to Eudoxus.

On the 15th: According to Euctemon Aquila rises, the south wind blows.

On the 16th: According to Democritus Aquila rises at the same time as the sun, and there tends to be a change in the weather with thunder, lightning, and either rain or wind, or usually both. According to Eudoxus Sirius rises acronychally, south winds. According to Callippus Gemini finishes setting, south winds.

On the 19th: According to Eudoxus [. . .] sets.

On the 21st: According to Eudoxus Scorpio rises in the morning, and it is stormy.

On the 23rd: According to Eudoxus Capella sets in the morning.

On the 26th: According to Eudoxus Aquila rises in the morning.

The sun traverses **Capricorn** in twenty-nine days.

On the 1st day: The winter solstice according to Euctemon, there is a change in the weather. According to Callippus Sagittarius finishes rising, winter solstice, it is stormy.

On the 2nd: According to Euctemon Delphinus rises, it is stormy.

On the 4th: Winter solstice according to Eudoxus, it is stormy.

On the 7th: According to Euctemon Aquila sets in the evening, and it is stormy.

On the 9th: According to Eudoxus Corona sets acronychally.

On the 12th: According to Democritus the south wind <usually> blows. [. . .] rises.

On the 14th: According to Euctemon the middle of winter, a strong south wind blows stormily at sea.

On the 15th: According to Callippus Capricorn begins to rise, south wind.

On the 16th: According to Euctemon a stormy south wind at sea.

On the 18th: [. . .] sets acronychally, and the south wind blows.

On the 27th: According to Euctemon Delphinus sets in the evening. According to Callippus Cancer finishes setting, it is stormy.

The sun traverses **Aquarius** in thirty days.

On the 2nd: According to Callippus Leo begins to set, rainy.

On the 3rd: According to Euctemon Lyra sets in the evening, rainy. Stormy according to Democritus.

On the 4th day: According to Eudoxus Delphinus sets acronychally.

On the 11th: According to Eudoxus Lyra sets acronychally, rain.

On the 14th: A nice day according to Eudoxus, sometimes the west wind also blows.

On the 16th: According to Democritus the west wind begins to blow forty-three days after the solstice, and it continues (from this date?).

On the 17th: According to Euctemon the season when the west wind blows. According to Callippus the middle of Aquarius rises, the west wind blows.

On the 25th: According to Euctemon [. . .] sets in the evening, and it is extremely stormy.

The sun traverses **Pisces** in thirty days.

On the 2nd: It is the season of the appearance of the swallow, and the bird-bringing winds blow. According to Callippus Leo finishes setting, and the swallow appears, there is a change in the weather.

On the 4th: According to Democritus the variable days, called the Halcyon (days) begin. According to Eudoxus Arcturus rises acronychally, and it rains, and the swallow appears, and the north winds blow for the next thirty days, especially the winds called the 'bird-bringers'.

On the 12th: According to Euctemon Arcturus rises in the evening, and Vindemiatrix is visible, the cold north wind blows.

On the 14th: According to Democritus cold winds called the 'bird-bringers' blow, for nine days especially. According to Euctemon Pegasus sets in the morning, the cold north wind blows.

On the 17th: According to Eudoxus it is stormy, and the kite appears. According to Callippus the southern fish rises, the north wind ends.

On the 21st: According to Eudoxus Corona rises acronychally, the bird-bringing winds begin to blow.

On the 22nd: According to Euctemon the kite appears, the bird-bringing winds blow until the equinox.

On the 29th: According to Euctemon the first stars in Scorpio set, the cold north wind blows.

On the 30th: According to Callippus the northern fish finishes rising, the kite appears, the north wind blows.

The sun traverses **Aries** in thirty-one days.

On the 1st: According to Callippus the binding (of the fish) in Pisces rises, vernal equinox. Equinox according to Euctemon, and light drizzle, it is very stormy, there is a change in the weather.

On the 3rd: According to Callippus Aries begins to rise, rain or snow.

On the 6th: Equinox according to Eudoxus, there is rain.

On the 10th: According to Euctemon the Pleiades disappear.

On the 13th: According to Eudoxus the Pleiades set acronychally and Orion begins to set after nightfall, there is rain. According to Democritus the Pleiades disappear at the same time as the sun rises and are invisible for forty nights.

On the 21st: According to Eudoxus the Hyades set acronychally.

On the 23rd: According to Euctemon the Hyades disappear, and hail follows, and the west wind blows. According to Callippus the claws begin setting, and a lot of hail.

On the 27th: According to Eudoxus Lyra rises acronychally.

The sun traverses **Taurus** in thirty-two days.

On the 1st day: According to Eudoxus Orion sets acronychally, rainy. According to Callippus Aries finishes rising, rainy, and a lot of hail.

On the 2nd: According to Euctemon Sirius disappears, and there is hail.

The 4th: <According to Euctemon?> Lyra rises. According to Eudoxus Sirius sets acronychally, and there is rain. According to Callippus the tail of Taurus rises, south winds.

On the 7th: There is rain according to Eudoxus.

On the 8th: According to Euctemon Capella rises in the morning, good weather, it rains with a wet south wind.

On the 9th: According to Eudoxus Capella rises in the morning.

On the 11th: According to Eudoxus Scorpio begins to set in the morning, and there is rain.

On the 13th: According to Euctemon the Pleiades rise, summer begins, and there is a change in the weather. According to Callippus the head of Taurus rises, there is a change in the weather.

On the 21st: According to Eudoxus the whole of Scorpio sets in the morning.

On the 22nd: According to Eudoxus the Pleiades rise and there is a change in the weather.

On the 25th: According to Euctemon Aquila sets in the evening.

On the 30th: According to Euctemon [. . .] rises in the evening.[81]

---

[81] Manitius adds '<Sagitta> rises in the evening.'

On the 31st: According to Euctemon Aquila rises in the evening.

On the 32nd: According to Euctemon Arcturus sets in the morning, there is a change in the weather. According to Callippus Taurus finishes rising. According to Euctemon the Hyades rise in the morning, there is a change in the weather.

The sun traverses **Gemini** in thirty-two days.

On the 2nd: According to Callippus Gemini begins to rise, south winds.

On the 5th: According to Eudoxus the Hyades rise in the morning.

On the 7th: According to Eudoxus Aquila rises acronychally.

On the 10th: According to Democritus it is wet.

On the 13th: According to Eudoxus Arcturus sets in the morning.

On the 18th: According to Eudoxus Delphinus rises acronychally

On the 24th: According to Euctemon the shoulder of Orion rises. According to Eudoxus Orion begins to rise.[82]

On the 29th: According to Democritus Orion begins to rise, and the weather tends to change after that.

A.iv. Puteoli parapegma (see p. 158).

A.v. Ovid's *Fasti* (astrometeorological excerpts)[83]

I will sing of time marked by the Latin year, and its causes, and of the constellations rising and dipping below the earth (1.1–2).

January

3rd day before the Nones of Jan. (1.311–14): Thus when the third night before the Nones comes and the soil is damp, spread with the sky's dew, you will search in vain for the arms of the eight-footed crab. Diving headfirst it will go beneath the waters.

---

[82] Manitius adds 'in the morning'.

[83] Latin text in Wormell, Alton, and Courtney, 1997. I am not entirely happy with excerpting the parapegmatic information in this way, but I do so on the grounds that a complete translation of Ovid would be too voluminous, and excerpts such as these will be useful for comparative purposes in a book like this. Among other things, the many indirect allusions in the poem to stellar phenomena and weather will be lost in such an exercise, as will a good deal of important contextual material.

Nones of Jan. (1.315–16): When the Nones approach, rains sent to you from dark clouds signal the rising of Lyra.

5th day before the Ides of Jan. (1.457–8): Meanwhile the bright constellation Delphinus is elevated above the horizon, raising its face above its watery home.

4th day before the Ides of Jan. (1.459): The next dawn marks the midpoint of winter, and what has passed will be equal to what remains.

16th day before the Kalends of Feb. (1.651–2): When this has passed you will run, Phoebus, leaving Capricorn behind, through the sign of the water-carrying youth.

10th day before the Kalends of Feb. (1.653–4): When from then the seventh sun has set in the waves, Lyra will no longer shine in any part of the sky.

9th day before the Kalends of Feb. (1.655–6): From that constellation, the fire that beats in the middle of Leo's breast will be buried at nightfall.

February

4th day before the Nones of Feb. (11.73–8): When the next Sun, about to set in the western waves, removes the jewelled collar from his purple horses, anyone raising his eyes to the stars that night will ask, 'where, today, is Lyra, which shone yesterday?' And while he looks for Lyra, he will see that the back of Leo has also gone, submerged beneath the churning water.

3rd day before the Nones of Feb. (11.79–80): Delphinus, which you saw thus decorated with stars, will flee your sight on the next night.

Nones of Feb. (11.145–8): Now Aquarius shines down to the middle of its belly, and pours flowing water mixed with nectar. If anyone is afraid of the north wind, let him be glad: a softer breeze comes from the west.

5th day before the Ides of Feb. (11.149–52): On the fifth day, when shining Venus has raised its light from the ocean waves, this is also the beginning of spring. But do not be mistaken, cold weather remains for you. Cold remains: winter, departing, leaves large footprints behind.

3rd day before the Ides of Feb. (11.153–4): When the third night arrives, you will immediately see that Boötes has thrust out his two feet.

16th day before the Kalends of March (11.243–6): Three constellations are in one area: Corvus, Draco, and Crater is in the middle. On the Ides they are hidden. They rise the next night.

15th day before the Kalends of March (11.453–8): When this day has risen, cease trusting the winds. At that season the winds lose reliability. Their blowing is inconstant, and for six days the door of the Aeolian prison stands wide open. Then light Aquarius sinks with his tilted urn, and you, Pisces, receive the aetherial horses.

March

5th day before the Nones of March (iii.399–400): When the third night of the month has brought about its arrival, one of the twin Fishes will have fled.

3rd day before the Nones of March (iii.403–7): When from golden cheeks Dawn has begun to spread dew, and the sun shines for the fifth time (in the month), Arctophylax – or lazy Boötes if you prefer – will have set, and will evade your sight. But Vindemiatrix will not be hidden.

Nones of March (iii.449–50): And when the stars colour the blue sky, look up, you will see the neck of Pegasus.

8th day before the Ides of March (iii.459): Right at the onset of night you will see Corona.

17th day before the Kalends of April (iii.711–12): The next day, when dawn has renewed the soft grass, the first part of Scorpio will be visible.

16th day before the Kalends of April (iii.793–4): The kite's star leans down to the Lycaonian Bear, and this night it becomes visible.

10th day before the Kalends of April (iii.851–2): Now, with face raised to the sun you can say, 'yesterday he reached the fleece of Aries'.

7th day before the Kalends of April (iii.877–8): When Venus has three times sent forth the dawn, you count the length of day equal to that of night.

April

Kalends of April (iv.163–4): While I speak, Scorpio, causing fear with the tip of his tail, is thrown into the green waters.

4th day before the Nones of April (iv.169): The Pleiades will begin to relieve their father's shoulders.

8th day before the Ides of April (iv.386): Dangling Libra brought heavenly rain.

5th day before the Ides of April (iv.388): Sword-bearing Orion will have sunk in the ocean.

18th day before the Kalends of May (iv.625–6): The next morning make for safe harbours, sailor, for the wind from the west will be mixed with hail.

15th day before the Kalends of May (iv.677–8): When Venus looks back at you the fourth time from the Ides, that night the Hyades will set in the sea.

12th day before the Kalends of May (iv.715–20): The sun leaves the leader of the woolly herd, as a larger victim awaits. Whether it is a cow or a bull is not quickly known, for the front part is visible, but the rear is hidden.

7th day before the Kalends of May (iv.901–4): When April has six sunrises left, the season of spring is in the middle of its course, and you will search for Aries in vain. The rains will be the sign, and Sirius rises.

May

Kalends of May (v.111–12): On the first night the star that was kind to Jupiter in his infancy is visible to me: the rainy constellation Capella is born.

6th day before the Nones of May (v.159–64): Next, when the Sun on dawn's horses raises its rosy lamp, chasing off the stars, the cold north-west wind will tickle the tips of the corn, and white sails will be set on the Calabrian waters. And just as dark twilight moves into night, none of the whole herd of the Hyades will be hidden.

5th day before the Nones of May (v.379–80): In less than four nights Centaurus, half human and half the body of a golden horse, will send out his stars.

3rd day before the Nones of May (v.415–16): Curved Lyra wants to follow (Centaurus), but the road is not yet ready. The right time will be the third night.

Day before the Nones of May (v.417–18): When we say that the Nones are about to arrive, Scorpio will be seen in the sky from his middle.

5th day before the Ides of May (v.493–4): If you look for Boeotian Orion during this time, you will be mistaken.

3rd day before the Ides of May (v.599–602): You will see all the Pleiades, the whole crowd of sisters, when it is one more night until the Ides. Reliable authorities tell me that summer begins, and the season of warm spring ends.

Day before the Ides of May (v.603–4): The time before the Ides signals the rising of Starry Taurus.

11th day before the Kalends of June (v.723): In the night of this day Sirius goes forth.

8th day before the Kalends of June (v.731): When the plentiful sea has received this (day) in her waters, you will see the beak of Auriga.

7th day before the Kalends of June (v.733–4): The coming of dawn will hide Boötes from your eyes, and towards day there will be the Hyades.

June

Kalends of June (vi.195–6): If you seek stars, taloned Aquila now rises.

4th day before the Nones of June (vi.197–8): The next dawn calls the Hyades, the horns of the forehead of Taurus, and the earth is soaked with much rain.

7th day before the Ides of June (vi.235–6): On the third day after the Nones Phoebe is seen to take away Arcturus, and Ursa has no fear for its back.

4th day before the Ides of June (vi.471–2): The sailor sitting in the stern says 'We shall see Delphinus when damp night comes forth and the day is pushed back.'

17th day before the Kalends of July (vi.711–12): The third night will come, when you, Thyone of Dodona (one of the Hyades), will stand visible on the forehead of Taurus.

16th day before the Kalends of July (vi.715–16): Insofar as the winds can be trusted, spread the sheets for the west wind, sailors. Tomorrow it will come steadily over your waters.

15th day before the Kalends of July (vi.717–20): When the Sun has bathed his rays in the waves, and the twin poles are wrapped in serene stars, Orion will raise his powerful shoulders above the earth, and Delphinus will be visible the next night.

13th day before the Kalends of July (vi.725–7): Now there are six and six days remaining in the month, to which, however, you should add one day: the sun leaves Geminus, and the sign of Cancer reddens.

11th day before the Kalends of July (vi.735–6): ... Ophiucus, struck by his grandfather's missiles, rises above the earth and stretches his snake-wrapped hand.

6th day before the Kalends of July (vi.785–8): Behold, returning from a suburban shrine a not-quite-sober man throws words such as these to the stars: 'Your belt is now invisible, Orion, and maybe tomorrow it will be so too, but then it will be visible to me.' If he were not drunk, he would have said that that same day was the solstice.

A.vi. Columella's parapegma[84]

RR xi.2.4–5 (January, from the Ides onward)

On the Ides of January the weather is windy and conditions unstable.

On the XVIII Kal. Feb. the weather is unstable.

On the XVII Kal. Feb. the sun moves into Aquarius. Leo begins to set in the morning. A south-west, or sometimes a south wind, with rain.

On the XVI Kal. Feb. Cancer finishes setting. It is wintry.

On the XV Kal. Feb. Aquarius begins to rise. A south-west wind signifies a storm.

On the XI Kal. Feb. Lyra[85] sets in the evening. Rainy day.

On the IX Kal. Feb. it signifies a storm from the setting of the previous star. And sometimes there is a storm.

On the VI Kal. Feb. the bright star that is in the breast of Leo sets. Sometimes it is signified.[86] Winter is halved.

On the V Kal. Feb. a south or a south-west wind. It is wintry. Rainy day.

On the III Kal. Feb. Delphinus begins to set. Likewise Lyra sets.

On the day before the Kalends of February the setting of the above stars cause a storm. Sometimes however it signifies. *[There follows a description of seasonal farming activities.]*

RR xi. 2.11

Both of these tasks [cutting stakes and felling lumber] are better done under a waning moon, from the 20th to the 30th, for all timber cut thus is thought to be free of rot.

---

[84] Latin text in Ash, Forster, and Heffner, 1948–55. Excerpting just the astrometeorological material as I have here leaves out much that is very closely related to this material, and the reader is directed to Columella's complete text (available in English translation in the Loeb series) for a full appreciation of the important larger contexts from which the passages in this section are excerpted.

[85] Columella sometimes refers to this constellation as *Fidicula* and sometimes as *Fides*. Whether he means two different stars by the two different names is unclear. Generally the two Latin words are taken as synonyms in astronomical contexts.

[86] *Nonnumquam significatur.* I have argued elsewhere (Lehoux, 2004c) that Columella's uses of the verb *significat* are probably literal translations of the Greek ἐπισημαίνει, and that this may account for some of the odd ways in which the language has been pushed by later scribes who did not understand that the original word was both intransitive and impersonal. In any case, Columella's uses of *significat* are frequently unclear, and I suspect something has become garbled. I translate them here literally.

RR xi.2.14–15 (**February**)

On the Kalends of Feb. Lyra begins to set. South-east wind, and sometimes there is a south wind with hail.

On the III Non. Feb. all of Lyra and the middle of Leo set. North-west or north wind. Occasionally a west wind.

On the Nones of Feb. the middle parts of Aquarius rise. Windy weather.

On the VII Id. Feb. the constellation Ursa Major sets. The west winds begin to blow.

On the VI Id. Feb., windy weather.

On the III Id. Feb. a south-east wind. *[There follows a description of seasonal farming activities.]*

RR xi.2.20–1

On the Ides of Feb. Sagittarius sets in the evening. It is extremely wintry.

On the XVI Kal. March Crater rises in the evening. Change of wind.

On the XV Kal. March the sun moves into Pisces, sometimes the weather is windy.

On the XIII and XII Kal. March west or south wind with hail and pouring rain.

On the X Kal. March Leo finishes setting. The north winds, which are called the 'bird-bringers', usually last for thirty days, then the swallow also appears.

On the IX Kal. March Arcturus rises at the onset of night. Cold day with north-east or north-west wind. Sometimes rain.

On the VIII Kal. March Sagitta begins to rise at dusk. Variable weather. The days are called the Halcyon days, and indeed on the Atlantic Ocean it is seen to be extremely calm.

On the VII Kal. March windy weather. The swallow is seen. *[Seasonal farming instructions follow.]*

RR xi.2.23–4 (**March**)

On the Kalends of March a south-west or a south wind with hail.

On the VI Non. March Vindemiatrix appears, which the Greeks call Τρυγητήρ. North winds.

On the IV Non. March a west or sometimes a south wind. It is wintry.

On the Nones of March Pegasus rises in the morning. North-east breeze.

On the III Id. March the northern fish finishes rising. North winds.

On the day before the Ides of March the ship Argo rises. West or south wind, sometimes north-east. *[Seasonal farming instructions follow.]*

RR XI.2.30–1

On the Ides of March, Scorpio begins to set. It signifies a storm.

On the XVII Kal. Apr. Scorpio sets. It is wintry.

On the XVI Kal. Apr. the sun moves into Aries. West wind or north-west.

On the XII Kal. Apr. Pegasus sets in the morning. North winds.

On the X Kal. Apr. Aries begins to rise. Rainy day. Sometimes it snows.

On the IX and the VIII Kal. Apr. the vernal equinox signifies a storm. *[Seasonal farming instructions follow.]*

RR XI.2.34–5 (April)

On the Kalends of April Scorpio sets in the morning. It signifies a storm.

On the Nones of April a west or a south wind with hail. Sometimes the same thing on the day before the Nones.

On the VIII Id. Apr. the Pleiades are hidden in the evening. Sometimes it is wintry.

On the VII Id. Apr. and the VI and the V, south and south-west winds. It signifies a storm.

On the IV Id. Apr. Libra begins to set at sunrise. Sometimes it signifies a storm.

On the day before the Ides of April the Hyades are hidden. It is wintry. *[Seasonal farming instructions follow.]*

RR XI.2.35–7

On the Ides of April, as above, Libra sets. It is wintry.

On the XVIII Kal. May, windy weather and rain, but this is not steady.

On the XV Kal. May the sun moves into Taurus. It signifies rain.

On the XIV Kal. May the Hyades hide in the evening. It signifies rain.

On the XI Kal. May spring is halved. Rain and sometimes hail.

On the X Kal. May the Pleiades rise with the sun. A south-west or south wind. Humid day.

On the IX Kal. May Lyra appears at nightfall. It signifies a storm.

On the IV Kal. May there is a south wind with rain.

On the III Kal. May Auriga rises in the morning. Southerly day. Sometimes rain.

On the day before the Kalends of May Sirius hides in the evening. It signifies a storm. *[Seasonal farming instructions follow.]*

RR xi.2.39–40 (May)

On the Kalends of May the sun is said to stay in the same degree for two days. The Hyades rise with the sun.

On the VI Non. May the north winds blow.

On the V Non. May the whole of Centaurus appears. It signifies a storm.

On the III Non. May the same star signifies rain.

On the day before the Nones of May the middle of Scorpio sets. It signifies a storm.

On the Nones of May the Pleiades rise in the morning. West wind.

On the VII Id. May, beginning of summer. West or north-west wind, and sometimes rain.

On the VI Id. all the Pleiades appear. West or north-west wind, sometimes also rain.

On the III Id. May Lyra rises in the morning. It signifies a storm. *[Seasonal farming instructions follow.]*

RR xi.2.43

On the Ides of May Lyra rises in the morning. South or south-east wind. Sometimes a humid day.

On the XVII Kal. June the same as above.

On the XVI and XV Kal. June a south-east or south wind with rain.

On the XIV Kal. June the sun makes its entry into Geminus.

On the XII Kal. June the Hyades rise. North winds. Sometimes south wind with rain.

On the XI and X Kal. June Arcturus sets in the morning. It signifies a storm.

On the VIII, VII, and VI Kal. June Auriga rises in the morning. North winds. *[Seasonal farming instructions follow.]*

RR xi.2.45 (June)

On the Kalends and the IV Non. June Aquila rises. Windy weather and sometimes rain.

On the VII Id. June Arcturus sets. West or north-west wind.

On the IV Id. June Delphinus rises in the evening. West wind. Sometimes it is damp. *[Seasonal farming instructions follow.]*

RR xi.2.49

On the Ides of June heat begins.

On the XIII Kal. July the sun makes its entry into Cancer. It signifies a storm.

On the XI Kal. July *Anguifer*, which the Greeks call Ophiucus sets in the
morning. It signifies a storm.

On the VIII, VII, and VI Kal. of July, solstice. West wind and heat.

On the III Kal. July windy weather. *[Seasonal farming instructions follow.]*

RR xi.2.51 (July)

On the Kalends of July, west or south wind and heat.

On the IV Non. July Corona sets in the morning.

On the day before the Nones of July the middle of Cancer sets. Heat.

On the VIII Id. July the middle of Capricorn sets.

On the VII Id. July Cepheus rises in the evening. It signifies a storm.

On the VI Id. July the *prodromoi* begin to blow. *[Seasonal farming instructions
follow.]*

RR xi.2.52

. . . and wooded land can be most efficiently cleared when the moon is
waning.

On the Ides of July Procyon rises in the morning. It signifies a storm.

On the XIII Kal. Aug. the sun moves into Leo. West wind.

On the VIII Kal. Aug. Aquarius begins to visibly set. West or south wind.

On the VII Kal. Aug. Sirius appears. Hot, sultry air.

On the VI Kal. Aug. Aquila rises.

On the IV Kal. Aug. the bright star in the breast of Leo rises. Sometimes it
signifies a storm.

On the III Kal. Aug. Aquila sets. It signifies a storm. *[Seasonal farming
instructions follow.]*

RR xi.2.57 (August)

On the Kalends of August, Etesian winds.

On the day before the Nones of August the middle of Leo rises. It signifies a
storm.

On the VII Id. Aug. the middle of Aquarius sets. Hazy heat.

On the day before the Ides of August Lyra sets in the morning, and autumn
begins. *[Seasonal farming instructions follow.]*

RR xi.2.57–8

On the Ides of August the setting of Delphinus signifies a storm.

On the XIX Kal. Sept. the morning setting of the same star signifies a storm.

On the XIII Kal. Sept. the sun moves into Virgo. This and the next day it signifies a storm. Sometimes it also thunders.

On the same day Lyra sets.

On the X Kal. Sept. from that same star, a storm often arises, and rain.

On the VI Kal. Sept. Vindemiatrix rises in the morning, and Arcturus begins to set. Sometimes rain.

On the III Kal. Sept. the shoulders of Virgo rise. The Etesian winds cease blowing, and sometimes it is wintry.

On the day before the Kalends of September Andromeda rises in the evening. Sometimes it is wintry. *[Seasonal farming instructions follow.]*

RR xi.2.63 (**September**)

On the Kalends of September, heat.

On the IV Non. Sept. the southern fish finishes setting. Heat.

On the Nones of September Arcturus rises. West wind, or north-west.

On the VII Id. Sept. the northern fish finishes setting and Auriga rises. It signifies a storm.

On the III Id. Sept. west wind or south-west. The middle of Virgo rises. *[Seasonal farming instructions follow.]*

RR xi.2.65–6

On the Ides of September it sometimes signifies a storm from the previous day's constellation.

On the XV Kal. Oct. Arcturus rises. West or south-west wind, sometimes the south-east wind, which some people call *Vulturnus*.

On the XIV Kal. Oct. Spica rises. West wind or north-west.

On the XIII Kal. Oct. the sun moves into Libra. Crater appears in the morning.

On the XI Kal. Oct. Pisces sets in the morning, and Aries begins to set. West wind or north-west. Sometimes a south wind with rains.

On the X Kal. Oct. the ship Argo sets. It signifies a storm, sometimes rain.

On the IX Kal. Oct. Centaurus begins to rise in the morning. It signifies a storm, sometimes with rain.

On the VIII Kal. Oct. and the VII and VI the autumn equinox signifies rain.

On the V Kal. Oct. the Haedi rise. West wind. Sometimes a south wind with rain.

On the IV Kal. Oct. Virgo finishes rising. It signifies a storm. *[Seasonal farming instructions follow.]*

RR xi.2.72–4 (October)

On the Kalends of October and the VI Non. it sometimes signifies a storm.

On the IV Non. Oct. Auriga sets in the morning. Virgo finishes setting. Sometimes it signifies a storm.

On the III Non. Oct. Corona begins to rise. It signifies a storm.

On the day before the Nones of Oct. the Haedi rise in the evening. The middle of Aries sets. North-east wind.

On the VIII Id. Oct. the bright star in Corona rises.

On the VI Id. Oct. the Pleiades rise in the evening. West wind, and sometimes a south-west wind with rain.

On the III Id. and the day before the Ides of Oct. the whole of Corona rises in the morning. A cold south wind, and sometimes rain. *[Seasonal farming instructions follow.]*

RR xi.2.76–8

On the Ides of October and the two days following, sometimes a storm. But sometimes it is only damp. The belt of Orion rises in the evening.

On the XIV Kal. Nov. the sun moves into Scorpio.

On the XIII and XII Kal. Nov. the Pleiades begin to set at sunrise. It signifies a storm.

On the XI Kal. Nov. the tail of Taurus sets. South wind, sometimes rain.

On the VIII Kal. Nov. Centaurus finishes rising in the morning. It signifies a storm.

On the VII Kal. Nov. the forehead of Scorpio rises. It signifies a storm.

On the V Kal. Nov. the Pleiades set. It is wintry, with cold and frosts.

On the IV Kal. Nov. Arcturus sets in the evening. Windy day.

On the III Kal. Nov. and the day before the Kalends Cassiopeia begins to set. It signifies a storm. *[Seasonal farming instructions follow.]*

RR xi.2.84 (November)

On the Kalends of November and the day after the head of Taurus sets. It signifies rain.

On the III Non. Nov. Lyra rises in the morning. It is wintry and it rains.

On the VIII Id. Nov. the whole of that same constellation rises. South wind or west. It is wintry.

On the VII Id. Nov. it signifies a storm and it is wintry.

On the VI Id. Nov. the Pleiades set in the morning. It signifies a storm. It is wintry.

On the V Id. Nov. the bright star in Scorpio rises. It signifies a storm. Or else a south-east wind. Sometimes it is damp.

On the IV Id. Nov. beginning of winter. South wind, or south-east. Sometimes it is damp. *[Seasonal farming instructions follow.]*

RR xi.2.85

And in particular on the day before the full moon (or if not then certainly on the day of the full moon itself) you will be sure that all the beans that are to be sown *are* sown on one day – you can cover them over afterwards as protection from birds and cattle. And if the course of the moon is agreeable have them harrowed before the Ides of November, when the area is very rich and newly tilled, or if not then when it is very well fertilized.

RR xi.2.88–9

On the Ides of November, unstable day, but calm.

On the XVI Kal. Dec. Lyra rises in the morning. South wind, sometimes a strong north-east wind.

On the XV Kal. Dec. a north-east wind, sometimes a south wind with rain.

On the XIV Kal. Dec. the sun moves into Sagittarius. The Hyades rise in the morning. It signifies a storm.

On the XII Kal. Dec. the horns of Taurus set in the evening. Cold north-east wind and rain.

On the XI Kal. Dec. the Hyades set in the morning. It is wintry.

On the X Kal. Dec. Lepus sets in the morning. It signifies a storm.

On the VII Kal. Dec. Sirius sets at sunrise. It is wintry.

On the day before the Kalends of December the whole of the Hyades set. West wind or south wind. Sometimes rain. *[Seasonal farming instructions follow.]*

RR xi.2.93 (December)

On the Kalends of December, unstable day, but usually calm.

On the VIII Id. Dec. the middle of Sagittarius sets. It signifies a storm.

On the VII Id. Dec. Aquila rises in the morning. South-east wind. Sometimes a south wind, and it is damp.

On the III Id. Dec. a north-west or north wind. Sometimes a south wind with rain. *[Seasonal farming instructions follow.]*

RR xi.2.93–4

On the Ides of December the whole of Scorpio rises in the morning. It is wintry.

On the XVI Kal. Jan. the sun moves into Capricorn. Winter solstice as Hipparchus reckons it, and so it often signifies a storm.

On the XV Kal. Jan. it signifies a change of winds.

On the X Kal. Jan. Auriga sets in the morning. It signifies a storm.

On the IX Kal. Jan. winter solstice as the Chaldaeans observe it.

On the VI Kal. Jan. Delphinus begins to rise in the morning. It signifies a storm.

On the IV Kal. Jan. Aquila sets in the evening. It is wintry.

On the III Kal. Jan. Sirius sets in the evening. It signifies a storm.

On the day before the Kalends of January, a windy storm. *[Seasonal farming instructions follow.]*

RR xi.2.97 (**January, Kalends to the Ides**)

On the Kalends of January, unstable day.

On the III Non. Jan. Cancer sets. Variable weather.

On the day before the Nones of Jan., the middle of winter. Strong south wind, sometimes rain.

On the Nones of Jan. Lyra rises in the morning. Variable weather.

On the VI Id. Jan. south wind, sometimes a west wind.

On the V Id. Jan. south wind. Sometimes rain.

On the day before the Ides of January, unstable condition of the sky. *[Seasonal farming instructions follow.]*

RR xi.3.22 (**further lunar instructions**)

But whenever we are going to sow garlic or lay it away in storage if it is mature, we must be careful that the moon is below the earth at the time when we either bury it or dig it up. For if it is thus sown or stored, it is thought not to have an overly sharp flavour, nor to give bad breath to those who eat it.

A.vii. Pliny's parapegma[87]

HN xviii.207f.

LVII. First of all the explanation of the days of the year and of the sun's motion is quite difficult, with a fourth of a day and night added to the 365 days by the intercalary year. This makes it impossible to set down definite times to the stars. Add to this the admitted obscurity of the subject, where sometimes a weather sign runs ahead by many days (which situation the Greeks call προχειμάζειν), and sometimes it trails behind (which they call ἐπιχειμάζειν), and often the effects of the heavens fall in one place sooner and in another later. So we hear from common folk that the 'star is finished' when the weather calms down again. And moreover even though all these things correspond to stars that are stationary and fixed in the heavens, the motions of planets interfere in rain and hail storms with substantial effect, as we have shown, and they confound the orderliness of the expectation held . . . Thus Vergil says to learn the order of the planets, recommending the observation of the passage of the cold star Saturn . . . The matter has two facets: first of all to seek a law for the heavens, and from there to investigate it by means of arguments. Above all is the difference due to the curvature of the Cosmos, and of the globe of the earth, where the same star shows itself at different times to different people, which means that its force is not manifest everywhere on the same days. Further difficulty is the writers' observing in different places, and then they write different things about the same place. There used to be three schools, the Chaldaean, the Egyptian, and the Greek, and among us Caesar the dictator (using the work of the experienced scholar Sosigenes) added a fourth, returning the individual years to the course of the sun. On the discovery of an error this system was afterwards corrected by not intercalating for twelve years in a row, since the year was beginning to fall behind the stars where it had previously gone too fast. Both Sosigenes himself in his three studies – who in spite of being more careful than others, did not shy away from raising doubts by correcting himself – and other authors (whom we prefixed to this volume) set out these things, but it is rare that the opinion of one agrees with another. This is less incredible in the rest of them – for they are excused by their different locations. But of those who differ in the same region, we shall set out one disagreement as an example: Hesiod – for there is a book on *Astrology* in his name – gives the morning setting of the Pleiades as happening at the end

---

[87] Latin text in Mayhoff, 1897–1906.

of the autumnal equinox, Thales on the twenty-fifth day after the equinox, Anaximander 31, Euctemon 48.[88] We shall mostly follow the observation of Caesar (this will be the account for Italy), but, since we are investigators of the whole of nature and not only one country, we will also state the opinions of others, set out by regions rather than by authors, for that would be too verbose. Readers should remember that, for the sake of brevity, when Attica is spoken of, they must also understand the Cyclades islands; when Macedonia: Magnesia, Thrace; when Egypt: Phoenicia, Cyprus, Cicilia; when Boeotia: Locris, Phocis, and the bordering territories too; when the Hellespont: Chersonese and neighbouring areas as far as Mount Athos; when Ionia: Asia and the islands of Asia; when the Peloponnese: Achaea and the adjoining lands to the west; *Chaldaeans* will mean Assyria and Babylon. *[There follow some definitions and explanations of visibility conditions]*

HN xviii.220f.

LIX. The cardinal points of the seasons correspond to the fourfold division of the year, according to the increases of daylight. This increases from the winter solstice, and it equals the night at the vernal equinox, in ninety days and three hours. From there it is greater than the night until the solstice, in ninety-four days and twelve hours. *[lacuna]* ... until the autumnal equinox, and then, having become equal to the day, the night then outstrips it from then to the winter solstice, in eighty-eight days and three hours, where the 'hours' in each increase mean equinoctial rather than the hours of any particular day. All of these changes happen at the eighth degree of the signs, the winter solstice happens in Capricorn on the eighth day before the Kalends of January, the vernal equinox in Aries, the summer solstice in Cancer, and the other equinox in Libra, and these days not uncommonly have signs of storms. On the other hand, these cardinal points are also divided by particular points in time, all of them in the middle of their run of days, since between the summer solstice and the autumnal equinox, the setting of Lyra on the forty-sixth day is the beginning of autumn; from that equinox to the winter solstice, the morning setting of the Pleiades on the forty-fourth day is the beginning of winter; between the winter solstice and the vernal equinox, the west wind on the forty-fifth day is the beginning of the season of spring; from the vernal equinox, the morning rising of the Pleiades on the

---

[88] Following all MSS. Böckh emends it to 'Euctemon [44, Eudoxus] 48'.

forty-eighth day is the beginning of summer. We will begin from the sowing of corn, that is, the morning setting of the Pleiades – and it is not wise to overanalyse and increase the difficulty by mentioning the less important stars, even though after a long period the violent constellation of Orion falls from sight on these days too.

LX. Most people prepare for the times for sowing, and sow crops on the eleventh day from the autumnal equinox (there is an almost certain promise of rain for nine days after the rising of Corona), but Xenophon (says to do so) not before the god has given the sign (which our Cicero explained as being done with rain), although the true reckoning is not to sow before the leaves have begun to fall. Some say this happens at the exact setting of the Pleiades, on the third day before the Ides of November as we have said. Even the peddlers of clothing pay attention to that star, and it is easy to see in the sky. Thus greedy dealers who have an eye for opportunities foretell how the winter will be from its setting, since a cloudy setting announces a rainy winter, and they immediately raise the prices of cloaks; a clear setting announces a cruel winter, and they raise the prices of the rest of their clothes. But the farmer, untutored about the heavens, receives this sign in his thorn bushes and looking at his soil, when he has seen the leaves fall. Thus the temperatures of the year can be determined, sometimes changing earlier, sometimes later. *[There follows some discussion of the providence of nature in giving us signs such as these.]*

HN xviii.228f.

LXI. Varro has advised following this observation, for sowing beans at least. Others say they are to be sown at the full moon, but lentils from the twenty-fifth to the thirtieth days of the moon, and vetch on these same lunar days, for then they will be free of slugs. Some say to sow fodder at that time, or in the spring for seed. *[There follows more on nature and some specific hints for certain crops and their uses.]*

HN xviii.231f.

[LXII . . .] From the Kalends of November until the winter solstice has happened, do not set eggs under the hens. All summer until that date, put thirteen under each but fewer in winter, though no less than nine. Democritus thinks that the coming winter will be like the day of the winter solstice and the three days around it, and the same with the summer solstice. Mostly the fourteen days around the winter solstice are the Halcyon-bringers that calm the sky with peaceful winds. But in this and in all other matters the stars must be understood from the outcome of the signs, not expecting a calling to court of the weather on predetermined dates.

LXIII. You should not tend vines at the winter solstice. Hyginus advises that wine be taken off the dregs then, or racked seven days after this if the moon is past its seventh day, and (he advises) the sowing of cherries around the winter solstice. *[There follows a list of farming activities proper to the winter generally.]*

HN xviii.234f. (late December to mid-February)

LXIV. According to Caesar, from the winter solstice to the season of the west wind, the prominent stars signify:[89] on the III Kal. Jan. the morning setting of Sirius. On that day Aquila is supposed to set in the evening for Attica and the neighbouring areas. On the day before the Nones of January, according to Caesar, Delphinus rises in the morning, and the next day Lyra, at which time Sagitta disappears in the evening for Egypt. Likewise on the VI Id. Jan., the next few days will be wintry in Italy during the evening setting of the aforementioned Delphinus, and also when the sun is seen to enter into Aquarius, which happens on the XVI Kal. Feb. On the VIII Kal. Feb. the star in the breast of Leo, called 'the Royal Star' by Tubero, sets in the morning, and on the day before the Nones of February Lyra sets in the evening. *[A description of seasonal farming tasks follows.]*

HN xviii.237f. (mid-February to late March)

LXV. From the west wind to the vernal equinox, the XIV Kal. March signifies three variable days according to Caesar, as does the VIII Kal. March, when the swallows are seen, as they are the next day when Arcturus rises in the evening, likewise on the III Non. March (Caesar observed this to happen at the rising of Cancer, but most authors at the sinking of Vindemiatrix). On the VIII Id. March, at the rising of the northern fish and the next day of Orion, in Attica the kite is supposed to appear. Caesar also noted the setting of Scorpio on the Ides of March, the date of his death. On the XV Kal. Apr. the kite appears for Italy. On the XII Kal. Apr. Pegasus sets in the morning. *[A description of seasonal farming tasks follows.]*

---

[89] *significant.* Possibly some misunderstanding of the Greek verb ἐπισημαίνει, 'there is a change the weather'. See Lehoux, 2004c.

HN xviii.246f. (late March to early May)

LXVI. The vernal equinox is seen to be fixed to the VIII Kal. Apr. From then to the morning rising of the Pleiades, the Kalends of April is significant, according to Caesar. On the III Non. Apr. the Pleiades disappear in the evening for Attica, which they do the next day for Boeotia, but for Caesar and the Chaldaeans on the Nones, when Orion and his sword begin to disappear for Egypt. On the VI Id. Apr. rain is signified, according to Caesar, when Libra sets. The Hyades, a violent constellation and disorderly for land and sea, set in the evening for Egypt on the XIV Kal. May; but on the XVI for Attica, and on the XV according to Caesar, which signifies for four continuous days, and on the XII Kal. May for Assyria. This star is commonly called *Parilicium* because the XI Kal. May, the birthday of the city of Rome and the day on which clear weather usually returns, offers clarity for observation. By the Greeks it is called *Hyas* (from the clouds) and we, because of the similarity of the Greek name, have called it *Suculae* in the ignorant belief that it was named for *sues*, pigs. According to Caesar, the VIII Kal. May is also a noted day. On the VII Kal. May the Haedi rise for Egypt. On the VI Kal. May Sirius disappears in the evening for Boeotia and Attica. Lyra rises in the morning. On the V Kal. May the whole of Orion vanishes for Assyria, and on the IV, Sirius. On the VI Non. May the Hyades rise in the morning according to Caesar, and on the VIII Id. May rainy Capella. On this day, however, Sirius disappears in the evening for Egypt. These are all the stars' motions until the VI Id. May, which is the rising of the Pleiades. *[A description of seasonal farming tasks follows.]*

HN xviii.255f. (early May to late June)

[LXVII . . .] From the rising of the Pleiades these are significant, according to Caesar: the morning setting of Arcturus the following day; the rising of Lyra on the III Id. May; Capella setting in the evening, and Sirius for Attica on the XII Kal. June. On the XI Kal. June the sword of Orion begins to set according to Caesar. On the IV Non. June Aquila rises in the evening according to Caesar and for Assyria. On the VII Id. June Arcturus sets in the morning for Italy. On the IV Id. June Delphinus rises in the evening. On the XVII Kal. July, the sword of Orion rises, which is four days later for Egypt. On the XI Kal. July that same sword of Orion begins to set according to Caesar. On the VIII Kal. July the very longest day and shortest night of the whole year bring about the summer solstice. *[A description of seasonal farming tasks follows.]*

HN xviii.268f. (late June to early August)

[LXVIII . . .] From the summer solstice to the setting of Lyra: On the VI Kal.
July Orion rises according to Caesar, and his belt on the IV Non. July for
Assyria, and in Egypt Procyon – which star does not have a name among the
Romans, unless we want to think of it as *Canicula* (that is, the 'Little Dog',
as it is usually drawn among the stars) – is burning hot in the morning, and
it has great strength for heating, as we will show just below. On the IV Non.
July Corona sets in the morning according to the Chaldaeans. For Attica, the
whole of Orion rises on this day. On the day before the Ides of July, Orion
finishes rising according to the Egyptians. On the XVI Kal. Aug., Procyon
rises for Assyria, and three days from then the remarkable star, universally
recognized by all, which we call the rising of Sirius, while the sun enters the
first degree of Leo. This happens on the twenty-third day after the summer
solstice. They feel it at sea and on land, and even many wild animals, as we
have said in those chapters. It is no less venerated than those stars assigned
to the gods, and it ignites the sun, and it is the cause of great heat. On the
XIII Kal. Aug. Aquila sets in the morning for Egypt, and the precursors of
the Etesian winds begin, which Caesar figures are felt in Italy on the X Kal.
Aug. Aquila sets in the morning for Attica. On the III Kal. Aug. the Royal
star in the breast of Leo comes up in the morning according to Caesar. On
the VIII Id. Aug. the middle of Arcturus sets. On the III Id. Aug. Lyra by its
setting begins autumn, but it has been found that a true calculation shows
that this happens on the VI Ides. *[A description of the seasonal effects on crops
of the moon's phases and of the stars follows.]*

HN xviii.283 (miscellaneous dates)

. . . We have said that Aquila rises for Italy on the XIII Kal. January, and the
order of Nature does not suffer anything to do with planting to be of sure
promise before this date.

HN xviii.285f.

[LXIX . . .] In the eleventh year of his reign, Numa established *Robigalia*,
which is now held on the VII Kal. May, because that is the time when rust
(*robigo*) attacks the fields. Varro has set this time as when the sun reaches
the tenth degree of Taurus, as calculation established it at that time. But the
real reason is that, on one of the four days (as observed by different nations)
from the nineteenth day from the vernal equinox to the IV Kal. May, Sirius
sets – a star both violent in itself, and which must be preceded by the setting of

Procyon. Those same people, following a Sybilline oracle, instituted *Floralia* on the IV Kal. in the 516th year AUC.[90] so that everything would blossom well. Varro has set this date as when the sun reaches the fourteenth degree of Taurus. Thus if the full moon happens in this four-day period, the crops and everything that is flowering will necessarily be damaged. The first *Vinalia*, which is before this on the IX Kal. May and instituted for the tasting of wines, does not coincide with the fruit, nor do those (feasts) mentioned above for the vines and olives, for their conception begins at the rising of the Pleiades on the VI Id. May, as we have shown. There is another four-day period when I would not want moisture damage, for the setting of the cold star of Arcturus the next day freezes things, and even less (do I want) a full moon to happen then. On the IV Non. June Aquila rises in the evening again, a decisive day for olive and vine flowers, if a full moon occurs on it. Indeed, I also think the solstice on the VIII Kal. June to be a similar occasion, as is the rising of Sirius on the twenty-third day after the solstice, but only if a new moon occurs, for then the heat is to blame for the grapes ripening too firmly. Again, a full moon is harmful on the IV Non. July when Sirius rises for Egypt, or anyway on the XVI Kal. Aug. when (it rises) for Italy, and on the XIII Kal. Aug. when Aquila sets, and also on the X Kal. of that month. The later *Vinalia*, which is held on the XIV Kal. Sept., is later than these forces. Varro has set this as when Lyra itself begins to set in the morning, which he thinks is the beginning of autumn, and on which date there is a festival established for calming the weather. Lyra is now observed to set on the VI Id. Aug. *[There follows more on the effects of the moon, and a description of seasonal farming tasks.]*

HN XVIII.309f. (early August to early November)

LXXIV. Next in the division of the seasons is autumn, from the setting of Lyra to the equinox and from there the setting of the Pleiades and the beginning of winter. In this interval the following are significant: on the day before the Ides of August Pegasus rising in the evening for Attica, and the setting of Delphinus for Egypt and Caesar. On the XI Kal. Sept., for Caesar and Assyria the star called Vindemiatrix begins to rise in the morning, predicting the maturity of the vintage (the change in colour of the grapes will be its sign). For Assyria on the V Kal. Sept. Sagitta sets and the Etesian winds stop. On the Nones of Sept. Vindemiatrix rises for Egypt, Arcturus does so in the morning

---

[90] *Ab urbe condita*, from the founding of Rome.

for Attica, and Sagitta sets in the morning. On the V Id. Sept. according to Caesar, Capella rises in the evening, and the middle of Arcturus on the day before the Ides of Sept., when there is the most violent signification for land and sea for five days. This reason is given for it: if there is rain at the setting of Delphinus, there will not be any at Arcturus. The sign of the rising of that star is the disappearance of the swallows, since those that are caught (by it) are killed. On the XVI Kal. Oct. Spica, which Virgo holds, rises in the morning for Egypt and the Etesian winds stop. This happens according to Caesar on the XIV Kal. Oct. For Assyria the XIII and the XI Kal. Oct. are significant, and according to Caesar the junction of Pisces setting, and the equinoctial constellation itself on the VIII Kal. Oct. From there, Philippus, Callippus, Dositheus, Parmensicus, Conon, Crito, Democritus, and Eudoxus all agree – which is rare – that on the IV Kal. Oct. Capella rises in the morning and on the III Kal. Oct. the Haedi. On the VI Non. Oct. Corona rises in the morning for Attica. For Asia and Caesar Auriga sets in the morning on the V. On the IV Non. Oct. Corona begins to rise according to Caesar, and the next day the Haedi set in the evening. On the VIII Id. Oct. the bright star in Corona rises, on the VI Id. Oct. the Pleiades do so in the evening, and on the Ides all of Corona. On the XVII Kal. Nov. the Hyades rise in the evening. On the day before the Kalends, Arcturus sets and the Hyades rise with the sun according to Caesar. On the IV Non. Nov. Arcturus sets in the evening. On the V Id. Nov., the sword of Orion begins to set. From there the Pleiades set on the III Id. Nov. *[There follow seasonal farming activities, as well as descriptions of beneficial lunar phases (321f.), and a long discussion of weather signs of the Theophrastan type (340f.).]*

HN xviii.321f. (lunar instructions)

LXXV. To this will be added what is necessary concerning the moon and the winds and predictions, so that the account of the heavens will be complete. Again, Vergil, following the claim of Democritus, reckons that things should be divided according to the lunar days. And our concern is also with laws, both here and in the work as a whole.

Everything that is cut, picked, or shorn is done more safely when the moon is waning than when it is waxing. Do not touch manure except when the moon is waning, and you will manure best at the new moon or half moon. Geld pigs, cattle, sheep, and goats when the moon is waning. Put eggs under the hens at the new moon. Dig ditches under the full moon at night. Cover the roots of trees at the full moon. Sow in damp ground at the new moon and the four days around the new moon. They say to air and store grains and pulses around the end of the moon's cycle, to prepare nurseries when the moon is above the earth, and to tread grapes when the moon is below the earth and to cut timber and all the other things we mentioned in those places. But the observation is not without difficulty, as we have already said in our second book. But, as even rustics can understand, during the period when it is seen at sunset and it shines in the earlier hours of the night it will be waxing and it will appear to your eyes to be halved. But when it rises opposite the setting of the sun so that it is seen at the same time, that will be a full moon. During the period when it rises at sunrise, withholding light in the earlier hours of the night but extending it into the daytime, it will be waning and again halved. And when it ceases to be visible that will be conjunction, which is called the new moon. At the new moon and during the whole first day it will be above the earth whenever the sun is, the second day for ten-twelfths plus one forty-eighth of an hour of the night, and then on the third and up to the fifteenth day, adding the same parts of an hour. On the fifteenth it will be above the earth all night and likewise below the earth all day. On the sixteenth it will be below the earth for ten twelfths plus one forty-eighth of the first hour of the night, and it will increase by the same portion of an hour each day until the new moon, and whatever it subtracts from its time above the earth in the first part of the night it will add above the earth to the beginning of the day. It will count to thirty every other month, but subtract one from that for the remaining months. This is the account of the moon. That of the winds is a little more difficult. *[There follows an account of wind directions and their importance in farming.]*

A.viii: Ptolemy's *Phaseis*[91]

φάσεις ἀπλανῶν ἀστέρων καὶ συναγωγὴ ἐπισημασιῶν.

Θώθ

α. ὡρῶν ιδ´ ∟´ ὁ ἐπὶ τῆς οὐρᾶς τοῦ Λέοντος ἐπιτέλλει. Ἱππάρχῳ ἐτησίαι
     παύονται. Εὐδόξῳ ὑετία, βρονταί, ἐτησίαι παύονται.

β. ὡρῶν ιδ´ ὁ ἐπὶ τῆς οὐρᾶς τοῦ Λέοντος ἐπιτέλλει, καὶ Στάχυς κρύπτεται.
     Ἱππάρχῳ ἐπισημαίνει.

γ. ὡρῶν ιγ´ ∟´ ὁ ἐπὶ τῆς οὐρᾶς τοῦ Λέοντος ἐπιτέλλει. ὡρῶν ιε´ ὁ καλού-
     μενος Αἴξ ἑσπέριος ἀνατέλλει. Αἰγυπτίοις ἐτησίαι παύονται. Εὐδόξῳ

---

[91] I have, for simplicity's sake, followed Heiberg's 1907 edition in my translation. A new edition of this text would be very highly desirable, however. The two standard editions, Heiberg's and Wachsmuth's, were both published by Teubner within a year of each other. Heiberg's edition is prepared from four MSS that frequently disagree with each other on minor details, and on

ἄνεμοι μεταπίπτοντες. Καίσαρι ἄνεμος, ὑετός, βρονταί. Ἱππάρχῳ ἀπηλιώτης πνεῖ.

δ.   ὡρῶν ιε΄ ὁ ἔσχατος τοῦ Ποταμοῦ ἑῷος δύνει. Καλλίππῳ χειμαίνει καὶ ἐτησίαι παύονται.

ε.   ὡρῶν ιγ΄ ∟΄ Στάχυς κρύπτεται. ὡρῶν ιε΄ ∟΄ ὁ λαμπρὸς τῆς Λύρας ἑῷος δύνει. Μητροδώρῳ δυσαερία. Κόνωνι ἐτησίαι λήγουσιν.

ϛ.   ὡρῶν ιε΄∟΄ ὁ λαμπρὸς τῆς νοτίου Χηλῆς κρύπτεται. Αἰγυπτίοις ὁμίχλη καὶ καῦμα ἢ ὑετὸς ἢ βροντή. Εὐδόξῳ ἄνεμος, βροντή, δυσαερία. Ἱππάρχῳ ἄνεμος, νοτία.

ζ.   Μητροδώρῳ δυσαερία. Καλλίππῳ, Εὐκτήμονι, Φιλίππῳ δυσαερία καὶ ἀταξία ἀέρος. Εὐδόξῳ ὑετός, βρονταί, ἄνεμος μεταπίπτων.

η.   Αἰγυπτίοις ὑετία, χειμὼν κατὰ θάλασσαν ἢ νότος. Καίσαρι ἄνεμοι μεταπίπτοντες, ὑετία, καὶ ἐτησίαι παύονται.

θ.   ὡρῶν ιδ΄ ὁ λαμπρὸς τοῦ Ὄρνιθος ἑῷος δύνει. Αἰγυπτίοις ζέφυρος ἢ ἀργεστὴς πνεῖ.

ι.   ὡρῶν ιγ΄ ∟΄ ὁ λαμπρὸς τοῦ Περσέως ἑσπέριος ἀνατέλλει. Φιλίππῳ δυσαερία. Δοσιθέῳ χειμαίνει.

ια.   Αἰγυπτίοις ἐπισημαίνει.

ιβ.   ὡρῶν ιε΄ ὁ λαμπρὸς τῆς νοτίου Χηλῆς κρύπτεται.

ιγ.   Δοσιθέῳ ἀκρασία ἀέρων.

ιδ.   ὡρῶν ιδ΄∟΄ ὁ καλούμενος Κάνωβος ἐπιτέλλει. Καίσαρι βορέαι παύονται πνέοντες.

ιε.   Εὐδόξῳ ἄνεμοι νότιοι.

ιϛ.   Καλλίππῳ καὶ Κόνωνι ἐπισημαίνει.

ιζ.   ὡρῶν ιδ΄∟΄ ὁ λαμπρὸς τοῦ Ὄρνιθος ἑῷος δύνει, καὶ ὁ λαμπρὸς τῆς νοτίου Χηλῆς κρύπτεται, καὶ ὁ ἔσχατος τοῦ Ποταμοῦ ἑῷος δύνει. Εὐδόξῳ βορέαι παύονται. Μητροδώρῳ ἐπισημαίνει. Δημοκρίτῳ Ἀβδηρίτῃ ἐπισημαίνει, καὶ χελιδὼν ἀφανίζεται.

ιη.   ὡρῶν ιε΄∟΄ ὁ κατὰ τὸ γόνυ τοῦ Τοξότου κρύπτεται. Αἰγυπτίοις ὑετία, ἐπισημαίνει, φθινοπώρου ἀρχή, χελιδὼν ἀφανίζεται. Δοσιθέῳ νοτία. Εὐκτήμονι μετοπώρου ἀρχή.

Climata in particular. Two of the MSS are so corrupt as to be all but useless for the editions of the text (see Heiberg's notes in the *praefatio* to Ptolemy's *Phaseis*, pp. iv–v). Wachsmuth's edition is prepared from four *different* MSS, only two of which he seems to have seen himself. Like all earlier editors and translators of this text, Wachsmuth and Heiberg have frequently corrected conflicting details in the text in one way or another. Trying to triangulate between Wachsmuth's and Heiberg's corrections of their respective MSS has proven too cumbersome for the purposes of this translation, which is, after all, primarily meant to make Ptolemy's text accessible to a wider audience.

ιθ´. ὡρῶν ιε´ ὁ λαμπρὸς τοῦ νοτίου Ἰχθύος ἑσπέριος ἀνατέλλει. Ἱππάρχῳ δυσαερία καὶ ὑετία κατὰ θάλασσαν καὶ φθινοπώρου ἀρχή.

κ´. Καίσαρι μετοπώρου ἀρχή, καὶ χελιδὼν ἀφανίζεται. Μητροδώρῳ ὑετία κατὰ θάλασσαν καὶ δυσαερία.

κα´. ὡρῶν ιδ´ ὁ λαμπρὸς τῆς νοτίου Χηλῆς κρύπτεται. ὡρῶν ιε´ ὁ ἐν τῷ ἑπομένῳ ὤμῳ τοῦ Ἡνιόχου ἑσπέριος ἀνατέλλει. Αἰγυπτίοις ζέφυρος ἢ λίψ, ὀψὲ ἀπηλιώτης. Εὐδόξῳ μετόπωρον μέσον.

κβ´. ὡρῶν ιδ´ Ľ ὁ καλούμενος Ἀντάρης κρύπτεται. Αἰγυπτίοις ζέφυρος ἢ ἀργεστὴς καὶ ψακάς. Εὐδόξῳ νοτία.

κγ´. ὡρῶν ιδ´ Ľ ὁ καλούμενος Αἲξ ἑσπέριος ἀνατέλλει. ὡρῶν ιε´ Ľ Ἀρκτοῦρος ἑῷος ἀνατέλλει. Αἰγυπτίοις ψακὰς καὶ ἄνεμος, ἐπισημαίνει. Καλλίππῳ καὶ Μητροδώρῳ ὑετία.

κδ´. ὡρῶν ιγ´ Ľ ὁ κοινὸς Ἵππου καὶ Ἀνδρομέδας ἑῷος δύνει.

κε´. ὡρῶν ιγ´ Ľ ὁ λαμπρὸς τῆς νοτίου Χηλῆς κρύπτεται. ὡρῶν ιε´ ὁ λαμπρὸς τοῦ Ὄρνιθος ἑῷος δύνει. Αἰγυπτίοις ζέφυρος ἢ νότος καὶ δι᾽ ἡμέρας ὄμβρος.

κϛ´. ὡρῶν ιε´ Ἀρκτοῦρος ἑῷος ἀνατέλλει. Εὐδόξῳ ὑετός. Ἱππάρχῳ ζέφυρος ἢ νότος.

κζ´. ὡρῶν ιδ´ ὁ κοινὸς Ἵππου καὶ Ἀνδρομέδας ἑῷος δύνει, καὶ ὁ ἔσχατος τοῦ Ποταμοῦ ἑῷος δύνει.

κη´. μετοπωρινὴ ἰσημερία. Αἰγυπτίοις καὶ Εὐδόξῳ ἐπισημαίνει.

κθ´. ὡρῶν ιδ´ ὁ καλούμενος Ἀντάρης κρύπτεται. ὡρῶν ιδ´ Ľ Ἀρκτοῦρος ἑῷος ἀνατέλλει. Εὐκτήμονι ἐπισημαίνει. Δημοκρίτῳ ὑετὸς καὶ ἀνέμων ἀταξία.

λ´. ὡρῶν ιδ´ Ľ ὁ κοινὸς Ἵππου καὶ Ἀνδρομέδας ἑῷος δύνει. Εὐκτήμονι καὶ Φιλίππῳ καὶ Κόνωνι ἐπισημαίνει

Φαωφί

α´. Αἰγυπτίοις ζέφυρος ἢ νότος. Ἱππάρχῳ ἐπισημαίνει.

β´. ὡρῶν ιε´ ὁ κοινὸς Ἵππου καὶ Ἀνδρομέδας ἑῷος δύνει. ὡρῶν ιε´ Ľ ὁ λαμπρὸς τῆς βορείου Χηλῆς κρύπτεται. Εὐδόξῳ καὶ Εὐκτήμονι ἐπισημαίνει. Ἱππάρχῳ νότος ἢ ζέφυρος.

γ´. ὡρῶν ιδ´ Ἀρκτοῦρος ἑῷος ἀνατέλλει. ὡρῶν ιε´ Ľ ὁ λαμπρὸς τοῦ Ὄρνιθος ἑῷος δύνει.

δ´. ὡρῶν ιε´ ὁ λαμπρὸς τῆς βορείου Χηλῆς κρύπτεται. Αἰγυπτίοις καὶ Καλλίππῳ χειμάζει, δυσαερία. Εὐκτήμονι καὶ Φιλίππῳ ὑετός.

ε´. ὡρῶν ιε´ Ľ ὁ κοινὸς Ἵππου καὶ Ἀνδρομέδας ἑῷος δύνει. Εὐδόξῳ ὑετός. Εὐκτήμονι χειμάζει. Μητροδώρῳ ὑετός.

ϛ´.   ὡρῶν ιγ´ ∟´ Ἀρκτοῦρος ἑῷος ἀνατέλλει, καὶ ὁ ἔσχατος τοῦ Ποταμοῦ
      ἑῷος δύνει. ὡρῶν ιδ´ ∟´ ὁ λαμπρὸς τῆς βορείου Χηλῆς κρύπτεται, καὶ ὁ
      καλούμενος Ἀντάρης κρύπτεται. ὡρῶν ιε´ ∟´ ὁ λαμπρὸς τοῦ βορείου
      Στεφάνου ἑῷος ἀνατέλλει. Αἰγυπτίοις καὶ Καίσαρι χειμών, ὑετός,
      βρονταί, ἀστραπαί.

ζ´.   ὡρῶν ιγ´ ∟´ Στάχυς ἐπιτέλλει. ὡρῶν ιδ´ ὁ καλούμενος Αἲξ ἑσπέριος
      ἀνατέλλει, καὶ ὁ λαμπρὸς τῆς βορείου Χηλῆς κρύπτεται. Αἰγυπ-
      τίοις ὑετοί, χειμαίνει. Εὐδόξῳ ὑετὸς καὶ ἄνεμος μεταπίπτων. Δοσιθέῳ
      ἐπισημαίνει.

η´.   ὡρῶν ιγ´ ∟´ ὁ λαμπρὸς τῆς βορείου Χηλῆς κρύπτεται. ὡρῶν ιδ´ ∟´ ὁ ἐν τῷ
      ἑπομένῳ ὤμῳ τοῦ Ἡνιόχου ἑσπέριος ἀνατέλλει, καὶ Στάχυς ἐπιτέλλει.
      Δημοκρίτῳ χειμάζει, σπόρου ὥρα.

θ´.   ὡρῶν ιε´ ∟´ Στάχυς ἐπιτέλλει. Αἰγυπτίοις βορρᾶς πνεῖ.

ι´.   ὡρῶν ιε´ ὁ λαμπρὸς τοῦ βορείου Στεφάνου ἑῷος ἀνατέλλει. Ἱππάρχῳ
      νότος.

ια´. ὡρῶν ιε´ ὁ κατὰ τὸ γόνυ τοῦ Τοξότου κρύπτεται.

ιβ´. ὡρῶν ιε´ ὁ καλούμενος Ἀντάρης κρύπτεται. Αἰγυπτίοις ζέφυρος ἢ λίψ.
      Εὐδόξῳ ἐπισημαίνει. Ἱππάρχῳ ἀπηλιώτης.

ιγ´.

ιδ´. Δοσιθέῳ καὶ Εὐδόξῳ ἐπισημαίνει.

ιε´. Αἰγυπτίοις ἀργεστής, ὑετός.

ιϛ´. ὡρῶν ιδ´ ∟´ ὁ λαμπρὸς τοῦ βορείου Στεφάνου ἑῷος ἀνατέλλει. Εὐδόξῳ
      βορέαι ἢ νότοι. Δοσιθέῳ ἄνεμος μεταπίπτων. Καλλίππῳ ἐπισημαίνει.
      Καίσαρι ἄνεμος ἄτακτος, ὑετός, βρονταί.

ιζ´. ὡρῶν ιγ´ ∟´ ὁ καλούμενος Ἀντάρης κρύπτεται. Αἰγυπτίοις βορέας ἢ
      λίψ. Εὐδόξῳ ἐπισημαίνει.

ιη´. ὡρῶν ιγ´ ∟´ Ἀρκτοῦρος ἑσπέριος δύνει.

ιθ´. Εὐδόξῳ ἀνέμων μεταβολαί, βρονταί.

κ´.   ὡρῶν ιδ´ ὁ ἐν τῷ ἑπομένῳ ὤμῳ τοῦ Ἡνιόχου ἑσπέριος ἀνατέλλει.
      Ἱππάρχῳ νότος ἢ βορέας.

κα´.   ὡρῶν ιγ´ ∟´ ὁ καλούμενος Αἲξ ἑσπέριος ἀνατέλλει.

κβ´. ὡρῶν ιδ´ ὁ λαμπρὸς τοῦ βορείου Στεφάνου ἑῷος ἀνατέλλει. Αἰγυπτίοις
      ζέφυρος ἢ νότος δι᾿ ἡμέρας· ὑετός. Δοσιθέῳ ἐπισημασία.

κγ´.

κδ´. ὡρῶν ιδ´ ∟´ ὁ καλούμενος Κάνωβος ἑῷος δύνει.

κε´. Αἰγυπτίοις πνεύματα ἄτακτα.

κϛ´. ὡρῶν ιδ´ Ἀρκτοῦρος ἑσπέριος δύνει. Εὐδόξῳ ἐπισημαίνει. Καίσαρι
      βορέας πνεῖ.

κζ´. ὡρῶν ιγ´ ∟´ ὁ λαμπρὸς τοῦ βορείου Στεφάνου ἑῷος ἀνατέλλει. ὡρῶν ιδ´
      ∟´ ὁ κατὰ τὸ γόνυ τοῦ Τοξότου κρύπτεται. Αἰγυπτίοις καὶ Καλλίππῳ

ἐπισημαίνει. Εὐκτήμονι καὶ Καλλίππῳ ἀμιξία ἀέρος, κατὰ θάλασσαν χειμὼν πολύς.

κη΄.   ὡρῶν ιγ΄ Ꞁ΄ ὁ ἐν τῷ ἑπομένῳ ὤμῳ τοῦ Ἡνιόχου ἑσπέριος ἀνατέλλει. Μητροδώρῳ ἐπισημαίνει.[92] Εὐκτήμονι καὶ Καλλίππῳ ἀέρος μίξις, καὶ κατὰ θάλασσαν χειμάζει.

λ΄.   Αἰγυπτίοις χειμάζει σφόδρα.

Ἀθύρ

α΄.   ὡρῶν ιγ΄ Ꞁ΄ ὁ λαμπρὸς τῆς νοτίου Χηλῆς ἐπιτέλλει.

β΄.   ὡρῶν ιδ΄ Ꞁ΄ ὁ λαμπρὸς τῆς νοτίου Χηλῆς ἐπιτέλλει. ὡρῶν ιε΄ τὸ αὐτό. Αἰγυπτίοις ἐπισημαίνει. Δοσιθέῳ χειμάζει. Δημοκρίτῳ ψύχη ἢ πάχνη. Ἱππάρχῳ νότος πυκνός.

γ΄.   ὡρῶν ιγ΄ Ꞁ΄ ὁ λαμπρὸς τῆς βορείου Χηλῆς ἐπιτέλλει. ὡρῶν ιε΄ Ꞁ΄ ὁ λαμπρὸς τῆς Λύρας ἑῷος ἀνατέλλει. Εὐκτήμονι καὶ Φιλίππῳ ἄνεμος μέγας πνεῖ.

δ΄.   ὡρῶν ιδ΄ ὁ λαμπρὸς τῆς βορείου Χηλῆς ἐπιτέλλει. ὡρῶν ιδ΄ Ꞁ΄ Ἀρκτοῦρος ἑσπέριος δύνει. Αἰγυπτίοις νότος ἢ λίψ. Καλλίππῳ καὶ Εὐκτήμονι πνεύματα σφοδρά. Καίσαρι καὶ Μητροδώρῳ ἄνεμοι, χειμάζει.

ε΄.   ὡρῶν ιδ΄ Ꞁ΄ ὁ λαμπρὸς τῆς βορείου Χηλῆς ἐπιτέλλει.

ϛ΄.   ὡρῶν ιδ΄ ὁ κατὰ τὸ γόνυ τοῦ Τοξότου κρύπτεται. Κόνωνι καὶ Εὐδόξῳ ἀκρασία πνευμάτων. Καλλίππῳ ἀκρασία ἀέρων. Καίσαρι καὶ Ἱππάρχῳ νότος ἢ βορρᾶς ψυχρός.

ζ΄.   ὡρῶν ιδ΄ ὁ λαμπρὸς τῶν Ὑάδων ἑσπέριος ἀνατέλλει. Αἰγυπτίοις νότος λάβρος. Μέτωνι ζέφυρος. Εὐδόξῳ βορέας ἢ νότος. Μητροδώρῳ ἀκρασία ἀέρος. Εὐκτήμονι καὶ Φιλίππῳ καὶ Ἱππάρχῳ ὑετός.

η΄.   ὡρῶν ιγ΄ Ꞁ΄ ὁ λαμπρὸς τῶν Ὑάδων ἑσπέριος ἀνατέλλει. Καλλίππῳ ὑετία. Εὐκτήμονι ἐπισημαίνει.

θ΄.   ὡρῶν ιε΄ Ꞁ΄ ὁ κοινὸς Ποταμοῦ καὶ ποδὸς Ὠρίωνος ἑῷος δύνει. Αἰγυπτίοις χειμών, ὑετός.

ι΄.   ὡρῶν ιδ΄ ὁ καλούμενος Κάνωβος ἑῷος δύνει. Αἰγυπτίοις νότος ἢ ζέφυρος. Δοσιθέῳ χειμών.

ια΄.   ὡρῶν ιε΄ ὁ λαμπρὸς τῆς Λύρας ἑῷος ἀνατέλλει. Μέτωνι ὑετὸς θυελλώδης. Ἱππάρχῳ ἀργεστὴς ψυχρός.

ιβ΄.   ὡρῶν ιε΄ Ἀρκτοῦρος ἑσπέριος δύνει, καὶ ὁ κοινὸς Ποταμοῦ καὶ ποδὸς Ὠρίωνος ἑῷος δύνει.

ιγ΄.   ὡρῶν ιγ΄ Ꞁ΄ ὁ κατὰ τὸ γόνυ τοῦ Τοξότου κρύπτεται. Αἰγυπτίοις νότος ἢ εὖρος δι᾽ ἡμέρας, ψακάζει. Μητροδώρῳ χειμάζει, θύελλα. Εὐκτήμονι ὑετοί, χειμάζει.

---

[92] Following Petavius.

ιδ.   ὡρῶν ιδ΄ ∟΄ ὁ κοινὸς Ποταμοῦ καὶ ποδὸς Ὠρίωνος ἑῷος δύνει. Φιλίππῳ καὶ Εὐκτήμονι χειμών, θύελλα. Ἱππάρχῳ βορέας ἢ νότος ψυχρὸς καὶ ὑετός.

ιε.   ὡρῶν ιγ ∟΄ ὁ λαμπρὸς τοῦ Περσέως ἑῷος δύνει, καὶ ὁ λαμπρὸς τοῦ βορείου Στεφάνου ἑσπέριος δύνει. ὡρῶν ιε΄ ∟΄ ὁ λαμπρὸς τῶν Ὑάδων ἑῷος δύνει. Αἰγυπτίοις καὶ Ἱππάρχῳ χειμῶνος ἀρχή. Μητροδώρῳ καὶ Καλλίππῳ καὶ Κόνωνι ἐπισημασία.

ιϛ΄.   ὡρῶν ιγ ∟΄ ὁ λαμπρὸς τῶν Ὑάδων ἑῷος δύνει. ὡρῶν ιδ΄ ∟΄ τὸ αὐτό. ὡρῶν ιε΄ τὸ αὐτό. Εὐκτήμονι καὶ Δοσιθέῳ χειμάζει.

ιζ΄.   ὡρῶν ιδ΄ ὁ κοινὸς Ποταμοῦ καὶ ποδὸς Ὠρίωνος ἑῷος δύνει. ὡρῶν ιε΄ ∟΄ ὁ ἐπὶ τῆς κεφαλῆς τοῦ ἡγουμένου Διδύμου ἑσπέριος ἀνατέλλει. Εὐδόξῳ χειμῶνος ἀρχὴ καὶ ἐπισημασία. Δημοκρίτῳ χειμὼν καὶ κατὰ γῆν καὶ κατὰ θάλασσαν.

ιθ.   ὡρῶν ιδ΄ ∟΄ ὁ λαμπρὸς τῆς Λύρας ἑῷος ἀνατέλλει. Αἰγυπτίοις νότος ἢ εὖρος δι᾽ ἡμέρας. Καίσαρι χειμάζει.

κ.   ὡρῶν ιγ ∟΄ ὁ κοινὸς Ποταμοῦ καὶ ποδὸς Ὠρίωνος ἑῷος δύνει. ὡρῶν ιδ΄ ὁ λαμπρὸς τοῦ Περσέως ἑῷος δύνει. ὡρῶν ιε΄ ∟΄ ὁ ἐν τῷ ἡγουμένῳ ὤμῳ τοῦ Ὠρίωνος ἑῷος δύνει, καὶ ὁ μέσος τῆς ζώνης τοῦ Ὠρίωνος ἑῷος δύνει. Καίσαρι χειμών.

κα.   ὡρῶν ιε΄ ὁ ἐν τῷ ἡγουμένῳ ὤμῳ τοῦ Ὠρίωνος ἑῷος δύνει, καὶ ὁ μέσος τῆς ζώνης τοῦ Ὠρίωνος ἑῷος δύνει. ὡρῶν ιε΄ ∟΄ Ἀρκτοῦρος ἑσπέριος δύνει. Αἰγυπτίοις βορέας δι᾽ ἡμέρας καὶ νυκτός. Εὐδόξῳ ὑετός. Καίσαρι χειμών.

κβ.   ὡρῶν ιδ΄ ∟΄ ὁ ἐν τῷ ἡγουμένῳ ὤμῳ τοῦ Ὠρίωνος ἑῷος δύνει.

κγ.   ὡρῶν ιγ ∟΄ ὁ καλούμενος Κάνωβος ἑῷος δύνει. ὡρῶν ιδ΄ ὁ λαμπρὸς τοῦ βορείου Στεφάνου ἑσπέριος δύνει, καὶ ὁ ἐν τῷ ἡγουμένῳ ὤμῳ τοῦ Ὠρίωνος ἑῷος δύνει. ὡρῶν ιε΄ ὁ ἐπὶ τῆς κεφαλῆς τοῦ ἡγουμένου Διδύμου ἑσπέριος ἀνατέλλει. Εὐδόξῳ χειμέριος περίστασις.

κδ.   ὡρῶν ιγ ∟΄ ὁ ἐν τῷ ἐμπροσθίῳ δεξιῷ βατραχίῳ τοῦ Κενταύρου ἐπιτέλλει. ὡρῶν ιδ΄ ∟΄ ὁ μέσος τῆς ζώνης τοῦ Ὠρίωνος ἑῷος δύνει. ὡρῶν ιε΄ ∟΄ Κύων ἑῷος δύνει. Αἰγυπτίοις χειμέριος περίστασις. Εὐδόξῳ βορέας ψυχρός.

κε.   ὡρῶν ιγ ∟΄ ὁ ἐν τῷ ἡγουμένῳ ὤμῳ τοῦ Ὠρίωνος ἑῷος δύνει, καὶ ὁ καλούμενος Ἀντάρης ἐπιτέλλει. ὡρῶν ιδ΄ ∟΄ ὁ λαμπρὸς τοῦ Περσέως ἑῷος δύνει. Εὐκτήμονι καὶ Δοσιθέῳ χειμὼν καὶ ὑετία. Καίσαρι ἀκρασία ἀέρος.

κϛ΄.   ὡρῶν ιγ ∟΄ ὁ ἐν τῷ ἡγουμένῳ ὤμῳ τοῦ Ὠρίωνος ἑσπέριος ἀνατέλλει, καὶ ὁ ἔσχατος τοῦ Ποταμοῦ ἑσπέριος ἀνατέλλει. ὡρῶν ιδ΄ ὁ λαμπρὸς τῆς Λύρας ἑῷος ἀνατέλλει, καὶ ὁ μέσος τῆς ζώνης τοῦ Ὠρίωνος ἑῷος δύνει, καὶ ὁ καλούμενος Ἀντάρης ἐπιτέλλει. Εὐδόξῳ χειμὼν σφοδρός.

κζ΄. ὡρῶν ιδ΄ ∠΄ ὁ καλούμενος Ἀντάρης ἐπιτέλλει. ὡρῶν ιε΄ Κύων ἑῷος δύνει. ὡρῶν ιε΄ ∠΄ ὁ λαμπρὸς τοῦ Ὄρνιθος ἑῷος ἀνατέλλει, καὶ ὁ ἐν τῷ ἑπομένῳ ὤμῳ τοῦ Ὠρίωνος ἑῷος δύνει. Αἰγυπτίοις καὶ Ἱππάρχῳ νότος πυκνός. Εὐδόξῳ καὶ Κόνωνι χειμέριος ὁ ἀήρ. Καλλίππῳ ὑετία.

κη΄. ὡρῶν ιδ΄ ὁ ἐν τῷ ἡγουμένῳ ὤμῳ τοῦ Ὠρίωνος ἑσπέριος ἀνατέλλει. ὡρῶν ιδ΄ ∠΄ ὁ ἐπὶ τῆς κεφαλῆς τοῦ ἡγουμένου Διδύμου ἑσπέριος ἀνατέλλει. ὡρῶν ιε΄ ὁ ἐν τῷ ἑπομένῳ ὤμῳ τοῦ Ὠρίωνος ἑῷος δύνει, καὶ ὁ καλούμενος Ἀντάρης ἐπιτέλλει. Αἰγυπτίοις ψακάς.

κθ΄. ὡρῶν ιγ΄ ∠΄ ὁ μέσος τῆς ζώνης τοῦ Ὠρίωνος ἑῷος δύνει. ὡρῶν ιε΄ ∠΄ ὁ καλούμενος Ἀντάρης ἐπιτέλλει.

λ΄. ὡρῶν ιγ΄ ∠΄ ὁ μέσος τῆς ζώνης τοῦ Ὠρίωνος ἑσπέριος ἀνατέλλει. ὡρῶν ιδ΄ ∠΄ ὁ ἐν τῷ ἑπομένῳ ὤμῳ τοῦ Ὠρίωνος ἑῷος δύνει, καὶ ὁ ἐν τῷ ἡγουμένῳ ὤμῳ τοῦ Ὠρίωνος ἑσπέριος ἀνατέλλει. ὡρῶν ιε΄ ∠΄ ὁ ἐπὶ τῆς κεφαλῆς τοῦ ἑπομένου Διδύμου ἑσπέριος ἀνατέλλει.

Χοιάκ

α΄. ὡρῶν ιδ΄ ∠΄ Κύων ἑῷος δύνει. ὡρῶν ιε΄ ὁ λαμπρὸς τοῦ Περσέως ἑῷος δύνει. Αἰγυπτίοις νότος καὶ ὑετός. Εὐδόξῳ ἀκρασία ἀέρος. Δοσιθέῳ ἐπισημασία. Δημοκρίτῳ οὐρανὸς ταραχώδης καὶ ἡ θάλασσα ὡς τὰ πολλά.

β΄. ὡρῶν ιγ΄ ∠΄ ὁ ἐν τῷ ἑπομένῳ ὤμῳ τοῦ Ὠρίωνος ἑσπέριος ἀνατέλλει, καὶ ὁ κοινὸς Ποταμοῦ καὶ ποδὸς Ὠρίωνος ἑσπέριος ἀνατέλλει. ὡρῶν ιδ΄ ὁ ἐπὶ τῆς κεφαλῆς τοῦ ἡγουμένου Διδύμου ἑσπέριος ἀνατέλλει, καὶ ὁ ἐν τῷ ἑπομένῳ ὤμῳ τοῦ Ὠρίωνος ἑῷος δύνει. ὡρῶν ιδ΄ ∠΄ ὁ λαμπρὸς τοῦ βορείου Στεφάνου ἑσπέριος δύνει.

γ΄. ὡρῶν ιγ΄ ∠΄ ὁ ἐν τῷ ἑπομένῳ ὤμῳ τοῦ Ὠρίωνος ἑῷος δύνει. ὡρῶν ιε΄ ὁ ἐν τῷ ἡγουμένῳ ὤμῳ τοῦ Ὠρίωνος ἑσπέριος ἀνατέλλει.

δ΄. ὡρῶν ιγ΄ ∠΄ ὁ λαμπρὸς τῆς Λύρας ἑῷος ἀνατέλλει. ὡρῶν ιδ΄ ὁ ἐν τῷ ἑπομένῳ ὤμῳ τοῦ Ὠρίωνος ἑσπέριος ἀνατέλλει, καὶ ὁ μέσος τῆς ζώνης τοῦ Ὠρίωνος ἑσπέριος ἀνατέλλει. ὡρῶν ιε΄ ὁ ἐπὶ τῆς κεφαλῆς τοῦ ἑπομένου Διδύμου ἑσπέριος ἀνατέλλει. Αἰγυπτίοις ζέφυρος ἢ νότος δι᾽ ἡμέρας, ὕει. Κόνωνι χειμάζει.

ε΄. ὡρῶν ιγ΄ ∠΄ ὁ καλουμένους Αἲξ ἑῷος δύνει, καὶ ὁ ἐπὶ τῆς κεφαλῆς τοῦ ἡγουμένου Διδύμου ἑσπέριος ἀνατέλλει. ὡρῶν ιδ΄ Κύων ἑῷος δύνει. ὡρῶν ιε΄ ∠΄ ὁ ἐν τῷ ἡγουμένῳ ὤμῳ Ὠρίωνος ἑσπέριος ἀνατέλλει. Καίσαρι καὶ Εὐκτήμονι καὶ Εὐδόξῳ καὶ Καλλίππῳ χειμών.

ς΄. ὡρῶν ιδ΄ ὁ ἐν τῷ ἐμπροσθίῳ δεξιῷ βατραχίῳ τοῦ Κενταύρου ἐπιτέλλει. ὡρῶν ιδ΄ ∠΄ ὁ ἐν τῷ ἑπομένῳ ὤμῳ τοῦ Ὠρίωνος ἑσπέριος ἀνατέλλει. Μητροδώρῳ χειμερία περίστασις. Εὐκτήμονι καὶ Φιλίππῳ καὶ Καλλίππῳ ἀνέμων ἀκρασία.

ζ΄.   ὡρῶν ιδ΄ ὁ κοινὸς Ποταμοῦ καὶ ποδὸς Ὠρίωνος ἑσπέριος ἀνατέλλει. ὡρῶν ιδ΄ L΄ ὁ ἐπὶ τῆς κεφαλῆς τοῦ ἑπομένου Διδύμου ἑσπέριος ἀνατέλλει, καὶ ὁ μέσος τῆς ζώνης τοῦ Ὠρίωνος ἑσπέριος ἀνατέλλει. ὡρῶν ιε΄ ὁ λαμπρὸς τοῦ Ὄρνιθος ἑῷος ἀνατέλλει. Αἰγυπτίοις ψακάζει. Καίσαρι καὶ Κόνωνι χειμάζει.

η΄.   ὡρῶν ιε΄ ὁ ἐν τῷ ἑπομένῳ ὤμῳ τοῦ Ὠρίωνος ἑσπέριος ἀνατέλλει. ὡρῶν ιε΄ L΄ ὁ λαμπρὸς τοῦ Περσέως ἑῷος δύνει. Αἰγυπτίοις ψακάζει. Καίσαρι καὶ Εὐκτήμονι καὶ Εὐδόξῳ χειμών.

θ΄.   ὡρῶν ιγ΄ L΄ Κύων ἑῷος δύνει. ὡρῶν ιδ΄ ὁ καλούμενος Αἴξ ἑῷος δύνει, καὶ ὁ ἐπὶ τῆς κεφαλῆς τοῦ ἑπομένου Διδύμου ἑσπέριος ἀνατέλλει, καὶ ὁ ἔσχατος τοῦ Ποταμοῦ ἑσπέριος ἀνατέλλει. Αἰγυπτίοις καὶ Δοσιθέῳ καὶ Δημοκρίτῳ χειμών.

ι΄.   ὡρῶν ιε΄ ὁ λαμπρὸς τοῦ βορείου Στεφάνου ἑσπέριος δύνει, καὶ ὁ μέσος τῆς ζώνης τοῦ Ὠρίωνος ἑσπέριος ἀνατέλλει. Αἰγυπτίοις λίψ ἢ νότος. Εὐδόξῳ καὶ Δοσιθέῳ χειμέριος ἀήρ.

ια΄.   ὡρῶν ιγ΄ L΄ ὁ ἐπὶ τῆς κεφαλῆς τοῦ ἑπομένου Διδύμου ἑσπέριος ἀνατέλλει. Ἱππάρχῳ βορέας πολύς. Εὐδόξῳ ὑετός.

ιβ΄.   ὡρῶν ιδ΄ L΄ ὁ κοινὸς Ποταμοῦ καὶ ποδὸς Ὠρίωνος ἑσπέριος ἀνατέλλει. Καίσαρι νοτία. Εὐκτήμονι καὶ Εὐδόξῳ καὶ Καλλίππῳ χειμῶνος ἀὴρ καὶ ὑετία.

ιγ΄.   ὡρῶν ιγ΄ L΄ ὁ ἐν τῷ ἑπομένῳ ὤμῳ τοῦ Ἡνιόχου ἑῷος δύνει. ὡρῶν ιε΄ L΄ ὁ μέσος τῆς ζώνης τοῦ Ὠρίωνος ἑσπέριος ἀνατέλλει. Καίσαρι νοτία. Εὐκτήμονι καὶ Εὐδόξῳ καὶ Καλλίππῳ χειμῶνος ἀὴρ καὶ ὑετία.

ιδ΄.   ὡρῶν ιδ΄ L΄ ὁ καλούμενος Αἴξ ἑῷος δύνει. Μητροδώρῳ καὶ Εὐκτήμονι καὶ Καλλίππῳ χειμῶνος περίστασις. Δημοκρίτῳ βρονταί, ἀστραπαί, ὕδωρ, ἄνεμοι.

ιε΄.   Αἰγυπτίοις ἀργεστὴς ψυχρὸς ἢ νότος καὶ ὄμβρος. Καλλίππῳ νότος καὶ ἐπισημασία. Εὐδόξῳ χειμῶνος ἀήρ.

ις΄.   ὡρῶν ιδ΄ L΄ ὁ λαμπρὸς τοῦ Ὄρνιθος ἑῷος ἀνατέλλει. ὡρῶν ιε΄ ὁ κοινὸς Ποταμοῦ καὶ ποδὸς Ὠρίωνος ἑσπέριος ἀνατέλλει. Αἰγυπτίοις χειμάζει.

ιζ΄.   Ἱππάρχῳ νότος πολὺς ἢ βορέας.

ιη΄.   ὡρῶν ιδ΄ ὁ ἐν τῷ ἑπομένῳ ὤμῳ τοῦ Ἡνιόχου ἑῷος δύνει. Αἰγυπτίοις ὑετία μετὰ πνευμάτων. Εὐδόξῳ χειμάζει.

ιθ΄.   ὡρῶν ιε΄ ὁ καλούμενος Αἴξ ἑῷος δύνει. ὡρῶν ιε΄ L΄ ὁ λαμπρὸς τοῦ βορείου Στεφάνου ἑσπέριος δύνει. Αἰγυπτίοις βορέας ψυχρὸς ἢ νότος καὶ ὑετία.

κ΄.   ὡρῶν ιε΄ L΄ Προκύων ἑῷος δύνει. Καίσαρι χειμάζει.

κα΄.   ὡρῶν ιε΄ L΄ ὁ κοινὸς Ποταμοῦ καὶ ποδὸς Ὠρίωνος ἑσπέριος ἀνατέλλει.

κβ́. ὡρῶν ιέ Προκύων ἑῷος δύνει. Ἱππάρχῳ νότος.

κγ́. ὡρῶν ιδ́ ∟′ ὁ ἐν τῷ ἑπομένῳ ὤμῳ τοῦ Ἡνιόχου ἑῷος δύνει, καὶ ὁ ἐν τῷ ἐμπροσθίῳ δεξιῷ βατραχίῳ τοῦ Κενταύρου ἐπιτέλλει. ὡρῶν ιέ ∟′ ὁ λαμπρὸς τοῦ Ἀετοῦ ἑῷος ἀνατέλλει. Αἰγυπτίοις καὶ Εὐδόξῳ καὶ Δοσιθέῳ λὶψ ἢ νότος.

κδ́. ὡρῶν ιδ́ ∟′ Προκύων ἑῷος δύνει, καὶ ὁ ἔσχατος τοῦ Ποταμοῦ ἑσπέριος ἀνατέλλει. Εὐδόξῳ χειμερινὸς ἀήρ.

κέ. ὡρῶν ιγ́ ∟′ Προκύων ἑσπέριος ἀνατέλλει. ὡρῶν ιδ́ Προκύων ἑῷος δύνει. ὡρῶν ιέ ὁ λαμπρὸς τοῦ Ἀετοῦ ἑῷος ἀνατέλλει. Αἰγυπτίοις ἐπισημασία.

κϛ́. χειμερινὴ τροπή. ὡρῶν ιγ́ ∟′ Προκύων ἑῷος δύνει, καὶ Κύων ἑσπέριος ἀνατέλλει. ὡρῶν ιέ ∟′ ὁ καλούμενος Αἲξ ἑῷος δύνει.

κζ́. ὡρῶν ιγ́ ∟′ ὁ λαμπρὸς τοῦ Ἀετοῦ κρύπτεται. ὡρῶν ιδ́· Προκύων ἑσπέριος ἀνατέλλει. ὡρῶν ιδ́ ∟′ ὁ λαμπρὸς τοῦ Ἀετοῦ ἑῷος ἀνατέλλει.

κή. ὡρῶν ιέ ὁ ἐν τῷ ἑπομένῳ ὤμῳ τοῦ Ἡνιόχου ἑῷος δύνει. ὡρῶν ιέ ∟′ ὁ λαμπρὸς τοῦ νοτίου Ἰχθύος κρύπτεται. Αἰγυπτίοις καὶ Καίσαρι χειμών. Ἱππάρχῳ καὶ Μέτωνι ἐπισημαίνει, ὄμβρος.

κθ́. ὡρῶν ιδ́ ∟′ Προκύων ἑσπέριος ἀνατέλλει. Αἰγυπτίοις καὶ Κόνωνι καὶ Μέτωνι καὶ Καλλίππῳ χειμών. Καίσαρι καὶ Μητροδώρῳ ἐπισημασία, ἀκρασία.

λ́. ὡρῶν ιδ́ ὁ λαμπρὸς τοῦ Ἀετοῦ ἑῷος ἀνατέλλει, καὶ ὁ λαμπρὸς τοῦ Ἀετοῦ ἑσπέριος δύνει. Αἰγυπτίοις λὶψ καὶ ἀκρασία ἀέρος. Εὐδόξῳ καὶ Μητροδώρῳ χειμῶνος ἀήρ. Ἱππάρχῳ χειμὼν ἑσπέριος.

Τυβί

ά. ὡρῶν ιδ́ Κύων ἑσπέριος ἀνατέλλει. ὡρῶν ιέ Προκύων ἑσπέριος ἀνατέλλει. Εὐδόξῳ ἐπισημαίνει. Δημοκρίτῳ χειμὼν μέσος.

β́. ὡρῶν ιγ́ ∟′ ὁ ἐπὶ τῆς κεφαλῆς τοῦ ἡγουμένου Διδύμου ἑῷος δύνει. Δοσιθέῳ χειμαίνει.

γ́. ὡρῶν ιγ́ ∟′ ὁ λαμπρὸς τοῦ Ἀετοῦ ἐπιτέλλει. ὡρῶν ιέ ∟′ Προκύων ἑσπέριος ἀνατέλλει. Εὐκτήμονι καὶ Φιλίππῳ καὶ Δημοκρίτῳ ἐπισημαίνει.

δ́. ὡρῶν ιγ́ ∟′ ὁ λαμπρὸς τοῦ Ὄρνιθος ἑῷος ἀνατέλλει, καὶ ὁ ἐπὶ τῆς κεφαλῆς τοῦ ἑπομένου Διδύμου ἑῷος δύνει. ὡρῶν ιδ́ ∟′ ὁ λαμπρὸς τοῦ Ἀετοῦ ἑσπέριος δύνει. ὡρῶν ιέ ὁ λαμπρὸς τοῦ νοτίου Ἰχθύος κρύπτεται. Αἰγυπτίοις χειμὼν κατὰ θάλασσαν. Εὐκτήμονι ἐπισημαίνει.

έ. ὡρῶν ιδ́ ὁ ἐπὶ τῆς κεφαλῆς τοῦ ἡγουμένου Διδύμου ἑῷος δύνει. ὡρῶν ιέ ∟′ ὁ ἐν τῷ ἑπομένῳ ὤμῳ τοῦ Ἡνιόχου ἑῷος δύνει.

ϛʹ.   ὡρῶν ιγʹ Lʹ ὁ κατὰ τὸ γόνυ τοῦ Τοξότου ἐπιτέλλει. ὡρῶν ιδʹ ὁ ἐπὶ τῆς
κεφαλῆς τοῦ ἑπομένου Διδύμου ἑῷος δύνει. ὡρῶν ιδʹ Lʹ Κύων ἑσπέριος
ἀνατέλλει.

ζʹ.   ὡρῶν ιεʹ ὁ λαμπρὸς τοῦ Ἀετοῦ ἑσπέριος δύνει. Δοσιθέῳ ἐπισημαίνει.

ηʹ.   ὡρῶν ιδʹ Lʹ ὁ ἐπὶ τῆς κεφαλῆς τοῦ ἡγουμένου Διδύμου ἑῷος δύνει, καὶ ὁ
ἐπὶ τῆς κεφαλῆς τοῦ ἑπομένου Διδύμου ἑῷος δύνει, καὶ ὁ λαμπρὸς τοῦ
νοτίου Ἰχθύος κρύπτεται. Αἰγυπτίοις ποικίλη κατάστασις.

θʹ.   ὡρῶν ιγʹ Lʹ ὁ λαμπρὸς τῆς Λύρας ἑσπέριος δύνει. ὡρῶν ιεʹ Lʹ ὁ λαμπρὸς
τοῦ Ἀετοῦ ἑσπέριος δύνει. Αἰγυπτίοις ἐπισημαίνει. Δημοκρίτῳ νότος
πνεῖ ὡς τὰ πολλά.

ιʹ.   ὡρῶν ιεʹ Κύων ἑσπέριος ἀνατέλλει.

ιαʹ. ὡρῶν ιεʹ ὁ ἐπὶ τῆς κεφαλῆς τοῦ ἑπομένου Διδύμου ἑῷος δύνει. Εὐκτήμονι καὶ Φιλίππῳ μέσος χειμών.

ιβʹ. ὡρῶν ιδʹ ὁ κατὰ τὸ γόνυ τοῦ Τοξότου ἐπιτέλλει. ὡρῶν ιεʹ ὁ ἐπὶ τῆς
κεφαλῆς τοῦ ἡγουμένου Διδύμου ἑῷος δύνει. Ἱππάρχῳ καὶ Εὐδόξῳ
χειμαίνει.

ιγʹ. ὡρῶν ιδʹ ὁ λαμπρὸς τοῦ νοτίου Ἰχθύος κρύπτεται. ὡρῶν ιεʹ ὁ ἔσχατος
τοῦ Ποταμοῦ ἑσπέριος ἀνατέλλει. Αἰγυπτίοις νότος ἢ ζέφυρος, χειμὼν
καὶ κατὰ γῆν καὶ κατὰ θάλασσαν. Μητροδώρῳ καὶ Εὐκτήμονι καὶ
Φιλίππῳ καὶ αλλίππῳ νότος.

ιδʹ. ὡρῶν ιεʹ Lʹ ὁ ἐπὶ τῆς κεφαλῆς τοῦ ἑπομένου Διδύμου ἑῷος δύνει, καὶ ὁ
λαμπρὸς τοῦ Ὑδροχόου[93] ἑῷος δύνει, καὶ Κύων ἑσπέριος ἀνατέλλει.
Αἰγυπτίοις καὶ Εὐδόξῳ νότος σφοδρὸς καὶ ὑετός.

ιεʹ. ὡρῶν ιεʹ. . . . Αἰγυπτίοις καὶ Καίσαρι νότος πολύς, καὶ ἐπισημαίνει
κατὰ θάλασσαν, βροντή, ψακάς.

ιϛʹ. ὡρῶν ιεʹ ὁ λαμπρὸς τοῦ Ὑδροχόου ἑῷος δύνει. ὡρῶν ιεʹ Lʹ ὁ ἐπὶ τῆς
κεφαλῆς τοῦ ἡγουμένου Διδύμου ἑῷος δύνει. Εὐδόξῳ καὶ Δοσιθέῳ
νότος, ἐπισημαίνει. Ἱππάρχῳ ἀνέμων ἀκρασία.

ιζʹ. ὡρῶν ιγʹ Lʹ ὁ λαμπρὸς τοῦ νοτίου Ἰχθύος κρύπτεται.

ιηʹ. ὡρῶν ιδʹ ὁ λαμπρὸς τῆς Λύρας ἑσπέριος δύνει. ὡρῶν ιδʹ Lʹ ὁ κατὰ τὸ
γόνυ τοῦ Τοξότου ἐπιτέλλει.

ιθʹ. ὡρῶν ιδʹ Lʹ ὁ λαμπρὸς τοῦ Ὑδροχόου ἑῷος δύνει. Ἱππάρχῳ νότος ἢ
βορέας, χειμάζει.

κʹ.   Αἰγυπτίοις χειμῶνος ἀήρ.

καʹ. ὡρῶν ιδʹ ὁ λαμπρὸς τοῦ Ὑδροχόου ἑῷος δύνει. ὡρῶν ιεʹ ὁ ἐπὶ τῆς
καρδίας τοῦ Λέοντος ἑσπέριος ἀνατέλλει. Ἱππάρχῳ ἀπηλιώτης πνεῖ.

---

[93] Here and over the next fourteen days, following all MSS. Wachsmuth and Heiberg follow
Bonaventure's emendation to ὕδρου.

κβ. ὡρῶν ιγ΄ Ĺ ὁ ἐπὶ τῆς καρδίας τοῦ Λέοντος ἑσπέριος ἀνατέλλει, καὶ ὁ λαμπρὸς τοῦ Ὑδροχόου ἑσπέριος ἀνατέλλει, καὶ ὁ καλούμενος Κάνωβος
ἑσπέριος ἀνατέλλει. ὡρῶν ιδ΄ ὁ ἐπὶ τῆς καρδίας τοῦ Λέοντος ἑσπέριος
ἀνατέλλει. ὡρῶν ιδ΄ Ĺ ὁ ἐν τῷ ἐμπροσθίῳ δεξιῷ βατραχίῳ τοῦ Κενταύρου ἑῷος δύνει. ὡρῶν ιδ΄ Ĺ ὁ ἐπὶ τῆς καρδίας τοῦ Λέοντος ἑσπέριος
ἀνατέλλει. Καίσαρι ἄνεμοι σφοδροί.

κγ. ὡρῶν ιγ΄ Ĺ ὁ λαμπρὸς τοῦ Ὑδροχόου ἑῷος δύνει. Εὐκτήμονι καὶ Φιλίππῳ χειμών. Μητροδώρῳ ἀκαταστασία ἀέρος.

κδ. ὡρῶν ιδ΄ ὁ λαμπρὸς τοῦ Ὑδροχόου ἑσπέριος ἀνατέλλει. Αἰγυπτίοις
ὕει ἢ πνίγη γίνεται. Καίσαρι καὶ Εὐκτήμονι χειμών.

κε. ὡρῶν ιδ΄ Ĺ ὁ λαμπρὸς τῆς Λύρας ἑσπέριος δύνει, καὶ ὁ λαμπρὸς τοῦ
Ὑδροχόου ἑσπέριος ἀνατέλλει. ὡρῶν ιε΄ ὁ κατὰ τὸ γόνυ τοῦ Τοξότου
ἐπιτέλλει. Αἰγυπτίοις καὶ Καλλίππῳ χειμών, ὑετός. Ἱππάρχῳ βορρᾶς
πνεῖ. Εὐκτήμονι καὶ Δημοκρίτῳ ἐφύει.

κϛ΄. ὡρῶν ιε΄ ὁ λαμπρὸς τοῦ Ὑδροχόου ἑσπέριος ἀνατέλλει. Εὐδόξῳ
χειμὼν μέσος.

κζ΄. Αἰγυπτίοις εὖρος ἢ νότος, ἐπισημαίνει.

κη΄. ὡρῶν ιε΄ Ĺ ὁ λαμπρὸς τοῦ Ὑδροχόου ἑσπέριος ἀνατέλλει. Αἰγυπτίοις
ὑετία. Ἱππάρχῳ ἐπισημασία.

κθ΄. Καλλίππῳ καὶ Εὐκτήμονι ἐφύει. Δημοκρίτῳ μέσος χειμών.

λ΄. Ἱππάρχῳ ἀπηλιώτης πνεῖ.

Μεχίρ

α. ὡρῶν ιε΄ Ĺ ὁ κατὰ τὸ γόνυ τοῦ Τοξότου ἐπιτέλλει. Εὐδόξῳ ὑετία.
Μητροδώρῳ ὑετία. Δοσιθέῳ χειμών.

β΄. Αἰγυπτίοις χειμὼν μέσος.

γ΄. Αἰγυπτίοις λὶψ ἢ νότος, ἐπισημαίνει.

δ΄. ὡρῶν ιγ΄ Ĺ ὁ λαμπρὸς τοῦ Ὄρνιθος ἑσπέριος δύνει. ὡρῶν ιε΄ ὁ λαμπρὸς
τῆς Λύρας ἑσπέριος δύνει. Ἱππάρχῳ νότος ἢ ἀργεστής.

ϛ΄. ὡρῶν ιγ΄ Ĺ ὁ ἐπὶ τῆς καρδίας τοῦ Λέοντος ἑῷος δύνει. ὡρῶν ιδ΄ ὁ καλούμενος Κάνωβος ἑσπέριος ἀνατέλλει. ὡρῶν ιε΄ Ĺ ὁ ἐπὶ τῆς οὐρᾶς τοῦ Λέοντος ἑσπέριος ἀνατέλλει, καὶ ὁ κατὰ τὸ γόνυ τοῦ Τοξότου ἐπιτέλλει.
Εὐδόξῳ ὑετός.

ζ΄. ὡρῶν ιδ΄ ὁ ἐπὶ τῆς καρδίας τοῦ Λέοντος ἑῷος δύνει. ὡρῶν ιε΄ ὁ ἐπὶ τῆς
οὐρᾶς τοῦ Λέοντος ἑσπέριος ἀνατέλλει.

η΄. ὡρῶν ιδ΄ Ĺ ὁ ἐπὶ τῆς καρδίας τοῦ Λέοντος ἑῷος δύνει, καὶ ὁ ἐπὶ τῆς
οὐρᾶς τοῦ Λέοντος ἑσπέριος ἀνατέλλει. Αἰγυπτίοις νότος ἢ ζέφυρος,
μεταξὺ χάλαζα.

θ.   ὡρῶν ιε΄ ὁ ἐπὶ τῆς καρδίας τοῦ Λέοντος ἑῷος δύνει. Εὐδόξῳ εὐδία, ἐνίοτε δὲ καὶ ζέφυρος πνεῖ.

ι.   ὡρῶν ιδ΄ ὁ ἐπὶ τῆς οὐρᾶς τοῦ Λέοντος ἑσπέριος ἀνατέλλει.

ια.   ὡρῶν ιε΄ Ϛ΄ ὁ ἐπὶ τῆς καρδίας τοῦ Λέοντος ἑῷος δύνει. Αἰγυπτίοις περίστασις χειμερινὴ ἢ ἔπομβρος καὶ ἀνέμων ἀκρασία. Δοσιθέῳ εὐδία, ἐνίοτε ζέφυρος πνεῖ.

ιβ.   ὡρῶν ιδ΄ ὁ λαμπρὸς τοῦ Ὄρνιθος ἑσπέριος δύνει. ὡρῶν ιε΄ ὁ ἔσχατος τοῦ Ποταμοῦ κρύπτεται. ὡρῶν ιε΄ Ϛ΄ ὁ λαμπρὸς τοῦ Περσέως ἑῷος ἀνατέλλει, καὶ ὁ λαμπρὸς τῆς Λύρας ἑσπέριος δύνει. Αἰγυπτίοις ἀνεμώδης κατάστασις. Καίσαρι ὑετία. Δημοκρίτῳ ζέφυρος ἄρχεται πνεῖν.

ιγ.   ὡρῶν ιγ΄ Ϛ΄ ὁ ἐπὶ τῆς οὐρᾶς τοῦ Λέοντος ἑσπέριος ἀνατέλλει. Αἰγυπτίοις καὶ Εὐδόξῳ ἔαρος ἀρχή, ζέφυρος ἄρχεται πνεῖν καὶ ἐνίοτε χειμών.

ιδ.   Αἰγυπτίοις καὶ Εὐδόξῳ ὑετία. Ἱππάρχῳ καὶ Καλλίππῳ καὶ Δημοκρίτῳ ζεφύρῳ ὥρα πνεῖν.

ιε.   Καίσαρι καὶ Μητροδώρῳ ἔαρος ἀρχή, καὶ ζέφυρος ἄρχεται πνεῖν.

ιζ.   Αἰγυπτίοις καὶ Εὐδόξῳ ζέφυροι πνέουσιν. Ἱππάρχῳ ἔαρος ἀρχή. Καλλίππῳ καὶ Μητροδώρῳ χειμών.

ιη.   Αἰγυπτίοις ἀπηλιώτης πνεῖ. Ἱππάρχῳ βορρᾶς ἢ ἀπηλιώτης πνεῖ.

ιθ.   ὡρῶν ιδ΄ ὁ ἐν τῷ ἐμπροσθίῳ δεξιῷ βατραχίῳ τοῦ Κενταύρου ἑῷος δύνει. ὡρῶν ιε΄ Ϛ΄ ὁ κοινὸς Ἵππου καὶ Ἀνδρομέδας ἑῷος ἀνατέλλει.

κα.   ὡρῶν ιδ΄ Ϛ΄ ὁ λαμπρὸς τοῦ Ὄρνιθος ἑσπέριος δύνει. Αἰγυπτίοις ἄνεμοι μεταπίπτουσιν. Ἱππάρχῳ νότος πνεῖ. Εὐκτήμονι καὶ Φιλίππῳ καὶ Δοσιθέῳ χειμών.

κβ.   Αἰγυπτίοις ἀνέμων ἀκαταστασία καὶ ὄμβροι.

κγ.   ὡρῶν ιδ΄ Ϛ΄ ὁ καλούμενος Κάνωβος ἑσπέριος ἀνατέλλει.

κδ.   Αἰγυπτίοις ζέφυρος ἢ νότος καὶ χάλαζα, ὑετός.

κε.   ὡρῶν ιδ΄ Ϛ΄ ὁ ἔσχατος τοῦ Ποταμοῦ κρύπτεται. ὡρῶν ιε΄ ὁ κοινὸς Ἵππου καὶ Ἀνδρομέδας ἑῷος ἀνατέλλει. Ἱππάρχῳ βορέας ψυχρὸς πνεῖ.

κϛ.   Αἰγυπτίοις ἀνεμώδης κατάστασις.

κη.   Ἱππάρχῳ καὶ Εὐκτήμονι ὀρνιθίαι ἄρχονται πνεῖν ψυχροί, καὶ χελιδόνι ὥρα φαίνεσθαι.

κθ.   ὡρῶν ιγ΄ Ϛ΄ ὁ κοινὸς Ἵππου καὶ Ἀνδρομέδας κρύπτεται. ὡρῶν ιε΄ ὁ λαμπρὸς τοῦ Ὄρνιθος ἑσπέριος δύνει. Αἰγυπτίοις καὶ Φιλίππῳ καὶ Καλλίππῳ χελιδὼν φαίνεται, καὶ ἀνεμώδης κατάστασις. Κόνωνι βορέαι ἄρχονται πνεῖν ψυχροί. Εὐδόξῳ ὑετὸς ἐπὶ χελιδόνι, καὶ ἐπὶ λ΄ ἡμέρας βορέαι πνέουσιν οἱ καλούμενοι ὀρνιθίαι.

λ.   Αἰγυπτίοις ὀρνιθίαι βορέαι, μεταξὺ ἀργεστής. Ἱππάρχῳ βορέαι ψυχροί. Μητροδώρῳ χελιδὼν φαίνεται, καὶ ἐπισημαίνει. Δημοκρίτῳ ποικίλαι ἡμέραι αἱ καλούμεναι ἀλκυονίδες.

Φαμενώθ

α. ὡρῶν ιδ´ Ľ ὁ κοινὸς Ἵππου καὶ Ἀνδρομέδας ἑῷος ἀνατέλλει. ὡρῶν ιέ Ľ Ἀρκτοῦρος ἑσπέριος ἀνατέλλει. Καίσαρι καὶ Δοσιθέῳ χειμών, ἐπιση- μαίνει.

β´. ὡρῶν ιδ´ ὁ κοινὸς Ἵππου καὶ Ἀνδρομέδας κρύπτεται.

γ´. ὡρῶν ιέ ὁ λαμπρὸς τοῦ Περσέως ἑῷος ἀνατέλλει.

δ´. ὡρῶν ιδ´ Ľ ὁ κοινὸς Ἵππου καὶ Ἀνδρομέδας ἑσπέριος δύνει.

ε´. ὡρῶν ιδ´ ὁ κοινὸς Ἵππου καὶ Ἀνδρομέδας ἐπιτέλλει. ὡρῶν ιέ Ἀρκτοῦρος ἑσπέριος ἀνατέλλει. Ἱππάρχῳ βορρᾶς ἢ νότος ψυχρὸς πνεῖ.

ϛ´. ὡρῶν ιδ´ ὁ ἔσχατος τοῦ Ποταμοῦ κρύπτεται. Αἰγυπτίοις λὶψ ἢ νότος, χάλαζα. Ἱππάρχῳ βορέας ψυχρὸς πνεῖ.

ζ´. ὡρῶν ιέ ὁ κοινὸς Ἵππου καὶ Ἀνδρομέδας ἑσπέριος δύνει. ὡρῶν ιέ Ľ ὁ λαμπρὸς τοῦ Ὄρνιθος ἑσπέριος δύνει.

η´. ὡρῶν ιδ´ Ľ Ἀρκτοῦρος ἑσπέριος ἀνατέλλει. Εὐκτήμονι βορρᾶς ψυχρὸς πνεῖ.

θ´. ὡρῶν ιέ Ľ ὁ λαμπρὸς τοῦ βορείου Στεφάνου ἑσπέριος ἀνατέλλει, καὶ ὁ κοινὸς Ἵππου καὶ Ἀνδρομέδας ἑσπέριος δύνει. Αἰγυπτίοις χειμάζει. Καίσαρι χελιδονίαι πνέουσιν ἐπὶ ἡμέρας ι´.

ι´. ὡρῶν ιγ´ Ľ ὁ κοινὸς Ἵππου καὶ Ἀνδρομέδας ἐπιτέλλει.

ια´. ὡρῶν ιγ´ Ľ ὁ λαμπρὸς τοῦ νοτίου Ἰχθύος ἐπιτέλλει, καὶ ὁ ἐν τῷ ἐμπροσθίῳ δεξιῷ βατραχίῳ τοῦ Κενταύρου ἑῷος δύνει. Αἰγυπ- τίοις ταραχώδης κατάστασις. Δημοκρίτῳ ἄνεμοι ψυχροὶ ὀρνιθίαι ἐπὶ ἡμέρας θ´.

ιβ´. ὡρῶν ιδ´ Ἀρκτοῦρος ἑσπέριος ἀνατέλλει. Εὐδόξῳ χειμών, καὶ ἰκτῖνος φαίνεται, καὶ ἐπισημαίνει. Μητροδώρῳ καὶ Εὐκτήμονι καὶ Φιλίππῳ βορέας ψυχρὸς πνεῖ. Ἱππάρχῳ ἔαρος ἀρχή.

ιγ´. ὡρῶν ιγ´ Ľ ὁ ἐπὶ τῆς οὐρᾶς τοῦ Λέοντος ἑῷος δύνει. Αἰγυπτίοις ψακάζει. Μητροδώρῳ καὶ Εὐκτήμονι βορέας πνεῖ. Δοσιθέῳ ἰκτῖνος ἄρχεται φαί- νεσθαι. Ἱππάρχῳ νότος πολύς.

ιδ´. ὡρῶν ιέ ὁ λαμπρὸς τοῦ βορείου Στεφάνου ἑσπέριος ἀνατέλλει. Αἰγυπ- τίοις καὶ Καλλίππῳ βορέας ψυχρὸς πνεῖ.

ιέ. ὡρῶν ιγ´ Ľ Ἀρκτοῦρος ἑσπέριος ἀνατέλλει.

ιϛ´. ὡρῶν ιγ´ Ľ ὁ ἔσχατος τοῦ Ποταμοῦ κρύπτεται. Καλλίππῳ βορρᾶς σύμμετρος πνεῖ.

ιζ´. ὡρῶν ιγ´ Ľ Στάχυς ἑσπέριος ἀνατέλλει. ὡρῶν ιδ´ Ľ Στάχυς ἑσπέριος ἀνατέλλει. Αἰγυπτίοις ἀνεμώδης κατάστασις. Εὐκτήμονι καὶ Φιλίππῳ ὀρνιθίαι ἄρχονται πνεῖν, καὶ ἰκτίνῳ ὥρα φαίνεσθαι.

ιη´. ὡρῶν ιδ´ ὁ ἐπὶ τῆς οὐρᾶς τοῦ Λέοντος ἑῷος δύνει. Αἰγυπτίοις ζέφυρος ἢ νότος πνεῖ. Εὐκτήμονι βορρᾶς ψυχρὸς πνεῖ. Δοσιθέῳ ὀρνιθίαι ἄρχονται πνεῖν. Ἱππάρχῳ βορρᾶς ἢ ἀργεστής.

ιθ´.   Αἰγυπτίοις καὶ Εὐκτήμονι βορρᾶς ψυχρὸς πνεῖ.

κ´.   ὡρῶν ιδ´ ὁ λαμπρὸς τοῦ νοτίου Ἰχθύος ἐπιτέλλει. ὡρῶν ιδ´ L´ ὁ λαμπρὸς τοῦ βορείου Στεφάνου ἑσπέριος ἀνατέλλει.

κα´.   ὡρῶν ιδ´ L´ ὁ λαμπρὸς τοῦ Περσέως ἑῷος ἀνατέλλει. Καλλίππῳ βορρᾶς πνεῖ, καὶ ἰκτῖνος φαίνεται.

κβ´.   Αἰγυπτίοις καὶ Δημοκρίτῳ χειμών, ἄνεμος ψυχρός.

κγ´.   Αἰγυπτίοις πνεύματα ψυχρὰ ἕως ἰσημερίας. Ἱππάρχῳ βορρᾶς πνεῖ.

κδ´.   Καίσαρι ἰκτῖνος φαίνεται, καὶ βορρᾶς πνεῖ.

κε´.   ὡρῶν ιδ´ L´ ὁ ἐπὶ τῆς οὐρᾶς τοῦ Λέοντος ἑῷος δύνει. Εὐδόξῳ ἰκτῖνος φαίνεται, καὶ βορρᾶς πνεῖ.

κϛ´.   ἐαρινὴ ἰσημερία. ὡρῶν ιδ´ ὁ λαμπρὸς τοῦ βορείου Στεφάνου ἑσπέριος ἀνατέλλει.

κζ´.   Καίσαρι βορρᾶς πνεῖ. Ἱππάρχῳ ὑετία.

κη´.   Αἰγυπτίοις βρονταί, ἐπισημασία. Φιλίππῳ καὶ Καλλίππῳ καὶ Εὐκτήμονι ὑετὸς ἢ ψακάς. Ἱππάρχῳ ἐπισημασία.

κθ´.   ὡρῶν ιε´ L´ ὁ καλούμενος Αἴξ ἑῷος ἀνατέλλει. Αἰγυπτίοις καὶ Κόνωνι καὶ Μέτωνι ἰσημερία. Εὐδόξῳ βορρᾶς πνεῖ.

λ´.   ὡρῶν ιγ´ L´ Στάχυς ἑῷος δύνει. Αἰγυπτίοις ἀργεστὴς ἄνεμος πνεῖ. Καλλίππῳ ὑετὸς ἢ νιφετός.

Φαρμουθί

α´.   ὡρῶν ιδ´ Στάχυς ἑῷος δύνει. Μέτωνι καὶ Καλλίππῳ καὶ Εὐδόξῳ ὑετός. Εὐκτήμονι καὶ Δημοκρίτῳ ἐπισημαίνει.

β´.   ὡρῶν ιγ´ L´ ὁ λαμπρὸς τοῦ βορείου Στεφάνου ἑσπέριος ἀνατέλλει. ὡρῶν ιδ´ L´ Στάχυς ἑῷος δύνει, καὶ ὁ καλούμενος Κάνωβος κρύπτεται. ὡρῶν ιε´ ὁ ἐπὶ τῆς οὐρᾶς τοῦ Λέοντος ἑῷος δύνει. Δοσιθέῳ καὶ Μέτωνι καὶ Καλλίππῳ ὑετία.

γ´.   ὡρῶν ιδ´ ὁ λαμπρὸς τοῦ Περσέως ἑῷος ἀνατέλλει. ὡρῶν ιδ´ L´ ὁ λαμπρὸς τοῦ νοτίου Ἰχθύος ἐπιτέλλει.

δ´.   ὡρῶν ιε´ L´ ὁ λαμπρὸς τῆς βορείου Χηλῆς ἑσπέριος ἀνατέλλει. Αἰγυπτίοις καὶ Κόνωνι ἐπισημαίνει. Εὐδόξῳ ὑετία γίνεται.

ε´.   ὡρῶν ιε´ Στάχυς ἑῷος δύνει.

ϛ´.   ὡρῶν ιε´ L´ ὁ λαμπρὸς τῆς νοτίου Χηλῆς ἑσπέριος ἀνατέλλει. Εὐδόξῳ ὑετός, ἐπισημαίνει.

ζ´.   ὡρῶν ιγ´ L´ ὁ λαμπρὸς τῆς νοτίου Χηλῆς ἑσπέριος ἀνατέλλει. ὡρῶν ιε´ L´ Στάχυς ἑῷος δύνει.

η´.   ὡρῶν ιε´ ὁ λαμπρὸς τῆς βορείου Χηλῆς ἑσπέριος ἀνατέλλει. Αἰγυπτίοις ζέφυρος καὶ χάλαζα. Κόνωνι ἐπισημαίνει. Εὐδόξῳ ὑετός.

θ´. ὡρῶν ιδ´ ∟´ ὁ λαμπρὸς τῆς βορείου Χηλῆς ἑσπέριος ἀνατέλλει. Αἰγυπτίοις καὶ Κόνωνι ζέφυρος ἢ νότος καὶ χάλαζα.

ι´. ὡρῶν ιδ´ ὁ λαμπρὸς τῆς βορείου Χηλῆς ἑσπέριος ἀνατέλλει. ὡρῶν ιε´ ∟´ ὁ λαμπρὸς τῆς Λύρας ἑσπέριος ἀνατέλλει. Ἱππάρχῳ νότος καὶ ἀνέμων συστροφή.

ια´. ὡρῶν ιγ´ ∟´ ὁ λαμπρὸς τῆς βορείου Χηλῆς ἑσπέριος ἀνατέλλει. Ἱππάρχῳ καὶ Δοσιθέῳ ἐπισημαίνει.

ιβ´. ὡρῶν ιε´ ∟´ ὁ ἐπὶ τῆς οὐρᾶς τοῦ Λέοντος ἑῷος δύνει.

ιγ´. ὡρῶν ιγ´ <∟´> . . . Αἰγυπτίοις νότος ἢ λίψ. Εὐδόξῳ ὑετία.

ιδ´. ὡρῶν ιγ´ ∟´ ὁ λαμπρὸς τοῦ Περσέως ἑῷος ἀνατέλλει. Αἰγυπτίοις ἀκρασία πνευμάτων. Ἱππάρχῳ ὑετία.

ιε´. Αἰγυπτίοις ἀέρος ἀκαταστασία καὶ ὑετός. Εὐκτήμονι καὶ Φιλίππῳ ἀκρασία πνευμάτων. Ἱππάρχῳ ὑετία.

ιϛ´. Εὐδόξῳ ζέφυρος καὶ ἀκρασία ἀέρος, μεταξὺ ψακάζει.

ιζ´. ὡρῶν ιε´ ∟´ ὁ κοινὸς Ποταμοῦ καὶ ποδὸς Ὠρίωνος κρύπτεται.

ιη´. ὡρῶν ιε´ ὁ καλούμενος Αἲξ ἑῷος ἀνατέλλει, καὶ ὁ λαμπρὸς τοῦ νοτίου Ἰχθύος ἐπιτέλλει. Δοσιθέῳ καὶ Καίσαρι ὑετία.

ιθ´. ὡρῶν ιε´ ὁ λαμπρὸς τῆς Λύρας ἑσπέριος ἀνατέλλει. Αἰγυπτίοις λευκόνοτος, βρονταί, ψακάς.

κ´. ὡρῶν ιδ´ ὁ καλούμενος Κάνωβος κρύπτεται. Αἰγυπτίοις ἀνέμων ἀκρασία.[94] Εὐδόξῳ καὶ Εὐκτήμονι ὑετία καὶ χάλαζα.

κα´. ὡρῶν ιε´ ὁ κοινὸς Ποταμοῦ καὶ ποδὸς Ὠρίωνος κρύπτεται. ὡρῶν ιε´ ∟´ ὁ λαμπρὸς τῶν Ὑάδων κρύπτεται. Μητροδώρῳ καὶ Καλλίππῳ χάλαζα. Εὐκτήμονι καὶ Φιλίππῳ ζέφυρος.

κβ´. ὡρῶν ιγ´ ∟´ ὁ λαμπρὸς τοῦ Περσέως ἑσπέριος δύνει. Αἰγυπτίοις καὶ Κόνωνι χάλαζα καὶ ζέφυρος. Καίσαρι καὶ Εὐδόξῳ ὑετία.

κγ´. ὡρῶν ιε´ ὁ λαμπρὸς τῶν Ὑάδων κρύπτεται. Αἰγυπτίοις ἀνεμώδης ψακάς.

κδ´. ὡρῶν ιδ´ ∟´ ὁ λαμπρὸς τῶν Ὑάδων κρύπτεται, καὶ ὁ κοινὸς Ποταμοῦ καὶ ποδὸς Ὠρίωνος κρύπτεται. ὡρῶν ιε´ ∟´ ὁ μέσος τῆς ζώνης τοῦ Ὠρίωνος κρύπτεται.

κε´. Αἰγυπτίοις λίψ ἢ νότος ἢ ἀργεστὴς καὶ ἀκρασία ἀέρος.

κϛ´. ὡρῶν ιδ´ ὁ λαμπρὸς τοῦ Περσέως ἑσπέριος δύνει, καὶ ὁ λαμπρὸς τῶν Ὑάδων κρύπτεται. ὡρῶν ιε´ ∟´ ὁ λαμπρὸς τοῦ Ὄρνιθος ἑσπέριος ἀνατέλλει, καὶ ὁ ἐν τῷ ἡγουμένῳ ὤμῳ τοῦ Ὠρίωνος κρύπτεται. Ἱππάρχῳ νότος ἢ ἀπαρκτίας ψυχρός.

---

[94] Following *C. Vat. gr.* 1594. Heiberg, following *C. Vat. gr.* 318, reads ἀκρισία, indistinctness.

κζ́.   ὡρῶν ιγ́ Ĺ ὁ λαμπρὸς τῶν Ὑάδων κρύπτεται, καὶ ὁ λαμπρὸς τῆς νοτίου
       Χηλῆς ἑῷος δύνει. ὡρῶν ιέ ὁ μέσος τῆς ζώνης τοῦ Ὠρίωνος κρύπτεται.
       Αἰγυπτίοις καὶ Καίσαρι χειμών. Εὐδόξῳ ὑετός.

κή.   ὡρῶν ιδ́ ὁ κοινὸς Ποταμοῦ καὶ ποδὸς Ὠρίωνος κρύπτεται. ὡρῶν ιδ́
       Ĺ ὁ λαμπρὸς τῆς Λύρας ἑσπέριος ἀνατέλλει. Αἰγυπτίοις λὶψ ἢ νότος,
       ὑετία.

κθ́.   ὡρῶν ιδ́ ὁ λαμπρὸς τῆς νοτίου Χηλῆς ἑῷος δύνει. ὡρῶν ιέ ὁ ἐν τῷ
       ἡγουμένῳ ὤμῳ τοῦ Ὠρίωνος κρύπτεται. Αἰγυπτίοις λὶψ ἢ νότος καὶ
       ὑετία. Μητροδώρῳ καὶ Καλλίππῳ ἐνίοτε χάλαζα. Δημοκρίτῳ ἐπιση-
       μαίνει.

λ́.   Αἰγυπτίοις καὶ Εὐδόξῳ ψακάς, ὑετός.

Παχών

ά.   ὡρῶν ιδ́ Ĺ ὁ λαμπρὸς τοῦ Περσέως ἑσπέριος δύνει, καὶ ὁ μέσος τῆς ζώνης
       τοῦ Ὠρίωνος κρύπτεται, καὶ ὁ λαμπρὸς τῆς νοτίου Χηλῆς ἑῷος δύνει.
       Αἰγυπτίοις ἀργεστὴς ἢ ζέφυρος, ἐπισημαίνει. Εὐκτήμονι καὶ Φιλίππῳ
       ὑετία ἢ χάλαζα.

β́.   ὡρῶν ιδ́ Ĺ ὁ καλούμενος Αἲξ ἑῷος ἀνατέλλει, καὶ ὁ ἐν τῷ ἡγουμένῳ
       ὤμῳ τοῦ Ὠρίωνος κρύπτεται. Αἰγυπτίοις ἀνεμώδης κατάστασις.
       Μητροδώρῳ καὶ Καλλίππῳ νοτία.

γ́.   ὡρῶν ιγ́ Ĺ ὁ κοινὸς Ποταμοῦ καὶ ποδὸς Ὠρίωνος κρύπτεται, καὶ ὁ
       καλούμενος Ἀντάρης ἑσπέριος ἀνατέλλει. ὡρῶν ιέ Ĺ Κύων κρύπτεται.
       Αἰγυπτίοις ἄνεμοι. Εὐδόξῳ ὑετός.

δ́.   ὡρῶν ιδ́ ὁ ἐν τῷ ἡγουμένῳ ὤμῳ τοῦ Ὠρίωνος κρύπτεται, καὶ ὁ
       μέσος τῆς ζώνης τοῦ Ὠρίωνος κρύπτεται, καὶ ὁ καλούμενος Ἀντάρης
       ἑσπέριος ἀνατέλλει. ὡρῶν ιδ́ Ĺ τὸ αὐτό. ὡρῶν ιέ τὸ αὐτό. Αἰγυπτίοις
       νηνεμία ἢ νότος καὶ ὑετία. Καίσαρι χειμών.

έ.   ὡρῶν ιγ́ Ĺ ὁ καλούμενος Κάνωβος κρύπτεται. ὡρῶν ιέ ὁ λαμπρὸς
       τῆς νοτίου Χηλῆς ἑῷος δύνει. Αἰγυπτίοις ἐπισημαίνει. Εὐκτήμονι καὶ
       Φιλίππῳ νηνεμία ἢ νότος, ψακάς.

ς́.   ὡρῶν ιγ́ Ĺ ὁ ἐν τῷ ἐμπροσθίῳ δεξιῷ βατραχίῳ τοῦ Κενταύρου ἑσπέριος
       ἀνατέλλει. ὡρῶν ιέ ὁ λαμπρὸς τοῦ Περσέως ἑσπέριος δύνει. ὡρῶν ιέ Ĺ ὁ
       ἐν τῷ ἑπομένῳ ὤμῳ τοῦ Ἡνιόχου ἑῷος ἀνατέλλει, καὶ ὁ ἐν τῷ ἑπομένῳ
       ὤμῳ τοῦ Ὠρίωνος κρύπτεται. Αἰγυπτίοις ψακάς.

ζ́.   ὡρῶν ιγ́ Ĺ ὁ ἐν τῷ ἡγουμένῳ ὤμῳ τοῦ Ὠρίωνος κρύπτεται, καὶ ὁ μέσος
       τῆς ζώνης κρύπτεται. ὡρῶν ιέ Κύων κρύπτεται.

ή.   ὡρῶν ιδ́ ὁ λαμπρὸς τῆς Λύρας ἑσπέριος ἀνατέλλει. ὡρῶν ιέ ὁ λαμ-
       πρὸς τοῦ Ὄρνιθος ἑσπέριος ἀνατέλλει, καὶ ὁ ἐν τῷ ἑπομένῳ ὤμῳ τοῦ

Ὠρίωνος κρύπτεται. ὡρῶν ιε΄ L΄ ὁ λαμπρὸς τῆς νοτίου Χηλῆς ἑῷος δύνει. Αἰγυπτίοις ἀργεστὴς καὶ ψακὰς ἢ νότος, βροντή.

θ. ὡρῶν ιδ΄ ὁ καλούμενος Αἴξ ἑῷος ἀνατέλλει. ὡρῶν ιε΄ L΄ ὁ λαμπρὸς τοῦ νοτίου Ἰχθύος ἐπιτέλλει. Αἰγυπτίοις ψακάς. Εὐδόξῳ ὑετός.

ι. ὡρῶν ιγ΄ L΄ ὁ λαμπρὸς τῆς βορείου Χηλῆς ἑῷος δύνει. Δοσιθέῳ ὑετία.

ια. ὡρῶν ιδ΄ L΄ ὁ ἐν τῷ ἑπομένῳ ὤμῳ τοῦ Ὠρίωνος κρύπτεται. Αἰγυπτίοις ἀνεμώδης κατάστασις.

ιβ. ὡρῶν ιγ΄ L΄ ὁ καλούμενος Αἴξ ἑῷος ἀνατέλλει. ὡρῶν ιδ΄ L΄ Κύων κρύπτεται. ὡρῶν ιε΄ L΄ ὁ λαμπρὸς τοῦ Περσέως ἑσπέριος δύνει. Αἰγυπτίοις ἀνεμώδης κατάστασις.

ιγ. Αἰγυπτίοις ζέφυρος ἢ ἀργεστὴς καὶ ὑετία. Εὐδόξῳ καὶ Δοσιθέῳ ὑετία.

ιδ. ὡρῶν ιδ΄ ὁ ἐν τῷ ἑπομένῳ ὤμῳ τοῦ Ὠρίωνος κρύπτεται, καὶ ὁ λαμπρὸς τῆς βορείου Χηλῆς ἑῷος δύνει. Αἰγυπτίοις ὄμβρος.

ιε. Αἰγυπτίοις ὑετός, θέρους ἀρχή. Εὐκτήμονι καὶ Φιλίππῳ ἐπισημαίνει.

ις΄. ὡρῶν ιγ΄ L΄ Ἀρκτοῦρος ἑῷος δύνει, καὶ ὁ ἐν τῷ ἑπομένῳ ὤμῳ τοῦ Ὠρίωνος κρύπτεται. Δοσιθέῳ ἐπισημαίνει.

ιζ. ὡρῶν ιγ΄ L΄ ὁ καλούμενος Αἴξ ἑσπέριος δύνει, καὶ ὁ λαμπρὸς τῆς Λύρας ἑσπέριος ἀνατέλλει. ὡρῶν ιδ΄ Κύων κρύπτεται, καὶ ὁ ἐν τῷ ἐμπροσθίῳ δεξιῷ βατραχίῳ τοῦ Κενταύρου ἑσπέριος ἀνατέλλει. Αἰγυπτίοις ζέφυρος ἢ ἀργεστής. Καίσαρι ὑετός. Μητροδώρῳ καὶ Εὐδόξῳ καὶ Ἱππάρχῳ ἐπισημαίνει· καὶ θέρους ἀρχή.

ιη. ὡρῶν ιγ΄ L΄ ὁ καλούμενος Ἀντάρης ἑῷος δύνει. ὡρῶν ιδ΄ L΄ ὁ λαμπρὸς τοῦ Ὄρνιθος ἑσπέριος ἀνατέλλει. ὡρῶν ιε΄ ὁ ἐν τῷ ἑπομένῳ ὤμῳ τοῦ Ἡνιόχου ἑῷος ἀνατέλλει. Αἰγυπτίοις ζέφυρος ἢ λίψ, ἐπισημασία. Εὐδόξῳ καὶ Κόνωνι ὑετία.

ιθ. ὡρῶν ιδ΄ L΄ ὁ καλούμενος Ἀντάρης ἑῷος δύνει. Αἰγυπτίοις καὶ Εὐδόξῳ καὶ Καλλίππῳ ἐπισημασία.

κ. ὡρῶν ιδ΄ ὁ καλούμενος Αἴξ ἑσπέριος δύνει. ὡρῶν ιε΄ ὁ καλούμενος Ἀντάρης ἑῷος δύνει. Καίσαρι ἐπισημασία, ὑετία.

κα. ὡρῶν ιε΄ L΄ ὁ καλούμενος Ἀντάρης ἑῷος δύνει. Καίσαρι ἐπισημαίνει.

κβ. Αἰγυπτίοις νότος ἢ ἀπηλιώτης. Εὐδόξῳ ὑετία. Ἱππάρχῳ νότος ἢ ἀπαρκτίας.

κγ. ὡρῶν ιγ΄ L΄ ὁ ἐν τῷ ἑπομένῳ ὤμῳ τοῦ Ἡνιόχου κρύπτεται, καὶ Κύων κρύπτεται. Αἰγυπτίοις ὄμβρος καὶ βροντή. Εὐδόξῳ θέρους ἀρχή, ὑετία.

κδ. ὡρῶν ιδ΄ L΄ ὁ καλούμενος Αἴξ ἑσπέριος δύνει, καὶ ὁ ἐν τῷ ἑπομένῳ ὤμῳ τοῦ Ἡνιόχου ἑῷος ἀνατέλλει. ὡρῶν ιε΄ L΄ ὁ λαμπρὸς τοῦ Ἀετοῦ ἑσπέριος ἀνατέλλει. Αἰγυπτίοις καὶ Ἱππάρχῳ ψακάζει καὶ ἐπισημαίνει.

κε. ὡρῶν ιδ΄ ὁ ἐν τῷ ἑπομένῳ ὤμῳ τοῦ Ἡνιόχου κρύπτεται. ὡρῶν ιε΄ ὁ λαμπρὸς τῆς βορείου Χηλῆς ἑῷος δύνει.

κϛʹ. ὡρῶν ιδʹ Ἀρκτοῦρος ἑῷος δύνει. Αἰγυπτίοις ἀργεστὴς ἢ ζέφυρος. Δοσιθέῳ νότος. Καίσαρι χειμάζει.

κζʹ. ὡρῶν ιεʹ ὁ λαμπρὸς τοῦ Ἀετοῦ ἑσπέριος ἀνατέλλει. ὡρῶν ιεΖʹ Προκύων κρύπτεται.

κηʹ. ὡρῶν ιδʹ Ζʹ ὁ ἐν τῷ ἑπομένῳ ὤμῳ τοῦ Ἡνιόχου ἑσπέριος δύνει. ὡρῶν ιεʹ ὁ καλούμενος Αἲξ ἑσπέριος δύνει.

κθʹ. ὡρῶν ιεʹ Ζʹ ὁ κατὰ τὸ γόνυ τοῦ Τοξότου ἑῷος δύνει. Αἰγυπτίοις ἀνεμώδης κατάστασις. Εὐκτήμονι καὶ Φιλίππῳ ἐπισημασία.

λʹ. ὡρῶν ιδʹ ὁ λαμπρὸς τοῦ Ὄρνιθος ἑσπέριος ἀνατέλλει. Εὐκτήμονι καὶ Φιλίππῳ καὶ Ἱππάρχῳ ἐπισημασία.

Παυνί

αʹ. ὡρῶν ιγʹ Ζʹ ὁ ἐν τῷ ἑπομένῳ ὤμῳ τοῦ Ἡνιόχου ἐπιτέλλει. ὡρῶν ιεʹ ὁ ἐν τῷ ἑπομένῳ ὤμῳ τοῦ Ἡνιόχου ἑσπέριος δύνει, καὶ Προκύων κρύπτεται. ὡρῶν ιεʹ Ζʹ ὁ λαμπρὸς τῆς βορείου Χηλῆς ἑῷος δύνει. Αἰγυπτίοις βορέας σφοδρός. Καλλίππῳ καὶ Εὐκτήμονι ἐπισημαίνει.

βʹ. ὡρῶν ιδʹ Ζʹ ὁ λαμπρὸς τοῦ Ἀετοῦ ἑσπέριος ἀνατέλλει. Αἰγυπτίοις ἐπισημασία. Μητροδώρῳ καὶ Καλλίππῳ νοτία.

γʹ. ὡρῶν ιγʹ Ζʹ ὁ λαμπρὸς τῶν Ὑάδων ἐπιτέλλει. ὡρῶν ιδʹ Ζʹ Προκύων κρύπτεται. Αἰγυπτίοις καὶ Δημοκρίτῳ ὑετία.

δʹ. Ἱππάρχῳ νότος ἢ ζέφυρος.

εʹ. ὡρῶν ιδʹ Ζʹ ὁ ἐν τῷ ἐμπροσθίῳ δεξιῷ βατραχίῳ τοῦ Κενταύρου ἑσπέριος ἀνατέλλει. ὡρῶν ιεʹ Ζʹ ὁ καλούμενος Αἲξ ἑσπέριος δύνει, καὶ ὁ ἐν τῷ ἑπομένῳ ὤμῳ τοῦ Ἡνιόχου ἑσπέριος δύνει. Καίσαρι νότος πνεῖ.

ϛʹ. ὡρῶν ιδʹ Προκύων κρύπτεται, καὶ ὁ λαμπρὸς τοῦ Ἀετοῦ ἑσπέριος ἀνατέλλει. ὡρῶν ιεʹ ὁ κατὰ τὸ γόνυ τοῦ Τοξότου ἑῷος δύνει.

ζʹ. ὡρῶν ιδʹ ὁ λαμπρὸς τῶν Ὑάδων ἐπιτέλλει. ὡρῶν ιδʹ Ζʹ Ἀρκτοῦρος ἑῷος δύνει. Αἰγυπτίοις ζέφυρος. Εὐδόξῳ καὶ Δοσιθέῳ νοτία.

ηʹ. Αἰγυπτίοις ἀργεστὴς ἢ ζέφυρος πνεῖ.

θʹ. ὡρῶν ιδʹ Ζʹ ὁ κατὰ τὸ γόνυ τοῦ Τοξότου ἑῷος δύνει. ὡρῶν ιεʹ Ζʹ ὁ λαμπρὸς τοῦ Ὕδρου κρύπτεται. Αἰγυπτίοις ἀργεστὴς καὶ ψακάς. Δημοκρίτῳ ὕδωρ γίνεται.

ιʹ. ὡρῶν ιγʹ Ζʹ ὁ λαμπρὸς τοῦ Ὄρνιθος ἑσπέριος ἀνατέλλει. ὡρῶν ιεʹ Ζʹ ὁ ἐπὶ τῆς κεφαλῆς τοῦ ἑπομένου Διδύμου κρύπτεται. Καίσαρι βρονταὶ καὶ ὑετός.

ιαʹ. ὡρῶν ιγʹ Ζʹ ὁ λαμπρὸς τοῦ Ἀετοῦ ἑσπέριος ἀνατέλλει, καὶ ὁ ἐπὶ τῆς κεφαλῆς τοῦ ἡγουμένου Διδύμου κρύπτεται. ὡρῶν ιεʹ ὁ ἐπὶ τῆς κεφαλῆς τοῦ ἑπομένου Διδύμου κρύπτεται. Αἰγυπτίοις ψακάζει. Καίσαρι βροντή, ὑετός.

ιβ΄. ὡρῶν ιδ΄ ∠΄ ὁ ἐπὶ τῆς κεφαλῆς τοῦ ἑπομένου Διδύμου κρύπτεται.

ιγ΄. ὡρῶν ιδ΄ ὁ ἐπὶ τῆς κεφαλῆς τοῦ ἡγουμένου Διδύμου κρύπτεται, καὶ ὁ κατὰ τὸ γόνυ τοῦ Τοξότου ἑῷος δύνει. ὡρῶν ιδ΄ ∠΄ ὁ ἐπὶ τῆς κεφαλῆς τοῦ ἡγουμένου Διδύμου κρύπτεται.

ιδ΄. ὡρῶν ιδ΄ ὁ ἐπὶ τῆς κεφαλῆς τοῦ ἑπομένου Διδύμου κρύπτεται. ὡρῶν ιδ΄ ∠΄ ὁ λαμπρὸς τῶν Ὑάδων ἐπιτέλλει. ὡρῶν ιε΄ ὁ ἐπὶ τῆς κεφαλῆς τοῦ ἡγουμένου Διδύμου κρύπτεται. ὡρῶν ιε΄ ∠΄ ὁ ἐπὶ τῆς κεφαλῆς τοῦ ἡγουμένου Διδύμου κρύπτεται.

ιε΄. ὡρῶν ιγ΄ ∠΄ ὁ ἐπὶ τῆς κεφαλῆς τοῦ ἑπομένου Διδύμου κρύπτεται, καὶ ὁ κατὰ τὸ γόνυ τοῦ Τοξότου ἑσπέριος ἀνατέλλει, καὶ ὁ κατὰ τὸ γόνυ τοῦ Τοξότου ἑῷος δύνει. ὡρῶν ιε΄ ὁ λαμπρὸς τοῦ Ὕδρου κρύπτεται. Αἰγυπτίοις ζέφυρος ἢ ἀργεστής, βροντή.

ιϛ΄. ὡρῶν ιγ΄ ∠΄ ὁ λαμπρὸς τοῦ βορείου Στεφάνου ἑῷος δύνει.

ιζ΄. ὡρῶν ιε΄ ὁ λαμπρὸς τῶν Ὑάδων ἐπιτέλλει. Αἰγυπτίοις δι᾽ ἡμέρας ψακάζει.

ιη΄. ὡρῶν ιδ΄ ὁ κατὰ τὸ γόνυ τοῦ Τοξότου ἑσπέριος ἀνατέλλει. ὡρῶν ιε΄ Ἀρκτοῦρος ἑῷος δύνει.

ιθ΄. Αἰγυπτίοις ζέφυρος ἢ ἀργεστής, ψακάζει.

κ΄. ὡρῶν ιδ΄ ∠΄ ὁ λαμπρὸς τοῦ Ὕδρου κρύπτεται, καὶ ὁ κατὰ τὸ γόνυ τοῦ Τοξότου ἑσπέριος ἀνατέλλει.

κα΄. ὡρῶν ιγ΄ ∠΄ ὁ ἐν τῷ ἡγουμένῳ ὤμῳ τοῦ Ὠρίωνος ἐπιτέλλει, καὶ ὁ ἔσχατος τοῦ Ποταμοῦ ἐπιτέλλει. Αἰγυπτίοις ψακάζει.

κβ΄. ὡρῶν ιε΄ ∠΄ ὁ λαμπρὸς τῶν Ὑάδων ἐπιτέλλει.

κγ΄. Αἰγυπτίοις καῦμα. Δοσιθέῳ ἐπισημασία.

κδ΄. ὡρῶν ιε΄ ὁ κατὰ τὸ γόνυ τοῦ Τοξότου ἑσπέριος ἀνατέλλει. Αἰγυπτίοις ζέφυρος ἢ νότος καὶ καῦμα.

κε΄. ὡρῶν ιδ΄ ὁ ἐν τῷ ἡγουμένῳ ὤμῳ τοῦ Ὠρίωνος ἐπιτέλλει, καὶ ὁ λαμπρὸς τοῦ Ὕδρου κρύπτεται. Αἰγυπτίοις ὑετός.

κϛ΄. Αἰγυπτίοις ζέφυρος, βροχή, βροντή.

κζ΄. ὡρῶν ιγ΄ ∠΄ ὁ ἐν τῷ ἑπομένῳ ὤμῳ τοῦ Ὠρίωνος ἐπιτέλλει. ὡρῶν ιδ΄ ὁ λαμπρὸς τοῦ βορείου Στεφάνου ἑῷος δύνει. ὡρῶν ιδ΄ ∠΄ ὁ ἐν τῷ ἐμπροσθίῳ δεξιῷ βατραχίῳ τοῦ Κενταύρου κρύπτεται.

κη΄. ὡρῶν ιγ΄ ∠΄ ὁ κοινὸς Ποταμοῦ καὶ ποδὸς Ὠρίωνος ἐπιτέλλει. Δημοκρίτῳ ἐπισημαίνει.

κθ΄. ὡρῶν ιε΄ ∠΄ ὁ κατὰ τὸ γόνυ τοῦ Τοξότου ἑσπέριος ἀνατέλλει. Ἱππάρχῳ ζέφυρος ἢ νότος πνεῖ.

λ΄. ὡρῶν ιγ΄ ∠΄ ὁ λαμπρὸς τοῦ Ὕδρου κρύπτεται. ὡρῶν ιδ΄ ∠΄ ὁ ἐν τῷ ἡγουμένῳ ὤμῳ τοῦ Ὠρίωνος ἐπιτέλλει. ὡρῶν ιε΄ ∠΄ Ἀρκτοῦρος ἑῷος δύνει.

Επιφί

α. θερινὴ τροπή. ὡρῶν ιγ´ Ϲ´ ὁ μέσος τῆς ζώνης τοῦ Ὠρίωνος ἐπιτέλλει. ὡρῶν ιδ´ ὁ ἐν τῷ ἑπομένῳ ὤμῳ τοῦ Ὠρίωνος ἐπιτέλλει. Αἰγυπτίοις ζέφυρος καὶ καῦμα.

β´. ὡρῶν ιε´ Ϲ´ ὁ λαμπρὸς τοῦ Περσέως ἑσπέριος ἀνατέλλει.

γ´. Αἰγυπτίοις καὶ Δημοκρίτῳ ζέφυρος πνεῖ.

δ´. Καλλίππῳ καὶ Δοσιθέῳ ἐπισημασία. Δημοκρίτῳ νότος καὶ ὕδωρ ἐξον, εἶτα βορέαι πρόδρομοι ἐπὶ ἡμέρας ζ.

ε´. ὡρῶν ιδ´ ὁ κοινὸς Ποταμοῦ καὶ ποδὸς Ὠρίωνος ἐπιτέλλει. ὡρῶν ιε´ ὁ ἐν τῷ ἡγουμένῳ ὤμῳ τοῦ Ὠρίωνος ἐπιτέλλει. Εὐδόξῳ ἐπισημαίνει.

ϛ´. ὡρῶν ιγ´ Ϲ´ ὁ ἐπὶ τῆς κεφαλῆς τοῦ ἡγουμένου Διδύμου ἐπιτέλλει. ὡρῶν ιδ´ ὁ μέσος τῆς ζώνης τοῦ Ὠρίωνος ἐπιτέλλει, καὶ ὁ ἔσχατος Ποταμοῦ ἐπιτέλλει, καὶ ὁ ἐπὶ τῆς κεφαλῆς τοῦ ἡγουμένου Διδύμου ἐπιτέλλει. Αἰγυπτίοις ἄνεμος καὶ ἀέρος ἀκρασία.

ζ´. ὡρῶν ιδ´ Ϲ´ ὁ λαμπρὸς τοῦ βορείου Στεφάνου ἑῷος δύνει.

η´. ὡρῶν ιε´ ὁ ἐπὶ τῆς κεφαλῆς τοῦ ἡγουμένου Διδύμου ἐπιτέλλει. ὡρῶν ιε´ Ϲ´ ὁ κοινὸς Ἵππου καὶ Ἀνδρομέδας ἑσπέριος ἀνατέλλει.

θ´. ὡρῶν ιε´ Ϲ´ ὁ ἐπὶ τῆς κεφαλῆς τοῦ ἡγουμένου Διδύμου ἐπιτέλλει. Αἰγυπτίοις καὶ Καίσαρι νότος καὶ καῦμα.

ι´. ὡρῶν ιδ´ Ϲ´ ὁ ἐν τῷ ἑπομένῳ ὤμῳ τοῦ Ὠρίωνος ἐπιτέλλει. ὡρῶν ιε´ Ϲ´ ὁ ἐπὶ τῆς καρδίας τοῦ Λέοντος κρύπτεται. Αἰγυπτίοις ἀργεστὴς καὶ ὑετία.

ια´. ὡρῶν ιδ´ Ϲ´ ὁ μέσος τῆς ζώνης τοῦ Ὠρίωνος ἐπιτέλλει. ὡρῶν ιε´ Ϲ´ ὁ ἐν τῷ ἡγουμένῳ ὤμῳ τοῦ Ὠρίωνος ἐπιτέλλει. Αἰγυπτίοις ζέφυρος ἢ ἀργεστὴς καὶ βροντή. Μητροδώρῳ ἀργεστής. Καλλίππῳ νότος. Ἱππάρχῳ νότος ἢ ζέφυρος.

ιβ´. ὡρῶν ιγ´ Ϲ´ ὁ ἐπὶ τῆς κεφαλῆς τοῦ ἑπομένου Διδύμου ἐπιτέλλει. ὡρῶν ιδ´ Ϲ´ ὁ κοινὸς Ποταμοῦ καὶ ποδὸς Ὠρίωνος ἐπιτέλλει. Αἰγυπτίοις ζέφυρος ἢ ἀργεστὴς καὶ καῦμα.

ιγ´. ὡρῶν ιε´ ὁ ἐπὶ τῆς καρδίας τοῦ Λέοντος κρύπτεται. Αἰγυπτίοις ἐπισημαίνει. Ἱππάρχῳ πρόδρομοι Κυνός.

ιδ´. ὡρῶν ιδ´ Ϲ´ ὁ ἐπὶ τῆς κεφαλῆς τοῦ ἑπομένου Διδύμου ἐπιτέλλει. Μέτωνι νοτία.

ιε´. ὡρῶν ιε´ Ϲ´ ὁ ἐν τῷ ἑπομένῳ ὤμῳ τοῦ Ὠρίωνος ἐπιτέλλει. Αἰγυπτίοις ἀργεστὴς ἢ ζέφυρος. Εὐκτήμονι καὶ Φιλίππῳ νοτία καὶ προδρόμων ἀρχή.

ιϛ´. ὡρῶν ιδ´ Ϲ´ ὁ ἐπὶ τῆς καρδίας τοῦ Λέοντος κρύπτεται. Αἰγυπτίοις ἐπισημαίνει, δυσαερία.

ιζ´. ὡρῶν ιε´ ὁ κοινὸς Ἵππου καὶ Ἀνδρομέδας ἑσπέριος ἀνατέλλει, καὶ ὁ μέσος τῆς ζώνης τοῦ Ὠρίωνος ἐπιτέλλει. ὡρῶν ιε´ Ϲ´ ὁ ἐπὶ τῆς κεφαλῆς τοῦ ἑπομένου Διδύμου ἐπιτέλλει.

ιη΄. ὡρῶν ιδ΄ ὁ ἐπὶ τῆς καρδίας τοῦ Λέοντος κρύπτεται. ὡρῶν ιε΄ ὁ λαμπρὸς τοῦ βορείου Στεφάνου ἑῷος δύνει, καὶ ὁ κοινὸς Ποταμοῦ καὶ ποδὸς Ὠρίωνος ἐπιτέλλει. Αἰγυπτίοις πρόδρομος ὥρᾳ α΄ πνεῖ. Μητροδώρῳ ζέφυρος ἢ ἀργεστής.

ιθ΄. ὡρῶν ιγ΄ ∠΄ Προκύων ἐπιτέλλει. Ἱππάρχῳ ἀνέμων ἀκρασία.[95]

κ΄. Αἰγυπτίοις καῦμα. Καίσαρι ἄνεμος πολύς. Ἱππάρχῳ βορέας ἄρχεται πνεῖν.

κα΄. ὡρῶν ιγ΄ ∠΄ ὁ ἐπὶ τῆς καρδίας τοῦ Λέοντος κρύπτεται.

κβ΄. ὡρῶν ιγ΄ ∠΄ Κύων ἐπιτέλλει. ὡρῶν ιδ΄ Προκύων ἐπιτέλλει. ὡρῶν ιδ΄ ∠΄ ὁ ἔσχατος τοῦ Ποταμοῦ ἐπιτέλλει. Αἰγυπτίοις ἄνεμος πολὺς καὶ ὑετία ἐνίοτε. Δημοκρίτῳ ὕδωρ, καταιγίδες.

κγ΄. ὡρῶν ιε΄ ὁ λαμπρὸς τοῦ Περσέως ἑσπέριος ἀνατέλλει. ὡρῶν ιε΄ ∠΄ ὁ μέσος τῆς ζώνης τοῦ Ὠρίωνος ἐπιτέλλει. Αἰγυπτίοις καὶ Δοσιθέῳ νότος καὶ καῦμα.

κδ΄. ὡρῶν ιδ΄ ∠΄ Προκύων ἐπιτέλλει. ὡρῶν ιε΄ ∠΄ ὁ κοινὸς Ποταμοῦ καὶ ποδὸς Ὠρίωνος ἐπιτέλλει. Ἱππάρχῳ ἐτησίαι ἄρχονται πνεῖν.

κε΄. Αἰγυπτίοις ζέφυρος ἢ ἀργεστὴς καὶ καῦμα.

κϛ΄. ὡρῶν ιδ΄ ∠΄ ὁ κοινὸς Ἵππου καὶ Ἀνδρομέδας ἑσπέριος ἀνατέλλει. ὡρῶν ιε΄ Προκύων ἐπιτέλλει. Αἰγυπτίοις ἀργεστὴς ἢ ζέφυρος.

κζ΄. ὡρῶν ιγ΄ ∠΄ ὁ λαμπρὸς τοῦ Ἀετοῦ ἑῷος δύνει. ὡρῶν ιε΄ ∠΄ ὁ λαμπρὸς τοῦ νοτίου Ἰχθύος ἑῷος δύνει. Μητροδώρῳ καὶ Εὐκτήμομι καὶ Φιλίππῳ ἐτησίαι πνέουσι, καὶ ὀπώρας ἀρχή. Καίσαρι πρόδρομοι πνέουσιν.

κη΄. ὡρῶν ιδ΄ Κύων ἐπιτέλλει. ὡρῶν ιε΄ ∠΄ ὁ λαμπρὸς τοῦ βορείου Στεφάνου ἑῷος δύνει, καὶ Προκύων ἐπιτέλλει. Αἰγυπτίοις δι᾽ ἡμέρας ζέφυρος καὶ καῦμα. Εὐκτήμονι καὶ Φιλίππῳ δυσαερία, πρόδρομοι πνέουσιν.

κθ΄. ὡρῶν ιδ΄ ὁ ἐν τῷ ἐμπροσθίῳ δεξιῷ βατραχίῳ τοῦ Κενταύρου κρύπτεται. Αἰγυπτίοις ἐτησίαι ἄρχονται πνεῖν. Μητροδώρῳ καὶ Καλλίππῳ ἀνεμώδης κατάστασις. Εὐκτήμονι χειμὼν κατὰ θάλασσαν.

λ΄. Εὐδόξῳ ἐτησίαι πνέουσιν. Μητροδώρῳ καὶ Καλλίππῳ ἀνεμώδης κατάστασις.

Μεσορί

α΄. Αἰγυπτίοις ζέφυρος ἢ νότος. Εὐδόξῳ καὶ Καίσαρι νότος.

β΄. ὡρῶν ιδ΄ ὁ λαμπρὸς τοῦ Ἀετοῦ ἑῷος δύνει. ὡρῶν ιε΄ ὁ λαμπρὸς τοῦ νοτίου Ἰχθύος ἑῷος δύνει. Μητροδώρῳ καὶ Καλλίππῳ καὶ Κόνωνι καὶ Δημοκρίτῳ καὶ Ἱππάρχῳ νότος καὶ καῦμα.

---

[95] Again, reading ἀκρασία for Heiberg's ἀκρισία (as at Pharmouthi 20, above).

γ΄.   Εὐκτήμονι καὶ Δοσιθέῳ νοτία καὶ πνίγη.

δ΄.   ὡρῶν ιγ΄ ∠΄ ὁ λαμπρὸς τῆς Λύρας ἑῷος δύνει. ὡρῶν ιδ΄ ὁ κοινὸς
     Ἵππου καὶ Ἀνδρομέδας ἑσπέριος ἀνατέλλει. ὡρῶν ιδ΄ ∠΄ Κύων
     ἐπιτέλλει.

ε΄.   Αἰγυπτίοις καῦμα. Εὐδόξῳ νοτία καὶ ὀπώρας ἀρχή. Δοσιθέῳ ἐτησίαι
     ἄρχονται.

ϛ΄.   ὡρῶν ιδ΄∠΄ ὁ λαμπρὸς τοῦ Ἀετοῦ ἑῷος δύνει, καὶ ὁ λαμπρὸς τοῦ νοτίου
     Ἰχθύος ἑῷος δύνει. Αἰγυπτίοις ἀργεστὴς ἢ ζέφυρος καὶ καῦμα. Εὐδόξῳ
     ἐτησίαι πνέουσιν.

ζ΄.   Καίσαρι νότος πνεῖ.

η΄.   Ἱππάρχῳ καῦμα.

θ΄.   ὡρῶν ιδ΄ ὁ λαμπρὸς τοῦ νοτίου Ἰχθύος ἑῷος δύνει. ὡρῶν ιε΄ Κύων
     ἐπιτέλλει.

ι΄.   ὡρῶν ιε΄ ὁ λαμπρὸς τοῦ Ἀετοῦ ἑῷος δύνει. ὡρῶν ιε΄ ∠΄ ὁ καλούμενος
     Αἴξ ἑσπέριος ἀνατέλλει. Καίσαρι ἐπισημασία. Εὐδόξῳ καὶ Δοσιθέῳ
     νοτία.

ια΄. ὡρῶν ιδ΄ ∠΄ ὁ λαμπρὸς τοῦ Περσέως ἑσπέριος ἀνατέλλει. ὡρῶν ιε΄ ὁ
     ἔσχατος τοῦ Ποταμοῦ ἐπιτέλλει. Εὐδόξῳ καῦμα μέγα.

ιβ΄. ὡρῶν ιγ΄∠΄ ὁ λαμπρὸς τοῦ νοτίου Ἰχθύος ἑῷος δύνει. Αἰγυπτίοις καῦμα.
     Δοσιθέῳ πνίγη καὶ μετὰ ταῦτα ἐτησίαι.

ιγ΄. ὡρῶν ιγ΄∠΄ ὁ κοινὸς Ἵππου καὶ Ἀνδρομέδας ἑσπέριος ἀνατέλλει. ὡρῶν
     ιδ΄ ὁ λαμπρὸς τῆς Λύρας ἑῷος δύνει.

ιδ΄. ὡρῶν ιε΄∠΄ Κύων ἐπιτέλλει.

ιε΄. Αἰγυπτίοις ἀργεστής, καῦμα μέγα καὶ πνιγετός.

ιϛ΄. Αἰγυπτίοις ἀργεστὴς ἢ νότος, ἀὴρ ὀμιχλώδης.

ιζ΄. Αἰγυπτίοις καῦμα μέγα καὶ πνιγετός.

ιη΄. ὡρῶν ιγ΄ ∠΄ ὁ ἐπὶ τῆς καρδίας τοῦ Λέοντος ἐπιτέλλει. Αἰγυπτίοις
     βρονταί. Εὐδόξῳ ἄνεμος μέγιστος. Ἱππάρχῳ ἀνέμων ταραχή.

ιθ΄. φθινοπώρου ἀρχή. ὡρῶν ιγ΄∠΄ ὁ λαμπρὸς τοῦ νοτίου Ἰχθύος ἑσπέριος
     ἀνατέλλει. ὡρῶν ιδ΄∠΄ ὁ ἐπὶ τῆς καρδίας τοῦ Λέοντος ἐπιτέλλει. Αἰγυπ-
     τίοις καῦμα.

κ΄.   ὡρῶν ιε΄ ὁ ἐπὶ τῆς καρδίας τοῦ Λέοντος ἐπιτέλλει. Καίσαρι ἐπισημαίνει.

κα΄. Καίσαρι ἐπισημαίνει, πνιγετός.

κβ΄. ὡρῶν ιγ΄∠΄ ὁ ἐπὶ τῆς οὐρᾶς τοῦ Λέοντος κρύπτεται, καὶ ὁ λαμπρὸς τοῦ
     Ὕδρου ἐπιτέλλει.

κγ΄. ὡρῶν ιγ΄∠΄ ὁ ἐν τῷ ἐμπροσθίῳ δεξιῷ βατραχίῳ τοῦ Κενταύρου κρύπτε-
     ται. ὡρῶν ιδ΄ ὁ ἐπὶ τῆς οὐρᾶς τοῦ Λέοντος κρύπτεται. Καίσαρι περίσ-
     τασις.

κδ΄. ὡρῶν ιδ΄ ὁ λαμπρὸς τοῦ Ὕδρου ἐπιτέλλει. Εὐδόξῳ ἐπισημαίνει.

κε΄. ὡρῶν ιε΄∠΄ ὁ ἐπὶ τῆς οὐρᾶς τοῦ Λέοντος κρύπτεται.

κϛ΄. ὡρῶν ιδ΄ ὁ λαμπρὸς τοῦ νοτίου Ἰχθύος ἑσπέριος ἀνατέλλει. Αἰγυπτίοις
    νότος ἢ ζέφυρος. Δημοκρίτῳ ἐπισημαίνει ὕδασι καὶ ἀνέμοις.

κζ΄. ὡρῶν ιδ΄ ∟΄ ὁ λαμπρὸς τοῦ Ὕδρου ἐπιτέλλει. Αἰγυπτίοις καῦμα καὶ
    ὁμίχλη.

κη΄. ὡρῶν ιδ΄ ὁ λαμπρὸς τοῦ Περσέως ἑσπέριος ἀνατέλλει.

κθ΄. ὡρῶν ιε΄ ὁ λαμπρὸς τοῦ Ὕδρου ἐπιτέλλει. Αἰγυπτίοις καὶ Καίσαρι
    ἐπισημαίνει, δυσαερία. Εὐδόξῳ βροντᾶν εἴωθεν.

λ΄. ὡρῶν ιε΄ ∟΄ ὁ ἐν τῷ ἑπομένῳ ὤμῳ τοῦ Ἡνιόχου ἑσπέριος ἀνατέλλει.
    Αἰγυπτίοις ζέφυρος ἢ ἀργεστής.

ἐπαγομένων

α΄. ὡρῶν ιε΄ ὁ λαμπρὸς τῆς Λύρας ἑῷος δύνει. ὡρῶν ιε΄ ∟΄ ὁ λαμπρὸς τοῦ
    Ὕδρου ἐπιτέλλει. Εὐδόξῳ καὶ Μητροδώρῳ ἐπισημαίνει.

β΄. ὡρῶν ιδ΄ ὁ καλούμενος Κάνωβος ἐπιτέλλει. ὡρῶν ιδ΄ ∟΄ ὁ λαμπρὸς τοῦ
    νοτίου Ἰχθύος ἑσπέριος ἀνατέλλει. Αἰγυπτίοις καῦμα. Εὐδόξῳ καὶ
    Καίσαρι ἐπισημαίνει. Ἱππάρχῳ νότος, καὶ ἐτησίαι παύονται.

γ΄. ὡρῶν ιδ΄ ∟΄ Στάχυς κρύπτεται. ὡρῶν ιε΄ ∟΄ ὁ ἐπὶ τῆς οὐρᾶς τοῦ Λέοντος
    ἐπιτέλλει. Ἱππάρχῳ ἀνέμων συστροφή.

δ΄. ὡρῶν ιε΄ ὁ ἐπὶ τῆς οὐρᾶς τοῦ Λέοντος ἐπιτέλλει. Καλλίππῳ ἐπισημαίνει.

ε΄. ὡρῶν ιγ΄ ∟΄ ὁ λαμπρὸς τοῦ Ὄρνιθος ἑῷος δύνει. Αἰγυπτίοις ζέφυρος ἢ
    ἀργεστής.

Phases of the Fixed Stars, and Collection of Weather Changes
Thoth

1. $14\frac{1}{2}$ hours:[96] the star on the tail of Leo rises. According to Hipparchus the
   Etesian winds stop. According to Eudoxus rainy, thundery, the Etesian
   winds stop.

2. 14 hours: the star on the tail of Leo rises, and Spica disappears. According
   to Hipparchus there is a change in the weather.

3. $13\frac{1}{2}$ hours: the star on the tail of Leo rises. 15 hours: the star called Capella
   rises in the evening. According to the Egyptians the Etesian winds stop.
   According to Eudoxus variable winds. According to Caesar wind, rain,
   thundery. According to Hipparchus the east wind blows.

4. 15 hours: the rearmost star of Eridanus sets in the morning. According to
   Callippus it is stormy and the Etesian winds stop.

---

[96] Meaning '(For the clima where the longest day is) $14\frac{1}{2}$ hours'.

5. $13\frac{1}{2}$ hours: Spica disappears. $15\frac{1}{2}$ hours: the bright star in Lyra sets in the morning. According to Metrodorus bad air. According to Conon the Etesian winds finish.

6. $15\frac{1}{2}$ hours: the bright star in the southern claw[97] disappears. According to the Egyptians mist and burning heat, or rain, or thunder. According to Eudoxus wind, thunder, bad air. According to Hipparchus wind, south winds.

7. According to Metrodorus bad air. According to Callippus, Euctemon, and Philippus bad air and unsettled air. According to Eudoxus rain, thundery, variable winds.

8. According to the Egyptians rainy, storms at sea or a south wind. According to Caesar variable winds, rainy, and the Etesian winds stop.

9. 14 hours: the bright star in Cygnus sets in the morning. According to the Egyptians the west wind or the north-west wind blows.

10. $13\frac{1}{2}$ hours: the bright star in Perseus rises in the evening. According to Philippus bad air. According to Dositheus it is stormy.

11. According to the Egyptians there is a change in the weather.

12. 15 hours: the bright star in the southern claw disappears.

13. According to Dositheus a bad mixture of airs.

14. $14\frac{1}{2}$ hours: the star called Canopus rises. According to Caesar the north winds stop blowing.

15. According to Eudoxus south winds.

16. According to Callippus and Conon there is a change in the weather.

17. $14\frac{1}{2}$ hours: the bright star in Cygnus sets in the morning, and the bright star in the southern claw disappears and the rearmost star of Eridanus sets in the morning. According to Eudoxus the north winds stop. According to Metrodorus there is a change in the weather. According to Democritus of Abdera there is a change in the weather, and the swallow is not visible.

18. $15\frac{1}{2}$ hours: the star on the knee of Sagittarius disappears. According to the Egyptians rainy, there is a change in the weather, autumn begins, the swallow is not visible. According to Dositheus south winds. According to Euctemon the beginning of autumn.

19. $15\frac{1}{2}$ hours: the bright star in the southern fish rises in the evening. According to Hipparchus bad air and rainy at sea and the beginning of autumn.

---

[97] The constellation Libra is regularly referred to by Ptolemy as 'the claws' (of Scorpio).

20. According to Caesar the beginning of autumn, and the swallow is not visible. According to Metrodorus rainy at sea and bad air.
21. 14 hours: the bright star in the southern claw disappears. 15 hours: the star in the rear shoulder of Auriga rises in the evening. According to the Egyptians west wind or south-west wind, later an east wind. According to Eudoxus the middle of autumn.
22. $14\frac{1}{2}$ hours: the star called Antares disappears. According to the Egyptians the west wind or the north-west wind, and drizzle. According to Eudoxus south winds.
23. $14\frac{1}{2}$ hours: the star called Capella rises in the evening. $15\frac{1}{2}$ hours: Arcturus rises in the morning. According to the Egyptians drizzle and wind, there is a change in the weather. According to Callippus and Metrodorus, rainy.
24. $13\frac{1}{2}$ hours: the star shared by Pegasus and Andromeda sets in the morning.
25. $13\frac{1}{2}$ hours: the bright star in the southern claw disappears. 15 hours: the bright star in Cygnus sets in the morning. According to the Egyptians west wind or south wind, and thunder storms throughout the day.
26. 15 hours: Arcturus rises in the morning. According to Eudoxus rain. According to Hipparchus west wind or south wind.
27. 14 hours: the star shared by Pegasus and Andromeda sets in the morning, and the rearmost star of Eridanus sets in the morning.
28. Autumnal equinox. According to the Egyptians and Eudoxus there is a change in the weather.
29. 14 hours: the star called Antares disappears. $14\frac{1}{2}$ hours: Arcturus rises in the morning. According to Euctemon there is a change in the weather. According to Democritus rain and unsettled winds.
30. $14\frac{1}{2}$ hours: the star shared by Pegasus and Andromeda sets in the morning. According to Euctemon, Philippus, and Conon there is a change in the weather.

Phaophi

1. According to the Egyptians west wind or south wind. According to Hipparchus there is a change in the weather.
2. 15 hours: the star shared by Pegasus and Andromeda sets in the morning. $15\frac{1}{2}$ hours: the bright star in the northern claw disappears. According to Eudoxus and Euctemon there is a change in the weather. According to Hipparchus south wind or west wind.

3. 14 hours: Arcturus rises in the morning. $15\frac{1}{2}$ hours: the bright star in Cygnus sets in the morning.

4. 15 hours: the bright star in the northern claw disappears. According to the Egyptians and Callippus it is stormy, bad air. According to Euctemon and Philippus rain.

5. $15\frac{1}{2}$ hours: the star shared by Pegasus and Andromeda sets in the morning. According to Eudoxus rain. According to Euctemon it is stormy. According to Metrodorus rain.

6. $13\frac{1}{2}$ hours: Arcturus rises in the morning, and the rearmost star in Eridanus sets in the morning. $14\frac{1}{2}$ hours: the bright star in the northern claw disappears, and the star called Antares disappears. $15\frac{1}{2}$ hours: the bright star in Corona Borealis rises in the morning. According to the Egyptians and Caesar stormy, rain, thundery, lightning.

7. $13\frac{1}{2}$ hours: Spica rises. 14 hours: the star called Capella rises in the evening, and the bright star in the northern claw disappears. According to the Egyptians rainy, it is stormy. According to Eudoxus rain and variable wind. According to Dositheus there is a change in the weather.

8. $13\frac{1}{2}$ hours: the bright star in the northern claw disappears. $14\frac{1}{2}$ hours: the star on the rear shoulder of Auriga rises in the evening, and Spica rises. According to Democritus it is stormy, time for sowing.

9. $15\frac{1}{2}$ hours: Spica rises. According to the Egyptians the north wind blows.

10. 15 hours: the bright star in Corona Borealis rises in the morning. According to Hipparchus south wind.

11. 15 hours: the star on the knee of Sagittarius disappears.

12. 15 hours: the star called Antares disappears. According to the Egyptians west wind or south-west wind. According to Eudoxus there is a change in the weather. According to Hipparchus east wind.

14. According to Dositheus and Eudoxus there is a change in the weather.

15. According to the Egyptians north-west wind, rain.

16. $14\frac{1}{2}$ hours: the bright star in Corona Borealis rises in the morning. According to Eudoxus north wind or south wind. According to Dositheus variable wind. According to Callippus there is a change in the weather. According to Caesar unsettled wind, rain, thundery.

17. $13\frac{1}{2}$ hours: the star called Antares disappears. According to the Egyptians north wind or south-west wind. According to Eudoxus there is a change in the weather.

18. $13\frac{1}{2}$ hours: Arcturus sets in the evening.

19. According to Eudoxus changing winds, thundery.
20. 14 hours: the star on the rear shoulder of Auriga rises in the evening. According to Hipparchus south wind or north wind.
21. $13\frac{1}{2}$ hours: the star called Capella rises in the evening.
22. 14 hours: the bright star in Corona Borealis rises in the morning. According to the Egyptians west wind or south wind throughout the day, rain. According to Dositheus a change in the weather.
24. $14\frac{1}{2}$ hours: the star called Canopus sets in the morning.
25. According to the Egyptians unsettled wind.
26. 14 hours: Arcturus sets in the evening. According to Eudoxus there is a change in the weather. According to Caesar the north wind blows.
27. $13\frac{1}{2}$ hours: the bright star in Corona Borealis rises in the morning. $14\frac{1}{2}$ hours: the star on the knee of Sagittarius disappears. According to the Egyptians and Callippus there is a change in the weather. According to Euctemon and Callippus unmixed air, very stormy at sea.
28. $13\frac{1}{2}$ hours: the star on the rear shoulder of Auriga rises in the evening. According to Metrodorus there is a change in the weather. According to Euctemon and Callippus mixed air, and it is stormy at sea.
30. According to the Egyptians it is very stormy.

Athyr

1. $13\frac{1}{2}$ hours: the bright star in the southern claw rises.
2. $14\frac{1}{2}$ hours: the bright star in the southern claw rises. 15 hours: the same. According to the Egyptians there is a change in the weather. According to Dositheus it is stormy. According to Democritus cold or frost. According to Hipparchus strong south wind.
3. $13\frac{1}{2}$ hours: the bright star in the northern claw rises. $15\frac{1}{2}$ hours: the bright star in Lyra rises in the morning. According to Euctemon and Philippus a strong wind blows.
4. 14 hours: the bright star in the northern claw rises. $14\frac{1}{2}$ hours: Arcturus sets in the evening. According to the Egyptians south wind or southwest wind. According to Callippus and Euctemon strong winds. According to Caesar and Metrodorus winds, it is stormy.
5. $14\frac{1}{2}$ hours: the bright star in the northern claw rises.
6. 14 hours: the star on the knee of Sagittarius disappears. According to Conon and Eudoxus a bad mixture of winds. According to Callippus a bad mixture of airs. According to Caesar and Hipparchus south wind or cold north wind.

7. 14 hours: the bright star in the Hyades rises in the evening. According to the Egyptians furious south wind. According to Meton west wind. According to Eudoxus north wind or south wind. According to Metrodorus a bad mixture of air. According to Euctemon, Philippus, and Hipparchus rain.

8. $13\frac{1}{2}$ hours: the bright star in the Hyades rises in the evening. According to Callippus rainy. According to Euctemon there is a change in the weather.

9. $15\frac{1}{2}$ hours: the star shared by Eridanus and the foot of Orion sets in the morning. According to the Egyptians stormy, rain.

10. 14 hours: the star called Canopus sets in the morning. According to the Egyptians south wind or west wind. According to Dositheus stormy.

11. 15 hours: the bright star in Lyra rises in the morning. According to Meton blustery rain. According to Hipparchus cold north-west wind.

12. 15 hours: Arcturus sets in the evening, and the star shared by Eridanus and the foot of Orion sets in the morning.

13. $13\frac{1}{2}$ hours: the star on the knee of Sagittarius disappears. According to the Egyptians south wind or east wind throughout the day, there is drizzle. According to Metrodorus it is stormy, blustery. According to Euctemon rainy, it is stormy.

14. $14\frac{1}{2}$ hours: the star shared by Eridanus and the foot of Orion sets in the morning. According to Philippus and Euctemon stormy, blustery. According to Hipparchus north wind or cold south wind and rain.

15. $13\frac{1}{2}$ hours: the bright star in Perseus sets in the morning, and the bright star in Corona Borealis sets in the evening. $15\frac{1}{2}$ hours: the bright star in the Hyades sets in the morning. According to the Egyptians and Hipparchus it is the beginning of winter. According to Metrodorus, Callippus, and Conon a change in the weather.

16. $13\frac{1}{2}$ hours: the bright star in the Hyades sets in the morning. $14\frac{1}{2}$ hours: the same. 15 hours: the same. According to Euctemon and Dositheus it is stormy.

17. 14 hours: the star shared by Eridanus and the foot of Orion sets in the morning. $15\frac{1}{2}$ hours: the star on the head of the leading twin rises in the evening. According to Eudoxus winter begins and a change in the weather. According to Democritus stormy both on land and at sea.

19. $14\frac{1}{2}$ hours: the bright star in Lyra rises in the morning. According to the Egyptians south wind or east wind throughout the day. According to Caesar stormy.

20. $13\frac{1}{2}$ hours: the star shared by Eridanus and the foot of Orion sets in the morning. 14 hours: the bright star in Perseus sets in the morning. $15\frac{1}{2}$ hours: the star on the leading shoulder of Orion sets in the morning, and the middle of Orion's belt sets in the morning. According to Caesar stormy.

21. 15 hours: the star on the leading shoulder of Orion sets in the morning, and the middle of Orion's belt sets in the morning. $15\frac{1}{2}$ hours: Arcturus sets in the evening. According to the Egyptians north wind throughout the day and the night. According to Eudoxus rain. According to Caesar stormy.

22. $14\frac{1}{2}$ hours: the star in the leading shoulder of Orion sets in the morning.

23. $13\frac{1}{2}$ hours: the star called Canopus sets in the morning. 14 hours: the bright star in Corona Borealis sets in the evening, and the star in the leading shoulder of Orion sets in the morning. 15 hours: the star on the head of the leading twin rises in the evening. According to Eudoxus wintry conditions.

24. $13\frac{1}{2}$ hours: the star in the right front hoof of Centaurus rises. $14\frac{1}{2}$ hours: the middle of Orion's belt sets in the morning. $15\frac{1}{2}$ hours: Sirius sets in the morning. According to the Egyptians wintry conditions. According to Eudoxus cold north wind.

25. $13\frac{1}{2}$ hours: the star in the leading shoulder of Orion sets in the morning, and the star called Antares rises. $14\frac{1}{2}$ hours: the bright star in Perseus sets in the morning. According to Euctemon and Dositheus stormy and rainy. According to Caesar a bad mixture of air.

26. $13\frac{1}{2}$ hours: the star in the leading shoulder of Orion rises in the evening, and the rearmost star in Eridanus rises in the evening. 14 hours: the bright star in Lyra rises in the morning, and the middle of Orion's belt sets in the morning, and the star called Antares rises. According to Eudoxus very stormy.

27. $14\frac{1}{2}$ hours: the star called Antares rises. 15 hours: Sirius sets in the morning. $15\frac{1}{2}$ hours: the bright star in Cygnus rises in the morning, and the star in the rear shoulder of Orion sets in the morning. According to the Egyptians and Hipparchus strong south wind. According to Eudoxus and Conon the air is wintry. According to Callippus rainy.

28. 14 hours: the star in the leading shoulder of Orion rises in the evening. $14\frac{1}{2}$ hours: the star on the head of the leading twin rises in the evening. 15 hours: the star in the rear shoulder of Orion sets in the morning, and the star called Antares rises. According to the Egyptians drizzle.

29. $13\frac{1}{2}$ hours: the middle of Orion's belt sets in the morning. $15\frac{1}{2}$ hours: the star called Antares rises.

30. $13\frac{1}{2}$ hours: the middle of Orion's belt rises in the evening. $14\frac{1}{2}$ hours: the star on the rear shoulder of Orion sets in the morning, and the star in the leading shoulder of Orion rises in the evening. $15\frac{1}{2}$ hours: the star on the head of the rear twin rises in the evening.

Choiak

1. $14\frac{1}{2}$ hours: Sirius sets in the morning. 15 hours: the bright star in Perseus sets in the morning. According to the Egyptians south wind and rain. According to Eudoxus a bad mixture of air. According to Dositheus a change in the weather. According to Democritus the sky is disturbed, as is the sea usually.

2. $13\frac{1}{2}$ hours: the star on the rear shoulder of Orion rises in the evening, and the star shared by Eridanus and the foot of Orion rises in the evening. 14 hours: the star on the head of the leading twin rises in the evening, and the star on the rear shoulder of Orion sets in the morning. $14\frac{1}{2}$ hours: the bright star in Corona Borealis sets in the evening.

3. $13\frac{1}{2}$ hours: the star in the rear shoulder of Orion sets in the morning. 15 hours: the star in the leading shoulder of Orion rises in the evening.

4. $13\frac{1}{2}$ hours: the bright star in Lyra rises in the morning. 14 hours: the star on the rear shoulder of Orion rises in the evening, and the middle of Orion's belt rises in the evening. 15 hours: the star on the head of the rear twin rises in the evening. According to the Egyptians west wind or south wind throughout the day, it rains. According to Conon it is stormy.

5. $13\frac{1}{2}$ hours: the star called Capella sets in the morning, and the star on the head of the leading twin rises in the evening. 14 hours: Sirius sets in the morning. $15\frac{1}{2}$ hours: the star on the leading shoulder of Orion rises in the evening. According to Caesar, Euctemon, Eudoxus, and Callippus stormy.

6. 14 hours: the star on the right front hoof of Centaurus rises. $14\frac{1}{2}$ hours: the star on the rear shoulder of Orion rises in the evening. According to Metrodorus wintry conditions. According to Euctemon, Philippus, and Callippus a bad mixture of winds.

7. 14 hours: the star shared by Eridanus and the foot of Orion rises in the evening. $14\frac{1}{2}$ hours: the star on the head of the rear twin rises in

the evening, and the middle of Orion's belt rises in the evening. 15 hours: the bright star in Cygnus rises in the morning. According to the Egyptians there is drizzle. According to Caesar and Conon it is stormy.

8. 15 hours: the star in the rear shoulder of Orion rises in the evening. $15\frac{1}{2}$ hours: the bright star in Perseus sets in the morning. According to the Egyptians there is drizzle. According to Caesar, Euctemon, and Eudoxus stormy.

9. $13\frac{1}{2}$ hours: Sirius sets in the morning. 14 hours: the star called Capella sets in the morning, and the star on the head of the rear twin rises in the evening, and the rearmost star in Eridanus rises in the evening. According to the Egyptians, Dositheus, and Democritus stormy.

10. 15 hours: the bright star in Corona Borealis sets in the evening, and the middle of Orion's belt rises in the evening. According to the Egyptians south-west wind or south wind. According to Eudoxus and Dositheus wintry air.

11. $13\frac{1}{2}$ hours: the star on the head of the rear twin rises in the evening. According to Hipparchus strong north wind. According to Eudoxus rain.

12. $14\frac{1}{2}$ hours: the star shared by Eridanus and the foot of Orion rises in the evening. According to Caesar south winds. According to Euctemon, Eudoxus and Callippus wintry air and rainy.

13. $13\frac{1}{2}$ hours: the star in the rear shoulder of Auriga sets in the morning. $15\frac{1}{2}$ hours: the middle of Orion's belt rises in the evening. According to Caesar south winds. According to Euctemon, Eudoxus, and Callippus wintry air and rainy.

14. $14\frac{1}{2}$ hours: the star called Capella sets in the morning. According to Metrodorus, Euctemon, and Callippus wintry conditions. According to Democritus thundery, lightning, wet, winds.

15. According to the Egyptians a cold north-west wind or south wind and thunder storms. According to Callippus south wind and a change in the weather. According to Eudoxus wintry air.

16. $14\frac{1}{2}$ hours: the bright star in Cygnus rises in the morning. 15 hours: the star shared by Eridanus and the foot of Orion rises in the evening. According to the Egyptians it is stormy.

17. According to Hipparchus strong south wind or north wind.

18. 14 hours: the star on the rear shoulder of Auriga sets in the morning. According to the Egyptians rainy with winds. According to Eudoxus it is stormy.

19. 15 hours: the star called Capella sets in the morning. $15\frac{1}{2}$ hours: the bright star in Corona Borealis sets in the evening. According to the Egyptians cold north wind or south wind and rainy.

20. $15\frac{1}{2}$ hours: Procyon sets in the morning. According to Caesar it is stormy.

21. $15\frac{1}{2}$ hours: the star shared by Eridanus and the foot of Orion rises in the evening.

22. 15 hours: Procyon sets in the morning. According to Hipparchus south wind.

23. $14\frac{1}{2}$ hours: the star on the rear shoulder of Auriga sets in the morning, and the star on the right front hoof of Centaurus rises. $15\frac{1}{2}$ hours: the bright star in Aquila rises in the morning. According to the Egyptians, Eudoxus, and Dositheus south-west wind or south wind.

24. $14\frac{1}{2}$ hours: Procyon sets in the morning, and the rearmost star in Eridanus rises in the evening. According to Eudoxus wintry air.

25. $13\frac{1}{2}$ hours: Procyon rises in the evening. 14 hours: Procyon sets in the morning. 15 hours: the bright star in Aquila rises in the morning. According to the Egyptians a change in the weather.

26. Winter solstice. $13\frac{1}{2}$ hours: Procyon sets in the morning, and Sirius rises in the evening. $15\frac{1}{2}$ hours: the star called Capella sets in the morning.

27. $13\frac{1}{2}$ hours: the bright star in Aquila disappears. 14 hours: Procyon rises in the evening. $14\frac{1}{2}$ hours: the bright star in Aquila rises in the morning.

28. 15 hours: the star in the rear shoulder of Auriga sets in the morning. $15\frac{1}{2}$ hours: the bright star in the southern fish disappears. According to the Egyptians and Caesar stormy. According to Hipparchus and Meton there is a change in the weather, thunderstorm.

29. $14\frac{1}{2}$ hours: Procyon rises in the evening. According to the Egyptians, Conon, Meton, and Callippus stormy. According to Caesar and Metrodorus a change in the weather, a bad mixture <of air>.

30. 14 hours: the bright star in Aquila rises in the morning, and the bright star in Aquila sets in the evening. According to the Egyptians south-west wind and a bad mixture of air. According to Eudoxus and Metrodorus wintry air. According to Hipparchus a storm in the evening.

Tybi

1. 14 hours: Sirius rises in the evening. 15 hours: Procyon rises in the evening. According to Eudoxus there is a change in the weather. According to Democritus the middle of winter.

2. $13\frac{1}{2}$ hours: the star on the head of the leading twin sets in the morning. According to Dositheus it is stormy.

3. $13\frac{1}{2}$ hours: the bright star in Aquila rises. $15\frac{1}{2}$ hours: Procyon rises in the evening. According to Euctemon, Philippus and Democritus there is a change in the weather.

4. $13\frac{1}{2}$ hours: the bright star in Cygnus rises in the morning, and the star on the head of the rear twin sets in the morning. $14\frac{1}{2}$ hours: the bright star in Aquila sets in the evening. 15 hours: the bright star in the southern fish disappears. According to the Egyptians there is a storm at sea. According to Euctemon there is a change in the weather.

5. 14 hours: the star on the head of the leading twin sets in the morning. $15\frac{1}{2}$ hours: the star on the rear shoulder of Auriga sets in the morning.

6. $13\frac{1}{2}$ hours: the star on the knee of Sagittarius rises. 14 hours: the star on the head of the rear twin sets in the morning. $14\frac{1}{2}$ hours: Sirius rises in the evening.

7. 15 hours: the bright star in Aquila sets in the evening. According to Dositheus there is a change in the weather.

8. $14\frac{1}{2}$ hours: the star on the head of the leading twin sets in the morning, and the star on the head of the rear twin sets in the morning, and the bright star in the southern fish disappears. According to the Egyptians complex conditions.

9. $13\frac{1}{2}$ hours: the bright star in Lyra sets in the evening. $15\frac{1}{2}$ hours: the bright star in Aquila sets in the evening. According to the Egyptians there is a change in the weather. According to Democritus a south wind usually blows.

10. 15 hours: Sirius rises in the evening.

11. 15 hours: the star on the head of the rear twin sets in the morning. According to Euctemon and Philippus the middle of winter.

12: 14 hours: the star on the knee of Sagittarius rises. 15 hours: the star on the head of the leading twin sets in the morning. According to Hipparchus and Eudoxus it is stormy.

13. 14 hours: the bright star in the southern fish disappears. 15 hours: the rearmost star in Eridanus rises in the evening. According to the Egyptians a south wind or west wind, stormy both on land and at sea. According to Metrodorus, Euctemon, Philippus, and Callippus a south wind.

14. $15\frac{1}{2}$ hours: the star on the head of the rear twin sets in the morning, and the bright star in Aquarius sets in the morning, and Sirius rises in the

evening. According to the Egyptians and Eudoxus a strong south wind and rain.

15. 15 hours: ... According to the Egyptians and Caesar much south wind, and there is a change in the weather at sea, thunder, drizzle.

16. 15 hours: the bright star in Aquarius sets in the morning. $15\frac{1}{2}$ hours the star on the head of the leading twin sets in the morning. According to Eudoxus and Dositheus a south wind, there is a change in the weather. According to Hipparchus a bad mixture of winds.

17. $13\frac{1}{2}$ hours: the bright star in the southern fish disappears.

18. 14 hours: the bright star in Lyra sets in the evening. $14\frac{1}{2}$ hours: the star on the knee of Sagittarius rises.

19. $14\frac{1}{2}$ hours: the bright star in Aquarius sets in the morning. According to Hipparchus south wind or north wind, it is stormy.

20. According to the Egyptians wintry air.

21. 14 hours: the bright star in Aquarius sets in the morning. 15 hours: the star on the heart of Leo rises in the evening. According to Hipparchus the east wind blows.

22. $13\frac{1}{2}$ hours: the star on the heart of Leo rises in the evening, and the bright star in Aquarius rises in the evening, and the star called Canopus rises in the evening. 14 hours: the star on the heart of Leo rises in the evening. $14\frac{1}{2}$ hours: the star on Centaurus' right front hoof sets in the morning. $14\frac{1}{2}$ hours: the star on the heart of Leo rises in the evening. According to Caesar strong winds.

23. $13\frac{1}{2}$ hours: the bright star in Aquarius sets in the morning. According to Euctemon and Philippus it is stormy. According to Metrodorus unstable air.

24. 14 hours: the bright star in Aquarius rises in the evening. According to the Egyptians it is rainy and it becomes stifling. According to Caesar and Euctemon stormy.

25. $14\frac{1}{2}$ hours: the bright star in Lyra sets in the evening, and the bright star in Aquarius rises in the evening. 15 hours: the star on the knee of Sagittarius rises. According to the Egyptians and Callippus stormy, rain. According to Hipparchus the north wind blows. According to Euctemon and Democritus rain falls.

26. 15 hours: the bright star in Aquarius rises in the evening. According to Eudoxus the middle of winter.

27. According to the Egyptians, an east wind or south wind, there is a change in the weather.

28. $15\frac{1}{2}$ hours: the bright star in Aquarius rises in the evening. According to the Egyptians rain. According to Hipparchus a change in the weather.

29. According to Callippus and Euctemon rain falls. According to Democritus the middle of winter.
30. According to Hipparchus the east wind blows.

Mecheir

1. $15\frac{1}{2}$ hours: the star on the knee of Sagittarius rises. According to Eudoxus rain. According to Metrodorus, rainy. According to Dositheus it is stormy.
2. According to the Egyptians, the middle of winter.
3. According to the Egyptians a south-west wind or a south wind, there is a change in the weather.
4. $13\frac{1}{2}$ hours: the bright star in Cygnus sets in the evening. 15 hours: the bright star in Lyra sets in the evening. According to Hipparchus a south wind or a north-west wind.
6. $13\frac{1}{2}$ hours: the star on the heart of Leo sets in the morning. 14 hours: the star called Canopus rises in the evening. $15\frac{1}{2}$ hours: the star on the tail of Leo rises in the evening, and the star on the knee of Sagittarius rises. According to Eudoxus rain.
7. 14 hours: the star on the heart of Leo sets in the morning. 15 hours: the star on the tail of Leo rises in the evening.
8. $14\frac{1}{2}$ hours: the star on the heart of Leo sets in the morning and the star on the tail of Leo rises in the evening. According to the Egyptians a south or west wind, in between hail.
9. 15 hours: the star on the heart of Leo sets in the morning. According to Eudoxus a nice day, sometimes also the west wind blows.
10. 14 hours: the star on the tail of Leo rises in the evening.
11. $15\frac{1}{2}$ hours: the star on the heart of Leo sets in the morning. According to the Egyptians stormy conditions, or downpour and a bad mixture of winds. According to Dositheus a nice day, sometimes the west wind blows.
12. 14 hours: the bright star in Cygnus sets in the evening. 15 hours: the rearmost star in Eridanus disappears. $15\frac{1}{2}$ hours: the bright star in Perseus rises in the morning, and the bright star in Lyra sets in the evening. According to the Egyptians windy conditions. According to Caesar rainy. According to Democritus the west wind begins to blow.
13. $13\frac{1}{2}$ hours: the star on the tail of Leo rises in the evening. According to the Egyptians and Eudoxus the beginning of spring. The west wind begins to blow and sometimes stormy.

14. According to the Egyptians and Eudoxus rainy. According to Hipparchus, Callippus, and Democritus it is the season for the west wind to blow.

15. According to Caesar and Metrodorus the beginning of spring. The west wind begins to blow.

17. According to the Egyptians and Eudoxus west winds blow. According to Hipparchus the beginning of spring. According to Callippus and Metrodorus it is stormy.

18. According to the Egyptians the east wind blows. According to Hipparchus the north or east wind blows.

19. 14 hours: the star on Centaurus' right front hoof sets in the morning. $15\frac{1}{2}$ hours: the star shared by Pegasus and Andromeda rises in the morning.

21. $14\frac{1}{2}$ hours: the bright star in Cygnus sets in the evening. According to the Egyptians variable winds. According to Hipparchus the south wind blows. According to Euctemon, Philippus, and Dositheus it is stormy.

22. According to the Egyptians unstable winds and thunderstorms.

23. $14\frac{1}{2}$ hours: the star called Canopus rises in the evening.

24. According to the Egyptians either a west wind or a south wind and hail, rain.

25. $14\frac{1}{2}$ hours: the rearmost star in Eridanus disappears. 15 hours: the star shared by Pegasus and Andromeda rises in the morning. According to Hipparchus a cold north wind blows.

26. According to the Egyptians windy conditions.

28. According to Hipparchus and Euctemon cold bird-bringing winds begin to blow, and it is the time for the swallows to appear.

29. $13\frac{1}{2}$ hours: the star shared by Pegasus and Andromeda disappears. 15 hours: the bright star in Cygnus sets in the evening. According to the Egyptians, Philippus, and Callippus the swallows appear, and windy conditions. According to Conon a cold north wind begins to blow. According to Eudoxus rain at the coming of the swallows, and for thirty days the north winds, called the 'bird-bringers' blow.

30. According to the Egyptians the northern bird-bringing winds blow, the north-west wind in between. According to Hipparchus cold north winds. According to Metrodorus the swallows appear, and there is a change in the weather. According to Democritus the variable days called the halcyon days.

Phamenoth

1. $14\frac{1}{2}$ hours: the star shared by Pegasus and Andromeda rises in the morning. $15\frac{1}{2}$ hours: Arcturus rises in the evening. According to Caesar and Dositheus stormy, there is a change in the weather.
2. 14 hours: the star shared by Pegasus and Andromeda disappears.
3. 15 hours: the bright star in Perseus rises in the morning.
4. $14\frac{1}{2}$ hours: the star shared by Pegasus and Andromeda sets in the evening.
5. 14 hours: the star shared by Pegasus and Andromeda rises. 15 hours: Arcturus rises in the evening. According to Hipparchus a north or a cold south wind blows.
6. 14 hours: the rearmost star of Eridanus disappears. According to the Egyptians a south-west or a south wind, hail. According to Hipparchus a cold north wind blows.
7. 15 hours: the star shared by Pegasus and Andromeda sets in the evening. $15\frac{1}{2}$ hours: the bright star in Cygnus sets in the evening.
8. $14\frac{1}{2}$ hours: Arcturus rises in the evening. According to Euctemon a cold north wind blows.
9. $15\frac{1}{2}$ hours: the bright star in Corona Borealis rises in the evening, and the star shared by Pegasus and Andromeda sets in the evening. According to the Egyptians it is stormy. According to Caesar the 'swallow-bringing' winds blow for ten days.
10. $13\frac{1}{2}$ hours: the star shared by Pegasus and Andromeda rises.
11. $13\frac{1}{2}$ hours: the bright star in the southern fish rises, and the star on Centaurus' right front hoof sets in the morning. According to the Egyptians disturbed conditions. According to Democritus cold bird-bringing winds for nine days.
12: 14 hours: Arcturus rises in the evening. According to Eudoxus stormy, and the kite appears, and there is a change in the weather. According to Metrodorus, Euctemon, and Philippus a cold north wind blows. According to Hipparchus the beginning of spring.
13: $13\frac{1}{2}$ hours: the star on the tail of Leo sets in the morning. According to the Egyptians there is drizzle. According to Metrodorus and Euctemon the north wind blows. According to Dositheus the kite begins to appear. According to Hipparchus a strong south wind.
14. 15 hours: the bright star in Corona Borealis rises in the evening. According to the Egyptians and Callippus a cold north wind blows.
15. $13\frac{1}{2}$ hours: Arcturus rises in the evening.
16. $13\frac{1}{2}$ hours: the rearmost star in Eridanus disappears. According to Callippus a calm north wind blows.

17. $13\frac{1}{2}$ hours: Spica rises in the evening. $14\frac{1}{2}$ hours: Spica rises in the evening. According to the Egyptians windy conditions. According to Euctemon and Philippus the bird-bringers begin to blow, and it is the season of the appearance of the kite.

18. 14 hours: the star on the tail of Leo sets in the morning. According to the Egyptians a west or a south wind blows. According to Euctemon a cold north wind blows. According to Dositheus the bird-bringers begin to blow. According to Hipparchus a north or a north-west wind.

19. According to the Egyptians and Euctemon a cold north wind blows.

20. 14 hours: the bright star in the southern fish rises. $14\frac{1}{2}$ hours: the bright star in Corona Borealis rises in the evening.

21. $14\frac{1}{2}$ hours: the bright star in Perseus rises in the morning. According to Callippus the north wind blows, and the kite appears.

22. According to the Egyptians and Democritus stormy, cold wind.

23. According to the Egyptians cold winds until the equinox. According to Hipparchus the north wind blows.

24. According to Caesar the kite appears, and the north wind blows.

25. $14\frac{1}{2}$ hours: the star on the tail of Leo sets in the morning. According to Eudoxus the kite appears and the north wind blows.

26. Vernal Equinox. 14 hours: the bright star in Corona Borealis rises in the evening.

27. According to Caesar the north wind blows. According to Hipparchus rainy.

28. According to the Egyptians thundery, a change in the weather. According to Philippus, Callippus, and Euctemon rain or drizzle. According to Hipparchus a change in the weather.

29. $15\frac{1}{2}$ hours: the star called Capella rises in the morning. The equinox, according to the Egyptians, Conon, and Meton. According to Eudoxus the north wind blows.

30. $13\frac{1}{2}$ hours: Spica sets in the morning. According to the Egyptians the north-west wind blows. According to Callippus rain or a snow-storm.

Pharmouthi

1. 14 hours: Spica sets in the morning. According to Meton, Callippus, and Eudoxus, rain. According to Euctemon and Democritus there is a change in the weather.

2. $13\frac{1}{2}$ hours: the bright star in Corona Borealis rises. $14\frac{1}{2}$ hours: Spica sets in the morning, and the star called Canopus disappears. 15 hours: the star on the tail of Leo sets in the morning. According to Dositheus, Meton, and Callippus rainy.

3. 14 hours: the bright star in Perseus rises in the morning. $14\frac{1}{2}$ hours: the bright star in the southern fish rises.

4. $15\frac{1}{2}$ hours: the bright star in the northern claw rises in the evening. According to the Egyptians and Conon there is a change in the weather. According to Eudoxus it is rainy.

5. 15 hours: Spica sets in the morning.

6. $15\frac{1}{2}$ hours: the bright star in the southern claw rises in the evening. According to Eudoxus rain, there is a change in the weather.

7. $13\frac{1}{2}$ hours: the bright star in the southern claw rises in the evening. $15\frac{1}{2}$ hours: Spica sets in the morning.

8. 15 hours: the bright star in the northern claw rises in the evening. According to the Egyptians a west wind and hail. According to Conon there is a change in the weather. According to Eudoxus rain.

9. $14\frac{1}{2}$ hours: the bright star in the northern claw rises in the evening. According to the Egyptians and Conon a west or a south wind and hail.

10. 14 hours: the bright star in the northern claw rises in the evening. $15\frac{1}{2}$ hours: the bright star in Lyra rises in the evening. According to Hipparchus a south wind and whirlwinds.

11. $13\frac{1}{2}$ hours: the bright star in the northern claw rises in the evening. According to Hipparchus and Dositheus there is a change in the weather.

12. $15\frac{1}{2}$ hours: the star on the tail of Leo sets in the morning.

13. 13 $<\frac{1}{2}>$ hours . . . According to the Egyptians a south or a south-west wind. According to Eudoxus rainy.

14. $13\frac{1}{2}$ hours: the bright star in Perseus rises in the morning. According to the Egyptians a bad mixture of winds. According to Hipparchus rainy.

15. According to the Egyptians unstable air and rain. According to Euctemon and Philippus a bad mixture of winds. According to Hipparchus rainy.

16. According to Eudoxus a west wind and a bad mixture of air, in between it drizzles.

17. $15\frac{1}{2}$ hours: the star shared by Eridanus and the foot of Orion disappears.

18. 15 hours: the star called Capella rises in the morning and the bright star in the southern fish rises. According to Dositheus and Caesar rainy.

19. 15 hours: the bright star in Lyra rises in the evening. According to the Egyptians the clearing south wind, thundery, drizzle.

20. 14 hours: the star called Canopus disappears. According to the Egyptians a bad mixture of winds. According to Eudoxus and Euctemon rainy and hail.

21. 15 hours: the star shared by Eridanus and the foot of Orion disappears. $15\frac{1}{2}$ hours: the bright star in the Hyades disappears. According to Metrodorus and Callippus hail. According to Euctemon and Philippus a west wind.

22. $13\frac{1}{2}$ hours: the bright star in Perseus sets in the evening. According to the Egyptians and Conon hail and a west wind. According to Caesar and Eudoxus rainy.

23. 15 hours: the bright star in the Hyades disappears. According to the Egyptians blustery drizzle.

24. $14\frac{1}{2}$ hours: the bright star in the Hyades disappears, and the star shared by Eridanus and the foot of Orion disappears. $15\frac{1}{2}$ hours: the middle of Orion's belt disappears.

25. According to the Egyptians a south-west, south, or a north-west wind and a bad mixture of air.

26. 14 hours: the bright star in Perseus sets in the evening, and the bright star in the Hyades disappears. $15\frac{1}{2}$ hours: the bright star in Cygnus rises in the evening, and the star in the leading shoulder of Orion disappears. According to Hipparchus a south or a cold arctic wind.

27. $13\frac{1}{2}$ hours: the bright star in the Hyades disappears and the bright star in the southern claw sets in the morning. 15 hours: the middle of Orion's belt disappears. According to the Egyptians and Caesar stormy. According to Eudoxus rain.

28. 14 hours: the star shared by Eridanus and the foot of Orion disappears. $14\frac{1}{2}$ hours: the bright star in Lyra rises in the evening. According to the Egyptians a south-west or a south wind, rainy.

29. 14 hours: the bright star in the southern claw sets in the morning. 15 hours: the star in the leading shoulder of Orion disappears. According to the Egyptians a south-west or a south wind and rainy. According to Metrodorus and Callippus sometimes hail. According to Democritus there is a change in the weather.

30. According to the Egyptians and Eudoxus drizzle, rain.

Pachon

1. $14\frac{1}{2}$ hours: the bright star in Perseus sets in the evening and the middle of Orion's belt disappears and the bright star in the southern claw sets in the morning. According to the Egyptians a north-west or a west wind,

there is a change in the weather. According to Euctemon and Philippus rainy or hail.

2. $14\frac{1}{2}$ hours: the star called Capella rises in the morning and the star in the leading shoulder of Orion disappears. According to the Egyptians windy conditions. According to Metrodorus and Callippus south winds.

3. $13\frac{1}{2}$ hours: the star shared by Eridanus and the foot of Orion disappears and the star called Antares rises in the evening. $15\frac{1}{2}$ hours Sirius disappears. According to the Egyptians windy. According to Eudoxus rain.

4. 14 hours: the star in the leading shoulder of Orion disappears and the middle of Orion's belt disappears and the star called Antares rises in the evening. $14\frac{1}{2}$ hours: the same. 15 hours: the same. According to the Egyptians calm or south wind and rainy. According to Caesar stormy.

5. $13\frac{1}{2}$ hours: the star called Canopus disappears. 15 hours: the bright star in the southern claw sets in the morning. According to the Egyptians there is a change in the weather. According to Euctemon and Philippus calm or south wind, drizzle.

6. $13\frac{1}{2}$ hours: the star on Centaurus' right front hoof rises in the evening. 15 hours: the bright star in Perseus sets in the evening. $15\frac{1}{2}$ hours: the star in the rear shoulder of Auriga rises in the morning and the star on the rear shoulder of Orion disappears. According to the Egyptians drizzle.

7. $13\frac{1}{2}$ hours: the star on the leading shoulder of Orion disappears and the middle of the belt disappears. 15 hours: Sirius disappears.

8. 14 hours: the bright star in Lyra rises in the evening. 15 hours: the bright star in Cygnus rises in the evening and the star in the rear shoulder of Orion disappears. $15\frac{1}{2}$ hours: the bright star in the southern claw sets in the morning. According to the Egyptians a north-west wind and drizzle, or a south wind, thunder.

9. 14 hours: the star called Capella rises in the morning. $15\frac{1}{2}$ hours: the bright star in the southern fish rises. According to the Egyptians drizzle. According to Eudoxus rain.

10. $13\frac{1}{2}$ hours: the bright star in the northern claw sets in the morning. According to Dositheus rainy.

11. $14\frac{1}{2}$ hours: the star in the rear shoulder of Orion disappears. According to the Egyptians windy conditions.

12. $13\frac{1}{2}$ hours: the star called Capella rises in the morning. $14\frac{1}{2}$ hours: Sirius disappears. $15\frac{1}{2}$ hours: the bright star in Perseus sets in the evening. According to the Egyptians windy conditions.

13. According to the Egyptians a west or a north-west wind and rainy. According to Eudoxus and Dositheus rainy.

14. 14 hours: the star in the rear shoulder of Orion disappears and the bright star in the northern claw sets in the morning. According to the Egyptians thunderstorm.

15. According to the Egyptians rain, beginning of summer. According to Euctemon and Philippus there is a change in the weather.

16. $13\frac{1}{2}$ hours: Arcturus sets in the morning and the star in the rear shoulder of Orion disappears. According to Dositheus there is a change in the weather.

17. $13\frac{1}{2}$ hours: the star called Capella sets in the evening and the bright star in Lyra rises in the evening. 14 hours: Sirius disappears and the star on the right front hoof of Centaurus rises in the evening. According to the Egyptians a west or a north-west wind. According to Caesar rain. According to Metrodorus, Eudoxus, and Hipparchus there is a change in the weather, and the beginning of summer.

18. $13\frac{1}{2}$ hours: the star called Antares sets in the morning. $14\frac{1}{2}$ hours: the bright star in Cygnus rises in the evening. 15 hours: the star on the rear shoulder of Auriga rises in the morning. According to the Egyptians a west or a south-west wind, a change in the weather. According to Eudoxus and Conon rainy.

19. $14\frac{1}{2}$ hours: the star called Antares sets in the morning. According to the Egyptians, Eudoxus and Callippus, a change in the weather.

20. 14 hours: the star called Capella sets in the evening. 15 hours: the star called Antares sets in the morning. According to Caesar a change in the weather, rainy.

21. $15\frac{1}{2}$ hours: the star called Antares sets in the morning. According to Caesar there is a change in the weather.

22. According to the Egyptians a south or an east wind. According to Eudoxus rainy. According to Hipparchus a south or an arctic wind.

23. $13\frac{1}{2}$ hours: the star on the rear shoulder of Auriga disappears and Sirius disappears. According to the Egyptians thunderstorm and thunder. According to Eudoxus the beginning of summer, rainy.

24. $14\frac{1}{2}$ hours: the star called Capella sets in the evening and the star on the rear shoulder of Auriga rises in the morning. $15\frac{1}{2}$ hours: the bright star in Aquila rises in the evening. According to the Egyptians and Hipparchus it drizzles and there is a change in the weather.

25. 14 hours: the star in the rear shoulder of Auriga disappears. 15 hours: the bright star in the northern claw sets in the morning.

26. 14 hours: Arcturus sets in the morning. According to the Egyptians a north-west or a west wind. According to Dositheus a south wind. According to Caesar it is stormy.

27. 15 hours: the bright star in Aquila rises in the evening. $15\frac{1}{2}$ hours: Procyon disappears.

28. $14\frac{1}{2}$ hours: the star in the rear shoulder of Auriga sets in the evening. 15 hours: the star called Capella sets in the evening.

29. $15\frac{1}{2}$ hours: the star on the knee of Sagittarius sets in the morning. According to the Egyptians windy conditions. According to Euctemon and Philippus a change in the weather.

30. 14 hours: the bright star in Cygnus rises in the evening. According to Euctemon, Philippus, and Hipparchus a change in the weather.

Payni

1. $13\frac{1}{2}$ hours: the star in the rear shoulder of Auriga rises. 15 hours: the star in rear shoulder of Auriga sets in the evening and Procyon disappears. $15\frac{1}{2}$ hours: the bright star in the northern claw sets in the morning. According to the Egyptians a strong north wind. According to Callippus and Euctemon there is a change in the weather.

2. $14\frac{1}{2}$ hours: the bright star in Aquila rises in the evening. According to the Egyptians a change in the weather. According to Metrodorus and Callippus south winds.

3. $13\frac{1}{2}$ hours: the bright star in the Hyades rises. $14\frac{1}{2}$ hours: Procyon disappears. According to the Egyptians and Democritus rainy.

4. According to Hipparchus a south or a west wind.

5. $14\frac{1}{2}$ hours: the star on the right front hoof of Centaurus rises in the evening. $15\frac{1}{2}$ hours: the star called Capella sets in the evening and the star on the rear shoulder of Auriga sets in the evening. According to Caesar the south wind blows.

6. 14 hours: Procyon disappears and the bright star in Aquila rises in the evening. 15 hours: the star on the knee of Sagittarius sets in the morning.

7. 14 hours: the bright star in the Hyades rises. $14\frac{1}{2}$ hours: Arcturus sets in the morning. According to the Egyptians a west wind. According to Eudoxus and Dositheus south winds.

8. According to the Egyptians a north-west or a west wind blows.

9. $14\frac{1}{2}$ hours: the star on the knee of Sagittarius sets in the morning. $15\frac{1}{2}$ hours: the bright star in Hydra disappears. According to the

Egyptians a north-west wind and drizzle. According to Democritus it is wet.

10. $13\frac{1}{2}$ hours: the bright star in Cygnus rises in the evening. $15\frac{1}{2}$ hours the star on the head of the rear twin disappears. According to Caesar thundery and rain.

11. $13\frac{1}{2}$ hours: the bright star on Aquila rises in the evening and the star on the head of the leading twin disappears. 15 hours: the star on the head of the rear twin disappears. According to the Egyptians it drizzles. According to Caesar thunder, rain.

12. $14\frac{1}{2}$ hours: the star on the head of the rear twin disappears.

13. 14 hours: the star on the head of the leading twin disappears and the star on the knee of Sagittarius sets in the morning. $14\frac{1}{2}$ hours: the star on the head of the leading twin disappears.

14. 14 hours: the star on the head of the rear twin disappears. $14\frac{1}{2}$ hours: the bright star in the Hyades rises. 15 hours: the star on the head of the leading twin disappears. $15\frac{1}{2}$ hours: the star on the head of the leading twin disappears.

15. $13\frac{1}{2}$ hours: the star on the head of the rear twin disappears and the star on the knee of Sagittarius rises in the evening and the star on the knee of Sagittarius sets in the morning. 15 hours: the bright star in Hydra disappears. According to the Egyptians a west or a north-west wind, thunder.

16. $13\frac{1}{2}$ hours: the bright star in Corona Borealis sets in the morning.

17. 15 hours: the bright star in the Hyades rises. According to the Egyptians it drizzles throughout the day.

18. 14 hours: the star on the knee of Sagittarius rises in the evening. 15 hours: Arcturus sets in the morning.

19. According to the Egyptians a west or a north-west wind, it drizzles.

20. $14\frac{1}{2}$ hours: the bright star in Hydra disappears and the star on the knee of Sagittarius rises in the evening.

21. $13\frac{1}{2}$ hours: the star on the leading shoulder of Orion rises and the rear-most star in Eridanus rises. According to the Egyptians it drizzles.

22. $15\frac{1}{2}$ hours: the bright star in the Hyades rises.

23. According to the Egyptians burning heat. According to Dositheus a change in the weather.

24. 15 hours: the star on the knee of Sagittarius rises in the evening. According to the Egyptians a west or a south wind and burning heat.

25. 14 hours: the star on the leading shoulder of Orion rises and the bright star in Hydra disappears. According to the Egyptians rain.

26. According to the Egyptians a west wind, showers, thunder.

27. $13\frac{1}{2}$ hours: the star on the rear shoulder of Orion rises. 14 hours: the bright star in Corona Borealis sets in the morning. $14\frac{1}{2}$ hours: the star on the right front hoof of Centaurus disappears.
28. $13\frac{1}{2}$ hours: the star shared by Eridanus and the foot of Orion rises. According to Democritus there is a change in the weather.
29. $15\frac{1}{2}$ hours: the star on the knee of Sagittarius rises in the evening. According to Hipparchus a west or a south wind blows.
30. $13\frac{1}{2}$ hours: the bright star in Hydra disappears. $14\frac{1}{2}$ hours: the star on the leading shoulder of Orion rises. $15\frac{1}{2}$ hours: Arcturus sets in the morning.

Epiphi

1. Summer solstice. $13\frac{1}{2}$ hours: the middle of Orion's belt rises. 14 hours: the star in the rear shoulder of Orion rises. According to the Egyptians a west wind and burning heat.
2. $15\frac{1}{2}$ hours: the bright star in Perseus rises in the evening.
3. According to the Egyptians and Democritus the west wind blows.
4. According to Callippus and Dositheus a change in the weather. According to Democritus a south wind and wet in the morning, then the northern *prodromoi* for 7 days.[98]
5. 14 hours: the star shared by Eridanus and the foot of Orion rises. 15 hours: the star in the leading shoulder of Orion rises. According to Eudoxus there is a change in the weather.
6. $13\frac{1}{2}$ hours: the star on the head of the leading twin rises. 14 hours: the middle of Orion's belt rises and the rearmost star in Eridanus rises and the star on the head of the leading twin rises. According to the Egyptians wind and a bad mixture of air.
7. $14\frac{1}{2}$ hours: the bright star in Corona Borealis sets in the morning.
8. 15 hours: the star on the head of the leading twin rises. $15\frac{1}{2}$ hours: the star shared by Pegasus and Andromeda rises in the evening.
9. $15\frac{1}{2}$ hours: the star on the head of the leading twin rises. According to the Egyptians and Caesar a south wind and burning heat.
10. $14\frac{1}{2}$ hours: the star on the rear shoulder of Orion rises. $15\frac{1}{2}$ hours: the star in the heart of Leo disappears. According to the Egyptians a north-west wind and rainy.
11. $14\frac{1}{2}$ hours: the middle of Orion's belt rises. $15\frac{1}{2}$ hours: the star on the leading shoulder of Orion rises. According to the Egyptians a west or a

[98] The *prodromoi* are precursors to the annual Etesian wind.

north-west wind and thunder. According to Metrodorus a north-west wind. According to Callippus a south wind. According to Hipparchus a south or a west wind.

12. $13\frac{1}{2}$ hours: the star on the head of the rear twin rises. $14\frac{1}{2}$ hours: the star shared by Eridanus and the foot of Orion rises. According to the Egyptians a west or a north-west wind and burning heat.

13. 15 hours: the star on the heart of Leo disappears. According to the Egyptians there is a change in the weather. According to Hipparchus the *prodromoi* of Sirius.

14. $14\frac{1}{2}$ hours: the star on the head of the rear twin rises. According to Meton south winds.

15. $15\frac{1}{2}$ hours: the star on the rear shoulder of Orion rises. According to the Egyptians a north-west or a west wind. According to Euctemon and Philippus south winds and the beginning of the *prodromoi*.

16. $14\frac{1}{2}$ hours: the star on the heart of Leo disappears. According to the Egyptians there is a change in the weather, bad air.

17. 15 hours: the star shared by Pegasus and Andromeda rises in the evening and the middle of Orion's belt rises. $15\frac{1}{2}$ hours: the star on the head of the rear twin rises.

18. 14 hours: the star on the heart of Leo disappears. 15 hours: the bright star in Corona Borealis sets in the morning and the star shared by Eridanus and the foot of Orion rises. According to the Egyptians the *prodromos* blows at the first hour. According to Metrodorus a west or a north-west wind.

19. $13\frac{1}{2}$ hours: Procyon rises. According to Hipparchus a bad mixture of winds.

20. According to the Egyptians burning heat. According to Caesar strong wind. According to Hipparchus the north wind begins to blow.

21. $13\frac{1}{2}$ hours: the star in the heart of Leo disappears.

22. $13\frac{1}{2}$ hours: Sirius rises. 14 hours: Procyon rises. $14\frac{1}{2}$ hours: the rearmost star in Eridanus rises. According to the Egyptians strong wind and rainy sometimes. According to Democritus wet, violent storm.

23. 15 hours: the bright star in Perseus rises in the evening. $15\frac{1}{2}$ hours the middle of Orion's belt rises. According to the Egyptians and Dositheus a south wind and burning heat.

24. $14\frac{1}{2}$ hours: Procyon rises. $15\frac{1}{2}$ hours: the star shared by Eridanus and the foot of Orion rises. According to Hipparchus the Etesian winds begin to blow.

25. According to the Egyptians a west or a north-west wind and burning heat.

26. $14\frac{1}{2}$ hours: the star shared by Pegasus and Andromeda rises in the evening. 15 hours: Procyon rises. According to the Egyptians a north-west or a west wind.

27. $13\frac{1}{2}$ hours: the bright star in Aquila sets in the morning. $15\frac{1}{2}$ hours: the bright star in the southern fish sets in the morning. According to Metrodorus, Euctemon, and Philippus the Etesian winds blow and the beginning of late summer. According to Caesar the *prodromoi* blow.

28. 14 hours: Sirius rises. $15\frac{1}{2}$ hours: the bright star in Corona Borealis sets in the morning and Procyon rises. According to the Egyptians a west wind throughout the day and burning heat. According to Euctemon and Philippus bad air, the *prodromoi* blow.

29. 14 hours: the star on the right front hoof of Centaurus disappears. According to the Egyptians the Etesian winds begin to blow. According to Metrodorus and Callippus windy conditions. According to Euctemon a storm at sea.

30. According to Eudoxus the Etesian winds blow. According to Metrodorus and Callippus windy conditions.

Mesore

1. According to the Egyptians a west or a south wind. According to Eudoxus and Caesar a south wind.

2. 14 hours: the bright star in Aquila sets in the morning. 15 hours: the bright star in the southern fish sets in the morning. According to Metrodorus, Callippus, Conon, Democritus, and Hipparchus a south wind and burning heat.

3. According to Euctemon and Dositheus south winds and stifling heat.

4. $13\frac{1}{2}$ hours: the bright star in Lyra sets in the morning. 14 hours: the star shared by Pegasus and Andromeda rises in the evening. $14\frac{1}{2}$ hours: Sirius rises.

5. According to the Egyptians burning heat. According to Eudoxus south winds and the beginning of late summer. According to Dositheus the Etesian winds begin.

6. $14\frac{1}{2}$ hours: the bright star in Aquila sets in the morning and the bright star in the southern fish sets in the morning. According to the Egyptians a north-west or a west wind and burning heat. According to Eudoxus the Etesian winds blow.

7. According to Caesar the south wind blows.

8. According to Hipparchus burning heat.

9. 14 hours: the bright star in the southern fish sets in the morning. 15 hours: Sirius rises.

10. 15 hours: the bright star in Aquila sets in the morning. $15\frac{1}{2}$ hours: the star called Capella rises in the evening. According to Caesar a change in the weather. According to Eudoxus and Dositheus south winds.

11. $14\frac{1}{2}$ hours: the bright star in Perseus rises in the evening. 15 hours: the rearmost star in Eridanus rises. According to Eudoxus a great burning heat.

12. $13\frac{1}{2}$ hours: the bright star in the southern fish sets in the morning. According to the Egyptians burning heat. According to Dositheus stifling heat and after this the Etesian winds.

13. $13\frac{1}{2}$ hours: the star shared by Pegasus and Andromeda rises in the evening. 14 hours: the bright star in Lyra sets in the morning.

14. $15\frac{1}{2}$ hours: Sirius rises.

15. According to the Egyptians a north-west wind, a great burning heat, and stifling.

16. According to the Egyptians a north-west or a south wind, misty air.

17. According to the Egyptians a great burning heat, and stifling.

18. $13\frac{1}{2}$ hours: the star in the heart of Leo rises. According to the Egyptians thundery. According to Eudoxus the strongest wind. According to Hipparchus a disturbance of the winds.

19. Beginning of autumn. $13\frac{1}{2}$ hours: the bright star in the southern fish rises in the evening. $14\frac{1}{2}$ hours: the star in the heart of Leo rises. According to the Egyptians burning heat.

20. 15 hours: the star in the heart of Leo rises. According to Caesar there is a change in the weather.

21. According to Caesar there is a change in the weather, stifling.

22. $13\frac{1}{2}$ hours: the star on the tail of Leo disappears and the bright star in Hydra rises.

23. $13\frac{1}{2}$ hours: the star in the right front hoof of Centaurus disappears. 14 hours: the star on the tail of Leo disappears. According to Caesar the winds veer around.

24. 14 hours: the bright star in Hydra rises. According to Eudoxus there is a change in the weather.

25. $15\frac{1}{2}$ hours: the star on the tail of Leo disappears.

26. 14 hours: the bright star in the southern fish rises in the evening. According to the Egyptians a south or a west wind. According to Democritus there is a change in the weather with respect to rains and winds.

27. $14\frac{1}{2}$ hours: the bright star in Hydra rises. According to the Egyptians burning heat and mist.

28. 14 hours: the bright star in Perseus rises in the evening.

29. 15 hours: the bright star in Hydra rises. According to the Egyptians and Caesar there is a change in the weather, bad air. According to Eudoxus there is usually thunder.
30. $15\frac{1}{2}$ hours: the star in the rear shoulder of Auriga rises in the evening. According to the Egyptians a west or a north-west wind.

Epagomenal days

1. 15 hours: the bright star in Lyra sets in the morning. $15\frac{1}{2}$ hours: the bright star in Hydra rises. According to Eudoxus and Metrodorus there is a change in the weather.
2. 14 hours: the star called Canopus rises. $14\frac{1}{2}$ hours: the bright star in the southern fish rises in the evening. According to the Egyptians burning heat. According to Eudoxus and Caesar there is a change in the weather. According to Hipparchus a south wind and the Etesian winds stop.
3. $14\frac{1}{2}$ hours: Spica disappears. $15\frac{1}{2}$ hours: the star on the tail of Leo rises. According to Hipparchus whirlwinds.
4. 15 hours: the star on the tail of Leo rises. According to Callippus there is a change in the weather.
5. $13\frac{1}{2}$ hours: the bright star in Cygnus sets in the morning. According to the Egyptians a west or a north-west wind.

A.ix. Polemius Silvius *Fasti* [99]

Ianuarius

dictus a Iano. habet dies xxxi. vocatur apud Hebraeos Sebet, apud Aegyptios Tybi, apud Athenienses Posideon, apud Graecos alios Edineus.

kalendae dictae ἀπὸ τοῦ καλεῖν, *hoc est a vocando, quia tum in rostris Romae ad contionem populus vocabatur.*	
IIII nonas.	*Circus privatus. auster, interdum cum pluvia.*
III	*dies auspicalium. natalis Ciceronis. Ludi.*
II	*Ludi Compitales.*
nonae dictae ideo, quia nonus dies eas discernit ab idibus. tempestatem significat.	
VIII idus.	*Epiphania, quo die interpositis temporibus stella magis [visa], quae Dominum natum nuntiabat. de aqua vinum fact[um et]*

---

[99] Latin text in Degrassi, 1963, vol. xiii.2, pp. 263–76.

	in amne Iordanis Salvator baptizatus est. auster interdum vel favonius.
VII	*prima consulis mappa, quae ideo sic vocatur, quia rex Tarquinius Romae, dum die circensium pranderet in Circo, de mensa suam mappam foras, ut aurigis post prandium currendi signum daret, abiecerat.*
VI	*auster interdum et imber.*
V	*senatus legitimus. suffecti consules designantur sive praetores.*
IIII	
III	*Carmentalia de nomine matris Evandri.*
II	*<natalis> d(omini) n(ostri) The(odosii) Aug(usti) pridie.*

idus dictae ἀπὸ τοῦ εἰδεῖν, *a videndo, quia, priusquam annus hic qui est fuerat, mense medio luna completa, quae incipiebat kalendis, de qua menses dictos accepimus, videbatur. secunda mappa. interdum ventus aut tempestas.*

XVIIII kalendas Februarias.	
XVIII	*natalis Honorii. Circenses. interdum auster et pluvia.*
XVII	
XVI	*Ludi Palatini.*
XV	*Ludi.*
XIIII	*Ludi.*
XIII	*natalis Gordiani. Circenses.*
XII	*Ludi. ventus africus. tempestatem significat.*
XI	*natalis sancti Vincentii martyris. dies pluvius.*
X	*senatus legitimus. quaestores Romae designantur.*
VIIII	*natalis Hadriani. Circenses.*
VIII	*interdum tempestas.*
<VII>	
VI	*Ludi Castorum Ostiis, quae prima facta colonia est.*
<V>	*Ludi. auster aut africus. interdum dies humidus.*
IIII	*Ludi.*
III	*tempestatem significat.*
II	*Circenses Adiabenis victis. interdum tempestas.*

Februarius

dictus a febro verbo, quod purgamentum veteres nominabant, quia tum Romae moenia lustrabantur. habet dies XXVIII. *vocatur apud Hebraeos Adar, apud Aegyptios Mechir, apud Athenienses Gamelion, apud Graecos Peritios.*

kalendae.	*Circenses. interdum auster cum grandine.*

IIII nonas.	
III nonas.	*senatus legitimus. corus aut africus.*
II	*Ludi.*
nonas.	*Ludi. ventus aut tempestas.*
VIII idus.	*Ludi.*
VII	*Ludi. favonius.*
VI	*ventus aut tempestas.*
V	*Circenses.*
IIII	*eurus.*
III	*Circenses.*
II	*Ludi.*
idus.	*parentatio tumulorum inc[ipit], quo die Roma liberata est de obsidione Gallorum.*
XVI kalendas Martii.	*venti mutatio.*
XV	*Lupercalia.*
XIIII	*natalis Faustinae uxoris Antonini.*
XIII	*Quirinalia, quo die Romulus, occisus a suis, Qui<rinus> ab hasta, quae a Sabinis curis vocatur, non apparuisse confictus est. favonius aut auster cum grandine.*
XII	
XI	
X	*venti septentrionales.*
VIIII	*frigidus dies de aquilone vel pluvia.*
VIII	*depositio sancti Petri et Pauli. Cara Cognatio ideo dicta, quia tunc, etsi fuerint vivorum parentum odia, tempore obitus deponantur. ventus aut tempestas.*
VII	*Terminalia. hoc die quarto bisextum anno vocamus, quem diem Λegyptii inter Augustum et Septembrem epagomeno quinque dierum mensi suo iungunt.*
VI	*Regifugium, cum Tarquinius Superbus fertur ab urbe expulsus.*
V	*Circenses.*
IIII	
III	*natalis Constantini.*
II	*Ludi.*

Martius

habet dies xxxi. vocatur apud Hebraeos Nisan, apud Aegyptios Famenoth, apud Athenienses Antesterion, apud Graecos Distros.

kalendae.	*interdum grando.*
VI nonas.	*septentrionales venti.*
V	
IIII	*interdum hiemat.*
III	
II	*aquilo*
nonas.	
VIII idus.	
VII	
VI	
V	*natalis Favonii.*
IIII	*septentrionales venti.*
III	
II	*flatus aquilonis.*
idus.	
XVII <kalendas> Aprilis. hiemat.	
XVI	*Circenses. favonius aut corus.*
XV	
XIIII	*Quinquatria.*
XIII	*Pelusia.*
XII	*septentrionales venti.*
XI	*dies pluvius, interdum ningit.*
X	
VIIII	*natalis calices.*
VIII	*aequinoctium. principium veris. crucimissio gentilium. Christus passus hoc die.*
VII	*interdum tempestas.*
VI	*Lavationem veteres nominabant. resurrectio.*
V	
IIII	
III	
II	*natalis Constanti[n]i. Circenses.*

Aprilis

habet dies xxx. vocatur apud Hebraeos Iar, apud Aegyptios Farmuti, apud Athenienses Elafybilion, apud Graecos Xanticus.

kalendae.	*tempestatem significant.*
IIII nonas.	

III	Circenses.
II	Ludi.
nonas.	favonius aut auster, interdum cum grandine.
VIII idus.	interdum hiemat.
VII	tempestatem significat.
VI	avium Alcyonum dicitur.
<V>	
IIII	Circenses. interdum tempestatem significat.
III	natalis <Se>veri. Circenses.
II	Ludi. hiemat.
idus.	senatus legitimus.
XVIII kalendas Maias.	Ludi. ventus ac tempestas.
XVII	Ludi.
XVI	Ludi.
XV	pluviam significat.
XIIII	Ludi.
XIII	Circenses. consulis tertiae mappae.
XII	
XI	natalis urbis Romae. consules ordinarii fasces deponunt. Parilia dicta de partu Iliae. interdum pluvia et grando.
X	africus aut auster. dies humidus.
VIIII	tempestatem significat.
VIII	
VII	
VI	natalis Antoni<n>i. Circenses.
V	Floria.
IIII	auster fere cum pluvia.
III	interdum dies humidus.
II	tempestatem significat.

Maius

dictus a maioribus. habet dies XXXI. vocatur apud Hebraeos Sivan, apud Aegyptios Pachon, apud Athenienses Munychion, apud Graecos Artemision.

kalendae.	
VI nonas.	septentrionales venti.
V	tempestatem significat.
IIII	Ludi.
III	pluvia.

II	*Ludi. <tempestatem> significat.*
nonas.	*Ludi. favonius.*
VIII idus.	*Ludi. interdum pluvia.*
VII	
VI	*natalis Claudii. tempestas interdum.*
V	
IIII	
III	*Ludi. tempestatem significat.*
II	*Ludi.*
idus.	*dies humidus.*
XVII kalendas Iunii. dies ut supra.	
XVI	*pluvia.*
XV	
XIIII	
XIII	
XII	*interdum pluvia.*
XI	*tempestatem significat.*
X	*ut supra.*
VIIII	
VIII	*venti.*
VII	
VI	
V	
IIII	*Ludi. interdum pluvia.*
III	
II	

Iunius

dictus a iunioribus. habet dies xxx. vocatur apud Hebraeos Tamuz, apud Aegyptios Pauni, apud Athenienses Thargilion, apud Graecos Desios.

kalendae.	*Circenses Fab<a>ricii.*
IIII nonas.	
III	
II	*Ludi.*
nonae.	
VIII idus.	
VII	*favonius aut corus.*
VI	

V
IIII   *interdum rorat.*
III
II
idus.   *natalis musarum.*
XVIII kalendas Iulii.
XVII
XVI   *tempestatem significat.*
XV
XIIII
XIII
XII
XI
X
VIIII
VIII   *solstitium. favonius.*
VII
VI
V   *initium aestatis.*
IIII   *dies ventosus.*
III
II

Iulius

dictus a Iulio Caesare. habet dies xxxi, cum Quintilis antea diceretur. vocatur apud Hebraeos Ab, apud Aegyptios Epifi, apud Athenienses Sciroforion, apud Graecos Panemos.

kalendae.
VI *nonas.*   *natalis genuinus d(omini) n(ostri) Valentiniani.*
V *nonas.*   *favonius.*
IIII
III
II   *Ludi.*
nonae.   *ancillarum feriae, quarum celebritas instituta est ideo, quia capta*
               Urbe a Gallis, cum finitimi prius victi tradi sibi Romanorum pro-
               cerum coniuges postularent et consilio Philotidis ancillae famulae
               dominarum vestibus adornatae datae illis fuissent, his nuntian-
               tibus praedictos somno sopitos et ebrios posse superari, facta vic-
               toria sic.

VIII idus.	*natalis Iulii Caesaris.*
VII	*tempestatem significat.*
VI	*Ludi.*
V	*Ludi. etesiae venti flare incipiunt.*
IIII	*Ludi.*
III	
II	
idus.	
XVII kalendas Augustas.	
XVI	
XV	
XIIII	
XIII	
XII	
XI	
X	
VIIII	*favonius vel auster. tempestatem significat.*
VIII	*natalis Constantini.*
VII	
VI	*dies canicularis.*
V	
IIII	*interdum tempestatem.*[100]
<III>	
<II>	

Augustus

habet dies XXXI. *prius Sextilis. dictus ab Octaviano Augusto. vocatur apud Hebraeos Elul, apud Aegyptios Mesore, apud Athenienses Ecatonbion, apud Graecos Loos.*

kalendae.	*natalis Pertinacis et martyrium Maccabaeorum.*
IIII nonas.	
III	
II	*tempestas.*
nonae.	
VIII idus.	*nebulosus aestus.*
VII	*natalis Constanti[n]i minoris.*

---

[100] Degrassi restores *<significat>*.

VI	*Ludi.*
V	
IIII	*natalis sancti Laurenti martyris.*
III	
II	
idus.	*natalis Hippolyti martyris. tempestatem significat.*
XVIIII *kalendas Septembris.*	
XVIII	
XVII	
XVI	
XV	*Ludi.*
XIIII	
XIII	
XII	*tempestas.*
XI	
X	*Circenses.*
VIIII	
VIII	
VII	*post hunc diem apud Aegyptios* v *dierum epagomenos tertius decimus mensis adiungitur.*
VI	*interdum pluvia.*
V	*Circenses.*
IIII	
III	
II	*interdum frigus.*

September

dictus a numero. habet dies xxx. *vocatur apud Hebraeos Tesri, apud Aegyptios Thoth, apud Athenienses Metageitnion, apud Graecos Gorpieos.*

kalendae.	
IIII nonas.	*favonius aut corus.*
III	*Ludi.*
II	
nonae.	
VIII idus.	
VII idus.	*tempestas.*
VI	
V	
IIII	

III	*favonius vel africus.*
II	*interdum tempestas.*
idus.	*hoc die Romae in aede Minervali per magistratum annis singulis ex aere clipei figebantur.*
XVIII kalendas Octobris.	
XVII	*favonius interdum sive vulturnus*
XVI	
XV	
XIIII	*favonius vel corus.*
XIII	
XII	*Ludi.*
XI	*favonius vel corus cum pluvia.*
X	*tempestatem significat.*
VIIII	
VIII	*aequinoctium.*
VII	
VI	*initium autumni.*
V	*favonius interdum.*
IIII	*tempestatem significat.*
III	*Ludi.*
II	

October

dictus a numero. habet dies xxxi. vocatur apud Hebraeos Maresuan, apud Aegyptios Faofi, apud Athenienses Boedromiion, apud Graecos Yperberetios.

kalendae.	*tempestatem.*[101]
VI nonas.	
V nonas.	
IIII	*tempestatem.*[102]
III	*ut supra.*
II	*Ludi. aquilo.*
nonae.	*Ludi.*
VIII idus.	
VII	
VI	*interdum pluvia.*
V	

---

[101] Degrassi restores *<significat>*.   [102] Degrassi restores *<significat>*.

IIII
III   *auster et pluvia.*
II   *natalis Virgilii.*
<*idus*>
XVII *kalendas Novembris.*
XVI
XV
XIIII   *tempestatem significat.*
<*XIII*>
XII
XI
X   *natalis Valentiniani purpurae.*
VIIII
VIII
VII   *tempestatem significat.*
VI
V   *interdum frigus.*
IIII   *ventosus dies.*
III   *tempestas.*
II   *ut supra.*

November

dictus a numero. habet dies xxx. vocatur apud Hebraeos Casleu, apud Aegyptios Atyr, apud Athenienses Pyanopsion, apud Graecos Dios.

kalendae.
IIII *nonas.*   *hiemat cum pluvia.*
III
II
nonae.
VIII *idus.*   *natalis Iuliani.*
VII   *hiemat.*
VI   *natalis Nervae. tempestas.*
V   *ut supra.*
IIII   *interdum rorat.*
III
II
idus.
XVIII *kalendas Decembris.*

XVII	
XVI	*aquilo.*
XV	*natalis Vespasiani. aquilo cum pluvia.*
XIIII	*tempestas.*
XIII	*aquilo frigidus.*
XII	
XI	*hiemat.*
X	*tempestas.*
VIIII	
VIII	*bruma.*
VII	
VI	
V	*Ludi.*
IIII	
III	*ut supra.*
II	*auster cum pluvia.*

December

dictus a numero. habet dies XXXI. *vocatur apud Hebraeos Tebet, apud Aegyptios Choiac, apud Athenienses Maimacterion, apud Graecos Apelleos.*

kalendae.	
IIII nonas.	
III	
II	
nonae.	
VIII idus.	*tempestas.*
VII	*interdum auster et rorat.*
VI	
V	
IIII	
III	*interdum pluvia.*
II	*Septimontium.*
idus.	*hiemat.*
XVIIII kalendas Ianuarias.	
XVIII	*natalis Veri. tempestatem.*[103]

[103] Degrassi restores *<significat>*.

XVII	
XVI	*feriae servorum.*
XV	
XIIII	
XIII	
XII	
XI	
X	*tempestas.*
VIIII	
VIII	*natalis Domini corporalis. solstitium et initium hiberni.*
VII	*natalis sancti Stephani martyris.*
VI	
V	*natalis Titi. tempestas.*
IIII	
III	
II	*ut supra.*

January

Named for Janus. It has thirty-one days. Called by the Hebrews *Sebet*, by the Egyptians *Tybi*, by the Athenians *Posideon*, but *Edineus* by the other Greeks.

The Kalends are so called from (the Greek) καλεῖν, which has to do with the announcement that was made to the assembly of the people at the Roman *Rostra*.

IIII Non. *Circus privatus.*[104] South wind, sometimes with rain.

III Non. *Auspicial* day. Birth of Cicero. *Ludi.*

II   *Ludi Compitales.*

The Nones are so called since they are nine days from the Ides. It signifies a storm.[105]

---

[104] On the festivals, rites and holidays throughout this calendar, see Salzman, 1990, chapter 4.

[105] Here as commonly in Latin parapegmata, the phrase *tempestatem significat* seems to be an imperfect translation of the Greek ἐπισημαίνει. I have argued elsewhere that *tempestatem significat* may be a corruption of *tempestas, significat*: 'Storm, and a change in the weather'. See Lehoux, 2004c, p. 85. The fact that a direct object has been added in Latin changes the meaning of the Greek original, and since the original meaning is lost in the Latin, I translate the Latin as it would have been understood by readers of this text. The derivation of the

VIII Ides Epiphany, on this day, after a time, the star that announced the birth of the Lord was seen by the Magi. Wine made from water, and the Saviour baptized in the river Jordan. South wind, or sometimes a west wind.

VII   The first *mappa* (cloth) for the consuls, which is so called because King Tarquin of Rome breakfasted on this day at a contest in the Circus, and threw out the cloth from his table in order that the charioteers could rush forth when the sign was given after his meal.

VI   South wind, sometimes with a downpour.

V   *Senatus legitimus.* Replacement consuls elected, or praetors.

IIII

III   *Carmentalia,* after the name of the mother of Evander.

II   Birth date of our lord Theodosius Augustus; *pridie.*

The Ides are so called from (the Greek) εἰδεῖν, for *seeing*, since in the old days the moon, which was new at the Kalends, was – and this is where we get our word – *seen* to be full in the middle of the month. Second *mappa*. Sometimes wind or storm.

XVIIII Kal. February.

XVIII   Birth of Honorius. *Circenses.* Sometimes a south wind and rain.

XVII

XVI   *Ludi Palatini.*

XV   *Ludi.*

XIIII   *Ludi.*

XIII   Birth of Gordianus. *Circenses.*

XII   *Ludi.* South-west wind. It signifies a storm.

XI   Birth of St Vincent the martyr. Rainy day.

X   *Senatus legitimus.* Quaestors elected at Rome.

VIIII   Birth of Hadrian. *Circenses.*

VIII   Sometimes stormy.

VII

VI   *Ludi Castorum* in Ostia, which was the first colony founded.

V   *Ludi.* South or south-west wind. Sometimes a humid day.

IIII   *Ludi.*

III   It signifies a storm.

II   *Circenses Adiabenis victis.* Sometimes stormy.

weather predictions in this text from an original (Columella) that included stellar phases as the implied subjects of the verbs will make some sense of the active verbs here.

February

Named from the word for purification, which was called *purgation* by our ancestors because of the purification of the walls of Rome. It has twenty-eight days. It is called *Adar* by the Hebrews, *Mechir* by the Egyptians, *Gamelion* by the Athenians, and *Peritios* by the Greeks.

Kalends	*Circenses.* Sometimes a south wind with hail.
IIII Non.	
III	*Senatus legitimus.* North-west or south-west wind.
II	*Ludi.*
Nones	*Ludi.* Wind or storm.
VIII Ides	*Ludi.*
VII	*Ludi.* West wind.
VI	Wind or storm.
V	*Circenses.*
IIII	South-east wind.
III	*Circenses.*
II	*Ludi.*
Ides	*Parentatio tumulorum* begins, on which day Rome was liberated from the Gallic siege.
XVI Kal. March.	Change of wind.
XV	*Lupercalia.*
XIIII	Birth of Faustina, wife of Antoninus.
XIII	*Quirinalia*, on which day Romulus, called Quirinus from the spear that by the Sabines is called a *curis*, was killed by his own people, though it was said that he 'disappeared'. West or south wind.
XII	
XI	
X	North winds.
VIIII	Cold day because of the north-east wind, or rain.
VIII	*Depositio* of Sts Peter and Paul. *Cara Cognatio*, which is named because at this time, if there is anger at living relatives, it is put aside. Wind or storm.
VII	*Terminalia.* On which day we have the intercalary day every fourth year. By the Egyptians that day is attached to the five days of the epagomenal month, between August and September.
VI	*Regifugium*, when Tarquinius Superbus was expelled from the city.

V	*Circenses.*
IIII	
III	Birth of Constantine
II	*Ludi.*

March

It has thirty-one days. It is called *Nisan* by the Hebrews, *Famenoth* by the Egyptians, *Antesterion* by the Athenians, and *Distros* by the Greeks.

Kalends	Sometimes hail.
VI Non.	North winds.
V	
IIII	Sometimes it is wintry.
III	
II	North-east wind.
Nones	
VIII Ides	
VII	
VI	
V	Birth of Favonius.
IIII	North winds.
III	
II	North-east breeze.
Ides	
XVII Kal. April. It is wintry.	
XVI	*Circenses.* West or north-west wind.
XV	
XIIII	*Quinquatria.*
XIII	*Pelusia.*
XII	North winds.
XI	Rainy day. Sometimes it snows.
X	
VIIII	*Natalis calices.*
VIII	Equinox. Beginning of spring. *Crucimissio gentilium.* Christ died on this day.
VII	Sometimes a storm.
VI	The ancients called it *Lavatio.* Resurrection.
V	

IIII

III

II   Birth of Constantius. *Circenses.*

April

It has thirty days. It is called *Iar* by the Hebrews, *Farmuti* by the Egyptians, *Elafybilion* by the Athenians, and *Xanticus* by the Greeks.

The Kalends signify a storm.

IIII Non.

III   *Circenses.*

II   *Ludi.*

Nones   West or south wind, sometimes with hail.

VIII Ides   Sometimes it is wintry.

VII   It signifies a storm.

VI   Called 'the time of the Halcyon birds'.[106]

V

IIII   *Circenses.* Sometimes it signifies a storm.

III   Birth of Severus. *Circenses.*

II   *Ludi.* It is wintry.

Ides   *Senatus legitimus.*

XVIII Kal. May. *Ludi.* Wind or storm.

XVII   *Ludi.*

XVI   *Ludi.*

XV   It signifies rain.

XIIII   *Ludi.*

XIII   *Circenses.* The third *mappae* for the consuls.

XII

XI   Anniversary of the founding of the city of Rome. The regular consuls lay down the *fasces*. Called the *Parilia* from parturition of *Ilia*.[107] Sometimes rain and hail.

X   South-west or south wind. Humid day.

VIIII   It signifies a storm.

VIII

VII

---

[106] I.e., the time when the Halcyon birds build their nests and the sea is supposed to be very calm.

[107] The mother of Romulus and Remus.

VI	Birth of Antoninus. *Circenses.*
V	*Floria.*
IIII	It causes a south wind with rain.
III	Sometimes a humid day.
II	It signifies a storm.

May

Named for the ancestors (*maiores*). It has thirty-one days. It is called *Sivan* by the Hebrews, *Pachon* by the Egyptians, *Munychion* by the Athenians, and *Artemision* by the Greeks.

Kalends	
VI Non.	North winds.
V	It signifies a storm.
IIII	*Ludi.*
III	Rain.
II	*Ludi.* It signifies [a storm].
Nones	*Ludi.* West wind.
VIII Ides	*Ludi.* Sometimes rain.
VII	
VI	Birth of Claudius. Sometimes a storm.
V	
IIII	
III	*Ludi.* It signifies a storm.
II	*Ludi.*
Ides	Humid day.
XVII Kal. June.	A day like the previous.
XVI	Rain.
XV	
XIIII	
XIII	
XII	Sometimes rain.
XI	It signifies a storm.
X	As above.
VIIII	
VIII	Winds.
VII	

VI
V
IIII   *Ludi.* Sometimes rain.
III
II

June

Named for young people (*iuniores*). It has thirty days. It is called *Tamuz* by the Hebrews, *Pauni* by the Egyptians, *Thargilion* by the Athenians, and *Desios* by the Greeks.

Kalends	*Circenses Fabaricii.*
IIII Non.	
III	
II	*Ludi.*
Nones	
VIII Ides	
VII	West or north-west wind.
VI	
V	
IIII	Sometimes it is damp.
III	
II	
Ides	Birth of the Muses.
XVIII Kal. July.	
XVII	
XVI	It signifies a storm.
XV	
XIIII	
XIII	
XII	
XI	
X	
VIIII	
VIII	Solstice. West wind.
VII	
VI	
V	Beginning of summer.

IIII	Windy day.
III	
II	

July

Named for Julius Caesar. It has thirty-one days, and was formerly called *Quintilis*. It is called *Ab* by the Hebrews, *Epifi* by the Egyptians, *Sciroforion* by the Athenians, and *Panemos* by the Greeks.

Kalends.	
VI Non.	True birth of our lord Valentinian.
V	West wind.
IIII	
III	
II	*Ludi.*
Nones	*Ancillarum feriae* (Feast of the handmaids), which festival was instituted for this reason: When the city was captured by the Gauls, the victorious Gauls demanded that the Roman chiefs hand over their wives to them. At the advice of Philotis, the handmaids, dressed in the clothes of their mistresses, went out to the Gauls instead. When the handmaids signalled that the Gauls were sound asleep, they were overtaken and victory was achieved.
VIII Ides	Birth of Julius Caesar.
VII	It signifies a storm.
VI	*Ludi.*
V	*Ludi.* The Etesian winds begin to blow.
IIII	*Ludi.*
III	
II	
Ides	
XVII Kal.	August.
XVI	
XV	
XIIII	
XIII	
XII	
XI	

X
VIIII   West or south wind. It signifies a storm.
VIII   Birth of Constantine.
VII
VI   Dog-days.
V
IIII   Sometimes a storm.
III
II

August

It has thirty-one days. Formerly *Sextilis*. Named for Octavian Augustus. It is called *Elul* by the Hebrews, *Mesore* by the Egyptians, *Ecatonbion* by the Athenians, and *Loos* by the Greeks.

Kalends.   Birth of Pertinax, and the martyrdom of the Maccabeans.
IIII Non.
III
II   Storm.
Nones
VIII Ides   Hazy heat.
VII   Birth of Constantius the younger.
VI   *Ludi.*
V
IIII   Birth of St Lawrence the martyr.
III
II
Ides   Birth of Hippolytus the martyr. It signifies a storm.
XVIIII Kal.   September.
XVIII
XVII
XVI
XV   *Ludi.*
XIIII
XIII
XII   Storm.
XI
X   *Circenses.*

VIIII	
VIII	
VII	After this day, according to the Egyptians, five epagomenal days are added to the thirtieth of the month.
VI	Sometimes rain.
V	*Circenses.*
IIII	
III	
II	Sometimes cold.

September

Named for the number (*septem*). It has thirty days. It is called *Tesri* by the Hebrews, *Thoth* by the Egyptians, *Metageitnion* by the Athenians, and *Gorpieos* by the Greeks.

Kalends	
IIII Non.	West or north-west wind.
III	*Ludi.*
II	
Nones	
VIII Ides	
VII Ides	Storm.
VI	
V	
IIII	
III	West or south-west wind.
II	Sometimes a storm.
Ides	On this day, at Rome in the temple of Minerva, the shields are hung up inside by the one-year magistrates.
XVIII Kal.	October.
XVII	Sometimes a west wind, or else a south-east wind.
XVI	
XV	
XIIII	West or north-west wind.
XIII	
XII	*Ludi.*
XI	West or north-west wind with rain.
X	It signifies a storm.
VIIII	

VIII	Equinox.
VII	
VI	Beginning of autumn.
V	Sometimes a west wind.
IIII	It signifies a storm.
III	*Ludi.*
II	

October

Named from the number (*octo*). It has thirty-one days. It is called *Maresuan* by the Hebrews, *Faofi* by the Egyptians, *Boedromiion* by the Athenians, and *Yperberetios* by the Greeks.

Kalends.	A storm.
VI Non.	
V Non.	
IIII	A storm.
III	As above.
II	*Ludi.* North-cast wind.
Nones	*Ludi.*
VIII Ides	
VII	
VI	Sometimes rain.
V	
IIII	
III	South wind and rain.
II	Birth of Vergil.
Ides	
XVII Kal. November.	
XVI	
XV	
XIIII	It signifies a storm.
XIII	
XII	
XI	
X	Anniversary of the accession of Valentinian.
VIIII	
VIII	
VII	It signifies a storm.

VI
V   Sometimes cold.
IIII   Windy day.
III   Storm.
II   As above.

November

Named from the number (*novem*). It has thirty days. It is called *Casleu* by the Hebrews, *Atyr* by the Egyptians, *Pyanopsion* by the Athenians, and *Dios* by the Greeks.

Kalends
IIII Non.   It is wintry, with rain.
III
II
Nones
VIII Ides   Birth of Julian.
VII   It is wintry.
VI   Birth of Nerva. Storm.
V   As above.
IIII   Sometimes it is damp.
III
II
Ides
XVIII Kal. December.
XVII
XVI   South wind.
XV   Birth of Vespasian. South wind with rain.
XIIII   Storm.
XIII   Cold south wind.
XII
XI   It is wintry.
X   Storm.
VIIII
VIII   Winter.
VII
VI
V   *Ludi.*
IIII

| III | As above. |
| II | South wind with rain. |

December

Named from the number (*decem*). It has thirty-one days. It is called *Tebet* by the Hebrews, *Choiac* by the Egyptians, *Maimacterion* by the Athenians, and *Apelleos* by the Greeks.

Kalends	
IIII Non.	
III	
II	
Nones	
VIII Ides	Storm.
VII	Sometimes a south wind, and it is damp.
VI	
V	
IIII	
III	Sometimes rain.
II	*Septimontium.*
Ides	It is wintry.
XVIIII Kal. January.	
XVIII	Birth of Verus. A storm.
XVII	
XVI	Feast of the slaves.
XV	
XIIII	
XIII	
XII	
XI	
X	Storm.
VIIII	
VIII	Bodily birth of the Lord. Solstice and the beginning of winter.
VII	Birth of St Stephan the martyr.
VI	
V	Birth of Titus. Storm.
IIII	
III	
II	As above.

A.x. Antiochus parapegma[108]

περὶ ἀστέρων ἀνατελλόντων καὶ δυνόντων ἐν τοῖς ιβ΄ μησὶ τοῦ ἐνιαυτοῦ σὺν τῷ ἡλίῳ. Ἀντιόχου ὁ περὶ ἀστέρων ἀνατελλόντων καὶ δυνόντων ἐν τοῖς ιβ΄ μησὶ τοῦ ἐνιαυτοῦ.

μὴν Ἰαννουάριος ♒

α΄. ὁ ἐπὶ τῆς κεφαλῆς τοῦ ἡγουμένου τῶν Διδύμων ἑῷος δύνει.

έ. Δελφὶς ἐπιτέλλει.

ζ΄. ὁ κατὰ τοῦ γόνατος τοῦ Τοξότου ἐπιτέλλει.

ια΄. ὁ λαμπρὸς τῆς Λύρας ἑσπέριος ἀνατέλλει· καὶ ποιεῖ ἐπισημασίαν ἀκίνδυνον.

ιε΄. ὁ λαμπρὸς τοῦ Ὑδροχόου ἑῷος δύνει.

ιη΄. ὁ ἐπὶ τῆς κεφαλῆς τοῦ Λέοντος ἑσπέριος ἀνατέλλει.

κβ΄. ὁ Ὑδροχόος ἀνατέλλει· ἐπισημασία.

κϛ΄. ὁ ἐπὶ τοῦ στήθους τοῦ Λέοντος δύνει· ἐπισημασία.

λα΄. ὁ Κάνωβος ἑσπέριος ἀνατέλλει.

μὴν Φεβρουάριος ♓

α΄. ὁ ἐπὶ τῆς καρδίας τοῦ Λέοντος ἑῷος δύνει.

δ΄. ὁ λαμπρὸς τῆς Λύρας ἑσπέριος δύνει· ἐπισημασία ἀκίνδυνος.

ϛ΄. ὁ λαμπρὸς τοῦ Ὄρνιθος ἑσπέριος ἀνατέλλει.

ιγ΄. ὁ ἐπὶ τοῦ δεξιοῦ βραχίονος τοῦ Κενταύρου ἑῷος δύνει.

ιθ΄. ἡ Παρθένος δύνει· ἐπισημασία.

κγ΄.[109] ὁ κοινὸς τοῦ Ἵππου καὶ τῆς Ἀνδρομέδας κρύπτεται· ἐπισημασία.

κε΄. ὁ ἔσχατος τοῦ Ποταμοῦ καὶ Ἰχθὺς νότιος ἀνατέλλει·[110] ἐπισημασία.

κζ΄. ὁ Ἀρκτοῦρος ἄρχεται ἀνατέλλων· ἐπισημασία.

μὴν Μάρτιος ♈

α΄. ὁ κοινὸς Ἵππου καὶ Ἀνδρομέδας ἑσπέριος ἀνατέλλει.

---

[108] Greek text from Boll, 1910a.

[109] Only the Munich MS has 23. The other two have 25 and 26.

[110] Boll thinks something must have dropped out in the middle: 'The rearmost star in Eridanus <...> and the southern fish rises', but compare 3 March. 'Rises' is actually in the singular in all three MSS, but this is not unusual in this parapegma. Compare 28 Apr. and 2 and 7 Oct. where pairs of stars also get singular verbs. 23 May seems to present a contrast to this, but one of the two subjects there, the Hyades, always takes a plural verb.

γ. ὁ ἔσχατος τοῦ Ποταμοῦ καὶ Ἰχθὺς νότιος ἀνατέλλει·[111] ἐπισημασία.

ε. Ἀρκτοῦρος ἀνατέλλει· ἐπισημασία.

ια. Ἵππος δύνει· ἐπισημασία.

ιβ. ὁ Στάχυς ἑσπέριος ἀνατέλλει.

ιδ. ὁ ἐπὶ τῆς οὐρᾶς τοῦ Λέοντος ἑῷος δύνει.

ιϛ. Ὑάδες ἀνατέλλουσιν· ἐπισημασία.

κβ. ἰσημερία ἐαρινή.

κζ. ὁ Στάχυς ἑῷος δύνει· ἐπισημασία.

λ. ὁ λαμπρὸς τοῦ Περσέως ἑῷος δύνει.

μὴν Ἀπρίλλιος ♉

α. ὁ λαμπρὸς τῆς νοτίας Χηλῆς τοῦ Σκορπίου ἑσπέριος ἀνατέλλει καὶ Πλειάδες κρύπτονται· ἐπισημασία.

ε. Πλειάδες τελείως κρύπτονται· ἐπισημασία.

ι. ὕψωμα ἡλίου.

ιδ. Κάνωβος κρύπτεται.

κ. ὁ λαμπρὸς τοῦ Περσέως δύνει.

κβ. ἀρχὴ παχνίτου.

κη. ὁ ἐπὶ τοῦ ἡγουμένου ποδὸς τοῦ Ὠρίωνος καὶ ὁ μέσος τῆς ζώνης κρύπτεται· ἐπισημασία.

λ. Κύων κρύπτεται· ἐπισημασία.

μὴν Μάιος ♊

α. Ὑάδων ἐπιτολή· ἐπισημασία.

β. ὁ λαμπρὸς τῆς Λύρας ἑσπέριος ἀνατέλλει· ἐπισημαίνει.

δ. Αἲξ ἑσπερία δύνει· ἐπισημασία.

θ. Πλειάδες ἀνατέλλουσιν ἅμα ἡλίῳ· ἐπισημαίνει.

ια. ὁ τοῦ ἡγουμένου ὤμου τοῦ Ὠρίωνος κρύπτεται.

ιγ. ὁ Κύων κρύπτεται καὶ ὁ ἐν τῷ δεξιῷ βραχίονι τοῦ Κενταύρου ἑσπέριος ἀνατέλλει.

ιζ. Ἀντάρης ἑῷος δύνει.

ιη. Αἲξ ἑσπερία ἀνατέλλει.

κα. Ἀρκτοῦρος ἑῷος δύνει· ἐπισημασία.

κγ. Ὑάδες καὶ Ἀετὸς ἀνατέλλουσιν· ἐπισημασία ἐν ἡμέραις ἑπτά.

κε. παχνίτου ἔκβασις.

---

[111] Boll again thinks something is missing in the middle. Compare 25 Feb.

κη΄. Προκύων κρύπτεται καὶ ὁ λαμπρὸς τοῦ Ἀετοῦ ἑσπέριος ἀνατέλλει.
λα΄. Πλειάδων ἐπιτολή.

μὴν Ἰούνιος ♋

α΄. ὁ λαμπρὸς τῶν Ὑάδων ἐπιτέλλει καὶ Ἀετὸς ἀνατέλλει· ἐπισημασία.
ς΄. ὁ ἐπὶ τῆς κεφαλῆς τῶν Διδύμων λαμπρὸς κρύπτεται·[112] ἐπισημασία.
ιγ΄. Ὠρίων ἄρχεται ἐπιτέλλων· καὶ ποιεῖ κλόνους καὶ ταραχὰς βροντῶν.
κα΄. ὁ λαμπρὸς τοῦ Ὑδροχόου κρύπτεται.
κδ΄. τροπὴ θερινή.
κε΄. Ὠρίωνος ζώνη ἐπιτέλλει· ἐπισημασία.

μὴν Ἰούλιος ♌

β΄. ὁ ἐπὶ τῆς κεφαλῆς τοῦ ἡγουμένου τῶν Διδύμων ἐπιτέλλει.
θ΄. Κηφεὺς ἀνατέλλει· ἐπισημασία.
ια΄. ὁ ἐπὶ τῆς καρδίας τοῦ Λέοντος κρύπτεται.
ιδ΄. Ὠρίων τελείως ἀνατέλλει ἅμα ἡλίῳ καὶ ποιεῖ ὕδατα καὶ ἀνέμους.
ιθ΄. Κυνὸς ἐπιτολὴ κατ' Αἰγυπτίους.
κα΄. Ὑδροχόος δύνει· ἐπισημασία.
κε΄. Κυνὸς ἐπιτολὴ ἐν τῷ δ΄ κλίματι.
κζ΄. ἡ Λύρα δύνει.
λ΄.[113] Κυνὸς ἐπιτολὴ ἐν τῷ ς΄ κλίματι.

μὴν Αὔγουστος ♍

α΄. Λέων ἀνατέλλει· ἐπισημασία.
β΄. γαυρίαμα Κυνὸς σὺν ἐξάλματι Λέοντος.
ζ΄. Ὑδροχόος μέσος δύνει· ἐπισημαίνει.
ια΄.[114] ὁ ἐπὶ τῆς καρδίας τοῦ Λέοντος ἀνατέλλει καὶ Λύρα δύνει· ἐπιση-
       μαίνει.
ιζ΄. ὁ ἐπὶ τῆς οὐρᾶς τοῦ Λέοντος κρύπτεται καὶ Δελφὶς δύνει· ἐπισημασία.
κα΄. ὁ λαμπρὸς τοῦ Ὑδροχόου ἐπιτέλλει.[115]
κγ΄. ὁ λαμπρὸς τοῦ νοτίου Ἰχθύος ἑσπέριος ἀνατέλλει.
κη΄.[116] ὁ ἐπὶ τῆς οὐρᾶς τοῦ Λέοντος ἐπιτέλλει.

---

[112] Boll thinks either 'leading' or 'following' has dropped out before 'Twins'.
[113] Vatican MS reads '29'.   [114] Munich MS has '8th'.
[115] Vatican MS only.   [116] Vatican MS has '27th'.

λ´.   Ἀνδρομέδα ἀνατέλλει· ἐπισημασία.
λα´. ὁ Στάχυς κρύπτεται· ἐπισημασία.

μὴν **Σεπτέμβριος** ♎

α´.   Ἰχθὺς νότιος ἐπιτέλλει· ἐπισημασία.
ε´.   ὁ λαμπρὸς τοῦ Ὄρνιθος ἑῷος δύνει.
θ´.   Ἀρκτοῦρος ἐπιτέλλει· ἐπισημασία.
ια´. Αἴξ ἀνατέλλει.
ιϛ´. ὁ λαμπρὸς τῆς νοτίας Χηλῆς τοῦ Σκορπίου κρύπτεται.
ιθ´. ὁ ἐν τῷ ἑπομένῳ ὤμῳ τοῦ Ἡνιόχου ἑσπέριος ἀνατέλλει.
κβ´. Ἰχθύες δύνουσιν· ἐπισημασία.
κε´. ἰσημερία μετοπωρινή.
κη´. Ἀρκτοῦρος ἑῷος ἀνατέλλει.

μὴν **Ὀκτώβριος**   ♏

β´.   Αἴξ ἑσπερία καὶ ὁ Στάχυς ἀνατέλλει.
ε´.   ὁ λαμπρὸς τοῦ Στεφάνου ἀνατέλλει· ἐν ἡμέραις θ´ ἐπισημασία.
ζ´.   Ταύρου οὐρὰ καὶ ὁ λαμπρὸς τῆς βορείας Χηλῆς τοῦ Σκορπίου ἀνατέλλει.
θ´.   Στέφανος τελείως ἀνατέλλει· ἐπισημασία.
ι´.   Πλειάδων δύσις καὶ Ἔριφοι ἀνατέλλουσι· ἐπισημασία.
ιγ´. Ταύρου οὐρὰ τελεία ἀνατέλλει· ἐπισημασία.
ιζ´. ὁ ἐν τῷ ἑπομένῳ ὤμῳ τοῦ Ἡνιόχου ἑσπέριος ἀνατέλλει.
ιθ´. ὁ λαμπρὸς τοῦ βορείου Ἰχθύος ἑσπέριος ἀνατέλλει.
κβ´. τὰ Νειλῷα·[117] ἐπισημαίνει.
κη´. Λύρα ἀνατέλλει ἅμα ἡλίῳ καὶ ποιεῖ χειμῶνα καὶ φυλορροεῖν τὰ δένδρα.
λα´. ὁ λαμπρὸς τῆς βορείας Χηλῆς τοῦ Σκορπίου ἀνατέλλει· ἐπισημασία.

μὴν **Νοέμβριος** ♐

α´.   ὁ κατὰ τὸ γόνυ τοῦ Τοξότου κρύπτεται.
γ´.   ὁ λαμπρὸς τῶν Ὑάδων ἑσπέριος ἀνατέλλει.
ε´.   Ὑάδες δύνουσιν· ἐν ἡμέραις ἑπτὰ ἐπισημασία.[118]

---

[117] All the MSS are corrupted here. The Vatican MS has τὰ νείλαια, Oxford τανηλῶνα, and Munich ἀνηλῶα.
[118] The Vatican MS omits the change in the weather for seven days, but includes a modified version of this the following day.

ϛ΄.   Ὠρίων ἄρχεται δύνειν ἅμα Ὑάσι καὶ Πλειάσιν.[119]

η΄.   Κάνωβος ἑῷος δύνει.

ια΄. Πλειάδων δύσις τελεία.

ιβ΄. Ὑάδες ἀνατέλλουσιν· ἐπισημασία.

ιζ΄. ὁ λαμπρὸς τοῦ Περσέως ἑῷος δύνει.[120]

κγ΄. ὁ μέσος τῆς ζώνης τοῦ Περσέως ἑῷος δύνει.[121]

κε΄. Κύων δύνει ἅμα ἡλίῳ· ἐπισημασία.

λ΄.   ὁ ἐν τῷ ἑπομένῳ ὤμῳ τοῦ Ὠρίωνος ἑσπέριος ἀνατέλλει.

μὴν **Δεκέμβριος**   ♑

β΄.   Κύων ἑῷος δύνει.

γ΄.   Ἀρκτοῦρος δύνει· ἐπισημασία.

δ΄.   Σκορπίος ἐπιτέλλει ἅμα ἡλίῳ· ἐπισημασία.

ζ΄.   Αἴξ ἑῷα δύνει.

θ΄.   Ἀετὸς ἐπιτέλλει ἅμα ἡλίῳ· ἐπισημαίνει.

κ΄.   Ταύρου κέρατα δύνει· ἐπισημαίνει.

κα΄. ὁ λαμπρὸς τοῦ Ὄρνιθος ἑῷος ἀνατέλλει.

κβ΄. τροπὴ χειμερινή.

κγ΄. Προκύων ἑῷος δύνει.

κε΄. ἡλίου γενέθλιον· αὔξει φῶς.

κϛ΄. Δελφὶς ἐπιτέλλει· ἐπισημαίνει ἐπὶ ἡμέραις ζ΄.

κη΄. Κύων ἑσπέριος ἀνατέλλει.

λ΄.   ὁ λαμπρὸς τοῦ Ἀετοῦ ἐπιτέλλει.

λα΄. ὁ ἐπὶ τῆς κεφαλῆς τοῦ ἡγουμένου τῶν Διδύμων ἑῷος δύνει.

On the Risings and Settings of the Stars with the Sun in the twelve Months of the Year. The Book of Antiochus, On the Risings and Settings of the Stars in the twelve Months of the Year.

Month of January   ♒

1: The star on the head of the leading twin sets in the morning.

5: Delphinus rises.

7: The star on the knee of Sagittarius rises.

---

[119] Vatican MS adds ἐπὶ ἡμέραις ιε΄ ἐπισημασία.

[120] Following the Oxford and Munich MSS. Vatican MS dates this phase to the 20th.

[121] Oxford MS has ἀνατέλλει.

11: The bright star in Lyra rises in the evening, and it causes a non-dangerous change in the weather.[122]

15: The bright star in Aquarius sets in the morning.

18: The star on the head of Leo rises in the evening.

22: Aquarius rises, change in the weather.

26: The star on the breast of Leo sets, change in the weather.

31: Canopus rises in the evening.

Month of February ♓

1: The star on the heart of Leo sets in the morning.

4: The bright star in Lyra sets in the evening, a non-dangerous change in the weather.

6: The bright star in Cygnus rises in the evening.

13: The star on the right arm of Centaurus sets in the morning.

19: Virgo sets, a change in the weather.

23: The star shared by Pegasus and Andromeda disappears, a change in the weather.

25: The rearmost star in Eridanus and the southern fish rise, a change in the weather.

27: Arcturus begins to rise, a change in the weather.

Month of March ♈

1: The star shared by Pegasus and Andromeda rises in the evening.

3: The rearmost star in Eridanus and the southern fish rise, a change in the weather.

5: Arcturus rises, a change in the weather.

11: Pegasus sets, a change in the weather.

12: Spica rises in the evening.

14: The star on the tail of Leo sets in the morning.

16: The Hyades rise, a change in the weather.

22: Vernal equinox.

27: Spica sets in the morning, a change in the weather.

30: The bright star in Perseus sets in the morning.

---

[122] καὶ ποιεῖ ἐπισημασίαν ἀκίνδυνον. This parapegma uses the noun ἐπισημασία rather than the verb ἐπισημαίνει throughout, except the Vatican MS which sometimes uses the verb. Here, as not infrequently elsewhere, only the Vatican MS mentions ἐπισημασία. The Munich and Oxford MSS omit it.

Month of April   ♉

1: The bright star in the southern claw of Scorpio rises in the evening and the Pleiades disappear, a change in the weather.
5: The Pleiades entirely disappear, a change in the weather.
10: ὕψωμα of the sun.[123]
14: Canopus disappears.
20: The bright star in Perseus sets.
22: The beginning of frostiness.[124]
28: The star on the leading foot of Orion and the middle of the belt disappear, a change in the weather.
30: Sirius disappears, a change in the weather.

Month of May   ♊

1: Rising of the Hyades, a change in the weather.
2: The bright star in Lyra rises in the evening, there is a change in the weather.
4: Capella sets in the evening, a change in the weather.
9: The Pleiades rise at the same time as the sun, there is a change in the weather.
11: The star on the leading shoulder of Orion disappears.
13: Sirius disappears and the star on the right arm of the Centaur rises in the evening.
17: Antares sets in the morning.
18: Capella rises in the evening.
21: Arcturus sets in the morning, a change in the weather.
23: The Hyades and Aquila rise, a change in the weather for seven days.
25: The end of frostiness.
28: Procyon disappears and the bright star in Aquila rises in the evening.
31: Rising of the Pleiades.

---

[123] This entry is unusual in a parapegma, but compare Oxford, 12 April and Clodius Tuscus 1 Jan. Ptolemy (*Tetr.* I.19) says that the sun has its exaltation in Aries, since it is then that it begins to grow warmer. I am unclear as to why this particular date should have been chosen, however. Weinstock, 1948, p. 39, proposes an unlikely solution.

[124] ἀρχὴ παχνίτου. This period ends on 25 May. The word is exceedingly rare, and I am deriving its meaning from πάχνη, 'frost', although, given the dates, I am not entirely happy with the semantic field of the word. Boll argues, plausibly, that it is sailor jargon for a stormy season, where the name is derived from the Egyptian month name *Pachon* (see Boll, 1910a, p. 22f.). Clodius Tuscus, C. Marcianus 335, and the Madrid and Oxford parapegma are the only other texts to use this word. Compare the Madrid parapegma, 26 April and 20 May, and C. Marcianus 335, 25 March. On 24 and 25 April Clodius has ἀρχὴ παχνήτου. Clodius does not have an entry in May for the end of this period.

Month of June ♋

 1: The bright star in the Hyades rises and Aquila rises, a change in the weather.
 6: The bright star on the head of the Twins disappears, a change in the weather.
13: Orion begins rising, and it causes agitation and disorder of thunder.
21: The bright star in Aquarius disappears.
24: Summer solstice.
25: Orion's belt rises, a change in the weather.

Month of July ♌

 2: The star on the head of the leading Twin rises.
 9: Cepheus rises, a change in the weather.
11: The star on the heart of Leo disappears.
14: Orion rises completely at the same time as the sun and causes rains and winds.
19: Rising of Sirius, according to the Egyptians.
21: Aquarius sets, a change in the weather.
25: Rising of Sirius in the fourth Clima.
27: Lyra sets.
30: Rising of Sirius in the sixth Clima.

Month of August ♍

1:[125] Leo rises, a change in the weather.
2:[126]
 7: The middle of Aquarius sets, there is a change in the weather.
11: The star on the heart of Leo rises and Lyra sets, a change in the weather.
17: The star on the tail of Leo disappears and Delphinus sets, a change in the weather.

---

[125] Munich MS only. The Vatican and Oxford MSS date this to the 3rd and omit the entry on the 2nd.

[126] The Munich MS has an entry here. It reads: γαυρίαμα Κυνὸς σὺν ἐξάλματι Λέοντος. The word for exultation is different here than at 10 April. The meaning of σὺν ἐξάλματι Λέοντος is opaque to me. As Alexander Jones pointed out to me: 'The Paris manuscript is a copy of a lost archetype, the Vienna manuscript is a copy of a lost copy of the same archetype, and the remaining manuscripts descend from a lost sister manuscript of the Vienna manuscript. The cryptic line is *only* in one of this last group, and hence it has to be spurious, since logically wherever P and V agree in error, the remaining manuscripts *must* share that error.'

21:    The bright star in Aquarius rises.
23:    The bright star in the southern Fish rises in the evening.
28:    The star on the tail of Leo rises.
30:    Andromeda rises, a change in the weather.
31:    Spica disappears, a change in the weather.

Month of September   ♎

 1:    The southern Fish rises, a change in the weather.
 5:    The bright star in Cygnus sets in the morning.
 9:    Arcturus rises, a change in the weather.
11:    Capella rises.
16:    The bright star in the southern claw of Scorpio disappears.
19:    The star in the rear shoulder of Auriga rises in the evening.
22:    Pisces sets, a change in the weather.
25:    Autumnal equinox.
28:    Arcturus rises in the morning.

Month of October   ♏

 2:    In the evening Capella, and Spica rises.
 5:    The bright star in Corona rises, a change of weather for nine days.
 7:    The tail of Taurus and the bright star in the northern claw of Scorpio rise.
 9:    Corona rises completely, a change in the weather.
10:    Setting of the Pleiades, and the Haedi rise, a change in the weather.
13:    The tail of Taurus rises completely, a change in the weather.
17:    The star in the rear shoulder of Auriga rises in the evening.
19:    The bright star in the northern Fish rises in the evening.
22:    Festival of the inundation of the Nile, there is a change in the weather.
28:    Lyra rises at the same time as the sun and causes stormy weather and (causes) the trees to shed their leaves.
31:    The bright star in the northern claw of Scorpio rises, a change in the weather.

Month of November   ♐

 1:    The star on the knee of Sagittarius disappears.
 3:    The bright star in the Hyades rises in the evening.
 5:    The Hyades set, a change in the weather for seven days.
 6:    Orion begins to set at the same time as the Hyades and the Pleiades.

8: Canopus sets in the morning.

11: Setting of the Pleiades completely.

12: The Hyades rise, a change in the weather.

17: The bright star in Perseus rises in the morning.

23: The middle of the belt of Perseus sets in the morning.

25: Sirius sets at the same time as the sun, a change in the weather.

30: The star in the rear shoulder of Orion rises in the evening.

Month of December   ♑

2: Sirius sets in the morning.

3: Arcturus sets, a change in the weather.

4: Scorpio rises at the same time as the sun, a change in the weather.

7: Capella sets in the morning.

9: Aquila rises at the same time as the sun, there is a change in the weather.

20: The horns of Taurus set, there is a change in the weather.

21: The bright star in Cygnus rises in the morning.

22: Winter solstice.

23: Procyon sets in the morning.

25: The birth of the sun, light increases.

26: Delphinus rises, there is a change in the weather for seven days.

28: Sirius rises in the evening.[127]

30: The bright star in Aquila rises.

31: The star on the head of the leading twin sets in the morning.

A.xi. Clodius Tuscus[128] parapegma

ἐφημερὶς τοῦ παντὸς ἐνιαυτοῦ ἐκ τῶν Κλοδίου τοῦ Θούσκου, καθ᾽ ἑρμηνείαν
πρὸς λέξιν.

Ἰανουάριος

α΄.   καλένδαις μακραὶ ἡμέραι ἄρχονται· ὁ ἥλιος ὑψοῦται, ὁ δ᾽ Ἀετὸς σὺν τῷ
Στεφάνῳ δύεται καὶ ποιεῖ χειμῶνας.

β΄.   τῇ πρὸ δ΄ νωνῶν Ἰανουαρίων ὁ μὲν ἥλιος πηδᾷ, τὸ δὲ μέσον τοῦ Καρκίνου δύεται, καὶ οἱ ἄνεμοι ἐναλλάττονται.

γ΄.   τῇ πρὸ γ΄ νωνῶν τὸ λοιπὸν τοῦ Καρκίνου δύεται, καὶ τροπὴ τοῦ ἀέρος
ποικίλη.

---

[127] Vatican MS only.   [128] Greek text from Wachsmuth, 1897, p. 117f.

δ. τῇ πρὸ αʹ τὸ μέσον τοῦ χειμῶνος καὶ νότος πολύς, εἶτα καὶ βορραῖ συνεχεῖς. καὶ ὁ Δελφὶν ἀνίσχει ἅμα τῷ κυνὶ περὶ τὸν ὄρθρον.

εʹ. νώναις Ἰανουαρίαις ἡ Λύρα ἀνίσχει, καὶ ὁ μὲν Ἀετὸς δύεται, ὁ δὲ Δελφὶν ὅλος ἐπιτέλλει. καὶ εἰκότως ἀνεμομαχία.

ϛʹ. τῇ πρὸ ηʹ εἰδῶν Ἰανουαρίων ἐν μὲν ἑσπέρᾳ ὁ Ἀετὸς δύεται, νότος δὲ πνεῖ.

ζʹ. τῇ πρὸ ζʹ εἰδῶν ἀπαρκτίας καὶ βορέας κατ᾽ ἀλλήλων μάχονται.

ηʹ. τῇ πρὸ ϛʹ εἰδῶν οἶκος Ἄρεως· νότος ἅμα καὶ ζέφυρος· ὁ δὲ Αἰγόκερως ἄρχεται [. . .] βροχή τε ἅμα καὶ ἐν ἑσπέρᾳ νότος πυκνός.

θʹ. τῇ πρὸ εʹ εἰδῶν νότος φυσᾷ σύνομβρος.

ιʹ. τῇ πρὸ δʹ εἰδῶν ὁμοίως· ὁ δὲ νότος βιαιότερος.

ιαʹ. τῇ πρὸ γʹ εἰδῶν ἀπαρκτίας μετὰ βροχῆς καὶ χιόνος.

ιβʹ. τῇ πρὸ αʹ εἰδῶν νότος πνεῖ.

ιγʹ. εἰδοῖς Ἰανουαρίαις ἄστρον κρυπτόν. καὶ πρῶτον μὲν λόγος τὸν Ὀϊστὸν δύεσθαι, ἐν δὲ τῇ νυκτὶ βρέχειν.

ιδʹ. τῇ πρὸ ιθʹ καλενδῶν Φεβρουαρίων ἄστρον κρυπτόν, καὶ ποικίλη τροπὴ βορέου ἅμα καὶ ἀπαρκτίου. καὶ ὁ μὲν Λέων ἄρχεται δύεσθαι, ἔσθ᾽ ὅτε δὲ βρέχει.

ιεʹ. τῇ πρὸ ιηʹ καλενδῶν ἀπαρκτίας καὶ βορρᾶς σφοδρός.

ιϛʹ. τῇ πρὸ ιζʹ καλενδῶν ὁ μὲν ἥλιος <ἐν> Ὑδροχόῳ, εὖρος δὲ μετὰ βροχῆς.

ιζʹ. τῇ πρὸ ιϛʹ καλενδῶν ἡ μὲν Λύρα ἄρχεται δύεσθαι, περὶ δὲ τὸν ὄρθρον ἀνεμομαχία.

ιηʹ. τῇ πρὸ ιεʹ καλενδῶν ἕωθεν μὲν δύεται ὁ Λέων. βορέου ἅμα καὶ νότου καὶ ἀπαρκτίου διαφορά, καὶ βροχή. ὁ δὲ Δελφὶν μετὰ Λέοντα δύεται.

ιθʹ. τῇ πρὸ ιδʹ καλενδῶν τροπή, καὶ τὸ μεσαίτατον τοῦ χειμῶνος.

κʹ. τῇ πρὸ ιγʹ καλενδῶν βορρᾶς καὶ νότος. καὶ τὸ μέσον μὲν τοῦ Καρκίνου δύεται, ὁ δὲ Ὑδροχόος ἄρχεται ἀνίσχειν.

καʹ. τῇ πρὸ ιβʹ καλενδῶν ὁ Ὑδροχόος παντελὴς ἀνίσχει, πνεῖ δὲ ὁ λίψ, καὶ ὕει.

κβʹ. τῇ πρὸ ιαʹ καλενδῶν ἡ Λύρα δύεται σὺν τῷ Καρκίνῳ, καὶ πρὸς ἑσπέραν ὕει.

κγʹ. τῇ πρὸ ιʹ καλενδῶν βορρᾶς πνεύσει μετὰ βροχῆς.

κδʹ. τῇ πρὸ θʹ καλενδῶν χειμὼν καὶ βορέου ἐπίτασις ἅμα καὶ εὔρου.

κεʹ. τῇ πρὸ ηʹ καλενδῶν ὡσαύτως.

κϛʹ. τῇ πρὸ ζʹ καλενδῶν χειμάζει, πνεῖ δὲ ὁ βορρᾶς σὺν τῷ εὔρῳ, καὶ ἄρχεται δύεσθαι ἡ Λύρα.

κζʹ. τῇ πρὸ ϛʹ καλενδῶν ἄστρον λαμπρὸν ἐν τῷ στήθει τοῦ Λέοντος ἄρχεται δύεσθαι, ἡ δὲ Λύρα ἐν ἑσπέρᾳ. καὶ βορρᾶς πνεῖ, ἔστι δ᾽ ὅτε καὶ ὕει.

κη΄. τῇ πρὸ ε΄ καλενδῶν ἀνεμομαχία μετὰ χιόνος.

κθ΄. τῇ πρὸ δ΄ καλενδῶν ὁ Δελφὶν δύεσθαι μελετᾷ.

λ΄. τῇ πρὸ γ΄ καλενδῶν ἡ Λύρα περὶ τὴν πρώτην φυλακὴν τῆς νυκτὸς ἄρχεται δύεσθαι ἐκ μέρους. καὶ συννέφεια ἔσται, καὶ σφοδρὸς βορρᾶς μετὰ βροχῆς.

λα΄. τῇ πρὸ α΄ καλενδῶν Φεβρουαρίων ὑετὸς χιόνι μεμιγμένος.

Φεβρουάριος

α΄. καλένδαις Φεβρουαρίαις κρυπτὸν ἄστρον, νότος καὶ εὖρος· καὶ ἡ Λύρα ἄρχεται δύεσθαι.

β΄. τῇ πρὸ δ΄ νωνῶν Φεβρουαρίων θυελλώδης ὁ ἀήρ, καὶ παραπνεύσει ὁ ζέφυρος.

γ΄. τῇ πρὸ γ΄ νωνῶν τὸ μέσον τοῦ Λέοντος σὺν τῇ Λύρᾳ δύεται. ἀπαρκτίας δὲ ἅμα καὶ βορρᾶς.

δ΄. τῇ πρὸ α΄ νωνῶν δύεται ὁ Δελφίν· καὶ ἐν ἑσπέρᾳ νότος ἐπιτείνει μετὰ βροχῆς.

ε΄. νώναις Φεβρουαρίαις τοῦ Ὑδροχόου τὰ μέσα ἀνίσχει· ταραχώδης δὲ ὁ ἀὴρ ἐκ τοῦ ζεφύρου.

ϛ΄. τῇ πρὸ η΄ εἰδῶν Φεβρουαρίων δύεται ἡ Λύρα, καὶ ὁ ζέφυρος ἀπὸ δυσμῶν πνεῖ.

ζ΄. τῇ πρὸ ζ΄ εἰδῶν ἀρχὴ ἔαρος, καθ᾽ ὃ ζέφυρος πνεῖ.

η΄. τῇ πρὸ ϛ΄ εἰδῶν ζέφυρος σὺν τῷ βορρᾷ πνεῖ.

θ΄. τῇ πρὸ ε΄ εἰδῶν ἄστρον κρυπτόν· καὶ ἀνίσχει ὁ Ὑδροχόος.

ι΄. τῇ πρὸ δ΄ εἰδῶν ἀπαρκτίας μετὰ ζεφύρου· ἔστι δ᾽ ὅτε καὶ βροχαί.

ια΄. τῇ πρὸ γ΄ εἰδῶν ἀπηλιώτης πνεῖ, καὶ ἀνίσχει ὁ Ἀρκτοῦρος.

ιβ΄. τῇ πρὸ α΄ εἰδῶν Φεβρουαρίων ἀνεμομαχία.

ιγ΄. εἰδοῖς Φεβρουαρίαις ὁ Τοξότης ἐν ἑσπέρᾳ δύεται, καὶ σφοδρὸς χειμών.

ιδ΄. τῇ πρὸ ιϛ΄ καλενδῶν Μαρτίων ὁ Κρατὴρ ἀνίσχει ἐν ἑσπέρᾳ, καὶ ἐναλλαγέντων τῶν ἀνέμων ὁ νότος ἐπικρατεῖ.

ιε΄. τῇ πρὸ ιε΄ καλενδῶν ὁ ἥλιος <ἐν> Ἰχθύσιν, καὶ ὁ ἀὴρ χειμαίνει.

ιϛ΄. τῇ πρὸ ιδ΄ καλενδῶν ἀπαρκτίας πνεύσει μετὰ τοῦ νότου· νέος δὲ ἥλιος.

ιζ΄. τῇ πρὸ ιγ΄ καλενδῶν δύεται ἡ Παρθένος· ῥεύσει δὲ νότος μετὰ ζεφύρου καὶ βορρᾶς.

ιη΄. τῇ πρὸ ιβ΄ καλενδῶν δύεται ὁ λεγόμενος Ὀιστός. ἐν ἑσπέρᾳ δὲ ζέφυρος πνεῖ, καὶ ἄρχεται ἡ Παρθένος δύεσθαι.

ιθ΄. τῇ πρὸ ια΄ καλενδῶν ἀπαρκτίας ῥεύσει μετὰ νότου· τὸν δὲ Ὀιστὸν φασι δύεσθαι.

κ. τῇ πρὸ ι´ καλενδῶν βορρᾶς μετὰ βροχῆς. καὶ ὁ Λέων δύεται· ἄρχονται δὲ οἱ βορραῖ οἱ λεγόμενοι χελιδόνιοι, οἳ πεφύκασιν ἐπὶ τριάκοντα
ἡμέρας πνεῖν, καὶ φαίνεται χελιδών.

κα. τῇ πρὸ θ´ καλενδῶν ὁ Ἀρκτοῦρος τῇ πρώτῃ φυλακῇ τῆς νυκτὸς ἄρχεται δύεσθαι, καὶ πνεῖ ζέφυρος. ἡ δὲ νὺξ συννεφής.

κβ´. τῇ πρὸ η´ καλενδῶν τὰ λεγόμενα Ἀλκυόνεια.

κγ´. τῇ πρὸ ζ´ καλενδῶν ὁ Ὑδροχόος ἄρχεται ἀνίσχειν, ἡ δὲ πρωινὴ
χειμάζει.

κδ. τῇ πρὸ ς´ καλενδῶν ἀργεστὴς πνεῖ ἅμα καὶ βορρᾶς.

κέ. τῇ πρὸ ε´ καλενδῶν ὁ Ἀρκτοῦρος ἀνίσχει, καὶ ὕει.

κς´. τῇ πρὸ δ´ καλενδῶν ὁ Ἀρκτοῦρος ἀνίσχει ἐν ἑσπέρᾳ.

κζ´. τῇ πρὸ γ´ καλενδῶν δύεται ἐν ἑσπέρᾳ ὁ παρ᾽ Ἕλλησι λεγόμενος Ὀϊστός.

κη´. τῇ πρὸ α´ καλενδῶν Μαρτίων ὁ ζέφυρος πλατύς, καὶ ἡ ἡμέρα πᾶσα
ἐαρινή.

Μάρτιος

α´. καλένδαις Μαρτίαις νότος καὶ λίψ.

β´. τῇ πρὸ ς´ νωνῶν Μαρτίων λίψ, καὶ ὁ Τρυγητὴς ἄρχεται φαίνεσθαι.
βορέας δὲ ψυχρὸς πνεῖ ἕως τῆς ἑωθινῆς δύσεως τοῦ Ἀρκτούρου.

γ´. τῇ πρὸ ε´ νωνῶν δυσαερία καὶ βροχή· καὶ ὁ Ἀρκτοῦρος ἀνίσχει ἡλίου
ἐγειρομένου, καὶ βορρᾶς πνεῖ.

δ´. τῇ πρὸ δ´ νωνῶν ὁ Ἀρκτοῦρος ἐν ἑσπέρᾳ ἀνίσχει.

έ. τῇ πρὸ γ´ νωνῶν ὡσαύτως.

ς´. τῇ πρὸ α´ νωνῶν ὑετὸς χιόνι μεμιγμένος.

ζ. νώναις Μαρτίαις δύεται ὁ Ἵππος ἀπὸ πρωί, καὶ βορρᾶς φυσᾷ. δύεται
δὲ καὶ ὁ Στέφανος ὄρθρου.

η´. τῇ πρὸ η´ εἰδῶν Μαρτίων ἄρχεται τὰ ὄρνεα φαίνεσθαι ἐπὶ τῆς
θαλάσσης, βορρᾶς δὲ καὶ ἀπαρκτίας φυσᾷ, καὶ προοίμιον τοῦ ἔαρος.
καὶ ὁ μὲν ἥλιος περὶ τὸ μέσον τῶν Ἰχθύων, ὁ δὲ Ἵππος δύεται.

θ´. τῇ πρὸ ζ´ εἰδῶν ἰκτῖνος ἄρχεται φαίνεσθαι, νότος δὲ πνεῖ, καὶ ὄρθρου
ὁ Ἰχθὺς ἀπὸ τοῦ νότου ἄρχεται κρύπτεσθαι.

ί. τῇ πρὸ ς´ εἰδῶν ὁ Ἵππος δύεται ὄρθρου, ὁ δὲ ἰκτῖνος ἀπὸ ὑψηλῶν ἐπὶ
τὰ χθαμαλὰ καθίπταται· καὶ ὁ μὲν Τρυγητὴς δύεται, ὁ δὲ Ἀρκτοῦρος
ἀνίσχει. βορρᾶς δὲ ψυχρὸς ῥεύσει.

ια´. τῇ πρὸ ε´ εἰδῶν χωρισμὸς μὲν τοῦ χειμῶνος, τροπὴ δὲ ἀπὸ βορρᾶ καὶ
ἀπαρκτίου.

ιβ´. τῇ πρὸ δ´ εἰδῶν παύεται μὲν ὁ Ἰχθὺς ἀπὸ τοῦ νότου ἀνίσχειν, ἀπαρκτίας δ᾽ ἢ νότος φυσᾷ.

ιγ´. τῇ πρὸ γ´ εἰδῶν ἡ μὲν Ἀργὼ ἀνίσχει ἐν ἑσπέρᾳ, ζέφυρος δὲ καὶ νότος πνεύσει καὶ ὁ ἐπὶ τῆς οὐρᾶς τοῦ Λέοντος <δύεται>· χειμάζει.

ιδ´. τῇ πρὸ α´ εἰδῶν βορρᾶς δι’ ὅλης τῆς ἡμέρας.

ιε´. εἰδοῖς Μαρτίαις ὁ Ἵππος δύεται, βορέας δὲ ψυχρὸς πνεῖ.

ιϛ´. τῇ πρὸ ιζ´ καλενδῶν Ἀπριλίων. . . . . .

ιζ´. τῇ πρὸ ιϛ´ καλενδῶν ἥλιος <ἐν> Κριῷ γίνεται· ὁ δὲ ζέφυρος κατὰ πλάτος πνεῖ. πέλαργος φαίνεται καὶ μέγα πέλαγος πλέεται.

ιη´. τῇ πρὸ ιε´ καλενδῶν ἀνεμομαχία, ὁ δὲ βορρᾶς ἐπικρατεῖ.

ιθ´. τῇ πρὸ ιδ´ καλενδῶν νότος πνεῖ, ὁ δὲ ἰκτῖνος φαίνεται ἕως τῆς ἰσημερίας.

κ´. τῇ πρὸ ιγ´ καλενδῶν βορέας πνεῖ εὔδιος.

κα´. τῇ πρὸ ιβ´ καλενδῶν ὁ μὲν Ἵππος ἕωθεν δύεται, βορέας δὲ ἢ ἀπαρκτίας πνεῖ.

κβ´. τῇ πρὸ ια´ καλενδῶν ὁ Κριὸς εἰς πλάτος ἀνίσχει, ἔστι δ’ ὅτε ἢ βρέχει ἢ νίφει.

κγ´. τῇ πρὸ ι´ καλενδῶν ὡσαύτως.

κδ´. τῇ πρὸ θ´ καλενδῶν ἰσημερία ἐαρινή, καὶ βροχὴ ἢ βροντώδης τροπή.

κε´. τῇ πρὸ η´ καλενδῶν ἀπαρκτίας ἢ βορρᾶς, καὶ ὁ Ἵππος ἕωθεν δύεται.

κϛ´. τῇ πρὸ ζ´ καλενδῶν οἱ Ἰχθύες ἀνίσχουσι, καὶ ἐκ τοῦ νότου βρέχει μετὰ χιόνος· ὁ δὲ Κριὸς ἕωθεν ἀνίσχει, κατὰ θάλασσαν ταραχὴ ἀέρος.

κζ´. τῇ πρὸ ϛ´ καλενδῶν ἰσημερία ἐαρινή. ἔστι δ’ ὅτε βρέχει καὶ βροντᾷ.

κη´. τῇ πρὸ ε´ καλενδῶν ἐξισοῦται ἡ ἡμέρα τῇ νυκτί.

κθ´. τῇ πρὸ δ´ καλενδῶν ὁ μὲν Σκορπίος δύεται, μέγας δὲ ῥέων ἄνεμος βροχὰς βροντώδεις συστρέφει.

λ´. τῇ πρὸ γ´ καλενδῶν ὁ μὲν Σκορπίος δύεται, ἀπαρκτίας δὲ φυσᾷ μετὰ βροχῆς.

λα´. τῇ πρὸ α´ καλενδῶν βρέχειν πέφυκε, καὶ ἐκ τοῦ νότου βροντᾷ.

Ἀπρίλιος

α´. καλένδαις Ἀπριλίαις ὁ μὲν Σκορπίος δύεται, καὶ ὁ ἥλιος μίαν <ἡμέρα> προστίθησι μοῖραν, καὶ συννέφεια γίνεται ἐκ τοῦ βορρᾶ, καὶ αἱ Πλειάδες ἄρχονται ἐπιτέλλειν καὶ ἐπισημαίνει.[129]

β´. τῇ πρὸ δ´ νωνῶν Ἀπριλίων συννέφεια καθ’ ὅλην τὴν ἡμέραν.

γ´. τῇ πρὸ γ´ νωνῶν ἐν ἑσπέρᾳ αἱ Πλειάδες δύονται.

δ´. τῇ πρὸ α´ νωνῶν ῥεύσει ὁ λίψ.

ε´. νώναις Ἀπριλίαις καταπνεῖ ὁ ζέφυρος.

ϛ´. τῇ πρὸ η´ εἰδῶν αἱ Ὑάδες ἀνίσχουσι, καὶ βροχαὶ ἐκ τοῦ νότου ἐπιτείνουσιν.

---

[129] Wachsmuth reads ἐπισημαίνειν.

ζ.   τῇ πρὸ ζ εἰδῶν νότος φυσᾷ, καὶ τὸ λοιπὸν τῶν Ὑάδων δύεται.

η.   τῇ πρὸ ϛ' εἰδῶν ὁ Ζυγὸς ὄρθρου ἄρχεται δύεσθαι.

θ.   τῇ πρὸ ε εἰδῶν ζάλη ἐκ τοῦ νότου.

ι.   τῇ πρὸ δ εἰδῶν βορρᾶς μὲν πᾶσαν καταπνεῖ τὴν ἡμέραν, ἡ δὲ ἑσπέρα ἔσται ἐπίβροχος.

ια. τῇ πρὸ γ εἰδῶν ψυχροὶ ἄνεμοι καὶ βροχαί.

ιβ. τῇ πρὸ α εἰδῶν αἱ Ὑάδες κρύπτονται.

ιγ. εἰδοῖς Ἀπριλίαις βορρᾶς ῥεύσει· μικροῦ παχνήτου ἀνατολή.

ιδ. τῇ πρὸ ιη καλενδῶν Μαίων κρύφιον ἄστρον καὶ ἄνεμοι καὶ ὄμβροι.

ιε. τῇ πρὸ ιζ καλενδῶν ἡ Ὑὰς δύεται καὶ ἄνεμοι ψυχροὶ πνέουσιν· ὁ δὲ Περσεὺς ἐπιτέλλει.

ιϛ'. τῇ πρὸ ιϛ' καλενδῶν αἱ Ὑάδες δύονται καὶ ζέφυρος πνεῖ.

ιζ. τῇ πρὸ ιε καλενδῶν ὁ μὲν ἥλιος <ἐν> Ταύρῳ, ἡ δὲ Ὑὰς δύεται.

ιη. τῇ πρὸ ιδ καλενδῶν λὶψ καταπνεύσει.

ιθ. τῇ πρὸ ιγ καλενδῶν καθόλου δύονται αἱ Ὑάδες, ὁ δὲ λὶψ <ἐν> ἑσπέρᾳ ἐπέρχεται.

κ.   τῇ πρὸ ιβ καλενδῶν ὁ ζέφυρος καταπνεῖ.

κα. τῇ πρὸ ια καλενδῶν ἡ κεφαλὴ τοῦ Ταύρου δύεται, καὶ οὐχ ἥκιστα ὕει.

κβ. τῇ πρὸ ι καλενδῶν αἱ Πλειάδες ἀνίσχουσι καὶ ζέφυρος πνεῖ.

κγ. τῇ πρὸ θ καλενδῶν ἡ Λύρα τῇ πρώτῃ φυλακῇ τῆς νυκτὸς φαίνεται, καὶ ἔσται τροπή.

κδ. τῇ πρὸ η' καλενδῶν φαίνεται ἡ Λύρα, καὶ βροχῶν σημασία· ἀρχὴ παχνήτου καὶ ἔαρος· νότος πνεῖ.

κε. τῇ πρὸ ζ καλενδῶν ὁ Σκορπίος δύεται· καὶ ὁ Κύων κρύπτεται καὶ ὁ ἐπὶ τῆς ζώνης τοῦ Ὠρίωνος· νότος τε πνεῖ· ἀρχή τε παχνήτου καὶ ἔαρος.

κϛ'. τῇ πρὸ ϛ' καλενδῶν δύεται καθόλου ἡ Ὑάς, καὶ ἀέρος ἔσται τροπή.

κζ. τῇ πρὸ ε καλενδῶν νότος καταπνεῖ.

κη'. τῇ πρὸ δ καλενδῶν ὕει ἐκ τοῦ νότου.

κθ'. τῇ πρὸ γ καλενδῶν οἱ Ἔριφοι ἀνίσχουσι, καὶ ἐξ ἑωθινῆς ὁ νότος πνεῖ.

λ'. τῇ πρὸ α καλενδῶν κρύπτεται ὁ Κύων ἐν ἑσπέρᾳ, καὶ τροπὴ ἐκ τοῦ νότου· ἅμα δὲ καὶ βορρᾶς ταράττει.

Μάιος

α.   καλένδαις Μαίαις ὁ μὲν Κύων κρύπτεται, δρόσος δὲ καταφέρεται.

β.   τῇ πρὸ ϛ' νωνῶν ἡ Ὑὰς μετὰ τοῦ ἡλίου ἀνίσχει.

γ.   τῇ πρὸ ε νωνῶν ὁ Κένταυρος ὅλος φαίνεται καὶ ζέφυρος πνεῖ.

δ.   τῇ πρὸ δ νωνῶν ὁ Σκορπίος ἔωθεν ἀνίσχει, καὶ βορρᾶς πνεῖ, καὶ δρόσος καταφέρεται.

ε.   τῇ πρὸ γ νωνῶν ἡ Λύρα ἔωθεν ἀνίσχει.

ϛʹ. τῇ πρὸ αʹ νωνῶν τὸ μέσον τοῦ Σκορπίου δύεται.

ζʹ. νώναις Μαίαις αἱ Πλειάδες ἀνίσχουσιν ἕωθεν, ζέφυρος δὲ πνεῖ.

ηʹ. τῇ πρὸ ηʹ εἰδῶν Μαίων προοίμιον θέρους, καὶ ζέφυρος ἐπικρατεῖ.

θʹ. τῇ πρὸ ζʹ εἰδῶν Μαίων ὡσαύτως.

ιʹ. τῇ πρὸ ϛʹ εἰδῶν ἡ μὲν Λύρα ἀνίσχει, ἡ δὲ Ὑὰς δύεται, καὶ ἡ κεφαλὴ τοῦ Ταύρου ἀναφαίνεται.

ιαʹ. τῇ πρὸ εʹ εἰδῶν αἱ Πλειάδες ἀναφαίνονται.

ιβʹ. τῇ πρὸ δʹ εἰδῶν ἀνίσχουσιν αἱ Πλειάδες, καὶ νότος πνεῖ.

ιγʹ. τῇ πρὸ γʹ εἰδῶν αἱ Ὑάδες ἀνίσχουσιν καὶ νότος πνεῖ.

ιδʹ. τῇ πρὸ αʹ εἰδῶν ὁ Σκορπίος δύεται, ἡ δὲ Λύρα ὄρθρου ἀνίσχει.

ιεʹ. εἰδοῖς Μαίαις ὁ Καρκίνος ἀνίσχει καὶ νότος πνεῖ.

ιϛʹ. τῇ πρὸ ιζʹ καλενδῶν Ἰουνίων· τὸ θέρος ἄρχεται.

ιζʹ. τῇ πρὸ ιϛʹ καλενδῶν δύεται ὁ Κύων.

ιηʹ. τῇ πρὸ ιεʹ καλενδῶν ὁ ἥλιος <ἐν> Διδύμοις.

ιθʹ. τῇ πρὸ ιδʹ καλενδῶν νότος ἐν ἑσπέρᾳ πνεῖ.

κʹ. τῇ πρὸ ιγʹ καλενδῶν αἱ Ὑάδες ἀνίσχουσι καὶ βορρᾶς πνεῖ.

καʹ. τῇ πρὸ ιβʹ καλενδῶν δύεται ὁ Ἀρκτοῦρος καὶ ταράττεται ὁ ἀήρ.

κβʹ. τῇ πρὸ ιαʹ καλενδῶν ὁ Τοξότης δύεται καὶ νότος πνεῖ.

κγʹ. τῇ πρὸ ιʹ καλενδῶν οἱ Δίδυμοι ἀνίσχουσιν καὶ ὁ Ἀετός.

κδʹ. τῇ πρὸ θʹ καλενδῶν ἄρχονται ἀνίσχειν αἱ Ὑάδες, καὶ εἰκότως βροχή.

κεʹ. τῇ πρὸ ηʹ καλενδῶν ὁ Ἔριφος ἀνίσχει ἕωθεν καὶ βορρᾶς πνεῖ.

κϛʹ. τῇ πρὸ ζʹ καλενδῶν δύεται ὁ Ἀρκτοῦρος, καὶ νότος μετὰ βορέου πνεῖ.

κζʹ. τῇ πρὸ ϛʹ καλενδῶν ἀνίσχει ὁ Ἀετός.

κηʹ. τῇ πρὸ εʹ καλενδῶν νότος πνεῖ, καὶ ἡ Λύρα ἕωθεν ἀνίσχει.

κθʹ. τῇ πρὸ δʹ καλενδῶν νότος πολύς.

λʹ. τῇ πρὸ γʹ καλενδῶν αἱ Πλειάδες ἀνίσχουσι, καὶ βροχὴ μετὰ βροντῶν.

λαʹ. τῇ πρὸ αʹ καλενδῶν Ἰουνίων χειμάζει ὁ ἀὴρ καὶ βροντᾷ, περὶ δὲ τὴν ἑσπέραν σφοδρότερον ταράττει.

Ἰούνιος

αʹ. καλένδαις Ἰουνίαις αἱ Ὑάδες ὅλαι ἀνίσχουσι καὶ νότος πνεῖ.

βʹ. τῇ πρὸ δʹ νωνῶν Ἰουνίων ὁ Ἀετὸς ἀνίσχει, ταραχή τε τοῦ ἀέρος, καὶ ζέφυρος πνεῖ.

γʹ. τῇ πρὸ γʹ νωνῶν βρονταὶ νότιοι.

δʹ. τῇ πρὸ αʹ νωνῶν νότος πνεῖ, καὶ ὡς εἰκὸς βροχαί.

εʹ. νώναις Ἰουνίαις ὁ Ἀετὸς ἀνίσχει, καὶ ὕει τε ἅμα καὶ νότος πνεῖ.

ϛʹ. τῇ πρὸ ηʹ εἰδῶν Ἰουνίων ὁ ὦμος τοῦ Ὠρίωνος ἀνίσχει.

ζʹ. τῇ πρὸ ζʹ εἰδῶν βορρᾶς πνεῖ καὶ ὕει.

ηʹ. τῇ πρὸ ϛʹ εἰδῶν ὁ Ἀρκτοῦρος ὄρθρου δύεται καὶ ζέφυρος πνεῖ.

θ. τῇ πρὸ ε´ εἰδῶν ὁ Δελφὶν ἄρχεται ἀνίσχειν, ὁ δὲ Ἀρκτοῦρος δύεται.

ι. τῇ πρὸ δ´ εἰδῶν ὡσαύτως· βορρᾶς δὲ πνεῖ, καὶ ἠρέμα ὕει.

ια. τῇ πρὸ γ´ εἰδῶν βροντώδης ὁ ἀὴρ καὶ ὑετώδης ἐκ τοῦ νότου.

ιβ. τῇ πρὸ α´ εἰδῶν ταραχώδης ὁ ἀὴρ μετὰ βροχῆς.

ιγ. εἰδοῖς Ἰουνίαις ζέφυρος ἢ ἀργέστης πνεύσει· εἰκότως δὲ καὶ βροντήσει.

ιδ. τῇ πρὸ ιη´ καλενδῶν Ἰουλίων ὁ Δελφὶν ἀνίσχει, καὶ νότος πνεῖ.

ιε. τῇ πρὸ ιζ´ καλενδῶν οἱ ὦμοι τοῦ Ὠρίωνος ἀνίσχουσι, καὶ προοίμια καυμάτων.

ις´. τῇ πρὸ ις´ καλενδῶν ἄστρον κρύφιον καὶ ζέφυρος σὺν τῷ νότῳ.

ιζ. τῇ πρὸ ιε´ καλενδῶν ζάλη μετρία τοῦ ἀέρος καὶ πνοὴ τοῦ βορέου.

ιη. τῇ πρὸ ιδ´ καλενδῶν ζέφυρος μετὰ νότου, καὶ οἱ ὦμοι τοῦ Ὠρίωνος φαίνονται.

ιθ. τῇ πρὸ ιγ´ καλενδῶν ὁ ἥλιος <ἐν> Καρκίνῳ, καὶ ὁ Ὠρίων ἀνίσχει ἕωθεν.

κ. τῇ πρὸ ιβ´ καλενδῶν νότος ἅμα καὶ ζέφυρος, καὶ βροντώδης ὑετός.

κα. τῇ πρὸ ια´ καλενδῶν ὁ Ὀφιοῦχος ὄρθρου δύεται.

κβ. τῇ πρὸ ι´ καλενδῶν νότος ἅμα καὶ βορέας.

κγ. τῇ πρὸ θ´ καλενδῶν ἐπιτολὴ τοῦ Ὠρίωνος.

κδ. τῇ πρὸ η´ καλενδῶν κρυφίου ἄστρου ἐπιτολὴ καὶ ἐπίτασις καυμάτων.

κε. τῇ πρὸ ζ´ καλενδῶν ἡ θερινὴ τροπή, αἰφνίδιός τε ταραχὴ τοῦ ἀέρος.

κς´. τῇ πρὸ ς´ καλενδῶν λὶψ ἅμα καὶ ζέφυρος.

κζ. τῇ πρὸ ε´ καλενδῶν βραχεῖα ἡ νύξ, καὶ ὁ μὲν Ὠρίων ἀνίσχει, νότος δὲ ῥεύσει.

κη. τῇ πρὸ δ´ καλενδῶν ἐν μὲν ἑσπέρᾳ βρέχει, ὁ δὲ Κύων ἀνίσχειν ἄρχεται.

κθ. τῇ πρὸ γ´ καλενδῶν ἀνεμομαχία.

λ. τῇ πρὸ α´ καλενδῶν ἕωθεν μὲν ὁ Κύων ἀνίσχει, ἡ δὲ ζώνη τοῦ Ὠρίωνος ἀναφαίνεται.

Ἰούλιος

α. καλένδαις Ἰουλίαις ταράττεται ὁ ἀὴρ ἐκ τοῦ βορρᾶ.

β. τῇ πρὸ ς´ νωνῶν Ἰουλίων κρύφιον ἄστρον καὶ ζέφυρος ἢ νότος.

γ. τῇ πρὸ ε´ νωνῶν ταραχαὶ τοῦ ἀέρος ἐκ τοῦ νότου.

δ. τῇ πρὸ δ´ νωνῶν ὁ Ὠρίων ἀνίσχει καὶ Στέφανος δύεται, καὶ ζέφυρος πνεῖ.

ε. τῇ πρὸ γ´ νωνῶν τὸ μέσον τοῦ Καρκίνου ἀνίσχει.

ς´. τῇ πρὸ α´ νωνῶν οἱ ἐτησίαι καὶ μετὰ νότον βορρᾶς.

ζ. νώναις Ἰουλίαις δύεται ὄρθρου ὁ Στέφανος, καὶ νότος πνεῖ.

η. τῇ πρὸ η´ εἰδῶν Ἰουλίων ὁ Κηφεὺς ἀνίσχει, καὶ τροπὴ τοῦ ἀέρος νοτία.

θ. τῇ πρὸ ζ´ εἰδῶν ὁ Ὠρίων ὅλος ἀνίσχει καὶ νότος πνεῖ.

ι. τῇ πρὸ ς´ εἰδῶν οἱ πρόδρομοι τῶν ἐτησίων πνέουσιν.

ια. τῇ πρὸ ε΄ εἰδῶν βροχὴ μετὰ βροντῶν καὶ βορρᾶς βίαιος.

ιβ. τῇ πρὸ δ΄ εἰδῶν ὁ Ὠρίων ὅλος ὄρθρου ἀνίσχει, καὶ ἐπιτείνουσιν οἱ λεγόμενοι πρόδρομοι.

ιγ. τῇ πρὸ γ΄ εἰδῶν λὶψ ταραχώδης.

ιδ. τῇ πρὸ α΄ εἰδῶν ἀνίσχει [. . .] καὶ πνεῖ ὁ βορρᾶς.

ιε. εἰδοῖς Ἰουλίαις ὁ Προκύων ἀνίσχει.

ιϛ΄. τῇ πρὸ ιζ΄ καλενδῶν Αὐγούστων ὁ Ὠρίων ἀνίσχει, καὶ βορρᾶς βίαιος πνεῖ.

ιζ. τῇ πρὸ ιϛ΄ καλενδῶν τὸ μεσαίτατον τοῦ θέρους, καὶ ψυχροτέρα ἡ ἡμέρα ἐκ τοῦ βορρᾶ.

ιη. τῇ πρὸ ιε΄ καλενδῶν ζέφυρος, ἴσως δὲ καὶ νότος. καὶ ὁ μὲν Κύων ὄρθρου ἀνίσχει, οἱ δὲ ἐτησίαι ἐπιτείνουσιν.

ιθ. τῇ πρὸ ιδ΄ καλενδῶν ὁ Ὠρίων ἀνίσχει, καὶ ἀργεστὴς φυσᾷ, καὶ ὅλος ὁ Ὠρίων φαίνεται.

κ. τῇ πρὸ ιγ΄ καλενδῶν ὁ ἥλιος <ἐν> Λέοντι, ἀργεστής τε φυσᾷ, καὶ οἱ πρόδρομοι τῶν ἐτησίων.

κα. τῇ πρὸ ιβ΄ καλενδῶν ἀπὸ ταύτης τῆς ἡμέρας οἱ ἐτησίαι σὺν καὶ τοῖς ἄλλοις ἀνέμοις ἐπὶ τεσσαράκοντα ἡμέρας πνέουσιν.

κβ. τῇ πρὸ ια΄ καλενδῶν οἱ πρόδρομοι καταφυσῶσιν.

κγ. τῇ πρὸ ι΄ καλενδῶν ὅλος ὁ Καρκίνος μετὰ τοῦ κυνὸς ἀνίσχει, καὶ ὁ Ἀετὸς δύεται.

κδ. τῇ πρὸ θ΄ καλενδῶν ὁ Λέων σὺν τῷ ἡλίῳ ἀνίσχει μετὰ τοῦ κυνός, ὁ δὲ Καρκίνος λήγει.

κε. τῇ πρὸ η΄ καλενδῶν ὁ Ὑδροχόος ἄρχεται δύεσθαι, ὁ δὲ Κύων περὶ ἀμφιλύκην ἀνίσχει, καὶ νότος φυσᾷ.

κϛ΄. τῇ πρὸ ζ΄ καλενδῶν ἀχλὺς καυσώδης. ὁ δὲ Ἀετὸς δύεται, ὁ δὲ Λέων ἀνίσχει, καὶ νότος πνεῖ.

κζ. τῇ πρὸ ϛ΄ καλενδῶν καῦμα ἐκ τοῦ κυνός· ἡ δὲ σταφυλὴ ἄρχεται περκάζειν.

κη. τῇ πρὸ ε΄ καλενδῶν καύσων βαρὺς καὶ οἱ ἐτησίαι σφοδρότεροι.

κθ. τῇ πρὸ δ΄ καλενδῶν ἐν τῷ στήθει τοῦ Λέοντος ἄστρον λαμπρὸν ἀνατέλλει, καὶ χλιαρὸς ὥσπερ ὁ βορρᾶς ἐκ τοῦ καύματος.

λ. τῇ πρὸ γ΄ καλενδῶν ἄρχονται ὀπῶραι φαίνεσθαι, ὁ δὲ Ἀετὸς ἕωθεν δύεται, καὶ ταραχώδης ὁ ἀήρ.

λα. τῇ πρὸ α΄ καλενδῶν Αὐγουστῶν νότος μετὰ λιβὸς πνεῖ.

Αὔγουστος

α. καλένδαις Αὐγούσταις ὁ Ἀετὸς δύεται ὄρθρου· λὶψ πνεύσει, καὶ καῦμα ξηρόν.

β. τῇ πρὸ δ΄ νωνῶν Αὐγούστων ἔτι δύεται ὁ Ἀετός, καὶ ὁ ἀὴρ νότιος.

γ.   τῇ πρὸ γ΄ νωνῶν τὸ αὐτὸ σημαίνει.

δ.   τῇ πρὸ α΄ νωνῶν τὸ μέσον τοῦ Λέοντος ἀνίσχει, φαίνεται δὲ καὶ τὸ δένδρον. ἐν τῷ διημέρῳ τούτῳ ὁ ἥλιος μίαν μοῖραν κρατεῖ· ὁ νότος δὲ ἀχλυώδης πνεῖ.

ε.   νώναις Αὐγούσταις δύεται μὲν ὁ Στέφανος, τὸ δὲ μέσον τοῦ Λέοντος ἀνίσχει καὶ γέρανοι φαίνονται, καὶ νότος πυκνὸς πνεῖ.

ϛ΄. τῇ πρὸ η΄ εἰδῶν Αὐγούστων ἡ Λύρα συστέλλεται, καὶ ἐκ τοῦ νότου καύσων.

ζ.   τῇ πρὸ ζ΄ εἰδῶν τὸ μέσον τοῦ Ὑδροχόου δύεται, καὶ ἐκ τοῦ νότου καῦμα ἀχλυῶδες.

η.   τῇ πρὸ ϛ΄ εἰδῶν ὁ Λέων ἀνίσχει, καὶ καῦμα βαρύ. ἐπινεφὴς ὁ ἀήρ, καὶ τὸ μέσον τοῦ Ὑδροχόου ἀνίσχει.

θ.   τῇ πρὸ ε΄ εἰδῶν ἄστρον κρυπτόν, ἀπαρκτίας πρᾶος καὶ μέτριον καῦμα.

ι.   τῇ πρὸ δ΄ εἰδῶν. ἐν ταύτῃ τῇ ἡμέρᾳ ἔκλειψις σεληνιακή, καὶ νότου ἅμα καὶ βορέου διαφορὰ καὶ καύματα.

ια΄. τῇ πρὸ γ΄ εἰδῶν δύεται ἡ Λύρα ὄρθρου. τὸ φθινόπωρον ἄρχεται, καὶ ἀνεμομαχία.

ιβ΄. τῇ πρὸ α΄ εἰδῶν ὡσαύτως.

ιγ΄. εἰδοῖς Αὐγούσταις ὁ Δελφὶν δύεται σὺν τῷ Λαγωῷ.

ιδ΄. τῇ πρὸ ιθ΄ καλενδῶν Σεπτεμβρίων αὐχμὸς καυσώδης.

ιε΄. τῇ πρὸ ιη΄ καλενδῶν ἄστρον κρυπτόν. ζέφυρος σὺν τῷ νότῳ πνεῖ.

ιϛ΄. τῇ πρὸ ιζ΄ καλενδῶν ὄρθρου ὁ Δελφὶν δύεται, καὶ νότος πνεῖ.

ιζ΄. τῇ πρὸ ιϛ΄ καλενδῶν ἀρχὴ φθινοπώρου.

ιη΄. τῇ πρὸ ιε΄ καλενδῶν δύεται ἡ Λύρα, καὶ νότος πνεῖ.

ιθ΄. τῇ πρὸ ιδ΄ καλενδῶν μέτριον τὸ καῦμα, καὶ δύεται ὁ Δελφίν.

κ.   τῇ πρὸ ιγ΄ καλενδῶν ἡ μὲν Λύρα δύεται ὄρθρου, ὁ δὲ ἥλιος <ἐν> Παρθένῳ. νότος ἄνεμος, καὶ βροντώδης ὑετός.

κα΄. τῇ πρὸ ιβ΄ καλενδῶν ὁ ἥλιος ἐφ᾽ ὅλης Παρθένου.

κβ΄. τῇ πρὸ ια΄ καλενδῶν ἀνίσχει ἡ Παρθένος.

κγ΄. τῇ πρὸ ι΄ καλενδῶν ἀπαρκτίας πνεύσει μέτριος, καὶ ὁλικὴ ἡ Παρθένος ἀνίσχει καὶ ἀὴρ καθαρός.

κδ΄. τῇ πρὸ θ΄ καλενδῶν τὸ μὲν αὐτὸ ζῴδιον, ἄνεμος δὲ βορρᾶς.

κε΄. τῇ πρὸ η΄ καλενδῶν παύονται οἱ ἐτησίαι, καὶ ψυχρότερος ἄρχεται πνεῖν βορρᾶς.

κϛ΄. τῇ πρὸ ζ΄ καλενδῶν ὁ μὲν Δελφὶν ἀνίσχει, νότος δὲ πνεῖ.

κζ΄. τῇ πρὸ ϛ΄ καλενδῶν ὁ Τρυγητὴς ἀνίσχει, νότος δὲ πνεῖ καὶ ζέφυρος ἅμα.

κη΄. τῇ πρὸ ε΄ καλενδῶν ἡ Παρθένος ἄρχεται φαίνεσθαι.

κθ΄. τῇ πρὸ δ΄ καλενδῶν μέτριος ὁ ζέφυρος.

λ.   τῇ πρὸ γ΄ καλενδῶν ὁμοίως, τῆς Παρθένου ἀνισχούσης.

λα΄. τῇ πρὸ α΄ καλενδῶν Σεπτεμβρίων ἡ Ἀνδρομέδα ἀνίσχει· εὖρος πνεῖ, καὶ ἐναλλάττονται οἱ ἄνεμοι.

Σεπτέμβριος

α΄. καλένδαις Σεπτεμβρίαις βροντώδης ὑετός, καὶ ἡ Ἀνδρομέδα ἀνίσχει. εὖρος πνεῖ, καὶ ἐναλλάττονται οἱ ἄνεμοι.

β΄. τῇ πρὸ δ΄ νωνῶν Σεπτεμβρίων ὁ Ἰχθὺς ἀπὸ τοῦ νότου παύεται δύεσθαι.

γ΄. τῇ πρὸ γ΄ νωνῶν βροντᾷ καὶ ὕει.

δ΄. τῇ πρὸ α΄ νωνῶν Ἀρκτοῦρος ἀνίσχει σὺν τῷ Τρυγητῇ, καὶ τὸν μὲν Ὀιστὸν ἀποκρύπτει.

ε΄. νώναις Σεπτεμβρίαις οἶκος Ἑρμοῦ. ζέφυρος δὲ πνεῖ, καὶ ἐκ τῆς τῶν ἀνέμων ἐναλλαγῆς ὑετός.

ϛ΄. τῇ πρὸ η΄ εἰδῶν Σεπτεμβρίων ἀνίσχει ὁ Ἵππος.

ζ΄. τῇ πρὸ ζ΄ εἰδῶν ἀνίσχει ἡ Αἴξ· ἐν δὲ τῇ ἑσπέρᾳ λὶψ μετὰ ὑετοῦ.

η΄. τῇ πρὸ ϛ΄ εἰδῶν ἀναφαίνεται ὁ Ἀρκτοῦρος, βορρᾶς δὲ φυσᾷ· ἔσθ᾽ ὅτε καὶ βροντᾷ.

θ΄. τῇ πρὸ ε΄ εἰδῶν τὸ μέσον τῆς Παρθένου ἀνίσχει, ζέφυρος δὲ μετὰ λιβὸς πνεῖ.

ι΄. τῇ πρὸ δ΄ εἰδῶν τὰ αὐτὰ σημαίνει.

ια΄. τῇ πρὸ γ΄ εἰδῶν ὡσαύτως.

ιβ΄. τῇ πρὸ α΄ εἰδῶν Σεπτεμβρίων ὁ Ἀρκτοῦρος ἀνίσχει.

ιγ΄. εἰδοῖς Σεπτεμβρίαις ὕει διὰ τῆς ἐπιτολῆς τοῦ Ἀρκτούρου.

ιδ΄. τῇ πρὸ ιη΄ καλενδῶν Ὀκτωβρίων αἱ Πλειάδες ἀνίσχουσι σὺν τῷ Ἵππῳ.

ιε΄. τῇ πρὸ ιζ΄ καλενδῶν ἡ χελιδὼν οὐδαμοῦ· ἀνίσχει δὲ καὶ ἡ Αἴξ, καὶ ὕει.

ιϛ΄. τῇ πρὸ ιϛ΄ καλενδῶν τὸ δωδεκατημόριον ἄρχεται τοῦ μετοπώρου.

ιζ΄. τῇ πρὸ ιε΄ καλενδῶν ζέφυρος μετὰ λιβὸς καὶ εὖρος πλατύς.

ιη΄. τῇ πρὸ ιδ΄ καλενδῶν ὁ Στάχυς ἀνίσχει, καὶ ἠρέμα ἀπαρκτίας.

ιθ΄. τῇ πρὸ ιγ΄ καλενδῶν ὁ ἥλιος <ἐν> Ζυγῷ· ὁ Κρατὴρ φαίνεται· νότος πνεῖ. τὸ μέσον τοῦ Ἀρκτούρου ὄρθρου φαίνεται.

κ΄. τῇ πρὸ ιβ΄ καλενδῶν ἐπιτολὴ τοῦ Ἀρκτούρου καὶ ὑετὸς πολύς.

κα΄. τῇ πρὸ ια΄ καλενδῶν ἡ μετοπωρινὴ ἰσημερία, καὶ οἱ Ἰχθύες δύονται.

κβ΄. τῇ πρὸ ι΄ καλενδῶν δύεται ἡ Ἀργώ, καὶ τροπὴ τοῦ ἀέρος ὑετώδης.

κγ΄. τῇ πρὸ θ΄ καλενδῶν δύονται οἱ Ἰχθύες, καὶ ὑετὸς ἐκ τοῦ νότου, καὶ γίνονται βροχαὶ καὶ ταραχαὶ ἀνέμων καὶ θαλάσσης.

κδ΄. τῇ πρὸ η΄ καλενδῶν ἔκλειψις σεληνιακὴ καὶ ἐπιτολὴ τοῦ Κενταύρου.

κε΄. τῇ πρὸ ζ΄ καλενδῶν συννεφὴς ὁ ἀὴρ καὶ ταραχώδης.

κϛ΄. τῇ πρὸ ϛ΄ καλενδῶν οἱ Ἔριφοι ἀνίσχουσι, καὶ νότος ψυχρός.

κζʹ. τῇ πρὸ ε΄ καλενδῶν παύεται ἡ Παρθένος ἀνίσχειν, αἱ δὲ Πλειάδες ἐν ἑσπέρᾳ φαίνονται, καὶ οἱ Ἔριφοι σὺν τῷ ἡλίῳ ἀνίσχουσιν, καὶ γίνονται βροχαὶ καὶ ταραχαὶ ἀνέμων καὶ θαλάσσης.

κη΄. τῇ πρὸ δ΄ καλενδῶν ἔτι μᾶλλον ὁ νότος.

κθ΄. τῇ πρὸ γ΄ καλενδῶν αἱ Πλειάδες ὄρθρου φαίνονται, καὶ πνεῖ νότος ὑετώδης.

λ΄. τῇ πρὸ α΄ καλενδῶν ἡ Αἲξ ἀνίσχει [. . .] μέχρι τῆς ἑσπερινῆς δύσεως τῶν Πλειάδων.

Ὀκτώβριος

α΄. καλένδαις Ὀκτωβρίαις αἱ Πλειάδες ἀπὸ τῆς ἀνατολῆς ἄρχονται φαίνεσθαι, καὶ νότος ὄρθρου πνεῖ.

β΄. τῇ πρὸ ϛ΄ νωνῶν Ὀκτωβρίων ὡσαύτως ὅ τε ἀὴρ αἵ τε Πλειάδες.

γ΄. τῇ πρὸ ε΄ νωνῶν ὁ Ἡνίοχος δύεται, καὶ ἀπὸ βορέου βροντᾷ.

δ΄. τῇ πρὸ δ΄ νωνῶν οἱ Ἔριφοι ἀνίσχουσι καὶ ὕει.

ε΄. τῇ πρὸ γ΄ νωνῶν ὁ Στέφανος ἀνίσχει καὶ τροπὴ βορεινή.

ϛ΄. τῇ πρὸ α΄ νωνῶν τὸ μέσον τοῦ Κριοῦ δύεται, καὶ ὁ Σκορπίος σὺν αὐτῷ.

ζ΄. νώναις Ὀκτωβρίαις ὡσαύτως.

η΄. τῇ πρὸ η΄ εἰδῶν Ὀκτωβρίων ὁ Στέφανος σὺν τοῖς Ἐρίφοις ἀνίσχων τρέπει τὸν ἀέρα.

θ΄. τῇ πρὸ ζ΄ εἰδῶν οἱ Ἔριφοι σὺν ταῖς Πλειάσιν ἀνίσχουσι, καὶ πνεῖ ἄνεμος λίψ.

ι΄. τῇ πρὸ ϛ΄ εἰδῶν ὁ Ζυγὸς ἄρχεται ἀνίσχειν, καὶ ζέφυρος πνεῖ.

ια΄. τῇ πρὸ ε΄ εἰδῶν ὁ Στέφανος ὄρθρου ἀνίσχων ἐναλλάττει τοὺς ἀνέμους.

ιβ΄. τῇ πρὸ δ΄ εἰδῶν αἱ Πλειάδες ἀνίσχουσι, καὶ νότος πνεῖ.

ιγ΄. τῇ πρὸ γ΄ εἰδῶν ὁ Στέφανος ἀνίσχων ἐναλλάττει τοὺς ἀνέμους, καὶ θαλάσσης ταραχὴ γίνεται.

ιδ΄. τῇ πρὸ α΄ εἰδῶν βορρᾶς πλατὺς πνεῖ.

ιε΄. εἰδοῖς Ὀκτωβρίαις τὸ μέσον τοῦ φθινοπώρου καὶ ἄνεμος νότος.

ιϛ΄. τῇ πρὸ ιζ΄ καλενδῶν Νοεμβρίων ὁ Ὠρίων ἀνίσχει, καὶ δροσώδης ὁ ἀήρ.

ιζ΄. τῇ πρὸ ιϛ΄ καλενδῶν ὡσαύτως.

ιη΄. τῇ πρὸ ιε΄ καλενδῶν ἡ πᾶσα ἡμέρα συννεφής.

ιθ΄. τῇ πρὸ ιδ΄ καλενδῶν ὁ ἥλιος <ἐν> Σκορπίῳ, καὶ ζέφυρος πνεῖ.

κ΄. τῇ πρὸ ιγ΄ καλενδῶν δύονται αἱ Πλειάδες, καὶ τροπὴ τοῦ ἀέρος.

κα΄. τῇ πρὸ ιβ΄ καλενδῶν ὡσαύτως· ἀλλὰ καὶ ὕει.

κβ΄. τῇ πρὸ ια΄ καλενδῶν [. . .]¹³⁰ ἡ οὐρὰ τοῦ Ταύρου, καὶ νότος πνεῖ ὑετώδης.

κγ΄. τῇ πρὸ ι΄ καλενδῶν δύεται ὁ Σκορπίος, καὶ βορρᾶς πνεῖ καὶ χειμὼν κατὰ θάλασσαν.

κδ΄. τῇ πρὸ θ΄ καλενδῶν δύονται αἱ Πλειάδες.

κε΄. τῇ πρὸ η΄ καλενδῶν ὁ Κένταυρος ὄρθρου δύεται.

κϛ΄. τῇ πρὸ ζ΄ καλενδῶν τὸ μέτωπον τοῦ Σκορπίου δύεται.

κζ΄. τῇ πρὸ ϛ΄ καλενδῶν αἱ Ὑάδες δύονται, καὶ βορρᾶς ψυχρὸς καὶ χειμὼν κατὰ θάλασσαν.

κη΄. τῇ πρὸ ε΄ καλενδῶν αἱ Πλειάδες καὶ ὁ Ὠρίων παντελῶς δύονται.

κθ΄. τῇ πρὸ δ΄ καλενδῶν ὁ Ἀρκτοῦρος δύεται, καὶ οἱ ἄνεμοι βιαιότεροι.

λ΄. τῇ πρὸ γ΄ καλενδῶν ἡ Κασσιόπεια ἄρχεται δύεσθαι.

λα΄. τῇ πρὸ α΄ καλενδῶν ὁ Ὠρίων ὅλος ἀποκρύπτεται, καὶ ὁ μὲν Ἀετὸς ἐν ἑσπέρᾳ, ἡ δὲ Λύρα ἀνίσχει.¹³¹

Νοέμβριος

α΄. καλένδαις Νοεμβρίαις αἱ Πλειάδες δύονται. ἕωθεν πάχνη, καὶ τοῦ Ἀρκτούρου δυομένου τροπὴ τοῦ ἀέρος ἐπὶ τὸ ψυχρότερον.

β΄. τῇ πρὸ δ΄ νωνῶν Νοεμβρίων ἄνεμοι ψυχροὶ καὶ βροχαί.

γ΄. τῇ πρὸ γ΄ νωνῶν ἡ Λύρα ὄρθρου ἀνίσχει, καὶ πρῶτα βορρᾶς, εἶτα καὶ νότος.

δ΄. τῇ πρὸ α΄ νωνῶν νότος σὺν τῷ ζεφύρῳ καὶ ὑετός.

ε΄. νώναις Νοεμβρίαις ἡ Λύρα ἀνίσχοντος ἡλίου φαίνεται, καὶ ἄνεμος βορέας.

ϛ΄. τῇ πρὸ η΄ εἰδῶν Νοεμβρίων ὄρθρου ὁ Ἀρκτοῦρος δύεται, καὶ συννέφεια.

ζ΄. τῇ πρὸ ζ΄ εἰδῶν αἱ Πλειάδες καὶ ὁ Ὠρίων δύονται, καὶ πνεῖ ὁ βορρᾶς.

η΄. τῇ πρὸ ϛ΄ εἰδῶν στυγνὸς ὁ ἀήρ.

θ΄. τῇ πρὸ ε΄ εἰδῶν λαμπρὸν ἄστρον τοῦ Σκορπίου, καὶ χειμερινὴ τροπή.

ι΄. τῇ πρὸ δ΄ εἰδῶν ἄρχεται ὁ χειμών.

ια΄. τῇ πρὸ γ΄ εἰδῶν αἱ Πλειάδες ὑποκρύπτονται.

ιβ΄. τῇ πρὸ α΄ εἰδῶν τοῦ Σκορπίου <τὸ> μέσον ἄστρον ἀνίσχει.

ιγ΄. εἰδοῖς Νοεμβρίαις αἱ Πλειάδες καὶ ὁ Ὠρίων ὄρθρου δύονται.

ιδ΄. τῇ πρὸ ιη΄ καλενδῶν Δεκεμβρίων ὄρθρου ὁ Σκορπίος δύεται.

ιε΄. τῇ πρὸ ιζ΄ καλενδῶν ἡ Λύρα ἀνίσχει ἕωθεν, εὖρός τε πνεῖ ἅμα καὶ νότος σὺν τῷ βορρᾷ.

---

¹³⁰ Wachsmuth restores δύεται.   ¹³¹ Wachsmuth inserts ὄρθρου here.

ιϛ'.   τῇ πρὸ ιϛ' καλενδῶν τὸ αὐτό.

ιζ'.   τῇ πρὸ ιε' καλενδῶν χειμάζει, καὶ νότος ἕπεται.

ιη'.   τῇ πρὸ ιδ' καλενδῶν ὁ ἥλιος <ἐν> Τοξότῃ. καὶ ὁ Ὠρίων ἀνίσχει σὺν τῇ Λύρᾳ, καὶ ταραχὴ τοῦ ἀέρος.

ιθ'.   τῇ πρὸ ιγ' καλενδῶν τὸ κέρας τοῦ Ταύρου σὺν τῷ ἡλίῳ δύεται, καὶ πνεῖ βορρᾶς.

κ'.   τῇ πρὸ ιβ' καλενδῶν ἀηδὴς χειμών.

κα'.   τῇ πρὸ ια' καλενδῶν αἱ Ὑάδες σὺν τῷ Λαγωῷ ὄρθρου δύονται.

κβ'.   τῇ πρὸ ι' καλενδῶν ὑετὸς ψυχρός.

κγ'.   τῇ πρὸ θ' καλενδῶν ὁ Ὠρίων καὶ τὰ κέρατα τοῦ Ταύρου δύονται.

κδ'.   τῇ πρὸ η' καλενδῶν προοίμια τῆς χειμερινῆς τροπῆς. δύεται ὁ Κύων, καὶ δρόσος φέρεται ψυχρά.

κε'.   τῇ πρὸ ζ' καλενδῶν ὁ ἥλιος ἐπὶ τῆς πρώτης μοίρας τοῦ Τοξότου.

κϛ'.   τῇ πρὸ ϛ' καλενδῶν ὡσαύτως.

κζ'.   τῇ πρὸ ε' καλενδῶν δύεται ὁ Κύων, καὶ βροχὴ ἔσται νοτία.

κη'.   τῇ πρὸ δ' καλενδῶν ἄρχεται ὁ Κύων δύεσθαι, καὶ νεφώδης ὁ ἀήρ.

κθ'.   τῇ πρὸ γ' καλενδῶν ὄρθρου δύεται ὁ Κύων, καὶ ἐκ τοῦ λιβὸς ὕει.

λ'.   τῇ πρὸ α' καλενδῶν δύεται ὁ Ὠρίων, καὶ ζέφυρος καὶ μετὰ νότου βροχή.

Δεκέμβριος

α'.   καλένδαις Δεκεμβρίαις σύγχυσις τοῦ ἀέρος· ἀπαρκτίας φυσᾷ, καὶ ὁ Ὠρίων ὅλος ὄρθρου δύεται.

β'.   τῇ πρὸ δ' νωνῶν Δεκεμβρίων ὁ Κύων δύεται· ἐν ἑσπέρᾳ ἀπαρκτίας ἐπιτείνει.

γ'.   τῇ πρὸ γ' νωνῶν χειμάζει σὺν ὑετῷ.

δ'.   τῇ πρὸ α' νωνῶν δύεται ὁ Τοξότης καὶ βορρᾶς πνεῖ.

ε'.   νώναις Δεκεμβρίαις δι᾽ ὅλης τῆς ἡμέρας βροχὴ καὶ πολὺς βορρᾶς.

ϛ'.   τῇ πρὸ η' εἰδῶν Δεκεμβρίων τὸ μέσον τοῦ Τοξότου ἀνίσχει.

ζ'.   τῇ πρὸ ζ' εἰδῶν ἀνίσχει ὁ Ἀετός, καὶ πνεῖ ὁ λίψ.

η'.   τῇ πρὸ ϛ' εἰδῶν ὁ Σκορπίος ὅλος ἀνίσχει.

θ'.   τῇ πρὸ ε' εἰδῶν ὁ Κύων ἐν ἑσπέρᾳ ἀνίσχει, καὶ ὁ νότος πνεῖ.

ι'.   τῇ πρὸ δ' εἰδῶν βορέας συχνός, καὶ ὁ οὐρανὸς μέλας ἐκ τοῦ χειμῶνος.

ια'.   τῇ πρὸ γ' εἰδῶν κρύφιον ἄστρον· ἀπαρκτίας ἅμα καὶ καικίας.

ιβ'.   τῇ πρὸ α' εἰδῶν βορρᾶς πρῶτα, ἔπειτα δὲ νότος βαρύς.

ιγ'.   εἰδοῖς Δεκεμβρίαις ὅλος ὁ Σκορπίος ἀνίσχει· βορρᾶς δὲ καὶ νότος ὑετὸν ἐπισπώμενοι.

ιδ'.   τῇ πρὸ ιθ' καλενδῶν Ἰανουαρίων δύεται ἡ Αἴξ.

ιε'.   τῇ πρὸ ιη' καλενδῶν νότος ἅμα καὶ βορρᾶς, καὶ σύγχυσις τοῦ ἀέρος.

ιϛ΄. τῇ πρὸ ιζ΄ καλενδῶν ὡσαύτως.

ιζ΄. τῇ πρὸ ιϛ΄ καλενδῶν ὁ ἥλιος ἐπὶ τοῦ Αἰγοκέρωτος.

ιη΄. τῇ πρὸ ιε΄ καλενδῶν ἄστρον κρυπτὸν καὶ ἀνεμομαχία.

ιθ΄. τῇ πρὸ ιδ΄ καλενδῶν ἡ Αἲξ ἀνίσχει.

κ΄. τῇ πρὸ ιγ΄ καλενδῶν ὁ Ἀετὸς ἀνίσχει καὶ ὁ Αἰγόκερως.

κα΄. τῇ πρὸ ιβ΄ καλενδῶν βορέας μὲν κατάρχεται, νότος δὲ τὴν πᾶσαν
ἡμέραν ἐπικρατεῖ.

κβ΄. τῇ πρὸ ια΄ καλενδῶν ὁ Ἀετὸς ἐν ἑσπέρᾳ ἀνίσχει.

κγ΄. τῇ πρὸ ι΄ καλενδῶν ὄρθρου ἡ Αἲξ δύεται· καὶ συμπληροῦται ἡ βροῦμα,
οἱονεὶ ἡ χειμερινὴ τροπή.

κδ΄. τῇ πρὸ θ΄ καλενδῶν χειμέριος ὁ ἀήρ.

κε΄. τῇ πρὸ η΄ καλενδῶν ὄρθρου ἡ Αἲξ δύεται.

κϛ΄. τῇ πρὸ ζ΄ καλενδῶν δύεται ἐν ἑσπέρᾳ ἡ Αἲξ, καὶ χειμάζει.

κζ΄. τῇ πρὸ ϛ΄ καλενδῶν ὁ Δελφὶν ἔωθεν ἀνίσχει, ἔχων τὸν ἀέρα.

κη΄. τῇ πρὸ ε΄ καλενδῶν ὁ ἥλιος ἀποστρέφεται ἀπὸ τοῦ νοτιαίου
καμπτῆρος.

κθ΄. τῇ πρὸ δ΄ καλενδῶν ὁ ἥλιος λαμπρός, ὁ δὲ Ἀετὸς δύεται, καὶ ὕει.

λ΄. τῇ πρὸ γ΄ καλενδῶν ὁ μὲν Δελφὶν ἀνίσχει, ὁ δὲ Κύων δύεται.

λα΄. τῇ πρὸ α΄ καλενδῶν Ἰανουαρίων κρυφία ταραχὴ καὶ ἀνεμώδης.
ἄρχεται ἀνίσχειν ὁ Δελφίν, καὶ ὁ ἥλιος εἰς ὕψος.

*Ephemeris for the whole year, by Clodius Tuscus, in translation, according to
his text.*

January

1. On the Kalends: the days begin to get long. The sun is elevated. Aquila
   sets with Corona and causes storms.

2. On the 4th day before the Nones of January: the sun 'leaps up'. The
   middle of Cancer sets, and the winds shift about.

3. On the 3rd day before the Nones: the remainder of Cancer sets, and a
   complex change in the air.

4. On the *pridie:* the middle of winter and a strong south wind, then
   continuous north winds. And Delphinus rises at the same time as Sirius
   around daybreak.

5. On the Nones of January: Lyra rises, Aquila sets, and the whole of
   Delphinus rises. Probably contrary winds.

6. On the 8th day before the Ides of January: in the evening Aquila sets,
   the south wind blows.

7. On the 7th day before the Ides: the arctic and the north wind fight each other.[132]

8. On the 6th day before the Ides: domicile of Mars.[133] South wind at the same time as the west wind. Capricorn begins to [...], showers,[134] along with a dense wind in the evening.

9. On the 5th day before the Ides: south wind blows gently, cloudy.

10. On the 4th day before the Ides: similar, the south wind is more violent.

11. On the 3rd day before the Ides: the arctic wind follows showers, and snow.

12. On the *pridie* of the Ides: the south wind blows.

13. On the Ides of January: hidden stars. Calculation[135] has the first part of Sagitta setting, and then at night it showers.

14. On the 19th day before the Kalends of February: hidden stars, and a variable change in the north wind and at the same time of the arctic wind. And Leo begins to set, and it showers then.

15. On the 18th day before the Kalends: arctic and strong north wind.

16. On the 17th day before the Kalends: the sun is in Aquarius, east wind with showers.

17. On the 16th day before the Kalends: Lyra begins to set, around dawn there are contrary winds.

18. On the 15th day before the Kalends: Leo sets in the morning. A changing about of the north and also the south and arctic winds, with showers. Delphinus sets after Leo.

19. On the 14th day before the Kalends: change, and the very middle of winter.

20. On the 13th day before the Kalends: north wind and south wind. And the middle of Cancer sets, and Aquarius begins to rise.

---

[132] ἀπαρκτίας καὶ βορέας: compare Aristotle, *Met.* 363b14.

[133] οἶκος Ἄρεως. The domiciles of Mars are Scorpio and Aries (Ptolemy, *Tetr.* 1.17), so this cannot mean that the sun is in the domicile of Mars since the sun is in Capricorn. I assume it means that Mars is in one of its domiciles (compare 5 Sept.). But, if this *is* a report of Mars moving into one of its domiciles then we may be tempted to combine it with the entry at 5 Sept. to try and determine the date of the calendar. The only year between 50 BC and AD 560 that fits the bill within narrowly determined parameters of accuracy is AD 110, but AD 110 does not agree with the entries for lunar eclipses on either 10 Aug. or 24 Sept. which in any case are incompatible with each other.

[134] In most parapegmata the more usual word for rain is ὑετός, but Clodius Tuscus often favours the less common βροχή here and throughout. To keep his terminology clear, I will translate ὑετός and its derivatives as 'rain', and βροχή and its derivatives as 'showers'.

[135] Or *a report*: λόγος.

21. On the 12th day before the Kalends: Aquarius rises completely, and the south-west wind blows, and it rains.
22. On the 11th day before the Kalends: Lyra sets with Cancer, and towards evening it rains.
23. On the 10th day before the Kalends: north wind will blow[136] with showers.
24. On the 9th day before the Kalends: storm and an increase in the north wind and in the east wind as well.
25. On the 8th day before the Kalends: similar.
26. On the 7th day before the Kalends: it is stormy, and the north wind blows with the east wind, and Lyra begins to set.
27. On the 6th day before the Kalends: the bright star in the breast of Leo begins to set, as (does) Lyra in the evening. And the north wind blows. And then it also rains.
28. On the 5th day before the Kalends: contrary winds with snow.
29. On the 4th day before the Kalends: Delphinus threatens to set.
30. On the 3rd day before the Kalends: Lyra begins to partly set, around the time of the first watch of the night. There are clouds, and a strong north wind with rain.
31. On the day before the Kalends of February: rain mixed with snow.

February

1. On the Kalends of February: hidden stars. South and east winds, and Lyra begins to set.
2. On the 4th day before the Nones: blustery air, and the west wind blows as well.
3. On the 3rd day before the Nones: the middle of Leo sets with Lyra. Arctic at the same time as a north wind.
4. On the day before the Nones: Delphinus sets, and in the evening a south wind increases, with showers.
5. On the Nones of February: the middle of Aquarius rises. The air is disturbed from the west.
6. On the 8th day before the Ides of February: Lyra sets, and the west wind blows because of its setting.

---

[136] It is unusual to see a future-tense verb (πνεύσει) in a parapegma, although they are common enough in this one. Compare 16, 17, 19 Feb., 10, 13 Mar. 4, 13, 23, 26 Apr. 13, 27 June, 1 Aug. I do not distinguish in my translation between the two verbs for *blow* (πνεύσει and ῥεύσει) used in this text. Compare also the Madrid parapegma, 8 April, and Lydus' *De mensibus*, xvi Kal. April and vi Non. Oct.

7. On the 7th day before the Ides: the beginning of spring, and the west wind blows from now.

8. On the 6th day before the Ides: the west wind blows with the north wind.

9. On the 5th day before the Ides: hidden stars. And Aquarius rises.

10. On the 4th day before the Ides: arctic with a west wind. And there are showers.

11. On the 3rd day before the Ides: the east *(apeliotes)* wind blows, and Arcturus rises.

12. On the day before the Ides of February: contrary winds.

13. On the Ides of February: Sagittarius sets in the evening and it is very stormy.

14. On the 16th day before the Kalends of March: the cup rises in the evening, and the south wind emerges the dominant among shifting winds.

15. On the 15th day before the Kalends: the sun is in Pisces and the air is stormy.

16. On the 14th day before the Kalends: the arctic wind will blow with a south wind, and the sun is fresh.[137]

17. On the 13th day before the Kalends: Virgo sets, and a south wind will blow with the west and north winds.

18. On the 12th day before the Kalends: The constellation called Sagitta sets, and in the evening the west wind blows, and Virgo begins to set.[138]

19. On the 11th day before the Kalends: an arctic wind will blow with a south wind. They say that Sagitta sets.

20. On the 10th day before the Kalends: north wind with showers, and Leo sets. The northern winds, the so-called 'swallow-bringers', begin, and these usually blow for thirty days, and the swallow appears.

21. On the 9th day before the Kalends: Arcturus begins to set at the first guard of the night, and the west wind blows. The night is cloudy.

22. On the 8th day before the Kalends: the so-called Halcyon days.

23. On the 7th day before the Kalends: Aquarius begins to rise, and the early morning is stormy.

24. On the 6th day before the Kalends: the north-west wind blows at the same time as the north wind.

---

[137] The phrase νέος δὲ ἥλιος is unique.

[138] This seems to contradict the entry for the previous day: Virgo sets. Entries like this point to either a complex intertwining of material gathered from different sources or from direct observation, or to carelessness.

25. On the 5th day before the Kalends: Arcturus rises and it rains.
26. On the 4th day before the Kalends: Arcturus rises in the evening.
27. On the 3rd day before the Kalends: the constellation the Greeks call Sagitta sets in the evening.
28. On the day before the Kalends of March: a diffuse west wind, and the day is completely springlike.

March

1. On the Kalends of March: south and south-west wind.
2. On the 6th day before the Nones of March: south-west wind, and Vindemiatrix begins to appear. A cold north wind blows until the morning setting of Arcturus.
3. On the 5th day before the Nones: bad air and showers. And Arcturus rises while the sun comes up, and the north wind blows.
4. On the 4th day before the Nones: Arcturus rises in the evening.
5. On the 3rd day before the Nones: similar.
6. On the day before the Nones: rain mixed with snow.
7. On the Nones of March: Pegasus sets early, and the west wind blows gently. Also Corona sets at daybreak.
8. On the 8th day before the Ides of March: birds begin to appear at sea, and north and arctic winds blow gently, and it is the prelude to spring. The sun is in the middle of Pisces, and Pegasus sets.
9. On the 7th day before the Ides: the kite begins to appear, and a south wind blows. The southern fish begins to disappear at daybreak.
10. On the 6th day before the Ides: Pegasus sets at daybreak, and the kite descends from on high to the ground. Vindemiatrix sets and Arcturus rises. A cold north wind will blow.
11. On the 5th day before the Ides: departure of winter, change from north and arctic winds.
12. On the 4th day before the Ides: the southern fish ceases to rise, an arctic or a south wind blows gently.
13. On the 3rd day before the Ides: Argo rises in the evening. A west and a south wind will blow and the star on the tail of Leo [. . .]. It is stormy.
14. On the day before the Ides: a north wind throughout the day.
15. On the Ides of March: Pegasus sets, and a cold north wind blows.
16. On the 17th day before the Kalends of April: [. . .].
17. On the 16th day before the Kalends: the sun is in Aries. A diffuse west wind blows. The stork appears and the open sea is sailable.

18. On the 15th day before the Kalends: contrary winds. The north wind triumphs.
19. On the 14th day before the Kalends: a south wind blows, and the kite appears until the equinox.
20. On the 13th day before the Kalends: a pleasant north wind blows.
21. On the 12th day before the Kalends: Pegasus sets in the morning. A north or an arctic wind blows.
22. On the 11th day before the Kalends: Aries rises in breadth, and so it either showers or is cloudy.
23. On the 10th day before the Kalends: similar.
24. On the 9th day before the Kalends: vernal equinox. Showers or a change to thunder.
25. On the 8th day before the Kalends: arctic or north wind, and Pegasus sets in the morning.
26. On the 7th day before the Kalends: the fishes rise, and there are showers from the south with snow. Aries rises in the morning. Disturbed air at sea.
27. On the 6th day before the Kalends: vernal equinox.[139] And so there are showers and thunder.
28. On the 5th day before the Kalends: the day is equal to the night.
29. On the 4th day before the Kalends: Scorpio sets, and a strong blowing wind gathers thunderous showers.
30. On the 3rd day before the Kalends: Scorpio sets, an arctic wind blows gently with showers.
31. On the day before the Kalends: showers came,[140] and thunder from the south.

April

1. On the Kalends of April: Scorpio sets, and the sun adds a degree. It is cloudy from the north. The Pleiades begin to rise and there is a change in the weather.
2. On the 4th day before the Nones of April: cloudy for the whole day.
3. On the 3rd day before the Nones: the Pleiades set in the evening.
4. On the day before the Nones: the south-west wind will blow.
5. On the Nones of April: the west wind blows.

---

[139] Compare three days earlier and one day later.   [140] βρέχειν πέφυκε. The perfect is unusual.

6. On the 8th day before the Ides: the Hyades rise, and showers increase from the south.

7. On the 7th day before the Ides: the south wind blows gently, and the rest of the Hyades set.

8. On the 6th day before the Ides: Libra begins to set at daybreak.

9. On the 5th day before the Ides: squalls from the south.

10. On the 4th day before the Ides: the north wind blows all day, and it will be very rainy in the evening.

11. On the 3rd day before the Ides: cold winds and showers.

12. On the day before the Ides: the Hyades dissapear.

13. On the Ides of April: the north wind will blow, sunrise a little frosty.

14. On the 18th day before the Kalends of May: hidden stars. Windy, and thunderstorms.

15. On the 17th day before the Kalends: the Hyades set[141] and cold winds blow. Perseus rises.

16. On the 16th day before the Kalends: the Hyades set and the west wind blows.

17. On the 15th day before the Kalends: the sun is in Taurus and the Hyades set.

18. On the 14th day before the Kalends: the south-west wind will blow.

19. On the 13th day before the Kalends: the Hyades generally set. The south-west wind comes up in the evening.

20. On the 12th day before the Kalends: the west wind blows.

21. On the 11th day before the Kalends: the head of Taurus sets, and there is not a little rain.

22. On the 10th day before the Kalends: the Pleiades rise and the west wind blows.

23. On the 9th day before the Kalends: Lyra appears at the first watch of the night, and there will be a change.

24. On the 8th day before the Kalends: Lyra appears. Sign of showers.[142] The beginning of the frost-season and of spring.[143] The south wind blows.

25. On the 7th day before the Kalends: Scorpio sets and Sirius disappears, as does the star on the belt of Orion. The south wind blows. The beginning of the frost-season and of spring.[144]

---

[141] This is the first of three days in a row with nearly identical entries for the setting of the Hyades.

[142] This entry is unique: βροχῶν σημασία.

[143] ἀρχὴ παχνήτου. See n. 124, above.   [144] Again, Clodius repeats himself.

26. On the 6th day before the Kalends: the Hyades generally set, and there will be a change in the air.
27. On the 5th day before the Kalends: the south wind blows.
28. On the 4th day before the Kalends: rain from the south.
29. On the 3rd day before the Kalends: the Haedi rise and from early morning the south wind blows.
30. On the day before the Kalends: Sirius disappears in the evening and a change from the south. At the same time the north wind also interferes.

May

1. On the Kalends of May: Sirius disappears, and dew settles.
2. On the 6th day before the Nones: the Hyades rise with the sun.
3. On the 5th day before the Nones: the whole of Centaurus appears and the west wind blows.
4. On the 4th day before the Nones: Scorpio rises in the morning, and the north wind blows, and dew settles.
5. On the 3rd day before the Nones: Lyra rises in the morning.
6. On the day before the Nones: the middle of Scorpio sets.
7. On the Nones of May: the Pleiades rise in the morning, and the west wind blows.
8. On the 8th day before the Ides of May: prelude to summer, and the west wind is dominant.
9. On the 7th day before the Ides: similar.
10. On the 6th day before the Ides: Lyra rises, and the Hyades set, and the head of Taurus appears.
11. On the 5th day before the Ides: the Pleiades appear.
12. On the 4th day before the Ides: the Pleiades rise and the south wind blows.
13. On the 3rd day before the Ides: the Hyades rise and the south wind blows.
14. On the day before the Ides: Scorpio sets, and Lyra rises at daybreak.
15. On the Ides of May: Cancer rises and the south wind blows.
16. On the 17th day before the Kalends of June: summer begins.
17. On the 16th day before the Kalends: Sirius sets.
18. On the 15th day before the Kalends: the sun in Gemini.
19. On the 14th day before the Kalends: the south wind blows in the evening.
20. On the 13th day before the Kalends: the Hyades rise and the north wind blows.

21. On the 12th day before the Kalends: Arcturus sets and the air is disturbed.
22. On the 11th day before the Kalends: Sagittarius sets and the south wind blows.
23. On the 10th day before the Kalends: Gemini rises, as does Aquila.
24. On the 9th day before the Kalends: the Hyades begin to rise and showers are likely.
25. On the 8th day before the Kalends: the Haedi rise in the morning and the north wind blows.
26. On the 7th day before the Kalends: Arcturus sets and the south wind blows with the north wind.
27. On the 6th day before the Kalends: Aquila rises.
28. On the 5th day before the Kalends: the south wind blows, and Lyra rises in the morning.
29. On the 4th day before the Kalends: a strong south wind.
30. On the 3rd day before the Kalends: the Pleiades rise and showers with thunder.
31. On the day before the Kalends of June: the air is stormy and it thunders. Around evening it is even more disturbed.

June

1. On the Kalends of June: the whole Hyades rise and the south wind blows.
2. On the 4th day before the Nones of June: Aquila rises. A disturbance in the air and the west wind blows.
3. On the 3rd day before the Nones: southerly thunder.
4. On the day before the Nones: the south wind blows, and showers are likely.
5. On the Nones of June: Aquila rises, and it rains at the same time as the south wind blows.
6. On the 8th day before the Ides of June: the shoulder of Orion rises.
7. On the 7th day before the Ides: the north wind blows and it rains.
8. On the 6th day before the Ides: Arcturus sets at daybreak, and the west wind blows.
9. On the 5th day before the Ides: Delphinus begins to rise, and Arcturus sets.
10. On the 4th day before the Ides: similar. And the north wind blows and it rains gently.
11. On the 3rd day before the Ides: the air is thundery and rainy from the south.
12. On the day before the Ides: the air is disturbed with showers.

13. On the Ides of June: the west or north-west wind will blow, and it will likely thunder.
14. On the 18th day before the Kalends of July: Delphinus rises and the south wind blows.
15. On the 17th day before the Kalends: the shoulders of Orion rise. Prelude to the hot season.
16. On the 16th day before the Kalends: hidden stars, and west with south wind.
17. On the 15th day before the Kalends: a moderate squall and a blowing of the north wind.
18. On the 14th day before the Kalends: a west with a south wind, and the shoulders of Orion appear.
19. On the 13th day before the Kalends: the sun in Cancer, and Orion rises in the morning.
20. On the 12th day before the Kalends: a south wind at the same time as a west wind, and thundery rain.
21. On the 11th day before the Kalends: Ophiucus sets at daybreak.
22. On the 10th day before the Kalends: south wind at the same time as a north wind.
23. On the 9th day before the Kalends: rising of Orion.
24. On the 8th day before the Kalends: rising of the hidden star, and intensification of heat.
25. On the 7th day before the Kalends: the summer solstice. Unforeseen disturbance in the air.
26. On the 6th day before the Kalends: a south-west wind at the same time as a west wind.
27. On the 5th day before the Kalends: the night is showery, and Orion rises. The south wind will blow.
28. On the 4th day before the Kalends: it showers in the evening, and Sirius begins to rise.
29. On the 3rd day before the Kalends: contrary winds.
30. On the day before the Kalends: Sirius rises in the morning, and the belt of Orion appears.

July

1. On the Kalends of July: the air is disturbed from the north.
2. On the 6th day before the Nones of July: hidden stars, and west or south wind.
3. On the 5th day before the Nones: disturbances in the air from the south.

4. On the 4th day before the Nones: Orion rises and Corona sets. The west wind blows.

5. On the 3rd day before the Nones: the middle of Cancer rises.

6. On the day before the Nones: the Etesian winds, after a south, a north wind.

7. On the Nones of July: Corona sets at daybreak, and the south wind blows.

8. On the 8th day before the Ides of July: Cepheus rises and a southerly change in the air.

9. On the 7th day before the Ides: the whole of Orion rises and the south wind blows.

10. On the 6th day before the Ides: the *prodromoi* ('precursors') of the Etesian winds blow.

11. On the 5th day before the Ides: showers with thunder and a violent north wind.

12. On the 4th day before the Ides: the whole of Orion rises at daybreak, and the winds called the *prodromoi* intensify.

13. On the 3rd day before the Ides: disturbed south-west wind.

14. On the day before the Ides: [. . .] rises and the north wind blows.

15. On the Ides of July: Procyon rises.

16. On the 17th day before the Kalends of August: Orion rises, and a violent north wind blows.

17. On the 16th day before the Kalends: the middle of summer, and the day becomes cooler from the north.

18. On the 15th day before the Kalends: west wind, and equally a south wind. And Sirius rises at daybreak, and the Etesian winds intensify.

19. On the 14th day before the Kalends: Orion rises, and the north-west wind blows gently, and the whole of Orion appears.[145]

20. On the 13th day before the Kalends: the sun in Leo. The north-west wind blows gently, and the *prodromoi* of the Etesian winds.

21. On the 12th day before the Kalends: from this day the Etesian winds will blow with the other winds for forty days.

22. On the 11th day before the Kalends: the *prodromoi* are discharged.

23. On the 10th day before the Kalends: the whole of Cancer rises with Sirius, and Aquila sets.

24. On the 9th day before the Kalends: Leo rises with the sun and with Sirius. Cancer ceases.

---

[145] Again, multiple sources seem likely.

25. On the 8th day before the Kalends: Aquarius begins to set, and Sirius rises around dawn, and the south wind blows gently.

26. On the 7th day before the Kalends: hot mist. Aquila sets and Leo rises, and the south wind blows.

27. On the 6th day before the Kalends: burning heat from Sirius, and the grapes begin to ripen.

28. On the 5th day before the Kalends: oppressive heat and an increase in the Etesian winds.

29. On the 4th day before the Kalends: the bright star in the breast of Leo rises, and it is moderately warm as the north wind (lessens) the burning heat.

30. On the 3rd day before the Kalends: fruits begin to appear, and Aquila sets in the morning, and the air is disturbed.

31. On the day before the Kalends of August: a south with a south-west wind blows.

August

1. On the Kalends of August: Aquila sets at daybreak. The south-west wind will blow, and dry burning heat.

2. On the 4th day before the Nones of August: Aquila still sets, and the air is southerly.

3. On the 3rd day before the Nones: it signifies the same thing.[146]

4. On the day before the Nones: the middle of Leo rises, and trees appear.[147] In this twenty-four-hour period the sun rules one degree. A misty south wind blows.

5. On the Nones of August: Corona sets, the middle of Leo rises, and cranes appear. A sharp south wind blows.

6. On the 8th day before the Ides of August: Lyra rises. Heat from the south.

7. On the 7th day before the Ides: the middle of Aquarius sets, and misty burning heat from the south.

8. On the 6th day before the Ides: Leo rises, and oppressive burning heat. The air is cloudy, and the middle of Aquarius rises.

9. On the 5th day before the Ides: stars hidden. Arctic wind early, and moderate heat.

---

[146] Or possibly 'the same signifies', τὸ αὐτὸ σημαίνει. If the text is correct, then comparison with 10 Sept. τὰ αὐτὰ σημαίνει, would seem to imply that τὸ αὐτό is the object of the verb here. My own suspicion is this entry is a corruption of τὸ αὐτό, ἐπισημαίνει. Compare 16 Nov.

[147] φαίνεται δὲ καὶ τὸ δένδρον. I am unsure what this may mean.

10. On the 4th day before the Ides: an eclipse of the moon on this day. And a shifting between the south and north winds, and burning heat.

11. On the 3rd day before the Ides: Lyra sets at daybreak. Autumn begins, and contrary winds.

12. On the day before the Ides: similar.

13. On the Ides of August: Delphinus sets with the hare.

14. On the 19th day before the Kalends of September: hot drought.

15. On the 18th day before the Kalends: hidden stars. The west wind blows with the south wind.

16. On the 17th day before the Kalends: Delphinus sets at daybreak, and the south wind blows.

17. On the 16th day before the Kalends: beginning of autumn.

18. On the 15th day before the Kalends: Lyra sets, and the south wind blows.

19. On the 14th day before the Kalends: the heat is moderate, and Delphinus sets.

20. On the 13th day before the Kalends: Lyra sets at daybreak. The sun in Virgo. South wind, and thundery rain.

21. On the 12th day before the Kalends: the sun is in Virgo completely.

22. On the 11th day before the Kalends: Virgo rises.

23. On the 10th day before the Kalends: a moderate arctic wind will blow, and Virgo is universal.[148]

24. On the 9th day before the Kalends: the same zodiacal sign, and the wind is northerly.

25. On the 8th day before the Kalends: the Etesian winds stop, and a colder north wind begins to blow.

26. On the 7th day before the Kalends: Delphinus rises, and the south wind blows.

27. On the 6th day before the Kalends: Vindemiatrix rises, and the south wind blows, as does the west wind at the same time.

28. On the 5th day before the Kalends: Virgo begins to appear.

29. On the 4th day before the Kalends: the west wind is moderate.

30. On the 3rd day before the Kalends: similar. Virgo rising.

31. On the day before the Kalends of September: Andromeda rises. The east wind blows, and the winds shift about.

---

[148] The Greek is obscure and may be corrupted. Most MSS have καὶ ὁλική ἡ Παρθένος, but Wachsmuth, following the translation in an eleventh-century Latin MS ('F'), emends it to καὶ ὁλική ἡ Παρθένος ἀνίσχει καὶ ἀὴρ καθαρός.

September

1. On the Kalends of September: thundery rain and Andromeda rises. The east wind blows, and the winds shift about.
2. On the 4th day before the Nones of September: the southern fish ceases to set.
3. On the 3rd day before the Nones: it thunders and rains.
4. On the day before the Nones: Arcturus rises with Vindemiatrix, and Sagitta disappears.
5. On the Nones of September: domicile of Mercury.[149] The west wind blows, and rain from the shiftiness of the winds.
6. On the 8th day before the Ides: Pegasus rises.
7. On the 7th day before the Ides: Capella rises. In the evening a south-west wind with rain.
8. On the 6th day before the Ides: Arcturus appears, and the north wind blows gently. And sometimes it thunders.
9. On the 5th day before the Ides: the middle of Virgo rises, and the west wind blows with a south-west wind.
10. On the 4th day before the Ides: it signifies the same things.[150]
11. On the 3rd day before the Ides: similar.
12. On the day before the Ides of September: Arcturus rises.
13. On the Ides of September: it rains because of the rising of Arcturus.
14. On the 18th day before the Kalends of October: the Pleiades rise with Pegasus.
15. On the 17th day before the Kalends: no more swallows, and also Capella rises, and it rains.
16. On the 16th day before the Kalends: the zodiacal sign of autumn begins.
17. On the 15th day before the Kalends: a west wind with a south-west and a diffuse east wind.
18. On the 14th day before the Kalends: Spica rises, and a gentle arctic wind.
19. On the 13th day before the Kalends: the sun is in Libra. The cup appears. The south wind blows. The middle of Arcturus appears at daybreak.
20. On the 12th day before the Kalends: rising of Arcturus and a lot of rain.
21. On the 11th day before the Kalends: autumnal equinox, and Pisces sets.

[149] The astrological 'houses' of Mercury are Gemini and Virgo (Ptolemy, *Tetr.* 1.17). Compare 8 Jan.
[150] Compare 3 Aug.

22. On the 10th day before the Kalends: Argo sets, and a rainy change in the air.
23. On the 9th day before the Kalends: Pisces sets, and rain from the south, and the winds and the sea become showery and disturbed.
24. On the 8th day before the Kalends: lunar eclipse,[151] and the setting of Centaurus.
25. On the 7th day before the Kalends: the air is cloudy and disturbed.
26. On the 6th day before the Kalends: the Haedi rise, and a cold south wind.
27. On the 5th day before the Kalends: Virgo ceases to rise, and the Pleiades appear in the evening, and the Haedi rise with the sun, and the winds and the sea become showery and disturbed.
28. On the 4th day before the Kalends: even more south wind.
29. On the 3rd day before the Kalends: the Pleiades appear at daybreak, and a rainy south wind blows.
30. On the day before the Kalends: Capella rises [. . .] until the evening setting of the Pleiades.

October

1. On the Kalends of October: the Pleiades begin to appear in the east, and a south wind blows at daybreak.
2. On the 6th day before the Nones of October: both the air and the Pleiades are similar.
3. On the 5th day before the Nones: Auriga sets, and it thunders from the north.
4. On the 4th day before the Nones: the Haedi rise and it rains.
5. On the 3rd day before the Nones: Corona rises and a northerly change.
6. On the day before the Nones: the middle of Aries sets, and Scorpio with it.
7. On the Nones of October: similar.
8. On the 8th day before the Ides of October: Corona, rising with the Haedi, changes the air.
9. On the 7th day before the Ides: the Haedi rise with the Pleiades, and the wind blows from the south-west.

[151] Compare 10 Aug. These two entries are not possible for the same year, and I have been unable to find any two years within reasonably close succession for which there are lunar eclipses on these two dates between 50 BC and AD 560.

10. On the 6th day before the Ides: Libra begins to rise, and the west wind blows.
11. On the 5th day before the Ides: Corona rising at daybreak shifts the winds about.
12. On the 4th day before the Ides: the Pleiades rise, and the south wind blows.
13. On the 3rd day before the Ides: Corona rising shifts the winds about, and a disturbance of the sea develops.
14. On the day before the Ides: a diffuse north wind blows.
15. On the Ides of October: the middle of autumn and a south wind.
16. On the 17th day before the Kalends of November: Orion rises, and the air is dewy.
17. On the 16th day before the Kalends: similar.
18. On the 15th day before the Kalends: the whole day is cloudy.
19. On the 14th day before the Kalends: the sun is in Scorpio, and the west wind blows.
20. On the 13th day before the Kalends: the Pleiades set, and a change in the air.
21. On the 12th day before the Kalends: similar, but it also rains.
22. On the 11th day before the Kalends: the tail of Taurus [. . .], and a rainy south wind blows.
23. On the 10th day before the Kalends: Scorpio sets, and the north wind blows, and stormy at sea.
24. On the 9th day before the Kalends: the Pleiades set.
25. On the 8th day before the Kalends: Centaurus sets at daybreak.
26. On the 7th day before the Kalends: the forehead of Scorpio sets.
27. On the 6th day before the Kalends: the Hyades set, and a cold north wind, and stormy at sea.
28. On the 5th day before the Kalends: the Pleiades and Orion completely set.
29. On the 4th day before the Kalends: Arcturus sets, and the winds are more fierce.
30. On the 3rd day before the Kalends: Cassiopeia begins to set.
31. On the day before the Kalends: the whole of Orion disappears, as does Aquila in the evening, and Lyra rises.

November

1. On the Kalends of November: the Pleiades set. Frost in the morning, and a change in the air for the colder from the setting of Arcturus.

2. On the 4th day before the Nones of November: cold winds and showers.
3. On the 3rd day before the Nones: Lyra rises at daybreak, and at first a north wind, then later a south wind.
4. On the day before the Nones: a south wind with a west wind and rain.
5. On the Nones of November: Lyra appears at sunrise, and the wind is northerly.
6. On the 8th day before the Ides of November: Arcturus sets at daybreak, and cloudy.
7. On the 7th day before the Ides: the Pleiades and Orion set, and the north wind blows.
8. On the 6th day before the Ides: gloomy sky.
9. On the 5th day before the Ides: the bright star in Scorpio [. . .],[152] and a stormy change.[153]
10. On the 4th day before the Ides: winter begins.
11. On the 3rd day before the Ides: the Pleiades completely disappear.
12. On the day before the Ides: the middle star of Scorpio rises.
13. On the Ides of November: the Pleiades and Orion set at daybreak.
14. On the 18th day before the Kalends of December: Scorpio sets at daybreak.
15. On the 17th day before the Kalends: Lyra rises in the morning, and the east wind blows at the same time as the south and the north winds.
16. On the 16th day before the Kalends: the same.
17. On the 15th day before the Kalends: it is stormy, and the south wind follows.
18. On the 14th day before the Kalends: the sun is in Sagittarius. Orion rises with Lyra, and a disturbance in the air.
19. On the 13th day before the Kalends: the horn of Taurus sets with the sun, and the north wind blows.
20. On the 12th day before the Kalends: unpleasant storm.[154]
21. On the 11th day before the Kalends: the Hyades set at daybreak with the hare.
22. On the 10th day before the Kalends: cold rain.
23. On the 9th day before the Kalends: Orion and the horns of Taurus set.
24. On the 8th day before the Kalends: prelude to the winter solstice. Sirius sets, and brings cold dew.

---

[152] λαμπρὸν ἄστρον τοῦ Σκορπίου. Columella V Id. Nov. has *stella clara Scorpionis exoritur*.

[153] χειμερινὴ τροπή, which can also mean 'winter solstice'.

[154] The imagery here is more vivid than is usual in a parapegma.

25. On the 7th day before the Kalends: the sun is in the first degree of Sagittarius.[155]
26. On the 6th day before the Kalends: similar.
27. On the 5th day before the Kalends: Sirius sets, and there are southerly showers.
28. On the 4th day before the Kalends: Sirius begins to set,[156] and the air is cloudy.
29. On the 3rd day before the Kalends: Sirius sets at daybreak, and it rains from the south-west.
30. On the day before the Kalends: Orion sets, and a west wind, and rain with a south wind.

December

1. On the Kalends of December: confusion of the air, the arctic wind blows gently, and the whole of Orion sets at daybreak.
2. On the 4th day before the Nones of December: Sirius sets. In the evening the arctic wind increases.
3. On the 3rd day before the Nones: it is stormy with rain.
4. On the day before the Nones: Sagittarius sets, and the north wind blows.
5. On the Nones of December: showers all day, and a strong north wind.
6. On the 8th day before the Ides of December: the middle of Sagittarius rises.
7. On the 7th day before the Ides: Aquila rises, and the south-west wind blows.
8. On the 6th day before the Ides: the whole of Scorpio rises.
9. On the 5th day before the Ides: Sirius rises in the evening, and the south wind blows.
10. On the 4th day before the Ides: much north wind, and the sky is black from storm.[157]
11. On the 3rd day before the Ides: hidden stars, and an arctic wind at the same time as a north-east wind.
12. On the day before the Ides: first a north wind, later an oppressive south wind.
13. On the Ides of December: the whole of Scorpio rises. A north wind and a rainy south wind following.

---

[155] Notice the strange seven-day gap between this and 'the sun is in Sagittarius' at 18 Nov.
[156] This disagrees with the previous day's entry, as well as 24 Nov.
[157] Perhaps the imagery here and on the 12th seems immediate rather than predictive?

14. On the 19th day before the Kalends of January: Capella sets.
15. On the 18th day before the Kalends: south wind at the same time as a north wind, and a confusion of the air.
16. On the 17th day before the Kalends: similar.
17. On the 16th day before the Kalends: the sun is in Capricorn.
18. On the 15th day before the Kalends: hidden stars, and contrary winds.
19. On the 14th day before the Kalends: Capella rises.
20. On the 13th day before the Kalends: Aquila rises, as does Capricorn.
21. On the 12th day before the Kalends: the north wind begins, but the south wind overtakes it all day.
22. On the 11th day before the Kalends: Aquila rises in the evening.
23. On the 10th day before the Kalends: Capella sets at daybreak, and the *bruma* is reached, which is the winter solstice.[158]
24. On the 9th day before the Kalends: the air is stormy.
25. On the 8th day before the Kalends: Capella sets at daybreak.
26. On the 7th day before the Kalends: Capella sets in the evening, and it is stormy.
27. On the 6th day before the Kalends: Delphinus rises in the morning, restraining the air.
28. On the 5th day before the Kalends: the sun turns back from its southernmost point.
29. On the 4th day before the Kalends: the sun is bright, and Aquila sets, and it rains.
30. On the 3rd day before the Kalends: Delphinus rises, and Sirius sets.
31. On the day before the Kalends of January: hidden disturbance, and windy. Delphinus begins to rise, and the sun is higher.

A.xii. Madrid parapegma[159]

μηνολόγιον ἀστρολογικόν.
μηνὸς Μαρτίου

δ.   ἡμέρα δύσκολος.
ἐν ε΄ Ἀρκτοῦρος ἑσπέρας ἀνατέλλει.
ς΄. χελιδόνες παίζουσι.
θ.   βορρᾶς ἀναπνεῖ.

---

[158] συμπληροῦται ἡ βροῦμα, οἱονεὶ ἡ χειμερινὴ τροπή. *Bruma* is just Latin for winter solstice.
[159] Greek text following Bianchi, 1914, p. 49f.

ιβ΄.   πελαργοὶ φαίνονται καὶ Ὄρνιθες ἑσπέριοι δύνουσι.[160]

ιβ΄.   βορρᾶς πνεῖ.

ιέ.   Στάχυς δύνει.

ιη΄.   Στέφανος ἑσπέρας ἀνατέλλει.

ιθ΄.   ὁ λεγόμενος Ἵππος ἑῷος ἀνατέλλει· καὶ νότος πνεῖ.

κ΄.   ἡμέρα δύσκολος· καὶ βορρᾶς στεγνός.

κά.   ἰσημερία ἐαρινή.

κγ΄.   Στάχυς ἑσπέρας ἀνατέλλει· καὶ γίνεται ταραχὴ τοῦ ἀέρος πρὸ ἡμέρας
       καὶ μεθ᾽ ἡμέραν.

κδ΄.   χειμὼν ἢ ἄνεμος.

κέ.   Ἀετὸς φαίνεται· καὶ πολλάκις ὑετός.

κϛ΄.   [. . .] ἑσπέρας δύνει.

κη΄.   χειμάζει.

λ΄.   ἄστατος βορρᾶς πνεῖ.

μηνὸς Ἀπριλλίου

α΄.   Πλειάδες ἀκρόνυχοι φαίνονται.

β΄.   ἄνεμος ἢ βροχή.

γ΄.   ἡμέρα δύσκολος.

δ΄.   νότος πνεῖ.

ϛ΄.   ζέφυρος.

η΄.   ἄστατος ἡμέρα.

ιθ΄.   Πλειάδες ἑσπέριαι κρύπτονται.

κ΄.   δύσκολος·[161] καὶ νότος πνεῖ.

κέ.   ὑετὸς ἢ ζέφυρος.

κϛ΄.   παχνίτης πνεῖ.

κϛ΄.[162]   Ὠρίων ἑσπέριος κρύπτεται.

λ΄.   ἄνεμος ἢ ὑετός.

μηνὸς Μαίου

α΄.   Ὑάδες ἅμα ἡλίου ἀνατολῇ ἐπιτέλλουσιν.

γ΄.   Κύων ἑσπέριος κρύπτεται.

---

[160] Cygnus is otherwise always in the singular. Bianchi assumes the entry is really about birds
rather than stars, and that a verb has dropped out (birds *[appear]* or some such thing), as has
the name of a star as a subject for δύνουσι. Three things incline me against Bianchi's reading:
(1) the repetition of an identical type of supposed omission at 16 May; (2) the agreement of
subject and verb in both places, and (3) the fact that comparison with Ptolemy shows that
Cygnus can be expected to be listed at about this time of year as setting in the evening, and
rising in the evening at about the time of the next entry for ὄρνιθες in this text, 16 May.

[161] Here and at 15 Aug. and 21 Feb., Bianchi restores <ἡμέρα>, which is very plausible.

[162] Bianchi emends the second κϛ΄ to κζ΄.

δ.   Λύρα ἑσπέριος ἐπιτέλλει· καὶ σφόδρα ἀλλοιοῦται ὁ ἀήρ.

ζ.   Πλείαδες ἑῷοι φαίνονται· καὶ ἄρχεται ὁ ἀὴρ καθίστασθαι.

η΄.   Αἲξ ἑσπέρας δύνει· καὶ ἔσονται βρονταί.

ιβ΄.   Πλείαδες ἐπιτέλλουσι.

ιδ΄.   ζέφυρος πνεῖ.

ιϛ΄.   Ὄρνιθες ἑσπέριαι ἀνατέλλουσι·[163] καὶ λὶψ πνεῖ.

ιθ΄.   Ὑάδες ἑῷοι φαίνονται· καὶ Κύων ἑσπέριος δύνει.

κ΄.   παχνίτης μέγας· καὶ δύσκολος ἡμέρα.

κβ΄.   νότος πνεῖ.

κγ΄.   Ἀετὸς ἀνατέλλει καὶ ταραχὴ τοῦ ἀέρος γίνεται πρὸ β΄ ἡμερῶν καὶ μεθ'
       ἡμέρας.

κδ΄.   Αἲξ ἑσπέριος κρύπτεται καὶ νότος ἀὴρ πνεῖ· καὶ οἱ παχνίται
       τελοῦνται.

κϛ΄.   Ὑάδες ἀνατέλλουσι.

κη΄.   ὁ τῶν Διδύμων α΄ ἀνατέλλει· ἄνεμος μέγας καὶ βρονταί.

λ΄.   Προκύων ἑσπέριος κρύπτεται καὶ νότος ἔσται.

μηνὸς Ἰουνίου

β΄.   ἀστὴρ ἑσπέριος ἀνατέλλει.

γ΄.   ἡμέρα δύσκολος· καὶ ὁ α΄ τῶν Διδύμων κρύπτεται· ἄνεμοι καὶ
       βρονταί.

ϛ΄.   Ἀρκτοῦρος δύνει.

θ΄.   Δελφὶς ἐπιτέλλει.

ιδ΄.   Αἲξ ἑσπερία δύνει.

ιϛ΄.   Κύων ἑσπέριος δύνει.

ιη΄.   ἡμέρα δύσκολος.

ιθ΄.   Ἀετὸς ἐπιτέλλει.

κα΄.   τροπὴ θερινή.

κε΄.   Ὠρίων ἑῷος ἐπιτέλλει· καὶ ἀλλοιοῦται ὁ ἀὴρ πρὸ ἡμερῶν.

λ΄.   Ἀριάδνης Στέφανος δύνει· καὶ βορρᾶς πνεῖ.

μηνὸς Ἰουλλίου

γ΄.   ἀρχὴ πρόδρομος ἀνέμων.

δ΄.   Ὠρίων ὅλος ἐπιτέλλει.

ϛ΄.   ἡμέρα δύσκολος.

ι΄.   ἐτήσιοι ἄρχονται πνεῖν.

---

[163] See note on 12 March, above.

ιε΄.   Ὠρίων ἐκφαίνεται.

ιθ΄.   Κύων ἐπιτέλλει.

κ΄.   ἡμέρα δύσκολος.

κε΄.   Λύρα δύνει· ἄνεμος ἔσται πρὸ ἡμερῶν, καὶ Ἀετὸς ἑῷος δύνει.

λ΄.   ὁ ἐπὶ τῆς οὐρᾶς τοῦ Λέοντος ἐπιτέλλει· καὶ βορρᾶς πνεῖ.

μηνὸς Αὐγούστου

α΄.   ὁ λαμπρὸς τοῦ Ὑδροχόου δύνει· σφοδρὸς ἄνεμος πρὸ ἡμερῶν καὶ μεθ᾽
       ἡμέρας.

δ΄.   ἡμέρα δύσκολος.

ζ΄.   ὁ ἐπὶ τῆς καρδίας τοῦ Λέοντος ἐπιτέλλει.

ιγ΄.   Δελφὶς ἐπιτέλλει· ὁμιχλώδης ὁ ἀήρ.

ιε΄.   δύσκολος· καὶ Λύρα ἑῷος δύνει.

ιθ΄.   Δελφὶς ἑῷος δύνει· καὶ τροπαὶ χειμεριναί.

κε΄.   ὁ λαμπρὸς τῆς Λύρας δύνει· ἄνεμος καὶ βροχὴ πρὸ ἡμερῶν καὶ μεθ᾽
       ἡμέρας.

κη΄.   Προτρυγητὴρ ἑῷος ἀνατέλλει· καὶ Ὀιστὸς δύνει· καὶ πληροῦνται αἱ
       μ΄ ἡμέραι ἀπὸ τῆς τοῦ Κυνὸς ἐπιτολῆς.

μηνὸς Σεπτεμβρίου

β΄.   Ἄνδρομος ἑσπέρας ἀνατέλλει· εὐκαιρία ἔσται.

γ΄.   ἡμέρα δύσκολος.

ς΄.   Αἲξ ἑσπερία ἀνατέλλει.

ιε΄.   Στάχυς δύνει· ἄνεμος ἢ βροχὴ τῇ ἑξῆς.

ιθ΄.   Περσεὺς ἄρχεται φαίνεσθαι· καὶ Στάχυς ἑῷος ἐπιτέλλει· καὶ ἀλλοιοῦται
       ὁ ἀὴρ πρὸ β΄ ἡμερῶν.

κς΄.   ἡμέρα δύσκολος· καὶ ἰσημερία φθινοπωρινή· καὶ γίνεται μεγίστη
       ταραχὴ τοῦ ἀέρος πρὸ β΄ ἡμερῶν ἢ μετὰ β΄ ἡμέρας.

κθ΄.   Ὑάδες ἅμα ἡλίῳ ἀνατέλλουσι· καὶ νότος πνεῖ.

μηνὸς Ὀκτωβρίου

β΄.   Ὑάδες ἀκρόνυχοι φαίνονται καὶ βορρᾶς πνεῖ· ὁ δὲ Στέφανος[164]
       ἀνατέλλει· καὶ γίνεται σφοδρὰ μεταβολὴ τοῦ ἀέρος.

ζ΄.   Ἔριφοι ἑσπέριοι ἀνατέλλουσι· καὶ γίνεται ταραχὴ τοῦ ἀέρος.

η΄.   ὁ τοῦ Στέφανου ἀστὴρ ἐπιτέλλει.

---

[164] Στέφανος, following Bianchi. The MS has πρόφανος.

ι΄.   Ταύρου οὐρὰ δύνει· καὶ ταραχὴ ἔσται τοῦ ἀέρος πρὸ ἡμέρας καὶ μεθ’ ἡμέραν.

ιδ΄. χειμάζει.

ιέ.   Ὠρίων δύνει· ἄνεμος ἢ βροχὴ πρὸ ἡμέρας ἢ μεθ’ ἡμέρας.

ιζ΄. Ὑάδες ἑσπέριοι ἐπιτέλλουσι· καὶ γίνεται ταραχὴ τοῦ ἀέρος.

ιη΄. χειμάζει· νότος.

κ΄.   ἀληθῶς ὁ ἐπὶ τοῦ Ἡνίοχου[165] ἀνατέλλει· σφόδρα ἄνεμος ἢ βροχὴ πρὸ ἡμερῶν ἢ μεθ’ ἡμέρας.

κα΄. ἡμέρα δύσκολος.

κδ΄. μέλας ἀνατέλλει.

κϛ΄. νότος πνεῖ.

κζ΄. ὁ Σκορπίος ἄρχεται ἀνατέλλειν· ταραχὴ τοῦ ἀέρος.

λ΄.   Ὑάδες ἄρχονται.

μηνὸς Νοεμβρίου

α΄.   Λύρα δύνει· νότος πνεῖ.

έ.   ἡμέρα δύσκολος.

ϛ΄.   Πλειάδες ἑῷαι δύνουσι· καὶ Λύρα ἐπιτέλλει.

ζ΄.   Σκορπίος κρύπτεται· καὶ βορρᾶς πνεῖ.

ια΄. ἡμέρα δύσκολος.

ιβ΄. ἄνεμος ἢ βροχή.

ιγ΄. Πλειάδες δύνουσι· καὶ γίνεται ταραχὴ τοῦ ἀέρος πρὸ ἡμερῶν καὶ μεθ’ ἡμέρας.

ιη΄. χειμάζει.

κ΄.   Ἡνίοχος[166] δύνει· νότος πνεῖ.

κα΄. Ὑάδες ἑῷοι δύνουσι· καὶ γίνεται ταραχὴ τοῦ ἀέρος τῇ αὐτῇ καὶ τῇ ἑξῆς.

κβ΄. χειμάζει.

κγ΄. Πλειάδων δύσις.

κϛ΄. Κύων δύνει.

κζ΄. Ὠρίων ἑσπέρας δύνει· καὶ Στέφανος ἐπιτέλλει.

μηνὸς Δεκεμβρίου

α΄.   Ὑάδες δύνουσιν· ἄνεμος ἢ βροχή.

γ΄.   ἡμέρα δύσκολος.

---

[165] ἀληθῶς ὁ ἐπὶ τοῦ Ἡνίοχου, following Bianchi. The MS has ἀληθὸς ὁ ἐπὶ τοῦ Ἰστιόχου.
[166] Following Bianchi. The MS has Ἔνιος.

ι΄.   Ἔριφοι ἑῷοι δύνουσιν.

ιε΄.   Αἲξ ἑσπερία δύνει· ἄνεμος σφοδρός.

ιδ΄.   Αἰγόκερως ἄρχεται ἀνατέλλειν· χειμὼν μέγας.

ιε΄.   ὁ λαμπρὸς τῶν Ὑάδων δύνει· ἄνεμος σφοδρός.

ιη΄.   χειμάζει.

κα΄.   ταραχὴ τοῦ ἀέρος.

κδ΄.   ἡμέρα δύσκολος.

κϛ΄.   βορρᾶς.

κθ΄.   ὁ ἐπὶ τῆς λαβίδος τοῦ Ὠρίωνος δύνει· χειμὼν ἔσται.

μηνὸς Ἰαννουαρίου

β΄.   ἡμέρα δύσκολος.

ϛ΄.   ζέφυρος ἀρκτώδης.

η΄.   εὔρου ἀρχή.

θ΄.   Ἀετὸς ἑσπέρας δύνει· ταραχὴ γίνεται πρὸ ἡμέρας ἢ μεθ' ἡμέραν.

ιβ΄.   ἀπηλιώτης πνεῖ.

ιδ΄.   ἡμέρα δύσκολος.

ιζ΄.   Παρθένος δύνει· καὶ γίνεται ταραχὴ τοῦ ἀέρος πρὸ ἡμερῶν ἢ μεθ' ἡμέρας.

κβ΄.   ἑσπέρας δύνει Τοξότης· ταραχὴ ἱκανὴ τοῦ ἀέρος.

κδ΄.   Ἀρκτοῦρος ἀνατέλλει.

κϛ΄.   χελιδὼν φαίνεται· καὶ ὁ λαμπρὸς ἀστὴρ ἐν τῷ Λέοντι δύνει· κινεῖται ὁ ἀὴρ πρὸ δύο ἡμερῶν.

κη΄.   Δελφὶς ἑσπέρας δύνει· καὶ χειμάζει.

μηνὸς **Φευρουαρίου**

α΄.   Λύρα δύνει· εὖρος πνεῖ.

β΄.   θολώδης ὁ ἀήρ.

γ΄.   Λέων σὺν τῇ Λύρᾳ δύνει.

δ΄.   ἀπαρκτίας πνεῖ.

θ΄.   βορρᾶς πνεῖ.

ι΄.   ἄστρον κρυπτὸν ἀνίσχει καὶ ὁ Δίδυμος.

ιβ΄.   ἀπηλιώτης πνεῖ.

ιγ΄.   ὁ Τοξότης δύνει.

κ΄.   ὁ Δίδυμος δύνει.

κα΄.   δύσκολος· καὶ οἱ λεγόμενοι χελιδόνιοι πνέουσι· καὶ φαίνονται χελιδόνες.

κγ΄.   τὰ λεγόμενα ἀλκυόνια.

κδ´. Ὑδροχόος ἀνίσχει.

κε´. Ἀρκτοῦρος ἑσπέρας ἀνατέλλει· καὶ γίνεται ταραχὴ ἱκανὴ τοῦ ἀέρος πρὸ ἡμερῶν τριῶν.

κϛ´. ἡμέρα δύσκολος· καὶ ἀπαρκτίας πνεῖ.[167]

κζ´. Ὀϊστὸς δύνει.

κη´. ζέφυρος πνεῖ.

Astronomical Menology
Month of March

4. Unpleasant day.
   On the 5th. Arcturus rises in the evening.
6. The swallows are playful.
9. The north wind blows.
12. Storks appear, and Cygnus sets in the evening.
12. The north wind blows.[168]
15. Spica sets.
18. Corona rises in the evening.
19. The star called Pegasus rises in the morning, and the south wind blows.
20. Unpleasant day, and a costive north wind.
21. Vernal equinox.
23. Spica rises in the evening, and there is a disturbance of the air for days before and the day after.[169]
24. Storm or wind.
25. Aquila rises, and a lot of rain.
26. [. . .] sets in the evening.
28. It is stormy.
30. An unstable north wind blows.

Month of April

1. The Pleiades appear acronychally.
2. Wind or showers.

---

[167] Bianchi prints Ἀρκτοῦρος πνεῖ without comment.
[168] Notice the repeated date.
[169] Bianchi thinks a number has dropped out before 'days' here and repeatedly in this parapegma, but I am not so sure a specific number is wanting. Compare also 23 May, 25 June, 25 July, 1 Aug., 25 Aug., 20 Oct., 13 Nov., 18 Jan.

3. Unpleasant day.
4. The south wind blows.
6. West wind.
8. Unstable day.
19. The Pleiades disappear in the evening.
20. Unpleasant, and the south wind blows.
25. Rain or a west wind.
26. Frosty (wind) blows.[170]
26. Orion disappears in the evening.[171]
30. Wind or rain.

Month of May

1. The Hyades rise at the same time as the rising of the sun.
3. Sirius disappears in the evening.
4. Lyra rises in the evening, and the air is much changed.
7. The Pleiades appear in the morning, and the air begins to be settled.
8. Capella sets in the evening, and it will be thundery.[172]
12. The Pleiades rise.
14. The west wind blows.
16. Cygnus rises in the evening,[173] and the south-west wind blows.
19. The Hyades appear in the morning, and Sirius sets in the evening.
20. A great frost-season, and an unpleasant day.
22. The south wind blows.
23. Aquila rises, and there is a disturbance of the air for two days before and after.
24. Capella disappears in the evening, and south air blows, and the frost-seasons end.
26. The Hyades rise.
28. The first of the twins rises, a strong wind and thundery.
30. Procyon disappears in the evening, and there will be a south wind.

---

[170] παχνίτης πνεῖ. The noun is rare. See note 124, above.   [171] Again, a repeated date.
[172] See note 136, above.   [173] See my note on 12 March.

Month of June

2. The star rises in the evening.[174]
3. Unpleasant day, and the first of the twins disappears, windy and thundery.
6. Arcturus sets.
9. Delphinus rises.
14. Capella sets in the evening.
16. Sirius sets in the evening.
18. Unpleasant day.
19. Aquila rises.
21. Summer solstice.
25. Orion rises in the morning, and the air is changed for days.
30 The crown of Ariadne sets,[175] and the north wind blows.

Month of July

3. Beginning of the *prodromoi* winds.
4. The whole of Orion rises.
6. Unpleasant day.
10. The Etesian winds begin to blow.
15. Orion appears.
19. Sirius rises.
20. Unpleasant day.
25. Lyra sets, there will be wind for days before, and Aquila sets in the morning.
30. The star on the tail of Leo rises, and the north wind blows.

Month of August

1. The bright star in Aquarius sets, strong wind for days before and days after.
4. Unpleasant day.
7. The star in the heart of Leo rises.
13. Delphinus rises, the air is misty.
15. Unpleasant, and Lyra sets in the morning.

---

[174] It seems that the specific star name has dropped out here. Bianchi proposes Aquila from comparison with the Quintilius parapegma. Compare also Oxford and Aëtius 2 Jun.

[175] Corona. See Bianchi, 1914, p. 52.

19. Delphinus sets in the morning, and stormy changes.
25. The bright star in Lyra sets, wind and showers for days before and days after.
28. Vindemiatrix rises in the morning, and Sagitta sets, and this completes the forty days after the rising of Sirius.

Month of September

2. Andromeda[176] rises in the evening: it will be favourable.[177]
3. Unpleasant day.
6. Capella rises in the evening.
15. Spica sets, wind or showers the day after.
19. Perseus begins to appear, and Spica rises in the morning, and the air is changed for two days before.
26. Unpleasant day, and autumnal equinox, and there is the greatest change in the air two days before or two days after.
29. The Hyades rise at the same time as the sun, and the south wind blows.

Month of October

2. The Hyades appear acronychally, and the north wind blows. Corona rises, and there is a great change in the air.
7. The Haedi rise in the evening, and there is a disturbance of the air.
8. The star of Corona rises.
10. The tail of Taurus sets and there will be a disturbance of the air for days before and the day after.
14. It is stormy.
15. Orion sets; wind or showers for days before or days after.
17. The Hyades rise in the evening, and there is a disturbance of the air.
18. It is stormy; south wind.
20. The star on Auriga rises truly; strong wind or showers for days before or days after.
21. Unpleasant day.
24. The black (star?)[178] rises.

---

[176] Ἄνδρομος.   [177] εὐκαιρία ἔσται, a unique entry in a parapegma.
[178] μέλας ἀνατέλλει. Bianchi proposes reading this as a paraphrase of Clodius Tuscus' common 'hidden stars' entry, but this still does not solve the puzzle of what the referent could be.

26. The south wind blows.
27. Scorpio begins to rise, a disturbance of the air.
30. The Hyades begin.

Month of **November**

1. Lyra sets, the south wind blows.
5. Unpleasant day.
6. The Pleiades set in the morning, and Lyra rises.
7. Scorpio disappears, and the north wind blows.
11. Unpleasant day.
12. Wind or showers.
13. The Pleiades set, and there is a disturbance of the air for days before and days after.
18. It is stormy.
20. Auriga sets; the south wind blows.
21. The Hyades set in the morning, and there is a disturbance of the air on the same day and the following.
22. It is stormy.
23. Setting of the Pleiades.
26. Sirius sets.
27. Orion sets in the evening, and Corona rises.

Month of **December**

1. The Hyades set; wind or showers.
3. Unpleasant day.
10. The Haedi set in the morning.
15. Capella sets in the evening; a strong wind.[179]
14. Capricorn begins to rise; a great storm.
15. The bright star in the Hyades sets; a strong wind.
18. It is stormy.
21. A disturbance of the air.
24. Unpleasant day.
26. North wind.
29. The star on the handle of Orion sets; it will be stormy.

---

[179] The string of dates here, 10th, 15th, 14th, 15th, 18th, is obviously corrupt.

Month of January

2. Unpleasant day.
6. Northerly west wind.[180]
8. Beginning of the east *(eurus)* wind.
9. Aquila sets in the evening; there is a disturbance for days before or the day after.
12. The east *(apeliotes)* wind blows.
14. Unpleasant day.
17. Virgo sets, and there is a disturbance of the air for days before or days after.
22. Sagittarius sets in the evening, a sufficient disturbance of the air.
24. Arcturus rises.
26. The swallow appears, and the bright star in Leo sets; the air moves for two days before.
28. Delphinus sets in the evening, and it is stormy.

Month of February

1. Lyra sets; the east wind *(eurus)* blows.
2. The air is blustery.
3. Leo sets with Lyra.
4. The arctic wind blows.
9. The north wind blows.
10. The stars are hidden and the twin rises.[181]
12. The east wind *(apeliotes)* blows.
13. Sagittarius sets.
20. The twin sets.
21. Unpleasant, and the so-called swallow-bringing winds blow, and swallows appear.
23. The so-called Halcyon days.
24. Aquarius rises.
25. Arcturus rises in the evening, and there is a sufficient disturbance of the air for three days before.
26. Unpleasant day, and an arctic wind blows.
27. Sagitta sets.
28. The west wind blows.

---

[180] ζέφυρος ἀρκτώδης.   [181] Or: the hidden star rises, as does the twin.

A.xiii. Lydus, *De mensibus*[182]

Extracts related to parapegmata:

Ἰανουάριος

(iv.8) ἰστέον δὲ κατὰ τὴν ἡμέραν τῶν καλενδῶν τὸν ἥλιον ἐφ᾽ ὕψους γίνεσθαι, τὸν δὲ Στέφανον δύεσθαι ὄρθρου.

(iv.16)[183] τῇ πρὸ δέκα ὀκτὼ καλενδῶν Φεβρουαρίων ἀνεμομαχίαν ὁ Βάρρων λέγει γίνεσθαι, Δημόκριτος δὲ τὸν λίβα μετὰ ὄμβρου φησὶ γίνεσθαι.

(iv.17) ἰστέον δὲ κατὰ ταύτην τὴν ἡμέραν τοὺς Ἐρίφους δύεσθαι καὶ τροπὴν γίνεσθαι κατὰ Φίλιππον.

(iv.18) πρὸ δεκαπέντε καλενδῶν Φεβρουαρίων ὁ Δημόκριτος λέγει δύεσθαι τὸν Δελφῖνα καὶ τροπὴν ὡς ἐπὶ πολὺ γίνεσθαι. πρὸ δεκατριῶν καλενδῶν Φεβρουαρίων ὁ μὲν Εὐκτήμων τὸν Καρκίνον δύεσθαι, ὁ δὲ Κάλλιππος τὸν Ὑδροχόον ἀνίσχειν λέγει, ὃν Δευκαλίωνα Ἵππαρχος καλεῖ [. . .] βήσσαις καὶ χαράδραις τῶν ὀρῶν. πρὸ δεκαδύο καλενδῶν Φεβρουαρίων Εὔδοξος τὸν Ὑδροχόον ἀνίσχειν λέγει. πρὸ δεκαμιᾶς καλενδῶν Φεβρουαρίων τὸν ἥλιον ἐν Ὑδροχόῳ γενέσθαι ὁ Καῖσαρ λέγει, ὁ δὲ Εὔδοξος ἀνίσχειν αὐτὸν καὶ βροχὰς σημαίνειν· τῇ δὲ πρὸ δέκα καλενδῶν Φεβρουαρίων ὁ Δημόκριτος ἄνεμον λίβα πνεῦσαι λέγει.

Μάρτιος

(iv.44) τῇ πρὸ τεσσάρων νωνῶν Μαρτίων ἄνεμον βιαιότερον ὡς ἐπίπαν πνεῖν προλέγει ὁ Εὔδοξος.

(iv.48) νώναις Μαρτίαις ὁ Βάρρων ὄρθρου τὸν Στέφανον δύεσθαι λέγει καὶ πνεῖν τὸν βορρᾶν.

(iv.49) (εἰδοῖς Μαρτίαις) . . . ταύτην τὴν ἡμέραν ὁ Μητρόδωρος κακὴν παραδίδωσιν.

---

[182] Greek text following Wünsch, 1898. Astrometeorological parts excerpted in Wachsmuth, 1897, pp. 295–9.

[183] I will insert book and chapter numbers when the excerpts pick up after skipping some of Lydus' text. Chapter numbers follow Wünsch's 1898 edition of the *De mensibus*, not the extracts at the end of Wachsmuth's 1897 edition of the *De ostentiis*.

(IV.50)   ἡ πρὸ δεκαεπτὰ καλενδῶν Ἀπριλίων ἀπάρκτιος.[184] ἐν ταύτῃ τῇ
ἡμέρᾳ Εὔδοξος τοὺς Ἰχθύας ἀνίσχειν καὶ βορρᾶν πνεῖν παραδίδωσιν.

(IV.51)   (πρὸ ἑκκαίδεκα καλενδῶν Ἀπριλίων) . . . ἐν δὲ τῇ ἡμέρᾳ τῶν Βακ-
χαναλίων Δημόκριτος δύεσθαι τοὺς Ἰχθύας λέγει, ὁ δὲ Βάρρων
ἀνεμομαχίαν ἔσεσθαι παραδίδωσιν.

(IV.54)   πρὸ δεκαπέντε καλενδῶν Ἀπριλίων ὁ Εὐκτήμων ποικίλους ἀνέμους πνέειν λέγει.

(IV.61)   ὁ Φίλιππος δύεσθαι τῇ πρὸ ἐννέα καλενδῶν Ἀπριλίων τὰς Ὑάδας
μετὰ νότου, ἀνίσχειν δὲ λέγει ὁ Μητρόδωρος. τῇ πρὸ ὀκτὼ
καλενδῶν ἰσημερία ἐαρινή.

Σεπτέμβριος

(IV.123)   τῇ νεομηνίᾳ ὁ Μητρόδωρος λέγει τὴν Ἀνδρομέδαν ἀνίσχειν καὶ
τῶν ἄλλων παυομένων ἀνέμων τὸν εὖρον ἐπικρατεῖν.

(IV.124)   (τῇ πρὸ τεσσάρων νονῶν Σεπτεμβρίων) . . . ἐν ταύτῃ τῇ ἡμέρᾳ
ὁ Δημόκριτος λέγει ἐναλλαγὴν ἀνέμων συμβαίνειν καὶ βροχῆς
ἐπικράτειαν.

(IV.126)   τῇ πρὸ ὀκτὼ εἰδῶν Σεπτεμβρίων Εὔδοξος τὸν Ἵππον δύεσθαι
καὶ ζέφυρον ἢ ἀργέστην πνεῖν σημειοῦται.

(IV.128)   πρὸ δέκα ὀκτὼ καλενδῶν Ὀκτωβρίων ὁ Δοσίθεος τὸν Ἀρκτοῦρον
ἀνίσχειν σημειοῦται. τῇ πρὸ δεκαδύο καλενδῶν Ὀκτωβρίων ὁ
Καῖσαρ τὰς χελιδόνας ἐκδημεῖν λέγει.

Ὀκτώβριος

(IV.135)   καλένδαις Ὀκτωβρίαις φησὶν ὁ Βάρρων τὰς Πλειάδας ἀπὸ
ἀνατολῶν ἀνίσχειν.

(IV.136)   τῇ πρὸ ἓξ νονῶν Ὀκτωβρίων Εὔδοξος περὶ τὴν ἑσπέραν βροχὴν
ἔσεσθαι ὑπολαμβάνει.

(IV.139)   τῇ πρὸ μιᾶς νονῶν Ὀκτωβρίων ὁ Δημόκριτος τοὺς Ἐρίφους
ἀνίσχειν καὶ βορρᾶν πνεῖν διισχυρίζεται, ὁ δὲ Εὔδοξος δύεσθαι τὸ
μέσον τοῦ Κριοῦ λέγει. νόναις Ὀκτωβρίαις ὁ Βάρρων ἐν ἑσπέρᾳ
τὰς Πλειάδας ἀνίσχειν καὶ ζέφυρον πνεῖν, εἶτα καὶ λίβα προλέγει.

(IV.141)   τῇ πρὸ μιᾶς εἰδῶν Ὀκτωβρίων ὁ Εὐκτήμων τὸ μεσαίτατον
τοῦ φθινοπώρου εἶναι νομίζει. τῇ πρὸ δεκαπέντε καλενδῶν

---

[184] Wünsch, following the MSS reads ἄπρακτος.

Νοεμβρίων ἥλιος <ἐν> Σκορπίῳ γίνεται, ὡς Κάλλιππός φησι. τῇ πρὸ δεκατεσσάρων καλενδῶν Νοεμβρίων ὁ Μητρόδωρος τὰς Ὑάδας ἐν ἑσπέρᾳ ἀνίσχειν λέγει καὶ ἄνεμον βίαιον.

(ιν.143)   τῇ πρὸ μιᾶς καλενδῶν Νοεμβρίων ὁ Βάρρων τὴν Λύραν ἅμα ἡλίῳ ἀνίσχειν λέγει.

Νοέμβριος

(ιν.148)   τῇ πρὸ τεσ<σά>ρων καὶ τριῶν νωνῶν Νοεμβρίων ἐν τῷ ναῷ τῆς Ἴσιδος συμ<πέ>ρασμα τῶν ἑορτῶν· ἐπετελ<εῖτο> δὲ καὶ ὁ λεγόμενος Δρεπαν [...] <κα>θ' ἣν ἑορτὴν ὁ Μητρόδωρος νό<τον> φυσῆσαι λέγει.

(ιν.149)   (<τῇ πρὸ ...[185] εἰδῶν Νοεμ>βρίων) ... ἐκ ταύτης τῆς ἡμέρας <Εὔ>δοξος χειμῶνα λέγε<ι>.

(ιν.152)   τῇ πρὸ ἑπτὰ καλενδῶν Δεκεμβρίων ὁ Δημόκριτος λέγει τὸν ἥλιον <ἐν> Τοξότῃ γίνεσθαι.

Δεκέμβριος

(ιν.154)   (<καλένδαις Δεκεμβρίαις>)[186] ... κα<τὰ> ταύτην τὴν ἡμέραν δύεσθαι τὰς Ὑάδας καὶ τὸ λοιπὸν χ<ειμῶ>να ὁ Βάρρων λέγει.

(ιν.155)   κατὰ δὲ τὴν ἑξῆς ὁ Εὔδοξος τὸν Τοξότη<ν ἀνί>σχειν καὶ χειμῶνα προλέγει·

(ιν.156)   τῇ πρὸ τριῶν νωνῶν Δεκεμ<βρίων ἡμέρα> ἀπάρκτιος,[187] καθ' ἣν ὁ Εὐκ<τή>μων τὸν Κύνα δύεσθαι καὶ <τὸν χειμῶ>να ἐνάρ<χες>θαι λ<έγει>.

January

(ιν.8)   And one should know that on the day of the Kalends the sun is higher, and Corona sets at daybreak.

(ιν.16)[188]   On the eighteenth day before the Kalends of February, Varro says there are contrary winds, and Democritus says there is a southwest wind with a thunderstorm.

---

[185] Wünsch restores ἑπτά.

[186] Restored by Wachsmuth based on the entry for 'the sixth [day before the Nones]' on the following day.

[187] Again the MSS have ἄπρακτος.

[188] I will insert book and chapter numbers when the excerpts pick up after skipping some of Lydus' text. Chapter numbers follow Wünsch's 1898 edition of the *De mensibus*, not the extracts at the end of Wachsmuth's 1897 edition of the *De ostentiis*.

(IV.17)   And one should know that on this day the Haedi set and there is a change[189] according to Philippus.

(IV.18)   On the fifteenth day before the Kalends of February, Democritus says Delphinus sets and there is usually a change. On the thirteenth day before the Kalends of February, Euctemon says Cancer sets, and Callippus that Aquarius rises, which Hipparchus calls *Deucalion* in the glens and streams of the mountains.[190] On the twelfth day before the Kalends of February, Eudoxus says Aquarius rises. On the eleventh day before the Kalends of February, Caesar says the sun is in Aquarius, and Eudoxus that this rises and signifies showers, and on the tenth day before the Kalends of February, Democritus says the south-west wind blows.

March

(IV.44)   On the fourth day before the Nones of March, Eudoxus predicts that a more violent wind blows, for the most part.

(IV.48)   On the Nones of March, Varro says Corona sets at daybreak, and the north wind blows.

(IV.49)   (The Ides of March) . . . Metrodorus teaches that this is a bad day.

(IV.50)   On the seventeenth day before the Kalends of April, an arctic wind. On this day, Eudoxus teaches that Pisces rises and the north wind blows.

(IV.51)   (The sixteenth day before the Kalends of April) . . . On the day of the Bacchanalia, Democritus says Pisces sets, and Varro predicts that there will be contrary winds.[191]

(IV.54)   On the fifteenth day before the Kalends of April, Euctemon says variable winds blow.

(IV.61)   Philippus says the Hyades set, with a south wind, on the ninth day before the Kalends of April, but Metrodorus that they rise. On the eighth day before the Kalends of April, the vernal equinox.

---

[189] Here and elsewhere in this parapegma: τροπὴν γίνεσθαι.

[190] I have no idea what this means. Wünsch puts in a lacuna before 'in the glens . . .', which makes things only slightly less disconcerting.

[191] The future tense is unusual in parapegmata, occurring only in this one, the Madrid parapegma, and Clodius Tuscus. See the footnote to Clodius Tuscus, 23 Jan.

September

(IV.123)   On the first of the month, Metrodorus says Andromeda rises and the east wind is dominant, all the other winds ceasing.

(IV.124)   (The fourth day before the Nones of September) . . . On this day, Democritus says there is a shifting of winds, and a dominion of showers.

(IV.126)   On the eighth day before the Ides of September, Eudoxus indicates that Pegasus sets and a west or a north-west wind blows.

(IV.128)   On the eighteenth day before the Kalends of October, Dositheus indicates that Arcturus rises. On the twelfth day before the Kalends of October, Caesar says the swallows depart.

October

(IV.135)   On the Kalends of October, Varro says the Pleiades rise after sunrise.

(IV.136)   On the sixth day before the Nones of October, Eudoxus supposes there will be showers around evening time.

(IV.139)   On the day before the Nones of October, Democritus is sure that the Haedi rise and the north wind blows. Eudoxus says the middle of Aries sets. On the Nones of October, Varro predicts that the Pleiades rise in the evening, and the west wind blows, or else the south-west.

(IV.141)   On the day before the Ides of October, Euctemon thinks it is the very middle of autumn. On the fifteenth day before the Kalends of November, the sun is in Scorpio, as Callippus says. On the fourteenth day before the Kalends of November, Metrodorus says the Hyades rise in the evening, and a violent wind.

(IV.143)   On the day before the Kalends of November, Varro says Lyra rises at sunrrise.

November

(IV.148)   On the f[o]urth and the third day before the Ides of November, the con[cl]usion of the festivals of Isis in the *naos*. The so-called *Drepan[. . .]* was finis[hed] [. . . a]t which festival, Metrodorus says, the so[uth] wind will blow.

(IV.149)   (The [. . . day before the Ides of Nove]mber) . . . From this day, [Eu]doxus say[s], winter.

(IV.152)   On the seventh day before the Kalends of December, Democritus says the sun is in Sagittarius.

December

(IV.154)   ([The Kalends of December])[192] . . . On this day Varro says the Hyades set, and the remainder stormy.

(IV.155)   On the sixth [day before the Nones of December], Eudoxus predicts that Sagittariu[s ri]ses, and stormy.

(IV.156)   On the third [day] before the Nones of Dec[ember], arctic wind, from which Euc[te]mon s[ays] Sirius sets and [winte]r be[gi]ns.

A.xiv. Oxford parapegma[193]

ἀπλανῶν ἐπιτολαὶ καὶ δύσεις

Φεβρουάριος,[194] καθ᾽ Ἕλληνας Περίτιος, κατ᾽ Αἰγυπτίους Μεχίρ.

α.

γ.   ὁ λαμπρὸς τοῦ Ὄρνιθος δύνει· θαλάσσης ταραχὴ γίνεται.

ϛ.   ζέφυρος πνεῖ.

ιθ.   Ἠριδανὸς ὑπὸ κρύψιν γίνεται καὶ ποιεῖ βροχάς.

κβ.   Ὀιστὸς δύνει ἑσπέριος καὶ ἔστι ταραχὴ τοῦ ἀέρος.

κδ.   Κηφεὺς ἐπιτέλλει μόνος.

κε.   Ἀρκτοῦρος ἑσπέριος ἐπιτέλλει.

κϛ.   ὁ ἐν γόνασιν ἀστὴρ ἐπιτέλλει· καὶ γίνονται ἀνεμομαχίαι· καὶ χελιδόνες φαίνονται. κεῖται ἄστρῳ Ὑδροχόῳ· νὺξ ὡρῶν ιγ΄, ἡμέρα ια΄.

Τούτῳ τῷ μηνὶ σελήνης οὔσης ἐν Ὑδροχόῳ, ἐὰν γένωνται βρονταί, πολέμους ἰσχυροὺς εἰς τὴν γῆν σημαίνει καὶ σάλους καὶ νόσους περὶ τοὺς ἀνθρώπους, σίτων δὲ καὶ τῶν ἄλλων καρπῶν φθορὰν καὶ χωρῶν τινῶν ἀπώλειαν. κατὰ δὲ Εὔδοξον χειμῶνας πολλούς. καὶ τὰ ἐσπαρμένα οὐκ ἔσται καλά· καὶ ἑρπετῶν ἀπώλεια. ἐὰν δὲ σεισμὸς γένηται, ἀναίρεσιν σημαίνει.[195]

---

[192] Restored by Wachsmuth based on the entry for 'the sixth [day before the Nones]' on the following day.

[193] Published in *CCAG* IX.1, pp. 128–37.   [194] January is missing.

[195] This entire paragraph is excised by Weinstock (in the *CCAG* edition). It is true that the material here is more or less what we should expect in a brontologia rather than what we

Μάρτιος, καθ᾽ Ἕλληνας Δύστρος, κατ᾽ Αἰγυπτίους Φαμενώθ.

α.   Ἄρεως ἑορτή.[196]
θ.   Πλοιαφέσια.
ιβ′.   Ἀρκτοῦρος ἐπιτέλλει· θαλάσσης ταραχή.
ιζ′.   πελαργοί.
ιθ′.   ὁ ἐπὶ τῆς οὐρᾶς τοῦ Λέοντος δύνει· θαλάσσης ταραχή· καὶ ὁ λεγόμενος
      Ἵππος ἑῷος ἐπιτέλλει.
κγ′.   ἰσημερία ἐαρινή· καὶ ἔστι μεγίστη ταραχὴ τοῦ ἀέρος.
κϛ′.   ὁ σύνδεσμος τῶν Ἰχθύων δύνει· καὶ γίνεται θαλάσσης ταραχὴ καὶ
      ἀνέμων κίνησις. οὗτος ὁ μὴν κεῖται ἄστρῳ Ἰχθύσι· νὺξ ὡρῶν ιβ′,
      ἡμέρα ιβ′.

Ἀπρίλλιος, καθ᾽ Ἕλληνας Ξανθικός.[197]

ιβ′.   ὕψωμα ἡλίου.
ιε′.   Περσεὺς ἄρχεται ἐπιτέλλειν· καὶ νότος πνεῖ.
κα′.   παχνίτης.
κε′.   ὁ ἐπὶ τῆς ζώνης τοῦ Ὠρίωνος κρύπτεται καὶ νότος γίνεται. οὗτος ὁ
      μὴν κεῖται ἄστρῳ Κριῷ· νὺξ ὡρῶν ια′, ἡμέρα ιγ′.

Μάιος, καθ᾽ Ἕλληνας Ἀρτεμίσιος, κατ᾽ Αἰγυπτίους Παχών.

α.   ὁ λαμπρὸς τῆς Λύρας ἀστὴρ ἑσπέριος ἐπιτέλλει· ἀὴρ ὁμιχλώδης· καὶ
      Ὑάδες ἅμα ἡλίου ἀνατολῇ ἐπιτέλλουσι.
ϛ′.   Αἲξ ἑῷος ἀνατέλλει καὶ σφόδρα ἀλλοιοῦται ὁ ἀήρ.
ζ′.   Πλειάδες ἑῷοι φαίνονται καὶ ἄρχεται ὁ ἀὴρ καθίστασθαι.
η′.   πάχνης μέγας.
ιδ′.   Προκύων ἐπιτέλλει· βρονταὶ καὶ νιφάδες.
ιθ′.   Ὑάδες ἑῷοι φαίνονται καὶ ἀλλοιοῦται πάνυ ὁ ἀὴρ πρὸ μιᾶς ἢ δύο.
κβ′.   ὁ τῶν Διδύμων νότιος Ἡρακλῆς ἐπιτέλλει· ἀέρες καθαροὶ καὶ παχνί-
      του ἔκβασις.
κδ′.   Αἲξ ἑσπέριος κρύπτεται.
      οὗτος ὁ μὴν κεῖται ἄστρῳ Ταύρῳ· νὺξ ὡρῶν ι′, ἡμέρα ιδ′.

should expect in a 'pure' parapegma. Nevertheless, parapegmata are flexible things, and it is
clear that the material was seen as closely enough related to warrant inclusion in this text by a
copyist. Far from ruining the urtext, the copyist has composed a new hybrid text of some
interest. The inclusion of the Eudoxus reference is particularly noteworthy. Unfortunately we
only have this type of entry for the month of February.

[196] The MS has ἔαρος.   [197] The Egyptian month name, Φαρμουθί, has dropped out.

Ἰούνιος, καθ᾽ Ἕλληνας Δαίσιος, κατ᾽ Αἰγυπτίους Παυνί.

α΄. Ἀρκτοῦρος δύνει· ζέφυρος πνεῖ.

β΄. Ἀετὸς ἑσπέριος ἐπιτέλλει.

ϛ΄. ὁ τῶν Διδύμων βόρειος Ἀπόλλων ἐπιτέλλει καὶ γίνεται καύματα· καὶ Ἀρκτοῦρος ἑῷος δύνει.

θ΄. Δελφὶς ἑσπέριος ἐπιτέλλει.

κβ΄. τροπὴ θερινή.

κδ΄. Ἀριάδνης Στέφανος δύνει καὶ γίνεται καύματα.

κε΄. Ὠρίων ἑῷος ἄρχεται ἐπιτέλλειν.
 οὗτος ὁ μὴν κεῖται ἄστρῳ Διδύμῳ· νὺξ ὡρῶν θ΄, ἡμέρα ιε΄.

Ἰούλιος, καθ᾽ Ἕλληνας Πάνεμος, κατ᾽ Αἰγυπτίους Ἐπιφί.

α΄. Ὠρίων ἄρχεται ἐπιτέλλειν ἀρχὴ προδρόμων.

ϛ΄. Προκύων ἐπιτέλλει ἄνεμοι ἐτήσιοι.

ιη΄. Κυνὸς ἐπιτολή.

κ΄. Ἀετὸς δύνει καὶ γίνεται καύματα.
 οὗτος ὁ μὴν κεῖται ἄστρῳ Καρκίνῳ· νὺξ ὡρῶν ι΄, ἡμέρα ιδ΄.

Αὔγουστος, καθ᾽ Ἕλληνας Λῷος, κατ᾽ Αἰγυπτίους Μεσωρί.

α΄. Ὕδρος ἑσπέριος δύνει καὶ πνέουσιν ἄνεμοι ἰσχυροί.

ε΄. Δελφὶς ἄρχεται ἐπιτέλλειν· ἀὴρ ὁμιχλώδης.

ιγ΄. ἡ προτομὴ τοῦ Ἵππου ἑῷα ἀνατέλλει καὶ ἐπισημαίνει.

ιε΄. Λύρα ἑῷα δύνει.

ιζ΄. ὁ ἐπὶ τῆς καρδίας τοῦ Λέοντος ἐπιτέλλει καὶ ἐπισημαίνει.

κ΄. νέον ἔτος Αἰγυπτίων.
 οὗτος ὁ μὴν κεῖται ἄστρῳ· νὺξ ὡρῶν ια΄, ἡμέρα ὡρῶν ιγ΄.[198]

Σεπτέμβριος, καθ᾽ Ἕλληνας Γορπιαῖος, κατ᾽ Αἰγυπτίους Θώθ.

α΄. ἡ Ἀνδρομέδα ἑσπέριος ἐπιτέλλει καὶ πνεῖ νότος.

ϛ΄. Αἲξ ἑσπέριος ἐπιτέλλει καὶ ζέφυρος πνεῖ.

ιε΄. Στάχυς δύνει· ὄμβροι σὺν ἀνέμοις.

κγ΄. νέον ἔτος.

κε΄. Ὑάδες ἄρχονται ἐπιτέλλειν καὶ γίνονται βροχαί.
 οὗτος ὁ μὴν κεῖται ἄστρῳ· νὺξ ὡρῶν ιβ΄, ἡμέρα ὡρῶν ιβ΄.

---

[198] From this point on the zodiacal constellations have dropped out of the text.

Ὀκτώβριος, καθ᾽ Ἕλληνας Ὑπερβερεταῖος, κατ᾽ Αἰγυπτίους Φαωφί.

α.　Πλειάδες ἀκρόνυχοι ἐπιτέλλουσι καὶ νότος πνεῖ.

ε.　Στέφανος ἐπιτέλλει καὶ ἐπισημαίνει.

ιη.　ὁ λαμπρὸς τοῦ βορείου Ἰχθύος ἐπιτέλλει καὶ νότος πνεῖ.

κθ.　Σκορπίος ἄρχεται ἐπιτέλλειν καὶ ποιεῖν ψύξεις.
　　　οὗτος ὁ μὴν κεῖται ἄστρῳ· νὺξ ὡρῶν ιγ´, ἡμέρα ὡρῶν ια´.

Νοέμβριος, καθ᾽ Ἕλληνας Δῖος, κατ᾽ Αἰγυπτίους Ἀθύρ.

α.　Λαμπτὴρ ἐπιτέλλει καὶ ποιεῖ χειμῶνα· ἀστρῷος Κύων ἐπιτέλλει καὶ
　　　ποιεῖ ἀνέμους ψυχροὺς καὶ βροχώδεις.

ϛ´.　Σκορπίος ὑπὸ κρύψιν γίνεται καὶ ποιεῖ βροχήν.

ιθ.　Ὀρίων δύνει καὶ νότος πνεῖ.

κβ´.　ἡλιοδύσια.

κη.　Κύων δύνει καὶ γίνεται θαλάσσης ταραχή.
　　　οὗτος ὁ μὴν κεῖται ἄστρῳ· νὺξ ὡρῶν ιδ´, ἡμέρα ί.

Δεκέμβριος, καθ᾽ Ἕλληνας Ἀπελλαῖος, κατ᾽ Αἰγυπτίους Χοιάκ.

α.　Ὑάδες δύνουσι· καὶ ποιεῖ χειμῶνα.

ε.　Αἲξ δύνει καὶ πνέουσιν ἄνεμοι σφοδροί.

κ.　τροπὴ χειμερινή.

κα.　Αἰγόκερως ἐπιτέλλει· θαλάσσης ταραχή.
　　　οὗτος ὁ μὴν κεῖται ἄστρῳ· νὺξ ὡρῶν ιε´, ἡμέρα θ´.

Risings and settings of the fixed stars

February, according to the Greeks *Peritios*. According to the Egyptians *Mechir*.

1.
3. The bright star in Cygnus sets. There is a disturbance at sea.
6. The west wind blows.
19. Eridanus comes out of hiding and it causes rains.
22. Sagitta sets in the evening and there is a disturbance in the air.
24. Cepheus rises alone.
25. Arcturus rises in the evening.

26. The star on the knees rises, and there are contrary winds.
    Also the swallows appear. (This month) is situated in the constellation of Aquarius. The night is 13 hours, and the day 11.[199]

This month, when the moon is in Aquarius: if there is thunder, it signifies terrible wars on earth, confusions and diseases among men, ruin of grain and other crops, and the destruction of some lands. According to Eudoxus, many storms. What is sown will be no good. Destruction of beasts. If there is an earthquake, it signifies death.

March, according to the Greeks *Dustros*. According to the Egyptians *Phamenoth*.

1. Feast of Ares.[200]
9. Festival of the launching of the ship of Isis.[201]
12. Arcturus rises, disturbance at sea.
17. Storks.
19. The star on the tail of Leo sets. Disturbance at sea, and the star called Pegasus rises in the morning.
23. Vernal equinox, and there is a great disturbance of the air.
26. The binding of the fish sets. There is a disturbance at sea and a movement of winds.
    This month is situated in the constellation Pisces. The night is 12 hours, the day 12.

April, according to the Greeks *Xanthicos*.

12. ὕψωμα of the sun.[202]
15. Perseus begins to rise and the south wind blows.
21. Frost-season.
25. The star on the belt of Orion disappears and there is a south wind.
    This month is situated in the constellation Aries. Night is 11 hours, day 13.

---

[199] The scheme for daylight in this text is simply to add or decrease the length of day and night by 1 hour per month up to a maximum length of 15 hours and a minimum of 9. For this scheme, see *HAMA*, pp. 706–11.

[200] The Mars festival is discussed in Weinstock, 1948, p. 38. Its celebration on 1 March seems to indicate an eastern provenance for the text as a whole.

[201] Weinstock notes that this festival marks the opening of shipping in the spring, and he therefore concludes that the text hails from a sea port.

[202] Compare Antiochus, 10 April, Clodius Tuscus, 1 Jan.

May, according to the Greeks *Artemisios*, according to the Egyptians *Pachon*.

1. The bright star in Lyra rises in the evening. Misty air, and the Hyades rise at the same time as the sun does.
6. Capella rises in the morning and the air is much changed.
7. The Pleiades appear in the morning and the air begins to settle.
8. Much frost.
14. Procyon rises. Thundery and snowy.
19. The Hyades appear in the morning and the air is especially changed for one or two (days) before.
22. Heracles, the southern twin, rises.[203] Clear airs and the completion of the frost-season.
24. Capella disappears in the evening.
    This month is situated in the constellation Taurus. Night is 10 hours, day 14.

June, according to the Greeks *Daisios*, according to the Egyptians *Payni*.

1. Arcturus sets. The west wind blows.
2. Aquila rises in the evening.
6. Apollo, the northern twin, rises and there is burning heat, and Arcturus sets in the morning.
9. Delphinus rises in the evening.
22. Summer solstice.
24. The crown of Ariadne (Corona) sets and there is burning heat.
25. Orion begins to rise in the morning.
    This month is situated in the constellation Gemini. The night is 9 hours, the day 15.

July, according to the Greeks *Panemos*, according to the Egyptians *Epeiph*.

1. Orion begins to rise. Beginning of the *prodromoi* winds.
6. Procyon rises. Etesian winds.
18. Rising of Sirius.
20. Aquila sets and there is burning heat.
    This month is situated in the constellation Cancer. The night is 10 hours, the day 14.

---

[203] There is a tradition where the twins in Gemini are thought of as Apollo and Heracles rather than the more usual Castor and Pollux. See, e.g., Ptolemy, *Tetr.* I.9.23.

August, according to the Greeks *Lōos*, according to the Egyptians *Mesore*.

1. Hydra sets in the evening and severe winds blow.
5. Delphinus begins to rise. Misty air.
13. The head of Pegasus rises in the morning and there is a change in the weather.
15. Lyra sets in the morning.
17. The star in the heart of Leo rises and there is a change in the weather.
20. Egyptian New Year.[204]

    This month is situated in the constellation . . . The night is 11 hours, the day 13 hours.

September, according to the Greeks *Gorpiaios*, according to the Egyptians *Thoth*.

1. Andromeda rises in the evening and the south wind blows.
6. Capella rises in the evening and the west wind blows.
15. Spica sets. Thunderstorms with winds.
23. New Year.[205]
25. The Hyades begin to rise and there are rains.

    This month is situated in the constellation . . . The night is 12 hours, the day 12 hours.

October, according to the Greeks *Hyperberetaios*, according to the Egyptians *Phaophi*.

1. The Pleiades rise acronychally and the south wind blows.
5. Corona rises and there is a change in the weather.
18. The bright star in the southern fish rises and the south wind blows.
29. Scorpio begins to rise and it causes chills.

    This month is situated in the constellation . . . The night is 13 hours, the day 11 hours.

---

[204] Weinstock, 1948, p. 40, shows that if this is a reference to the first day of Thoth in the Egyptian calendar and if the Julian date is correct then we can date the composition of this parapegma (or at least this entry) to about AD 15, since we know that the first of Thoth fell on 19 July in AD 139 and that the Egyptian calendar slipped by one day every four years relative to the Julian.

[205] Weinstock, 1948, p. 40, plausibly connects this with the birthday of Augustus, which was marked as the New Year by law in Asia Minor from 9 BC.

November, according to the Greeks *Dios*, according to the Egyptians *Athyr*.

1. The Lantern rises and causes storms.[206] The Starry Dog[207] rises and causes cold and rainy winds.
6. Scorpio comes out of hiding and causes rain.
19. Orion sets and the south wind blows.
22. Sunset.[208]
28. Sirius sets and there is a disturbance at sea.
    This month is situated in the constellation . . . The night is 14 hours, the day 10.

December, according to the Greeks *Apellaios*, according to the Egyptians *Choiak*.

1. The Hyades set and it causes storms.
5. Capella sets and much wind blows.
20. Winter solstice.
21. Capricorn rises. Disturbance at sea.
    This month is situated in the constellation . . . The night is 15 hours, the day 9.

A.xv. Aëtius parapegma[209]

περὶ ἐπισημασιῶν

ἐπεὶ δὲ οἱ κατ᾿ οὐρανὸν ἀστέρες ἀνατέλλοντες κατὰ τοὺς τεταγμένους αὐτοῖς ὑπὸ τοῦ θεοῦ καιροὺς καὶ δύνοντες ὁμοίως τὸν ἀέρα ἀλλοιοῦσιν, ὡς συμβαίνειν ἐκ τούτου καὶ τοὺς ἀνέμους ἄλλοτε ἄλλους πνεῖν, ἀναγκαῖον ἐνόμισα ἐνταῦθα δηλῶσαι καὶ τοὺς καιρούς, ἐν οἷς αἱ τῶν σαφῶς ἀλλοιούντων τὸν ἀέρα ἀνατολαὶ καὶ δύσεις γίγνονται· καὶ γὰρ καὶ τῶν ὑγιαινόντων τὰ σώματα καὶ πολλῷ μᾶλλον τῶν νοσούντων ἀλλοιοῦται πρὸς

---

[206] A unique entry. Star unknown. Perhaps a corruption of λαμπρός.
[207] Again, unique. The reference is to Sirius.
[208] May be referring to the diminishing of the sun's strength at this time of year generally, but it is unclear why this particular date should have been chosen, unless it is a corruption of the entry we find at Clodius Tuscus, 19 Nov. or Antiochus, 25 Nov. For speculation on the matter, see Weinstock, 1948, p. 41, who connects it with the festival of *Brumalia* on 24 Nov. in, e.g., Polemius Silvius.
[209] Greek text from Olivieri, 1935, book iii.164.

τὴν τοῦ ἀέρος κατάστασιν. μηνὶ **Δύστρῳ**, ὅ ἐστι **Μαρτίῳ**, ιθ´ ὁ λεγόμενος Ἵππος ἑῷος ἀνατέλλει. μηνὶ Μαρτίῳ κγ´ ἰσημερία ἐαρινὴ καὶ ἐστὶ μεγίστη ταραχὴ τοῦ ἀέρος. μηνὶ **Ἀπριλλίῳ** πρώτῃ Πλειάδες ἀκρόνυχοι φαίνονται· μηνὶ Ἀπριλλίῳ ιθ´ Πλειάδες ἑσπέριοι κρύπτονται· μηνὶ Ἀπριλλίῳ κα´ Πλειάδες ἅμα ἡλίου ἀνατολῇ ἐπιτέλλουσι καί ἐστὶ μεγίστη ἡ περὶ τὸν ἀέρα ταραχή· μηνὶ **Ξανθικῷ** κζ´ Ὠρίων ἑσπέριος κρύπτεται. μηνὶ **Ἀρτεμισίῳ**· ὅ ἐστι **Μαΐῳ**, α´ Ὑάδες ἅμα ἡλίου ἀνατολῇ ἐπιτέλλουσι. μηνὶ Μαΐῳ δ´ Λύρα ἑσπέριος ἐπιτέλλει καὶ ἀλλοιοῦται ὁ ἀὴρ ἱκανῶς· μηνὶ Μαΐῳ ς´ Αἲξ ἑῷος ἀνατέλλει καὶ σφόδρα ἀλλοιοῦται ὁ ἀήρ· μηνὶ Μαΐῳ ζ´ Πλειάδες ἑῷαι φαίνονται καὶ ἄρχεται ὁ ἀὴρ καθίστασθαι· μηνὶ Μαΐῳ ιθ´ Ὑάδες ἑῷαι φαίνονται καὶ ἀλλοιοῦται πάνυ ὁ ἀὴρ πρὸ μιᾶς ἢ δύο· μηνὶ Μαΐῳ κδ´ Αἲξ ἑσπέριος κρύπτεται· κινεῖται ὁ ἀὴρ πρὸ δύο ἡμερῶν. μηνὶ **Ἰουνίῳ** β´ Ἀετὸς ἑσπέριος ἐπιτέλλει· μηνὶ Ἰουνίῳ ς´ Ἀρκτοῦρος ἑῷος δύνει· μηνὶ Ἰουνίῳ θ´ Δελφὶς ἑσπέριος ἐπιτέλλει. μηνὶ Ἰουνίῳ κε´ Ὠρίων ἑῷος ἄρχεται ἐπιτέλλειν· εἰσὶ δὲ τροπαὶ θεριναὶ καὶ ἀλλοιοῦται σφόδρα ὁ ἀὴρ πρὸ τριῶν ἡμερῶν. μηνὶ **Ἰουλίῳ** γ´ Ὠρίων ὅλος ἐπιτέλλει· μηνὶ Ἰουλίῳ ιδ´ Προκύων ἑῷος ἐπιτέλλει· μηνὶ Ἰουλίῳ ιθ´ Κύων ἑῷος ἐπιτέλλει καὶ γίγνεται μεγίστη τοῦ ἀέρος ταραχή, ἐνίοτε πρὸ δύο ἡμερῶν· μηνὶ Ἰουλίῳ κε´ Ἀετὸς ἑῷος δύνει, κινεῖται δὲ ὁ ἀὴρ πρὸ τριῶν ἡμερῶν. μηνὶ **Αὐγούστῳ** ιε´ Λύρα ἑῷος δύνει· μηνὶ Αὐγούστῳ ιθ´ Δελφὶς ἑῷος δύνει καί ἐστι τροπὴ τοῦ θέρους ἐπὶ τὸ ψυχρόν· μηνὶ Αὐγούστῳ κη´ Προτρυγητὴρ ἑῷος ἐπιτέλλει καὶ Ὀιστὸς δύνει· ἐστὶ δὲ τὸ τέλος τῶν μετὰ τὴν ἐπιτολὴν τοῦ Κυνὸς ἡμερῶν μ´. μηνὶ **Σεπτεμβρίῳ** ζ´ Αἲξ ἑσπέριος ἐπιτέλλει· μηνὶ Σεπτεμβρίῳ ιδ´ Ἀρκτοῦρος ἐπιτέλλει καὶ ἀλλοιοῖ τῇ ἑξῆς τὸν ἀέρα· μηνὶ Σεπτεμβρίῳ ιθ´ Στάχυς ἑῷος ἐπιτέλλει, ἀλλοιοῦται δὲ ὁ ἀὴρ πρὸ β´ ἡμερῶν· μηνὶ Σεπτεμβρίῳ κε´ ἰσημερία φθινοπωρινὴ καὶ ἐστὶ μεγίστη ταραχὴ τοῦ ἀέρος πρὸ γ´ ἡμερῶν· διὸ παραφυλάττεσθαι χρὴ μηδὲ φλεβοτομεῖν μηδὲ καθαίρειν μηδ᾽ ἄλλως τὸ σῶμα κινεῖν σφοδρᾷ κινήσει ἀπὸ ιε´ τοῦ Σεπτεμβρίου μέχρι κδ´. μηνὶ **Ὀκτωβρίῳ** ς´ Στέφανος ἑῷος ἐπιτέλλει καί ἐστι σφόδρα μεταβολὴ τοῦ ἀέρος· μηνὶ Ὀκτωβρίῳ ζ´ Ἔριφοι ἑσπέριοι ἐπιτέλλουσιν· μηνὶ Ὀκτωβρίῳ ιζ´ Ὑάδες ἑσπέριοι ἐπιτέλλουσι καὶ ἱκανὴ ταραχὴ τοῦ ἀέρος γίγνεται· μηνὶ Ὀκτωβρίῳ κγ´ ἅμα ἡλίου ἀνατολῇ Πλειάδες δύνουσι καί ἐστι μεγίστη ταραχὴ τοῦ ἀέρος πρὸ α´ ἡμέρας. μηνὶ **Νοεμβρίῳ** ς´ Πλειάδεις ἑῷαι δύνουσι καὶ ἄρχεται καθίστασθαι ὁ ἀήρ· μηνὶ Νοεμβρίῳ ιγ´ Λύρα ἑῷος ἐπιτέλλει· μηνὶ Νοεμβρίῳ κα´ Ὑάδες ἑῷαι δύνουσι καὶ ταραχὴ περὶ τὸν ἀέρα γίγνεται τῇ ἑξῆς· μηνὶ Νοεμβρίῳ κζ´ Ὠρίων ἐπιτέλλει καὶ Στέφανος δύνει. μηνὶ **Δεκεμβρίῳ** α´ Κύων ἑῷος δύνει· τετήρηται δὲ τοῖς πολλοῖς, ὡς εἴγε χειμάσειε ταύτῃ τῇ ἡμέρᾳ, ἐπιμένει ὡς ἐπίπαν ἡ ταραχὴ τοῦ ἀέρος μέχρις ἡμερῶν λζ´· εἰ δὲ εὐδιάσει, τὸ

αὐτὸ συμβαίνει· μηνὶ Δεκεμβρίῳ ι´ Ἔριφοι ἑῷοι δύνουσι· μηνὶ Δεκεμβρίῳ κα´
Αἲξ ἑῷος δύνει, ταραχὴ δὲ γίγνεται μετὰ μίαν ἡμέραν· μηνὶ Δεκεμβρίῳ κγ´
τροπαὶ χειμεριναί. μηνὶ Ἰαννουαρίῳ δ´ Δελφὶς ἐπιτέλλει· μηνὶ Ἰαννουαρίῳ ε´
Ἀετὸς ἑσπέριος δύνει, ταραχὴ δὲ σφοδρὰ γίγνεται μετὰ δύο ἡμέρας· μηνὶ
Ἰαννουαρίῳ κε´ ὁ λαμπρὸς ἀστὴρ ἐν τῷ Λέοντι δύνει, κινεῖ πρὸ γ´ ἡμερῶν
τὸν ἀέρα· μηνὶ Ἰαννουαρίῳ κη´ Δελφὶς ἑσπέριος δύνει. μηνὶ **Φευρουαρίῳ** ϛ´
ζέφυρος πνεῖ· μηνὶ Φευρουαρίῳ κβ´ Ὀιστὸς δύνει ἑσπέριος καί ἐστι ταραχὴ
ἱκανὴ τοῦ ἀέρος· μηνὶ Φευρουαρίῳ κε´ Ἀρκτοῦρος ἑσπέριος ἐπιτέλλει· μηνὶ
Φευρουαρίῳ κϛ´ χελιδόνες φαίνονται.

On Weather Changes

Because the stars in the sky change the air, rising and also setting according
to their order at the seasons ordained by god (so it follows from this that
the different winds blow at different times as well), I thought it necessary to
clarify the seasons here, those in which the risings and settings pertaining
to the more prominent changes of the air occur. For the bodies of healthy
people, and especially those of sick people, change according to the condition
of the air.

In the month of **Dustros,** which is **March**

19th: the star called Pegasus rises in the morning.
23rd of March: vernal equinox, and there is the greatest disturbance of the
　　air.

In the month of **April**

1st: the Pleiades appear acronychally.
19th of April: the Pleiades disappear in the evening.
21st of April: the Pleiades rise at the same time as the sun does, and there is
　　the greatest disturbance in the air.
27th of **Xanthicos:** Orion disappears in the evening.

In the month of **Artemision,** which is **May**

1st: the Hyades rise at the same time as the sun does.
4th of May: Lyra rises in the evening, and the air is sufficiently changed.
6th of May: Capella rises in the morning and the air is much changed.
7th of May: the Pleiades appear in the morning and the air begins to settle.

19th of May: the Hyades appear in the morning and the air is especially changed for one or two (days) before.

24th of May: Capella disappears in the evening, the air is moved for two days before.

In the month of **June**

2nd: Aquila rises in the evening.

6th of June: Arcturus sets in the morning.

9th of June: Delphinus rises in the evening.

25th of June: Orion begins to rise in the morning. It is the summer solstice, and the air is much changed for three days before.

In the month of **July**

3rd: the whole of Orion rises.

14th of July: Procyon rises in the morning.

19th of July: Sirius rises in the morning and there is the greatest disturbance of the air, sometimes for two days before.

25th of July: Aquila sets in the morning, and the air is moved for three days before.

In the month of **August**

15th: Lyra sets in the morning.

19th of August: Delphinus sets in the morning, and this is the change of the summer towards cold.

28th of August: Vindemiatrix rises in the morning, and Sagitta sets. It is the end of the forty days after the rising of Sirius.

In the month of **September**

7th: Capella rises in the evening.

14th of September: Arcturus rises and it changes the air on the following day.

19th of September: Spica rises in the morning, and the air is changed for two days before.

25th of September: autumnal equinox, and there is the greatest change of the air for three days before. Because of this it is necessary to be careful neither to phlebotomize, nor purge, nor otherwise to change the body violently from the 15th of September through the 24th.

In the month of **October**

6th: Corona rises in the morning and there is a great change in the air.

7th of October: the Haedi rise in the evening.

17th of October: the Hyades rise in the evening and there is a sufficient disturbance of the air.

23rd of October: the Pleiades set at the same time as the sun rises, and there is the greatest disturbance of the air for one day before.

In the month of **November**

6th: the Pleiades set in the morning, and the air begins to settle.

13th of November: Lyra rises in the morning.

21st of November: the Hyades set in the morning and there is a disturbance in the air afterwards.

27th of November: Orion rises and Corona sets.

In the month of **December**

1st: Sirius sets in the morning. It has been observed by many that if it is stormy on this day, a disturbance of the air will, for the most part, remain for thirty-seven days. And if it is a nice day, it signifies the same thing.

10th of December: the Haedi set in the morning.

21st of December: Capella sets in the morning, and there is a disturbance for one day after.

23rd of December: winter solstice.

In the month of **January**

4th: Delphinus rises.

5th of January: Aquila sets in the evening, and there is a great disturbance for two days after.

25th of January: the bright star in Leo sets, it moves the air for three days before.

28th of January: Delphinus sets in the evening.

In the month of **February**

6th: the west wind blows.

22nd of February: Sagitta sets in the evening, and there is a sufficient disturbance of the air.

25th of February: Arcturus rises in the evening.

26th of February: the swallows appear.

A.xvi. Quintilius parapegma[210]

ἐπειδὴ καὶ οἱ κατ᾽ οὐρανὸν ἀστέρες ἀνατέλλοντες κατὰ τοὺς τεταγμένους αὐτοῖς ὑπὸ τοῦ θεοῦ καιροὺς καὶ δύνοντες ὁμοίως τὸν ἀέρα ἀλλοιοῦσιν, ὡς συμβαίνειν ἐκ τούτων καὶ τοὺς ἀνέμους ἄλλοτε ἄλλους πνεῖν, δηλώσομεν καὶ τοὺς καιροὺς ἐν οἷς καὶ τῶν σαφῶς ἀλλοιούντων τὸν ἀέρα ἀνατολαὶ καὶ δύσεις γίνονται. καὶ γὰρ καὶ τῶν ὑγιαινόντων τὰ σώματα καὶ πολλῷ μᾶλλον τῶν νοσούντων ἀλλοιοῦνται πρὸς τὴν τοῦ ἀέρος κατάστασιν.

ἀστέρων ἐπιτολαὶ καὶ δύσεις κατὰ Κυιντίλλιον

μηνὶ Ξανθικῷ

πρὸ τρισκαιδεκάτης καλανδῶν Ἀπριλλίων Ἵππος ἐπιτέλλει ἑῷος.

πρὸ ἐννέα ἰσημερία ἐαρινή.

καλάνδαις Ἀπριλλίαις Πλειάδες ἀκρόνυχοι κρύπτονται· καὶ ἔσται ταραχὴ
    μεγάλη τοῦ ἀέρος.

μηνὶ Ἀρτεμισίῳ

πρὸ δώδεκα καλανδῶν Μαίων Πλειάδες ἑσπέριοι κρύπτονται.

πρὸ δώδεκα καλανδῶν Μαίων Ὡρίων ἑσπέριος κρύπτεται.

κα΄ Πλειάδες ἐπιτέλλουσι καὶ ἔσται ταραχὴ τοῦ ἀέρος.

πρὸ ἓξ νοννῶν Μαίων Λύρα ἑσπέριος ἐπιτέλλει.

Μαίῳ δ΄ καὶ ἀλλοιοῦται ὁ ἀὴρ ἱκανῶς.[211]

νόνναις Μαίαις Αἲξ ἑῷος ἐπιτέλλει ἢ Μαίῳ ε΄.

πρὸ ἑπτὰ εἰδῶν Μαίων Πλειάδες ἑῷαι φαίνονται· καὶ ὁ ἀὴρ ἄρχεται καθίστασθαι.

μηνὶ Δαισίῳ

πρὸ τρισκαιδεκάτης καλανδῶν Ἰουνίων Ὑάδες ἑῷαι φαίνονται· καὶ
    ἀλλοιοῦται ὁ ἀὴρ πρὸ μιᾶς ἢ δύο.

---

[210] Greek text following Boll, 1910b.

[211] Here as elsewhere, Boll appends the anomalously numbered entry to the previous one, as
    probably an interpolation.

πρὸ ὀκτὼ καλανδῶν Ἰουνίων Αἲξ ἑσπέριος κρύπτεται· καὶ κινεῖται ὁ ἀήρ.
πρὸ τριῶν νοννῶν Ἰουνίων Ἀετὸς ἑσπέριος ἐπιτέλλει.
πρὸ ἑπτὰ εἰδῶν Ἰουνίων Ἀρκτοῦρος ἑῷος δύνει.
πρὸ τεσσάρων εἰδῶν Ἰουνίων Δελφὶς ἑσπέριος ἐπιτέλλει.

μηνὶ Πανέμῳ

πρὸ ἓξ καλανδῶν Ἰουλίων τροπαὶ θεριναὶ καὶ Ὠρίων ἄρχεται ἐπιτέλλειν·
　　καὶ ἀλλοιοῦται ὁ ἀὴρ πρὸ δύο ἢ τριῶν ἡμερῶν.
πρὸ τεσσάρων νοννῶν Ἰουλίων Κύων ἑῷος ἐπιτέλλει.²¹²
Ἰουλίῳ ιθʹ καὶ γίνεται οὐχ ἡ τυχοῦσα ταραχὴ περὶ τὸν ἀέρα.
Ἰουλίῳ γʹ Ὠρίων ἑῷος ὅλος ἐπιτέλλει.

μηνὶ Λώῳ

πρὸ ἐννέα καλανδῶν Αὐγούστων οἱ ἐτησίαι ἄρχονται πνεῖν.
Αὐγούστῳ ιεʹ Λύρα ἑῷος δύνει.
Ἰουλίῳ κεʹ ὁ Ἀετὸς ἑῷος δύνει καὶ κινεῖται ὁ ἀὴρ πρὸ δύο ἡμερῶν.
πρὸ τριῶν καλανδῶν Αὐγούστων ὁ λαμπρὸς ἀστὴρ ἐν τῷ στήθει τοῦ
　　Λέοντος ἐπιτέλλει.
Αὐγούστῳ ιθʹ Δελφίν.²¹³
πρὸ τριῶν νοννῶν Αὐγούστων Ἀετὸς ἑῷος δύνει, Αὐγούστῳ κεʹ.
πρὸ ἐννέα καλανδῶν Σεπτεμβρίων Δελφὶς ἑῷος δύνει· καὶ ἐστὶ τὸ τέλος τῶν
　　μετὰ τὴν ἐπιτολὴν τοῦ Κυνὸς ἡμερῶν μʹ.

μηνὶ Γορπιαίῳ

πρὸ τεσσάρων καλανδῶν Σεπτεμβρίων Προτρυγητὴρ ἑῷος ἐπιτέλλει καὶ
　　Ὀιστὸς δύνει.
πρὸ ἓξ [. . .] Σεπτεμβρίων Ἵππος ἑῷος δύνει.
ζʹ Σεπτεμβρίῳ Αἲξ ἑσπέριος ἐπιτέλλει.
πρὸ τεσσάρων εἰδῶν Σεπτεμβρίων Αἲξ ἑσπέριος ἐπιτέλλει.
Σεπτεμβρίῳ ιδʹ Ἀρκτοῦρος ἐπιτέλλει καὶ τῇ ἑξῆς ἀλλοιοῖ τὸν ἀέρα.

---

²¹² Boll splits this into two entries to try and correct the date: πρὸ τεσσάρων νοννῶν [. . .]
　　followed in the next line by [. . .] Ἰουλίων Κύων ἑῷος ἐπιτέλλει.
²¹³ Boll restores ἑῷος δύνει.

μηνὶ Ὑπερβερεταίῳ

πρὸ δυοκαίδεκα καλανδῶν Ὀκτωβρίων Στάχυς ἑῷος ἐπιτέλλει, ιθ΄ καὶ ἀλλοιοῦται ὁ ἀὴρ πρὸ δύο ἡμερῶν.

πρὸ ἓξ καλανδῶν Ὀκτωβρίων ἰσημερία φθινοπώρου.

Σεπτεμβρίῳ κγ΄ καὶ ἐστὶ ταραχὴ τοῦ ἀέρος· διὸ παραφυλάττεσθαι προσήκει μήτε φλεβοτομεῖν μήτε καθαίρειν, μηδ᾽ἄλλως τὸ σῶμα κινεῖν σφοδρᾷ κινήσει ἀπὸ ιε΄ Σεπτεμβρίου μεχρὶ κδ.

Ὀκτωβρίῳ ϛ΄ ἐστι σφοδρὰ μεταβολή.

πρὸ ἐννέα νοννῶν Ὀκτωβρίων Στέφανος ἑῷος ἐπιτέλλει.

πρὸ ὀκτὼ εἰδῶν Ὀκτωβρίων Ἔριφοι ἑσπέριοι ἐπιτέλλουσι.

πρὸ πεντεκαίδεκα καλανδῶν Νοεμβρίων Ὑάδες ἑσπέριαι ἐπιτέλλουσι καὶ πολλὴ τοῦ ἀέρος ταραχή.

μηνὶ Δίῳ

πρὸ ἐννέα καλανδῶν Νοεμβρίων πρώτῃ ἡλίου ἀνατολῇ Πλειάδες δύνουσι καὶ ἐστί ταραχὴ τοῦ ἀέρος.

πρὸ τριῶν εἰδῶν Νοεμβρίων Πλειάδες ἑῷαι δύνουσι καὶ Ὡρίων ἄρχεται δύνειν· καὶ ἄρχεται καθίστασθαι ὁ ἀήρ.

πρὸ ὀκτωκαίδεκα καλανδῶν Δεκεμβρίων Λύρα ἐπιτέλλει.

μηνὶ Ἀπελλαίῳ

πρὸ δέκα καλανδῶν Δεκεμβρίων Ὑάδες ἑῷαι δύνουσι· καὶ ἐστὶ ταραχὴ περὶ τὸν ἀέρα, Νοεμβρίῳ ιζ.

πρὸ τεσσάρων καλανδῶν Δεκεμβρίων Ὡρίων ἐπιτέλλει καὶ Στέφανος δύνει, Νοεμβρίου κζ.

πρώτῃ Δεκεμβρίου καλάνδαις Δεκεμβρίαις Κύων ἑῷος δύνει. τετήρηται δὲ τοῖς πολλοῖς ὡς εἴγε χειμάσει ἐν ταύτῃ τῇ ἡμέρᾳ ἐπιμένει ὡς ἐπίπαν ἡ ταραχὴ τοῦ ἀέρος ἄχρι ἡμέρας λζ. εἰ δὲ εὐδιάσει, τὸ αὐτὸ συμβαίνει.

πρὸ τριῶν εἰδῶν Δεκεμβρίων Ἔριφοι ἑῷαι δύνουσι. δεκάτη.

μηνὶ Αὐδυναίῳ

πρὸ ἕνδεκα καλανδῶν Ἰαννουαρίων Αἲξ ἑῷος δύνει, κα, ταραχὴ δὲ γίνεται μετὰ μίαν ἡμέραν.

πρὸ ἐννέα καλανδῶν Ἰαννουαρίων τροπαὶ χειμεριναί.

νόνναις Ἰαννουαρίαις Δελφὶς ἐπιτέλλει.

πρὸ ὀκτὼ εἰδῶν Ἰαννουαρίων Ἀετὸς ἑσπέριος δύνει, ε´ Ἰαννουαρίου.²¹⁴

κε´ Ἰαννουαρίου μηνὶ **Περιτίῳ**

πρὸ ἑπτὰ καλανδῶν Φεβρουαρίων ὁ λαμπρὸς ἀστὴρ ἐν τῷ στήθει τοῦ
    Λέοντος δύνει· καὶ γίνεται κίνησις ἀνέμου πρὸ τριῶν ἡμερῶν.

κζ´ Ἰαννουαρίου²¹⁵ πρὸ τεσσάρων καλανδῶν Φεβρουαρίων Δελφὶς ἑσπέριος
    δύνει.

πρὸ τριῶν νοννῶν **Φεβρουαρίων** Λύρα ἑσπέριος δύνει· καὶ ζέφυρος πνεῖ.

μηνὶ **Δύστρῳ**

πρὸ ἑπτὰ καλανδῶν Μαρτίων Ὀιστὸς δύνει ἑσπέριος καὶ ὄρνιθες πνέουσι·
    καὶ γίνεται ταραχὴ μεγάλη τοῦ ἀέρος.

πρὸ τεσσάρων καλανδῶν Μαρτίων Ἀρκτοῦρος ἑσπέριος ἐπιτέλλει.

πρὸ τριῶν καλανδῶν Μαρτίων χελιδόνες φαίνονται.

Because the stars in the sky change the air, rising and also setting according to their order at the seasons ordained by god (so it follows from this that the different winds blow at different times as well), we shall clarify the seasons in which the risings and settings pertaining to the more prominent changes of the air occur. For the bodies of healthy people, and especially those of sick people, change according to the condition of the air.

The risings and settings of the stars, according to Quintillius

In the month of **Xanthicos**

The thirteenth day before the Kalends of April, Pegasus rises in the morning. The ninth day before, vernal equinox.

On the Kalends of **April**, the Pleiades disappear acronychally, and there is a great disturbance of the air.

In the month of *Artemision*

The twelfth day before the Kalends of May, the Pleiades disappear in the evening.

---

²¹⁴ The MSS have ε´ Φεβρουαρίου καὶ Ἰαννουαρίου.
²¹⁵ MSS have Ἰαννουαρίου καὶ Φεβρουαρίῳ.

The twelfth day before the Kalends of May, Orion disappears in the evening.[216]

21st: The Pleiades rise, and there will be a disturbance of the air.

The sixth day before the Nones of **May**, Lyra rises in the evening.

4 May, the air is also changed sufficiently.

On the Nones of May, Capella rises in the morning, or 5 May.

The seventh day before the Ides of May, the Pleiades appear in the morning, and the air begins to be settled.

In the month of **Daisios**

The thirteenth day before the Kalends of June, the Hyades appear in the morning, and the air is changed for one or two days before.

The eighth day before the Kalends of June, Capella disappears in the evening, and the air is moved.

The third day before the Nones of **June**, Aquila rises in the evening.

The seventh day before the Ides of June, Arcturus sets in the morning.

The fourth day before the Ides of June, Delphinus rises in the evening.

In the month of **Panemos**

The sixth day before the Kalends of July, summer solstice, and Orion begins to rise, and the air is changed for two or three days before.

The fourth day before the Nones of **July**, Sirius rises in the morning.

19 July: and there is not the coincident disturbance in the air.[217]

3 July: the whole of Orion rises in the morning.

In the month of **Lōos**

The ninth day before the Kalends of August: the Etesian winds begin to blow.

15 **August**: Lyra sets in the morning.[218]

25 July: Aquila sets in the morning, and the air is moved for two days before.

The third day before the Kalends of August, the bright star in the breast of Leo rises.

19 August: Delphinus.

---

[216] Notice the doubling of the date.   [217] Something has gone awry with the dates here.
[218] The date seems right, but it should be two lines lower.

The third day before the Nones of August: Aquila sets in the morning,[219]
 25 August.
The ninth day before the Kalends of September, Delphinus sets in the morning, and it is the end of the forty days after the rising of Sirius.

In the month of **Gorpiaios**

The fourth day before the Kalends of September, Vindemiatrix rises in the morning, and Sagitta sets.
The sixth day before the [. . .] of **September**, Pegasus sets in the morning.
7 September: Capella rises in the evening.
On the fourth day before the Ides of September, Capella rises in the evening.
14 September: Arcturus rises, and it changes the air on the following day.

In the month of **Hyperberetaios**

The twelfth day before the Kalends of October, Spica rises in the morning, nineteenth, and the air is changed for two days before.
The sixth day before the Kalends of October, autumnal equinox.
23 September: There is also a disturbance of the air. Because of this it is necessary to be careful neither to phlebotomize, nor purge, nor otherwise to change the body violently from the 15th of September through the 24th.
6 **October**: there is a great change.
The ninth day before the Nones of October, Corona rises in the morning.
The eighth day before the Ides of October, the Haedi rise in the evening.
The fifteenth day before the Kalends of November, the Hyades rise in the evening, and much disturbance of the air.

In the month of **Dios**

The ninth day before the Kalends of November, at the first rising of the sun, the Pleiades set, and there is a disturbance of the air.
The third day before the Ides of **November**, the Pleiades set in the morning, and Orion begins to set, and the air begins to be settled.
The eighteenth day before the Kalends of December, Lyra rises.

---

[219] Again, something has gone awry with the dates.

In the month of **Apellaios**

The tenth day before the Kalends of December, the Hyades set in the morning, and there is a disturbance in the air, 17 November.

The fourth day before the Kalends of December, Orion rises and Corona sets, 27 November.

On the first of **December**, the Kalends of December, Sirius sets in the morning. It has been observed by many that if it is stormy on this day, a disturbance of the air will, for the most part, remain for thirty-seven days. And if it is a nice day, the same thing happens.

The third day before the Ides of December, the Haedi set in the morning, tenth.

In the month of **Audunaios**

The eleventh day before the Kalends of January, Capella sets in the morning, 21st, there is a disturbance for one day after.

The ninth day before the Kalends of January, winter solstice.

The Nones of **January**, Delphinus rises.

The eighth day before the Ides of January, Aquila sets in the evening, 5 January.

25 January, in the month of **Peritios**

The seventh day before the Kalends of February, the bright star in the breast of Leo sets, and there is a movement of wind for three days before.

27 January,[220] the fourth day before the Kalends of February, Delphinus sets in the evening.

The third day before the Nones of **February**, Lyra sets in the evening, and the west wind blows.

In the month of **Dustros**

The seventh day before the Kalends of March, Sagitta sets in the evening, and the bird-bringers blow, and there is a great disturbance of the air.

The fourth day before the Kalends of March, Arcturus rises in the evening.

The third day before the Kalends of March, swallows appear.

---

[220] MSS have '27 January, and in February . . .'.

A.xvii. Paris parapegma[221]

<Σεπτέμβριος>

α. βρονταὶ καὶ διαλλαγαὶ ἀνέμων.

β´. ὅλη ἡ Ἀνδρομέδα ἀνίσχει.

γ´. Ἀρκτοῦρος σὺν τῷ Τρυγητῇ ἀνατέλλει.

δ´. ὁ Ὀιστὸς πᾶσιν φαίνεται.

ε´. ἄρρητοι πάλιν βρονταὶ καὶ ἀλλαγαὶ ἀνέμων.

ϛ´. ὁ Ἵππος ὁ ἀστὴρ ἀνίσχει.

ζ´. ἡ Αἲξ ἀνίσχει· καὶ λὶψ ἄνεμος πνεῖ.

η´. ὁ Ἀρκτοῦρος ἀναφέρεται.

θ´. αὐγώδης Ἀρκτοῦρος ἀνίσχει σὺν Παρθένῳ.

ι´. ὁ Ἀρκτοῦρος καθόλου φαίνεται.

ια´. Ἵππος σὺν τῇ Πλειάδι ἀνίσχει.

ιβ´. χελιδὼν οὐδαμῶς φαίνεται.

ιγ´. τὸ πλεῖον τοῦ Ἵππου φαίνεται.

ιδ´. ὅλος ὁτηλεύκη περιφαίνεται.

ιε´. ἡ Αἲξ καθαρὴ φαίνεται.

ιϛ´. ζέφυρος μετὰ λιβός, εὖρος ἐύς.

ιζ´. Ἀπαρκτίας φυσᾷ ἠρέμα.

ιη´. ὁ ἥλιος ἐν τῷ Ζυγῷ καλῶς ἐπιποιῶν νότον.

ιθ´. ἡ μετοπωρινὴ ἰσημερία.

κ´. ὁ Κρατὴρ φαίνεται.

κα´. Ἀργὼ δύνει.

κγ´. ἐπιτολὴ τοῦ Κενταύρου· ταραχὴ ἀέρος καὶ θαλάσσης.

κδ´. συννεφὴς ὁ ἀὴρ καὶ ταραχώδης ἐκ τοῦ Κενταύρου.

κε´. οἱ Ἔριφοι ἀνίσχουσιν καὶ νότος πνεῖ· καὶ ἐστὶν μεγίστη ταραχὴ τοῦ ἀέρος πρὸ τριῶν ἡμερῶν· διὸ παραφυλάττεσθαι χρὴ τοῦ φλεβοτομεῖν μηδὲ καθαίρειν, μηδ᾽ ὅλως τὸ σῶμα κινεῖν σφοδρᾷ κινήσει· ὁμοίως δὲ καὶ ἀπὸ τὰς ιε ἕως τὰς κδ αὐτοῦ.

κϛ´. οἱ Ἔριφοι σὺν τῷ ἡλίῳ ἀνίσχουσιν.

κζ´. ἐκ τῶν Ἐρίφων νότος ὡς ἐπὶ τὸ πολύ.

κη´. ὁ Ἡνίοχος ἄρχεται φαίνεσθαι.

κθ´. αἱ Πλειάδες φαίνονται ἐν ταῖς δυσὶν ἡμέραις.

λ´. ἡ Αἲξ ἀνατέλλει· καὶ τὸ Δελτωτὸν ἑσπέριον ἐπιτέλλει· καὶ κινεῖται ὁ ἀὴρ πρὸ μιᾶς ἡμέρας· ποτὲ δὲ καὶ ταραχὴ γίνεται.

---

[221] Greek text from Bianchi, 1914, pp. 22–48.

Ὀκτώβριος

α΄.   αἱ Πλειάδες ὄρθρου φαίνονται.

β΄.   καὶ τὴν β΄ τὸ αὐτό, ἀλλὰ καὶ τὴν τριτάτην.

γ΄.   Ἡνίοχος δυόμενος· πρὸς δὲ τετάρτην πάλιν.

δ΄.   οἱ Ἔριφοι ἀνίσχουσιν καὶ Στέφανος τὴν πέμπτην.

ε΄.   Στέφανος ἑῷος ἐπιτέλλει καί ἐστιν σφοδρὰ μεταβολὴ τοῦ ἀέρος.

ϛ΄.   Ἔριφοι ἑσπέριοι ἐπιτέλλουσιν.

η΄.   καὶ τὴν η΄ Στέφανος ἀνίσχει σὺν Ἐρίφοις.

θ΄.   καὶ τὴν θ΄ Ἔριφοι φαίνονται σὺν Πλειάσιν.

ι΄.   καὶ τὴν ι΄ ὁ Ζυγὸς ἄρχεται προανίσχων.

ια΄.   τὴν ια΄ Στέφανος ἀνίσχει πρὸς ὄρθρον.

ιβ΄.   τὴν ιβ΄ τὸ αὐτὸ καὶ τὴν ιγ΄.
       ὁμοίως καὶ τὴν μετ᾽ αὐτήν· πρὸς δὲ τρίτην τοιουτῶδες.

ιϛ΄.   καὶ τὴν ιϛ΄ Ὠρίων ἐπανίσχει.

ιζ΄.   Ὑάδες ἑσπέριοι ἐπιτέλλουσιν καὶ ἱκανὴ ταραχὴ τοῦ ἀέρος γίνεται.

κ΄.   τὴν κ΄ δὲ δύνουσιν Πλειάδες καὶ τὴν πρώτην.

κβ΄.   καὶ τὴν δευτέραν ἡ οὐρὰ τοῦ Ταύρου δύνει ὅλη.
       Πλειάδες δύνουσιν· καί ἐστιν μεγίστη ταραχὴ τοῦ ἀέρος πρὸ μιᾶς ἡμέρας.[222]

κδ΄.   πρὸς δὲ καὶ τὴν δ΄ αἱ Πλειάδες ἀνίσχουσιν· ἀλλὰ καὶ τὴν ζ΄.

κη΄.   δυτικὸς ὁ Ἀρκτοῦρος.

κθ΄.   καὶ τὴν κθ΄ τὸ αὐτό.
       τριακοντάδα πάλιν Ὠρίων ὅλος κρύπτεται καὶ Λύρα προανίσχει.

Νοέμβριος

α΄.   δύνουσιν αἱ Πλειάδες.

β΄.   Ἀρκτοῦρος· καὶ τροπὴ τοῦ ἀέρος.

γ΄.   ἡ Λύρα ὀρθρία· πρὸς δὲ καὶ τὴν δ΄.

ε΄.   ἡ Λύρα σύνεγγυς ἡλίῳ ἀνατέλλει.

ϛ΄.   Ἀρκτοῦρος δὲ ὄρθριος ἐπιδύνει· καὶ Πλειάδες ἑῷαι δύνουσιν καὶ ἄρχεται καθίστασθαι ὁ ἀήρ.

ζ΄.   δύνουσιν Πλειάδες καὶ Ὠρίων μετ᾽ αὐτῶν.

θ΄.   τὸ αὐτό, τὴν θ΄ Ἀντάρης φαίνεται· καὶ χειμέριος τροπὴ λέγεται εἶναι.

ι΄.   καὶ τὴν δεκάτην τὸ αὐτό.

ια΄.   τὴν ια΄ Λύρα ἑῷος ἐπιτέλλει.

---

[222] Date omitted in MS. Bianchi restores κγ΄.

ιβ΄. αἱ Πλειάδες· πρὸς καὶ τὴν ιβ΄ τοῦ Ταύρου τὸ μεσότατον ἀστρίτζην ἐπανίσχει.

ιγ΄. πάλιν δὲ Πλειάδες καὶ Ὠρίων πρὸς ὄρθρον ἐπιδύνουσιν· καὶ πρὸς δὲ ταῦτα ἡ Λύρα προανίσχει.

ιε΄. καὶ Λύρα σὺν Ὠρίωνι καὶ ταραχὴ ἀνέμου.

ιθ΄. τοῦ Ταύρου κέρας δύνει.

κ΄. ὁμοίως δέ· τὴν πρωτίστην ταύτης Ὑάδες σὺν τῷ Λαγωῷ δύνουσιν πρὸς τὸν ὄρθρον.

κβ΄. καὶ τὴν β΄ τοῦ αὐτοῦ· τὴν γ΄ ὁ Ὠρίων.

κδ΄. τὴν δ΄ καὶ ε΄ σὺν αὐτῷ· καὶ τὴν ς΄ προοίμια χειμερινὰ κατάρχονται τῆς ὥρας.

κζ΄. τὴν κζ΄ μέχρι τῆς τριαντάδος τὸ αὐτὸ γίνεται.

Δεκέμβριος

α΄. ὁ Κύων ὄρθριος δύνει· χύσεις τοῦ ἀέρος· τετήρηται δὲ τοῖς πολ- λοῖς εἴτε χειμάσει ταύτῃ τῇ ἡμέρᾳ· ἐπιμένει δὲ ὡς ἐπὶ τὸ πολὺ ἡ ταραχὴ τοῦ ἀέρος μέχρι ἡμερῶν λζ΄· εἰ δὲ εὐδιάσει, τὸ αὐτὸ σημαίνει.

β΄. δύνει δὲ πρὸς τὴν ἑσπέραν Κύων· πάντα δὲ ταῦτα γίνεται ἕως ις΄.

ι΄. Ἔριφοι ἑῷοι δύνουσιν.

ια΄. Αἲξ ἑῷα δύνει· καὶ ταραχὴ δὲ γίνεται μετὰ α΄ ἡμέραν.

ιζ΄. τῇ ιζ΄ τὸ ζῷδιον Αἰγόκερως προσδέχεται τὸν ἥλιον.

ιη΄. ἡ ιη΄· πάλιν ἡ Αἲξ ἀνίσχει.

κγ΄. ἡ κγ΄· Αἲξ ὄρθριος δύνει· καί εἰσιν τροπαὶ χειμεριναί.

κη΄. καὶ τὴν ις΄ ἑσπέριος [. . .] τὴν δ΄ εἰκοστὴν ὀγδόην ἥλιος ἀναστρέφεται.

λ΄. ἀλλὰ καὶ λ΄ καὶ λα΄ προεπανίσχει ὁ Δελφὶν καὶ ἥλιος ὑψοῦται.

Ἰανουάριος

α΄. ὑψοῦται ὁ ἥλιος· Ἀετὸς σὺν Στεφάνῳ.

β΄. Ὠρίων δύνει.

γ΄. δύνει τὸ μέσον τοῦ Ἀετοῦ. δύνει Ἀετός, πτηνὸς ὄρνις· καὶ γίνεται τροπὴ τοῦ ἀέρος ποικίλη.

δ΄. τῇ δὲ δ΄ ὁ Δελφὶν ὄρθριος ἐπιτέλλει καὶ πολὺς νότος πνεῖ.

ε΄. τῇ ε΄ Ἀετὸς Λύρα Κόραξ δύνουσι καὶ ταραχὴ δὲ γίνεται μετὰ β΄ ἡμέρας.

ς΄. τῇ ς΄ δύνει δὲ Ἀετός. Δίδυμος ἀνατέλλει.

ζ΄. τῇ ζ΄ τὸ μέσον τῆς Λύρας ἐπιτέλλει.

η΄. τῇ η΄ ὁ Δελφὶν δύνει· καὶ πνεῖ νότος σὺν ἀπαρκτίᾳ.

θ.   καὶ τὴν θ᾽ δὲ καὶ τὴν ι᾽ τὸ αὐτὸ σημαίνουσιν.

ια.   τῇ ια᾽ ταραχὴ τοῦ ἀέρος ὑπάρχει.

ιβ.   τῇ ιβ᾽ Λύρα δύνει ἑσπερία.

ιγ.   τῇ τρίτῃ καὶ δεκάτῃ ὁ Ὀιστὸς καὶ τῇ νυκτὶ βροχίτζα.

ιε.   τῇ ιε᾽ Λύρα δύνει περὶ τὸν ὄρθρον.

ιθ.   τῇ ιθ᾽ ὁ Δελφὶν ἀνίσχει· πνεῖ δὲ καὶ βορρᾶς.

κβ.   τῇ κβ᾽ Λύρα σὺν τῷ Ἀετῷ δύνει.

κε.   Λύρα δύνει καὶ κινεῖ τὸν ἀέρα πρὸ γ᾽ ἡμερῶν.

κϛ.   Λύρα ἑσπερία δύνει καὶ Βασιλίσκος.

κζ.   τῇ κζ᾽ Δελφὶν ἑσπέριος δύνει καὶ ποιεῖ ἀνέμους καὶ χειμῶνας.

κθ.   τῇ κθ᾽ ὁ Δελφὶν δύνει.

λ.   Λύρα δύνει καὶ συννέφεια γίνεται καὶ βροχή.

λα.   τῇ λα᾽ ἄνεμος χειμωνώδης.

Φευρουαρίῳ

α.   ἄστρον κρυπτὸν ἀνίσχει.

β.   ζέφυρος προσπνέει.

γ.   τῇ γ᾽ Λύρα δύνει.

δ.   τῇ δ᾽ Δελφὶν ἑσπέριος δύνει.

ε.   τῇ ε᾽ τὰ μέσα Ὑδροχόου ἀνίσχουσι.

ϛ.   τῇ ϛ᾽ Λύρα δύεται καὶ ζέφυρος πνεῖ.

ζ.   ζέφυρος ἢ βορρᾶς πνέει.

θ.   τῇ θ᾽ τὸ κρυπτὸν ἄστρον φαίνει.

ι.   ἄνεμοι ῥαγδαῖοι.

ια.   τῇ ια᾽ ὁ Ἀρκτοῦρος καὶ νότος πνεῖ.

ιβ.   ὁ Τοξότης ἑσπέριος δύνει καὶ χειμάζει.

ιδ.   ὁ Κρατὴρ προσανίσχων.

ιε.   ὁ ἥλιος ἐν Ἰχθύσιν.

ιϛ.   ἀπαρκτίας ἐκ τούτου.

ιζ.   ἡ Παρθένος πρὸς δύσιν.

ιη.   ὁ Ὀιστὸς ἐπιδύνων.

ιθ.   ἀπαρκτίας δύνει ἤγουν πνεῖ.

κ.   τοῦ Λέοντος τὰ μέσα δύνει.

κα.   Ἀρκτοῦρος δύνει.

κβ.   Ὀιστὸς δύνει καί ἐστι ταραχὴ ἐαρινή.

κγ.   ὁ Ὑδροχόος ἀνίσχει καὶ χειμῶνα παραπέμπει.

κδ.   βορρᾶς πνεῖ.

κε.   ὁ Ἀρκτοῦρος αὐγινός.

κϛ΄.   χελιδόνες φαίνονται.
κζ΄.   Ὀιστὸς δύνει.

Μάρτιος

α΄.   νότος καὶ λὶψ πνέει.
β΄.   Προτρυγητὴρ ἀνίσχει.
γ΄.   Ἀρκτοῦρος ἀνίσχει.
ζ΄.   Ἵππος ἀπὸ πρωίας δύεται.
η΄.   ὄρνεα φαίνουνται <ἐπὶ> τοῦ Ὑδροχόου.
θ΄.   ἰκτῖνος ἄρχεται ἀνίσχειν· ὁ δὲ Ἰχθὺς μὲν ὄρθριος δύνει.
ι΄.   ὄρθριος Ἵππος ἐπὶ δύσιν.
ια΄. χωρισμὸν τοῦ χειμῶνος λεκτέον.
ιβ΄. παύεται Ἰχθὺς ἐκ τοῦ νότου.
ιγ΄. Αἲξ ἑσπέρᾳ προανίσχει.
ιδ΄. βορρᾶς ὅλως πνέει.
ιε΄. Ἵππος δύνει ὅλος.
ιϛ΄.²²³   τὰς ιζ΄ ὁμοίως· καὶ ψεκάζει.
ιζ΄. τῷ Κριῷ σελασφόρος.
ιη΄. ἰκτῖνος ἐπιφέρεται ἕως ἰσημερίας.
ιθ΄. τὰς δὲ ιθ΄ Ἵππος ἑῷος ἀνατέλλει.
κ΄.   εὔδιον τὸ κατάστημα ὅλον.
κα΄. Ἵππος ἕωθεν δύνει καί ἐστιν ταραχὴ τοῦ ἀέρος.
κβ΄. ὁ Κριὸς ἀνίσχει καὶ βρέχει.
κε΄. Ἵππος ἄρχεται δύνειν.
κϛ΄. Ἰχθύες ἀνίσχουσιν ἐκ νότου.
λα΄. ταραχὴ τοῦ ἀέρος.²²⁴

Ἀπρίλλιος

α΄.   Πλειάδες ἑσπέριοι κρύπτονται.
δ΄.   τὴν δ΄ λὶψ προσπνέει.
ε΄.   προανίσχουσιν Ὑάδες· καὶ βρονταί.
ζ΄.   Πλειάδες ἀνίσχουσιν.
η΄.   ζέφυρος ὄρθρου πνεῖ.
θ΄.   ζάλη ἀπὸ τοῦ νότου πέλει.

---

²²³ The MS has ιϛ΄ corrected to ιζ΄.
²²⁴ Following Bianchi. The MS has ἀρχὴ τοῦ ἀέρος.

ί.   βορρᾶς πνεῖ ὅλην τὴν ἡμέραν.

ια.   ψυχρὸς ἄνεμος καὶ βορρᾶς.

ιγ.   βορρᾶς πνεῖ ἐξ Ὑάδων.

ιδ.   κρυπτὸν ἄστρον προσδύνει.

ιε.   Ὑάδες δύνοντες φαίνουσιν.

ιη.   ὁ λὶψ προσπνέει.

ιθ.   Ὑάδες δύνουν.

κ.   ζέφυρος πνεῖ μετὰ βορρᾶ.

κα.   ἡ κάρα τοῦ Ταύρου δύνει.

κβ.   ἀνίσχουσιν αἱ Πλειάδες.

κγ.   ἡ Λύρα φαίνεται πρώτης νυκτός.

κε.   δύνει ὁ Κύων.

κζ.   νότος πνεῖ καὶ βρέχει.

λ.   ὁ Κύων κρύπτεται.

Μάιος

α.   Ὑάδες ἅμα ἡλίου ἀνατολῇ ἐπιτέλλουσιν.

γ.   Κένταυρος ὅλος φαίνεται· καὶ ζέφυρος πνεῖ.

δ.   Λύρα ἑσπέριος ἐπιτέλλει· καὶ ἀλλοιοῦται ὁ ἀήρ.

ϛ.   ἔωθεν ἀνίσχουσιν αἱ Πλειάδες.

ζ.   Πλειάδες.

θ.   ἡ Λύρα ἀνίσχει· Ὑὰς ἐπιδύνει.

ια.   Πλειὰς ἀναφαίνεται πάλιν.

ιβ.   νότος ἐπιπνέει.

ιγ.   Ὑάδες δύνουν· καὶ νότος πνεῖ.

ιδ.   Λύρα ἔωθεν ἀνίσχει· καὶ νότος πνεῖ.

ιζ.   ὁ Κύων ἐπιδύνει.

ιη.   νότος πνέει τὴν νύκτα· καὶ ἀλλοιοῦται ὁ ἀήρ.

κ.   Ὑάδες ἀνίσχουσιν· καὶ βορρᾶς πνεῖ.

κα.   Ἀρκτοῦρος ἐπιδύνει.

κδ.   Αἴξ ἑσπέριος κρύπτεται καὶ κινεῖται ὁ ἀὴρ πρὸ β´ ἡμερῶν.

κη.   Λύρα ἀνίσχει· καὶ νότος πνεῖ.

λ.   ἀνίσχουσιν Ὑάδες· καὶ βρονταί.

Ἰούνιος

α.   Ἀετὸς ἑσπέριος ἐπιτέλλει· καὶ βρέχει.

β.   Ἀετὸς ἀνίσχει· καὶ βρέχει καὶ πνεῖ νότος.

ζ. Ὠρίων ἀνίσχει.

η. ὄρθριος Ἀρκτοῦρος ἐπιδύνει.

θ. Δελφὶς ἑσπέριος ἐπιτέλλει.

ιε. ὦμοι Ὠρίωνος ἀνίσχουσιν.

ιζ. ἄστρον κρυπτὸν ἐκφαίνεται.

κα. Ὀφιοῦχος ἀνίσχει.

κγ. ἐπιτολὴ Ὠρίωνος· καί εἰσι τροπαὶ θεριναί· καὶ ἀλλοιοῦται σφόδρα ὁ ἀὴρ πρὸ τριῶν ἡμερῶν.

Ἰούλλιος

α. Ὠρίων ἀνίσχει.

ζ. Στέφανος πρὸς ὄρθρον δύνει.

η. ὁ Κηφεὺς δύνει.

ι. ὁλόκληρος Ὠρίων ἐπανίσχει.

ιδ. Προκύων ἐπανίσχει.

Αὔγουστος

α. ἰκτῖνος ὄρθρου ἀνίσχει.

    Στέφανος δύνει.

    Λύρα κρύπτεται ἑῴα.

    ὁ Δελφὶν καὶ ὁ Λαγωὸς ἅμα ἑῷοι πρὸς ὄρθρον δύνουσιν.

    ὄρθρου Δελφὶν ἀνίσχει.

    Προτρυγητὴς ἑῷος ἐπιτέλλει· καὶ ὁ Ὀιστὸς δύνει. ἔστιν τέλος τῶν μετὰ τὴν ἐπιτολὴν τοῦ Κυνὸς ἡμερῶν μ´.

September[225]

1. Thundery and interchanging of winds.
2. The whole of Andromeda rises.
3. Arcturus rises with Vindemiatrix.
4. Sagitta appears for all.
5. Horrible thunders and interchanging of winds again.
6. The star Pegasus rises.

---

[225] Bianchi, in order to compare this text to Clodius Tuscus, rearranges the parapegma to begin the year in January.

7. Capella rises, and a south-west wind blows.
8. Arcturus is carried up.
9. Arcturus rises at dawn with Virgo.
10. Arcturus generally appears.
11. Pegasus rises with the Pleiades.
12. Swallows are no longer seen.
13. Most of Pegasus appears.
14. The whole of ὁτηλεύκη appears.
15. Clear Capella appears.
16. West wind with the south-west, a fortunate east wind.[226]
17. The arctic wind blows very gently.
18. The sun is in Libra, causing a good south wind.
19. Autumnal equinox.
20. The cup appears.
21. *Argo sets.*[227]
23. Rising of Centaurus, disturbance of the air and the sea.
24. The air is cloudy, and disturbed from Centaurus.
25. The Haedi rise and the south wind blows, and there is the greatest disturbance of the air three days before. Because of this it is necessary to be careful neither to phlebotomize, nor purge, nor to change the body violently at all. Similarly, from the 15th through the 24th of this (month).
26. The Haedi rise with the sun.
27. From the Haedi, a south wind usually.
28. Auriga begins to appear.
29. The Pleiades appear at the close of the days.
30. Capella rises, and the triangle rises in the evening, and the air is moved one day before; sometimes there is also a disturbance.

October

1. The Pleiades appear at daybreak.
2. And on the 2nd the same,[228] but also the 3rd.
3. Auriga setting, and towards the 4th again.
4. The Haedi rise, and Corona on the 5th.
5. Corona rises in the morning, and there is a great change of the air.

---

[226] Here and in the previous entry, Bianchi notes a tendency toward poetic metre and language.
[227] Written in a second hand.
[228] I follow the vagaries of the MS, with its inconsistent use of 'on the xth'.

6. The Haedi rise in the evening.
8. And on the 8th, Corona rises with the Haedi.
9. And on the 9th, the Haedi appear with the Pleiades.
10. And on the 10th, Libra begins to be about to rise.
11. On the 11th, Corona rises towards daybreak.
12. On the 12th the same, and the 13th.
    Similar, and the day after it. Towards the 3rd, likewise.[229]
16. And on the 16th, Orion rises completely.
17. The Hyades rise in the evening, and there is a sufficient disturbance of the air.
20. On the 20th, the Pleiades set, and the (2)1st.
22. And on the (2)2nd, the whole tail of Leo sets.
    The Pleiades set, and there is the greatest disturbance of the air one day before.[230]
24. Towards the (2)4th, the Pleiades rise, but also on the (2)7th.
28. Arcturus is setting.[231]
29. And on the 29th the same. On the 30th, again, the whole of Orion disappears, and Lyra is about to rise.

November

1. The Pleiades set.
2. Arcturus, and a change of the air.
3. Lyra at daybreak, and towards the 4th.
5. Lyra rises close to the sun.
6. Arcturus sets completely at daybreak, and the Pleiades set in the morning, and the air begins to be settled.
7. The Pleiades set, and Orion with them.
9. The same. On the 9th, Antares appears, and they say there is a stormy change.[232]
10. And on the 10th, the same.
11. On the 11th, Lyra rises in the morning.
12. The Pleiades, and towards the 12th the middlemost part of Taurus rises completely.
13. Again, the Pleiades and Orion set completely towards daybreak, and towards the same, Lyra is about to rise.

---

[229] Text following Bianchi. Date omitted in MS. Bianchi restores thirteen.
[230] Date omitted in MS. Bianchi restores twenty-three.   [231] δυτικός, an unusual wording.
[232] χειμέριος τροπή. Compare Clodius Tuscus.

15. And Lyra with Orion, and a disturbance of the air.
19. The head of Taurus sets.
20. Similar. On the (2)1st,[233] the Hyades set with Lepus towards daybreak.
22. And on the (2)2nd of (the month). The (2)3rd, Orion.
24. On the (2)4th and (2)5th as well, and the (2)6th, the prelude to winter begins the season.
27. From the 27th until the 30th it is the same.

December

1. Sirius sets at daybreak. Melting of air. It has been observed by many that if it is stormy on this day, then a disturbance of the air will, for the most part, remain for thirty-seven days. And if it is a nice day, it signifies the same thing.[234]
2. Sirius sets toward evening, and all these things happen until the 16th.
10. The Hyades set in the morning.
11. Capella sets in the morning, and there is a disturbance one day after.
17. On the 17th, the sign of Capricorn admits the sun.
18. The 18th, Capella rises again.
23. The 23rd, Capella sets at daybreak, and it is the winter solstice.
28. And on the 16th, in the evening [. . .]. On the 28th, the sun turns back.
30. But on both the 30th and the 31st, Delphinus is about to rise, and the sun is elevated.

January

1. The sun is elevated. Aquila with Corona.
2. Orion sets.
3. The middle of Aquila sets. Aquila sets, winged bird,[235] and there is a variable change in the air.
4. On the 4th, Delphinus rises at daybreak, and much south wind blows.
5. On the 5th, Aquila, Lyra, Corvus set, and there is a disturbance two days after.

---

[233] ὁμοίως δέ· τὴν πρωτίστην ταύτης. I think Bianchi is right to read the ταύτης as referring to the ten-day period, and the shortening the formula for dates like 'the 22nd' to just 'the 2nd' is common in this calendar. Compare e.g. 21, 22, 24 Oct.

[234] Compare the Aëtius, Quintilius, and Oxford parapegmata.

[235] πτηνὸς ὄρνις. The entry is unique in a parapegma and I am not sure what to do with it.

6. On the 6th, Aquila sets. The twin rises.
7. On the 7th, the middle of Lyra rises.
8. On the 8th, Delphinus sets, and the south wind blows with the arctic.
9. Both the 9th and the 10th signify the same thing.
11. On the 11th, a disturbance of the air prevails.
12. On the 12th, Lyra sets in the evening.
13. On the 13th, Sagitta, and showery at night.
15. On the 15th, Lyra sets around daybreak.
19. On the 19th, Delphinus rises, and the north wind blows.
22. On the 22nd, Lyra sets with Aquila.
25. Lyra sets and moves the air three days before.
26. Lyra sets in the evening, as does Regulus.
27. On the 27th, Delphinus sets in the evening, and it causes winds and storms.
29. On the 29th, Delphinus sets.
30. Lyra sets, and there are clouds and showers.
31. On the 31st, stormy wind.

February

1. The hidden star rises.[236]
2. The west wind is about to blow.
3. On the 3rd, Lyra sets.
4. On the 4th, Delphinus sets in the evening.
5. On the 5th, the middle of Aquarius rises.
6. On the 6th, Lyra sets and the west wind blows.
7. A west or a north wind blows.
9. On the 9th, the hidden star appears.
10. Violent winds.
11. On the 11th, Arcturus, and the south wind blows.
12. Sagittarius sets in the evening, and it is stormy.
14. The cup is about to rise.
15. The sun is in Pisces.
16. An arctic wind from this.
17. Virgo towards setting.

---

[236] Compare Clodius Tuscus, 1 Feb.: 'The stars are hidden.' This text commonly adds a stellar phase to the phrase κρυπτὸν ἄστρον.

18. Sagitta setting completely.
19. The arctic wind sets, or rather it blows.
20. The middle of Leo sets.
21. Arcturus sets.
22. Sagitta sets, and there is a spring disturbance.
23. Aquarius rises, and storms accompany it.
24. The north wind blows.
25. Arcturus is bright.
26. The swallows appear.
27. Sagitta sets.

March

1. The south wind and the south-west wind blow.
2. Vindemiatrix rises.
3. Arcturus rises.
7. Pegasus sets early.
8. Birds appear at the time of Aquarius.
9. The kite begins to rise, and the fish sets at daybreak.
10. At daybreak, Pegasus at the point of setting.
11. It must be said to be the departure of winter.
12. The southern fish finishes.
13. Capella is about to rise in the evening.
14. The north wind blows completely.
15. The whole of Pegasus sets.
16.[237] On the 17th, similar, and it is drizzly.
17. The light-bringer is in Aries.
18. The kite appears until the equinox.
19. On the 19th, Pegasus rises in the morning.
20. The weather is all good.
21. Pegasus sets in the morning, and there is a disturbance of the air.
22. Aries rises, and it showers.
25. Pegasus begins to set.
26. Pisces rises from the south.
31. Disturbance of the air.

[237] Something has gone awry with the dates here.

April

1. The Pleiades disappear in the evening.
4. On the 4th, the south-west wind is about to blow.
5. The Hyades are about to rise, and showery.
7. The Pleiades rise.
8. The west wind blows at daybreak.
9. A squall comes from the south.
10. The north wind blows all day.
11. A cold wind, and a north wind.
13. The north wind blows from the Hyades.
14. The hidden star is about to set.
15. The Hyades appear setting.
18. The south-west wind is about to blow.
19. The Hyades set.
20. The west wind blows after a north wind.
21. The head of Taurus sets.
22. The Pleiades rise.
23. Lyra appears at first night.
25. Sirius sets.
27. The south wind blows, and it showers.
30. Sirius disappears.

May

1. The Hyades rise at sunrise.
3. The whole of Centaurus appears, and the west wind blows.
4. Lyra rises in the evening, and the air is changed.
6. The Pleiades rise in the morning.
7. Pleiades.
9. Lyra rises. The Hyades set completely.
11. The Pleiades appear again.
12. The south wind blows.
13. The Hyades set, and the south wind blows.
14. Lyra rises in the morning, and the south wind blows.
17. Sirius sets completely.
18. The south wind blows at night, and the air is changed.
20. The Hyades rise, and the north wind blows.
21. Arcturus sets completely.

24. Capella disappears in the evening, and the air is moved two days before.
28. Lyra rises, and the south wind blows.
30. The Hyades rise, and thundery.

June

1. Aquila rises in the evening, and it showers.
2. Aquila rises, and it showers, and the south wind blows.
7. Orion rises.
8. Arcturus sets completely at daybreak.
9. Delphinus rises in the evening.
15. Shoulders of Orion rise.
17. The hidden star appears.
21. Ophiucus rises.
23. Rising of Orion, and it is the summer solstice, and the air is especially changed three days before.

July

1. Orion rises.
7. Corona sets before daybreak.
8. Cepheus sets.
10. All of Orion rises completely.
14. Procyon rises completely.

August

1. The kite rises at daybreak.
   Corona sets.[238]
   Lyra disappears in the morning.
   Delphinus and Lepus set in the morning, towards daybreak.
   Delphinus rises at daybreak.
   Vindemiatrix rises in the morning, and Sagitta sets. It is the end of the 40 days after the rising of Sirius.

---

[238] There are no dates for the next handful of entries. Bianchi borrows dates for these phases from Clodius Tuscus and Aëtius. The format of Bianchi's edition makes it unclear whether the MS indicates lacunae where the dates should be or whether these entries are treated as continuations of the entry for 1 Aug.

A.xviii. Iudicia parapegma[239]

*prima die mensis **Septembris,** Icarus custos plaustri apparet cum solis ortu, et mutatur aer in .vii. horis, hoc fit inter diem et noctem.*

septima die eiusdem mensis <. . .> vespertinus apparet, et mutatur aer in ventum.

quarta decima die eiusdem Arcturus – id est Septemtrion – apparet cum solis ortu. mutatur aer in crastinum.

decima nona die eiusdem mensis, Sichis – id est Subole<s> quam Virgo tenet in manu – apparet. tunc mutatur aer ante duos dies.

vicesima quinta die eiusdem mensis Alferat – id est Equus – ponet, et erit calida mutatio aeris.

*sexta die mensis **Octobris** Stephanon – id est Corona – apparet, et est nimia mutatio aeris.*

septima die eiusdem Erifi – id est Edi – vespertini apparent, et tunc fit magna turbatio aeris.

vicesima tertia die mensis illius Pliades – id est Virgilie – cum solis ortu ponunt, et tunc fit magna turbatio aeris.

*sexta die mensis **Novembris** Lapsidis – id est Lucidus – ponit, et incipit tunc obscurari aer.*

tertia decima die eiusdem Lira apparet.

vicesima prima die Yades – id est Facule – ponunt et mutatur aer in crastinum.

vicesima septima die eiusdem <. . .> vespertinus apparet et Stephanon – id est Corona – ponet, et mutatur aer.

*prima die mensis **Decembris** Cion – id est Canis – apparet, et tunc fit magna turbatio aeris usque ad aliquantos dies.*

decima die mensis eiusdem Erifi – id est Edi – ponunt.

vicesima prima die Ectos – id est Aquila – apparet, et Esion – id est Eridanus – ponet, et erit turbatio aeris post unam diem.

vicesima tertia die Egera – id est Capra – oritur, et mutatur aer.

*quarta die mensis **Ianuarii** Delphinus apparet.*

quinta die Ectos vespertinus ponet.

vicesima quinta die Ectos – id est Aquila – ponet, et stella regia appellata Tuberoni in pectore Leonis occidit matutina, et turbatur aer ante .iii. dies.

---

[239] Latin text from Burnett, 1993.

vicesima octava Delphinus vespertinus apparet.
vicesima nona die Lira vespere ponet, et mutatur aer.

*sexta die **Februarii** zephirus flat.*
vicesima secunda die Ipos – id est Equus – vespertinus ponet.
vicesima tertia die Arcturus – id est Septemtrion – apparet.
vicesima quinta die Kalende .v. Libra apparet, et mutatur aer.

*quinta die mensis **Marcii** Cancer apparet.*
octava die eiusdem Piscis Aquilonis.
nona die eiusdem Orion apparet.
octava decima die eiusdem Esion – id est Eridanus – apparet.
vicesima prima die Ipos – id est Equus – <. . .> et est turbatio aeris.

*prima die **Aprilis** Pliades – id est Vergilie – paulum apparet.*
19 die mensis eiusdem Pliades vespere ponunt.
21 die Pliades apparent, et fit magna turbatio aeris.
vicesima septima die Orion vespertinus ponet, et mutatur aer usque in 9 die<i>
horas.

*prima die **Maii** Yades – id est Succule – cum solis ortu appare<n>t, et mutatur*
aer usque in 4 dies.
quarta die Lira vespere apparet, et mutatur aer nimis.
die 7 cum solis ortu Pliades cum Esion apparent, et incipit aer obscurari.
19 die eiusdem Lapsidis – id est Lucidus – apparet et mutatur aer ante duos
dies.
vicesima quarta die Exeon vespertinus ponet, et movetur aer ante unam diem.

*die secunda **Iunii** Ectos – id est Aquila – vespertinus apparet et fit caliditas et*
mutatur aer.
die 5 eiusdem Arct<ur>us ponet, et mutatur aer in duos dies.
die 9 vespertinus apparet Delfinus, et mutatur <aer>[240] *usque in 10 hora<m>*
diei.
25 die Orion incipit apparere, et mutatur aer nimis ante 3 dies.

*die tertia **Iulii** Orion plenus apparet, et fit caliditas in Ariete.*
die quarta Prochion – id est Antichanus – apparet, et est mutatio aeris.
octava decima die Cion – id est Canis – apparet plenus, et fit magna turbatio
aeris ante duos dies.

---

[240] Subject inadvertently omitted by Burnett.

vicesima quinta die Teros ponet, et movetur <aer> ante tres dies.

*quintadecima die **Augusti** Lira apparet, et mutatur aer.*
decima nona die Delphinus ponet et Friccos et Ydre pars prior – id est Oridus
*   – apparet.*
vicesima octava die Protriguis – id est Antevendemor – apparet et Cystos ida
*   Esie desunt, et est finis Cionis – id est Canis – ante unam diem.*

The first day of the month of **September,** Bootes, guardian of the Wain,
    appears at sunrise, and the air is changed for seven hours, which hap-
    pens between day and night.
The seventh day of this month [. . .] appears in the evening, and the air is
    changed with respect to wind.
The fourteenth day of this, Arcturus – that is, *Septemtrion* – appears at
    sunrise. The air is changed on the next day.
The nineteenth day of this month, Spica[241] – that is, the shoot that Virgo
    holds in her hand – appears. At this time the air is changed two days
    before.
The twenty-fifth day of this month, *Alferat* – that is, Pegasus – sets,[242] and
    there will be a hot change in the air.

The sixth day of the month of **October,** *Stephanon* – that is, Corona – appears,
    and there is an excessive change in the air.
The seventh day of this, *Erifi* – that is, the Haedi[243] – appear in the evening,
    and at this time there is a great disturbance in the air.
The twenty-third day of that month, the Pleiades – that is, *Vergilie* –
    set at sunrise, and at this time there is a great disturbance in the
    air.

---

[241] The text has *Sichis*, which is probably a corruption of the Greek Στάχυς, Spica. Here
    and commonly in the text, we see the transcription of Greek and even Arabic star names
    (see 25 Sept., for example) alongside their Latin equivalents. The formula in this text is always
    'on such-and-such a date, [Greek or Arabic star name] – that is, [Latin star name] –
    appears/sets'. In English, our star names are confusingly sometimes from the Latin, sometimes
    from the Greek, and sometimes from the Arabic. This means that in my translation, the
    English star name will sometimes be before the ' – that is', and sometimes after, depending
    on whether our name is taken from Latin or not. Where English uses neither of the
    names offered in the text, I have translated the Latin names into English, and left the
    non-Latin names in italics. I have preserved the quirks of spelling in the text where
    possible.
[242] This text regularly translates the Greek δύνειν as *ponere*. See Burnett, 1993, p. 31.
[243] The text has *Edi*.

The sixth day of the month of **November,** *Lapsidis* – that is, the 'Bright one'[244] – sets, and at this time the air begins to be darkened.

The thirteenth day of this, Lyra appears.

The twenty-first day, the Hyades – that is, *Facule* – set, and the air is changed on the next day.

The twenty-seventh day of this, [. . .] appears in the evening, and *Stephanon* – that is, Corona – sets, and the air is changed.

The first day of the month of **December,** *Cion* – that is, Sirius – appears, and at this time there is a great disturbance in the air for several days.

The tenth day of this month, *Erifi* – that is, the Haedi – set.

The twenty-first day, *Ectos* – that is, Aquila – appears, and *Esion* – that is Eridanus – sets, and there will be a disturbance in the air one day after.

The twenty-third day, *Egera* – that is, *Capra* – rises,[245] and the air is changed.

The fourth day of the month of **January,** Delphinus appears.

The fifth day, *Ectos* sets in the evening.

The twenty-fifth day, *Ectos* – that is, Aquila – sets, and the star in the breast of Leo, called 'the Royal Star' by Tubero,[246] sets in the morning, and the air is disturbed three days before.

The twenty-eighth, Delphinus appears in the evening.

The twenty-ninth day, Lyra sets in the evening, and the air is changed.

The sixth day of **February,** the west wind blows.

The twenty-second day, *Ipos* – that is, Pegasus – sets in the evening.

The twenty-third day, Arcturus – that is, *Septemtrion* – appears.

The twenty-fifth day (the fifth before the Kalends), Libra appears, and the air is changed.[247]

The fifth day of the month of **March,** Cancer appears.

The eighth day of this, Pisces Borealis.

The ninth day of this, Orion appears.

The eighteenth day of this, *Esion* – that is, Eridanus – appears.

---

[244] I am unaware of either of these terms as star names. Burnett, 1993, is right that *Lucidus* is probably a translation of λαμπρός, as in 'the bright star [in such-and such]'.

[245] Unusually for this text, the verb here is *oritur*. Burnett thinks, plausibly, that the star names are a corruption of *Aigokeros – id est Capricornus*.

[246] Compare Pliny. *HN* xviii.235.

[247] Burnett thinks that the dating formula here, combined with the lack of an entry for this date in either Pliny or Aëtius, points to a third source for this parapegma.

The twenty-first day, *Ipos* – that is, Pegasus – and there is a disturbance in the air.[248]

The first day of **April,** the Pleiades – that is, *Vergilie* – appear a little.[249]

The nineteenth day of this month, the Pleiades set in the evening.

The twenty-first day, the Pleiades appear, and there is a great disturbance in the air.

The twenty-seventh day, Orion sets in the evening, and the air is changed for nine hours of the day.

The first day of **May,** the Hyades – that is, *Succule* – appear at sunrise, and the air is changed for four days.

The fourth day, Lyra appears in the evening, and the air is changed excessively.

The seventh day, the Pleiades with *Esion*[250] appear at sunrise, and the air begins to darken.

The nineteenth day of this, *Lapsidis* – that is, the 'Bright One' – appears and the air is changed two days before.

The twenty-fourth day, *Exeon* sets in the evening, and the air is moved one day before.

The second day of **June,** *Ectos* – that is, Aquila – appears in the evening, and there is heat and the air is changed.

The fifth day of this, Arcturus sets, and the air is changed for two days.

The ninth day, *Delfinus* appears in the evening, and the air is changed until the tenth hour of the day.

The twenty-fifth day, Orion begins to appear, and the air is excessively changed three days before.

The third day of **July,** Orion appears completely, and there is heat in Aries.

The fourth day, Procyon – that is, *Antichanus* – appears, and there is a change in the air.

The eighteenth day *Cion* – that is, Sirius – appears completely, and there is a great disturbance in the air two days before.

The twenty-fifth day, *Teros* sets, and the air is moved three days before.

The fifteenth day of **August,** Lyra appears, and the air is changed.

The nineteenth day, Delphinus sets and *Friccos* and the first part of Hydra – that is, *Oridus* – appears.

---

[248] Burnett restores the lost verb as 'appears', although Pliny has 'Pegasus sets in the morning' on this date.

[249] *paulum apparet.*   [250] Eridanus: see 18 March.

The twenty-eighth day, Vindemiatrix[251] – that is, *Antevendemor* – appears, and *Cystos ida Esie desunt*,[252] and it is the end of *Cion* – that is, Sirius – one day before.

A.xix. al-Bīrūnī parapegma[253]

Chapter XIII

On the days of the Greek calendar as known both among the Greeks and other nations.

The Greek year agrees with the solar year; its seasons retain their proper places like the natural seasons of the solar year; it revolves parallel with the latter, and its single parts never cease to correspond with those of the latter, except by that quantity of time (the *portio intercalationis*) which, before it becomes perceptible, is appended to the year and added to it as one whole day (in every fourth year) by means of intercalation. Therefore the Greeks and Syrians and all who follow their example fix and arrange by this kind of year all annual, consecutive occurrences, and also the meteorological and other qualities of the single days that experience has taught them in the long run of time, which are called *anwā'* and *bawāriḥ*.

Regarding the cause of these *anwā'*,[254] scholars do not agree among each other. Some derive them from the rising and setting of the fixed stars, among them the Arabs. (Some poet says):

Those are my people (a bad set) like the *Banāt-Na'sh*,
Who do not bring rain like the other stars;

that is, they are good-for-nothing people like the *Banāt-Na'sh*, whose rising and setting do not bring rain.

Others, again, derive them from the days themselves, maintaining that they are peculiarities of them, that such is their nature, at least, on an average, and that besides they are increased or diminished by other causes. They say,

---

[251] *Protriguis*
[252] Burnett is right that this looks to be a corruption of Pliny's *et etesiae desiunt*, 'and the Etesian winds cease'.
[253] Translation is Sachau's, 1879, with minor modifications. Common changes are that I translate ἐπισημασία and ἐπισημαίνει where Sachau just transliterates them, and I have tried to update Sachau's system of Arabic transliteration as best I can.
[254] In a note, Sachau points out that he translates *anwā'* as ἐπισημασία, although it can also mean 'lunar mansion'. See Sachau, 1879, p. 428.

for instance: The nature of the season of summer is heat, the nature of the season of winter is cold, sometimes in a higher degree, sometimes less. The excellent Galen says:

To decide between these parties is only possible on the basis of experiment and examination. But to examine this difference of opinion is not possible except in a long space of time, because the motion of the fixed stars is very little known and because in a short space of time we find very little difference in their rising and setting.

Now, this opinion has filled Sinān ibn Thābit ibn Qurra with surprise. He says in his book on the *anwā'*, which he composed for the Khalif Almu'taḍid:

I do not know how Galen came to make such a mistake, skilled as he was in astronomy. For the rising and setting of the stars differ greatly and evidently in different countries. E.g., *Suhail* rises at Baghdad on the 5th of Ilūl, at Wāsiṭ two days later, at Basra somewhat earlier than at Wāsiṭ. People say: 'the *anwā'* differ in different countries'. But that is not the case. On the contrary, they occur always on one and the same day (everywhere); which proves that the stars and their rising and setting have nothing to do with this matter.

Afterwards he has given the lie to himself, though it is correct what he said, *viz.*, that the rising and setting of the stars are not to be considered as forming one of the causes of the *anwā'*, if you limit his assertion by certain conditions and do not understand it in that generality in which he has proclaimed it.

Further he (Sinān ibn Thābit) says:

The *anwā'* of the Arabs are mostly correct for al-Ḥijāz and the neighbourhood, those of the Egyptians for Egypt and the coasts of the sea, those of Ptolemy for Greece and the neighbouring mountains. If anybody would go to one of those countries and examine them there, he would find correct what Galen says regarding the difficulty of an examination of the *anwā'* in a short space of time.

In this respect he (Sinān) is right. Galen mentions and believes only what he considers as a truth, resting on certain arguments, and keeps aloof from everything that is beset with doubt and obscurity.

Sinān relates of his father, that he examined the *anwā'* in Iraq about thirty years with the view of finding certain principles with which to compare the *anwā'* of other countries. But fate overtook him before he could accomplish his plan.

Whichever of the two theories may be correct, whether the *anwā'* are to be traced back to the days of the year or to the rising and setting of the Lunar Stations, in any case there is no room for a third theory. To each of these theories, whichever you may hold to be correct, certain conditions attach, on which the correctness of the *anwā'* depends, that is, to prognosticate the character of the year, the season, the month, whether it will be dry or moist, whether it will answer to the expectations of people or not, to prognosticate it by means of the signs and proofs, of which the astronomical books on meteorology are full. For if the *anwā'* agree with those signs and proofs, they are true and will be fulfilled in their entire extent; if they do not agree, something different will occur.

Thus the matter stands between these two theories.

Sinān ibn Thābit prescribes that we should take into regard whether the Arabs and Persians agree on a *nau'*. If they do agree, its probability is strengthened and it is sure to take place; if they do not agree, the contrary is the case.

I shall mention in this book the comprehensive account of Sinān in his book on the *anwā'* and the proper times for secular affairs occurring in the Greek months. Of the rising and setting of the Lunar Stations I shall speak in a special chapter at the end of this book. For since the astronomers have found that their rising and setting proceed according to one and the same uniform order in these months, they have assigned them to their proper days, in order to unite them and prevent them from getting into confusion. God lends support and help!

Tishrīn I

1. People expect rain (Euctemon and Philippus); turbid air (Egyptians and Callippus).
2. Turbid wintry air (Callippus, Egyptians, and Euctemon); rain, (Eudoxus and Metrodorus).
3. Nothing mentioned.
4. Wearing wind (Eudoxus); wintry air (Egyptians).
5. Wintry air (Democritus); beginning of the time of sowing.
6. North wind (Egyptians).
7. South wind (Hipparchus).
8. Nothing mentioned. Wintry air, according to Sinān.
9. There is a change in the weather (Eudoxus); east wind (Hipparchus); west wind (Egyptians).
10. Nothing mentioned.
11. A change in the weather (Eudoxus and Dositheus).
12. Rain (the Egyptians).
13. Unsteady wind, a change in the weather, thunder, and rain (Callippus); north wind or south wind (Eudoxus and Dositheus).

Sinān attests that this is frequently true. On this day the waves of the sea are sure to be in great commotion.

14. A change in the weather and north wind (Eudoxus).
15. Change of the winds (Eudoxus).
16. Nothing mentioned.
17. Rain and a change in the weather (Dositheus); west wind or south wind (Egyptians).
18. Nothing mentioned.
19. Rain and a change in the weather (Dositheus); west wind or south wind (Egyptians).
20, 21. Nothing mentioned.
22. Unsteady, changing winds (Egyptians). On this day the air begins to get cold. It is no longer time for drinking medicine and for phlebotomy except in case of need. For the *Favourable Times* for such things are always then, when you intend thereby to preserve the health of the body. For if you are compelled to use such means, you cannot wait for a night or day, for heat or cold, for a lucky or unlucky day. On the contrary, you use it as soon as possible, before the evil takes root, when it would be difficult to eradicate it.
23. A change in the weather (Eudoxus); north wind or south wind (Caesar).
24. A change in the weather (Callippus and Egyptians).
25. A change in the weather (Metrodorus); change in the air (Callippus and Euctemon).
26. Nothing mentioned.
27. Wintry air (Egyptians).
28. Nothing mentioned. It is a favourable day for taking a warm bath and for eating things that are of a sharp, biting taste, nothing that is salt or bitter.
29. Hail or frost (Democritus); continual south wind (Hipparchus); tempest and wintry air (Egyptians).
30. Heavy wind (Euctemon and Philippus). The kites, the white carrion-vultures *(vultur percnopterus)*, and the swallows migrate to the lowlands, and the ants go into their nest.
31. Violent winds (Callippus and Euctemon); wind and wintry air (Metrodorus and Caesar); south wind (Egyptians). God knows best!

Tishrīn II

1. Clear (*lit.* unmixed) winds (Eudoxus and Conon).

2. Clear air with cold north wind and south wind.
3. South wind blows (Ptolemy); west wind (Egyptians); north or south wind (Eudoxus); rain (Euctemon, Philippus, and Hipparchus).
4. A change in the weather (Euctemon); rain (Philippus).
5. Wintry air and rain (Egyptians).
6. South or west wind (Egyptians); wintry air (Dositheus). Sinān says that this is borne out by practical experience.
7. Rain with whirlwind (Meton); cold wind (Hipparchus). This is the first day of the rainy season, when the sun enters the 21st degree of Cancer. Astrologers take the horoscope of this time and derive therefrom an indication as to whether the year will have much rain or little. Herein they rely upon the condition of Venus at the times of her rising and setting, I believe, however, that this is only peculiar to the climate of Iraq and Syria, not to other countries, for very frequently it rains with us in Khwārizm even before this time. Abū al-Qāsim 'Ubaid Allāh ibn Abdallāh ibn Khurdādhbih relates in his *Kitāb al-masālik wa-l-mamālik* that in Ḥijāz and Yaman it rains during Ḥazīrān, Tammūz, and part of Īlūl. I myself have been dwelling in Jurjān during the summer months, but there never passed ten consecutive days during which the sky was clear and free from clouds, and when it did not rain. It is a rainy country. People relate that one of the khalifs, I think it was al-Ma'mūn, stayed there during forty days whilst it rained without any interruption. So he said: 'Lead us out of this pissing, splashing country!'

The nearer a district is to Ṭabaristān, the more its air is moist, the more rainy it is. The air of the mountains of Ṭabaristān is so moist that if people break and pound garlic on the tops of the mountains, rain is sure to set in. As the cause of this subject, the vice-judge, al-Āmūlī, the author of the *Kitāb al-ghurra*, mentions this, that the air of the country is moist and dense with stagnant vapours. If, now, the smell of garlic spreads among these vapours, it dissolves the vapours by its sharpness and compresses the density of the air, in consequence of which rain follows.

Granted, now, that this be the cause of this appearance produced by the pounding of garlic, how do you, then, account for the famous well in the mountains of Farghāna, where it begins to rain as soon as you throw something dirty into this well?

And how do you account for the place called 'the shop of Solomon the son of David', in the cave called Ispahbadhān in the mountain of Ṭāq in

Ṭabaristān, where heaven becomes cloudy as soon as you defile it by filth or by milk, and where it rains until you clean it again?

And how do you account for the mountain in the country of the Turks? For if the sheep pass over it, people wrap their feet in wool to prevent their touching the rock of the mountain. For if they touch it, heavy rain immediately follows. Pieces of this rock the Turks carry about, and contrive to defend themselves thereby against all evil coming from the enemy, if they are surrounded by them. Now, those who are not aware of these facts consider this as a bit of sorcery on the part of the Turks.

Of a similar character is a fountain called 'the pure one' in Egypt in the lowest part of a mountain which adjoins a church. Into this fountain sweet, nicely smelling water is flowing out of a source in the bottom of the mountain. If, now, an individual that is impure through pollution or menstruation touches the water, it begins at once to stink, and does not cease until you pour out the water of the fountain and clean it; then it regains its nice smell.

Further, there is a mountain between Herāt and Sijistān, in a sandy country, somewhat distant from the road, where you hear a clear murmur and a deep sound as soon as it is defiled by human excrements or urine.

These things are natural peculiarities of the created beings, the causes of which are to be traced back to the simple elements and to the beginning of all composition and creation. And there is no possibility that our knowledge should ever penetrate to subjects of this description.

There are other districts of quite another character from that of the mountains of Ṭabaristān, e.g. Fusṭāṭ in Egypt, and the adjacent parts, for there it rains very seldom. And if it rains, the air is infected, becomes pestilential and hurts both animals and plants. Such things (i.e., such climactic differences) depend upon the nature of the place and its situation, whether it lies in the mountains or on the sea, whether it is a place of great elevation or a low country; further, upon the degree of northern or southern latitude of the place.

8. Rain and wintry air (Euctemon); wintry air and whirlwinds (Metrodorus); south wind or *Eurus*, i.e., south-east wind (Euctemon); east wind (Egyptians).

9. Nothing mentioned.

10. Wintry air and whirlwinds (Euctemon and Philippus); north wind, or cold south wind and rain (Hipparchus).

11. A change in the weather (Callippus, Conon, and Metrodorus). Sinān says that this is borne out by experience.
12. Wintry air (Eudoxus and Dositheus).
13. A change in the weather (Eudoxus); wintry air on land and sea (Democritus). Ships that are at sea on this day put in to shore, and navigation to Persia and Alexandria is suspended. For the sea has certain days when it is in uproar, when the air is turbid, the waves roll, and thick darkness lies over it. Therefore navigation is impracticable. People say that at this time there arises the wind at the bottom of the sea that puts the sea in motion. This they conclude from the appearance of a certain sort of fishes which then swim in the upper regions of the sea and on its surface, showing thereby that this storm is blowing at the bottom.

    Frequently, people say, this submarine storm rises a day earlier. Every sailor recognizes this by certain marks in his special sea. For instance, in the Chinese sea this submarine storm is recognized by the fishing-nets rising of themselves from the bottom of the sea to its surface. On the contrary, they conclude that the sea bottom is quiet if a certain bird sits hatching her eggs – for they hatch in a bundle of chips and wood on the sea, if they do not go on land nor sit down there. They lay their eggs only at that time when the sea is quiet.

    Further, people maintain that any wood which is cut on this day does not get worm-eaten, and that the white ant does not attack it. This peculiarity perhaps stands in connection with the nature of the mixture of the air on this special day.
14. Wintry air (Caesar); south wind or Eurus, i.e., south-east wind (Egyptians).
15. Nothing mentioned.
16. Wintry air (Caesar).
17. Rain (Eudoxus); wintry air (Caesar); north wind during night and day (Caesar).
18. Nothing registered.
19. Sharp wintry air (Eudoxus).
20. North wind (Eudoxus); severe wintry air (Egyptians). People say that on this day all animals that have no bones perish. This, however, is different in different countries. For I used to be molested by the gnats, that is, animals without bones, in Jurjān, whilst the sun was moving in the sign of Capricorn.
21. Wintry air and rain (Euctemon and Dositheus).

22. Very wintry air (Eudoxus). On this day people forbid to drink cold water during the night, for fear of the Yellow Water.

23. Rain (Philippus); wintry air (Eudoxus and Conon); continual south wind (Hipparchus and Egyptians). On this day falls the feast of gathering the olives, and the fresh olive-oil is pressed.

24. Light rain (Egyptians).

25, 26. Nothing mentioned.

27. In most cases a disturbance of the air on land and sea (Democritus); a change in the weather (Dositheus); south wind and rain (Egyptians).

28. Nothing mentioned. People say that on this day the waves of the sea roll heavily and that there is very little fishing.

29. Wintry air (Eudoxus and Conon); west or south wind and rain (Egyptians).

30. Nothing mentioned by the authorities hitherto quoted, nor by others.

Kānūn I

1. Wintry air (Callippus, Eudoxus, and Caesar). On this day people hold a fair in Damascus, which is called 'the fair of the cutting of the *ben*-nut', i.e., *nux unguentaria*.

2. Pure winds (*lit.* not mixed) (Euctemon and Philippus); sharp, wintry air (Metrodorus).

3. Wintry air (Conon and Caesar); light rain (Egyptians).

4. (Missing.)

5. Wintry air (Democritus and Dositheus). The same is confirmed by Sinān.

6. Wintry air (Eudoxus); vehement north wind (Hipparchus).

8. Nothing mentioned.

9. Wintry air and rain (Callippus, Euctemon, and Eudoxus).

10. Sharp wintry air (Callippus, Euctemon, and Metrodorus); thunder and lightning, wind and rain (Democritus).

11. South wind and a change in the weather (Callippus); wintry air and rain (Eudoxus and Egyptians). According to Sinān this is borne out by practical experience. Continued sexual intercourse on this day is objected to, which I do not quite understand. For sexual intercourse is not approved of in autumn, in the beginning of winter, and at the times of epidemic disease; on the contrary, at such times it is most noxious and pernicious to the body. Although we must say that the conditions of sexual intercourse depend upon a great many other things, as, for

example, age, time, place, custom, character, nourishment, the fullness or emptiness of the stomach, the desire, the female genitals, etc.

12. Wintry air (Egyptians).
13. Vehement south wind or north wind (Hipparchus).
14. Wintry air (Eudoxus); rain and wind (Egyptians).
15. Cold north wind or south wind and rain (Egyptians).
16. Wintry air (Caesar).
17. Nothing mentioned. People forbid on this day to take of the flesh of cows, of oranges, and mountain balm, to drink water after you lie down to sleep, to smear the camels with *nūra* (a depilatory unguent made of arsenic and quicklime), and to bleed anybody except him whose blood is feverish. The reason of all this is the cold and the moistness of the season. This day people call the 'Great Birth', meaning the winter-solstice. People say that on this day the light leaves those limits within which it decreases, and enters those limits within which it increases, that human beings begin growing and increasing, whilst the demons begin withering and perishing.

Ka'b the Rabbi relates that on this day the sun was kept back for Yosua the son of Nūn during three hours on a clouded day. The same story is told by the simpletons among the Shī'a regarding the prince of the believers, 'Alī ibn Abī Ṭālib. Whether, now, this story has any foundation or not, we must remark that those who are beset by calamity find its duration to be very long and think that the moment of liberation is very slow in coming. So, for example, 'Alī ibn al-Jahm said in a sleepless night, when he had gone out to war against the Greeks, oppressed by wounds and fatigue:

Has a stream swept away the morning,
Or has another night been added to the night?

Afterwards on being released he indulged in hallucinations and lying reports.

Something similar frequently happens on fast-days, if heaven during the latter part of them be clouded and dark; then people break their fast, whilst shortly afterwards, when the sky or part of it clears up, the sun appears still standing above the horizon, having not yet set. The charm-mongers say that it is a good omen on this day to rise from sleeping on the right side, and to fumigate with frankincense in the morning before speaking. It is also considered desirable to walk twelve consecutive steps towards the east at the moment of sunrise.

Yaḥyā ibn 'Alī, the Christian writer of Anbār, says that the rising-place of the sun at the time of the winter-solstice is the true east, that he rises from the very midst of paradise; that on this day the sages lay the foundations of the altars. It was the belief of this man that paradise is situated in the southern regions. But he had no knowledge of the difference of the zeniths. Besides, the dogma of his own religion proves his theory to be erroneous, for their law orders them to turn in praying towards the *east* (i.e., the rising-place of the sun), whilst he told them that the sun rises in paradise (i.e., in the *south* according to *his* theory). Therefore the Christians turn to no other rising-place but to that one of the equator, and they fix the direction of their churches accordingly.

This theory is not more curious than his view of the sun. For he maintains that the degrees through which the sun ascends and descends are 360 in number, corresponding to the days of the year; that during the five days which are the complement of the year the sun is neither ascending nor descending. Those are two-and-a-half days of Ḥazīrān and two-and-a-half days of Kānūn I.

A similar idea hovered in the mind of Abū al-Abbās al-Āmulī when he said in his book *On the Proofs for the Qibla* that the sun has 177 rising and setting places, thinking evidently that the solar year has got 354 days. He, however, who undertakes what he does not understand, incurs ignominy. Those crotchets of his are brought into connection with the argument regarding the five supernumerary days of the solar year and the six deficient days of the lunar year, of which we have already spoken.

18. Nothing mentioned.
19. South wind (Eudoxus, Dositheus, and Egyptians).
20. Wintry air (Eudoxus).
21. A change in the weather (Egyptians).
22. Nothing mentioned.
23. Nothing mentioned.
24. Wintry air (Caesar and Egyptians); a change in the weather and rain (Hipparchus and Meton).
25. Middling wintry air (Democritus).
26. (Missing.)
27. Nothing mentioned.
28. Wintry air (Dositheus).
29. A change in the weather (Callippus, Euctemon, and Democritus). People forbid on this day the drinking cold water after rising from sleep. They say that the demons vomit into the water, and that therefore he

who drinks of it is affected by stupidity and phlegm. This serves as a warning to people against that which they dread most. The cause of all this is the coldness and moisture of the air.

30. Wintry air on the sea (Egyptians).
31. Wintry air (Euctemon).

Kānūn II

1. Nothing mentioned by the parapegmatists.
2. A change in the weather (Dositheus). Some people say that wood which is cut on this day will not soon get dry.
3. Changeable air (Egyptians).
4. A change in the weather (Egyptians); south wind (Democritus), which observation is confirmed by Sinān.
5, 6. Nothing mentioned. People say that on the sixth there is an hour during which all salt water of the earth is getting sweet. All the qualities occurring in the water depend exclusively upon the nature of that soil by which the water is enclosed, if it be standing, or over which the water flows, if it be running. Those qualities are of a stable nature, not to be altered except by a process of transformation from degree to degree by means of certain *media*. Therefore this statement of the waters getting sweet in this one hour is entirely unfounded. Continual and leisurely experimentation will show to any one the futility of this assertion. For if the water were sweet it would remain sweet for some space of time. Nay, if you would place – in this hour or any other – in a well of salt water some pounds of pure dry wax, possibly the saltishness of the water would diminish. This has been mentioned by the experimenters, who go so far as to maintain that if you make a thin vase of wax and place it in sea water, so that the mouth of the vase emerges above the water, those drops of water which splash over into the vase become sweet. If all salt water were mixed with so much sweet water as would overpower its nature, in that case their theory would be realized (i.e., all salt waters would become sweet). An example of this process is afforded by the lake of Tinnīs, the water of which is sweet in autumn and winter in consequence of the great admixture of the water of the Nile, whilst at the other seasons it is salt, because there is very little admixture of Nile water.
7. Wintry air (Eudoxus and Hipparchus).
8. South wind (Callippus, Euctemon, Philippus, and Metrodorus); south wind and west wind and wintry air on the sea (Egyptians).

9. Violent south wind and rain (Eudoxus and Egyptians).

The authors of talismans say that if you draw the figures of grapes on a table, between the ninth and the sixteenth of the month, and place it among the vines as a sort of offering at the time of the setting of the Tortoise, that is, *al-nasr al-wāqiʿ*, the fruit will not be injured by anything.

10. Violent south wind and a change in the weather (Caesar and Egyptians).
11. South wind (Eudoxus and Dositheus); mixed winds (Hipparchus).
12. Nothing mentioned.
13. Wintry air (Hipparchus); a north wind or a south wind blows (Ptolemy).
14. Nothing mentioned.
15. East wind (Hipparchus).
16. Nothing mentioned.
17. Violent wind (Caesar).
18. Wintry air (Euctemon and Philippus); change of the air (Metrodorus).
19. Wintry air (Eudoxus and Caesar); suffocating air (Egyptians).
20. Clear sky (Euctemon and Democritus); north wind (Hipparchus); wintry air and rain (Egyptians).
21. Middling wintry air (Eudoxus).
22. A change in the weather (Hipparchus); rain (Egyptians).
23. Nothing mentioned. On this day people do not smear the camels with *nūra* (a depilatory unguent of arsenic and quicklime), nor bleed anybody except in cases of special need.
24. Clear sky (Callippus and Euctemon); middling wintry air (Democritus). Besides, the rule of the preceding day as regards the use of *nūra* and phlebotomy refers also to this day.
25. East wind (Hipparchus).
26. Rain (Eudoxus and Metrodorus); wintry air (Dositheus).
27. Severe winter (Egyptians).
28. South wind blows and a change in the weather (Ptolemy).
29. Nothing mentioned.
30. South wind (Hipparchus).
31. Nothing mentioned.

Shubāṭ

It is the leap-month. It appears to me that the following is the reason – but God knows best! – why people have shortened this month in particular so that it has only twenty-eight days, and why it has not had assigned to it twenty-nine or thirty or thirty-one days: If it were assigned twenty-nine days and were then to be increased by the leap-day, it would have 30 days and would no longer be distinguishable from the other months in a leap-year. The same would be the case if it had thirty days, whether the year be a leap-year or not. Likewise if it had thirty-one days, the same similarity with the other months in all sorts of years would exist. For this reason the leap-month has been assigned twenty-eight days, that it might be distinguished from the other months both in leap and common years.

For the same reason it was necessary that in the Greek year two months of more than thirty days should follow each other. For at the beginning they intentionally gave to each month thirty days and took away two from Shubāṭ. So they got seven supernumerary days (i.e., the five epagomenal days and the two days of Shubāṭ), which they had to distribute over eleven months, because Shubāṭ had to be left out. Now, it was not possible to distribute the complete months of thirty days so as to fall each of them between two months of thirty-one days, for the latter (i.e., the months of thirty-one days) are more in number than the former. Therefore it was necessary to let several months of more than thirty days follow each other. But the most important subject of their deliberation was to add them in the places which would be the most suitable to them, so that the sum of the days of both spring and summer is more than the sum of the days of autumn and winter, a fact which is the result of both ancient and modern observations.

Further, their months are proportional to each other in most cases; I mean to say: the sum of each month and of the seventh following one is sixty-one days, which is nearly equal to the time of the sun's mean motion through two signs of the zodiac. However, the sum of the days of Āb and Shubāṭ is fifty-nine days. This could not have been otherwise, for the reason we have mentioned for Shubāṭ. For if Āb had been assigned more than thirty-one days, it would have been different from all the other months, and people would have thought that this in particular was the leap-month. As for Tammūz and Kānūn the Last, the sum of their days is sixty-two. This, again, was necessary, because the number of the months of more than thirty days is greater than that of the months of thirty days. Wherever the supernumerary day is placed the circumstances are always the same. And, further, intercalation has been applied to Shubāṭ to the exclusion of the other months only for this reason, that Adhār I, which is the leap-month in the Jewish leap-year, falls on Shubāṭ and near it.

1. Rain (Eudoxus). The cold decreases a little.
2. West wind or south wind intermixed with hail (Egyptians). Sinān says that this is frequently the case.
3. Clear sky and frequently the west wind blows (Eudoxus).
4. Clear sky and frequently the west wind blows (Dositheus); severe wintry air, rain and unmixed winds (Egyptians).
5. Nothing mentioned. People say that the four winds are in uproar.
6. Rain (Caesar); winds (Egyptians); the west wind begins blowing (Democritus).
7. Beginning of the blowing of the west wind, frequently the air is wintry (Eudoxus and Egyptians). On this day the first *Coal* falls, called the minor one.[255]
8. The time of the blowing of the west wind (Callippus, Metrodorus, and Hipparchus); rain (Eudoxus and Egyptians). This is confirmed by Sinān as borne out by his observations.
9, 10. Nothing mentioned.
11. Wintry air (Philippus and Metrodorus); west wind (Eudoxus and Egyptians).
12. North and east wind (Hipparchus); east wind alone (Egyptians).
13, 14. Nothing mentioned. On the fourteenth falls the second *Coal*, called the middle one. As the poet says:

When Christmas has passed and Epiphany after it,
And ten days and ten days and five complete days,
And five days and six and four of Shubāṭ,
Then, no doubt, the greatest cold vanishes.
That is the time of the falling of the two *Coals*; afterwards the cold
   remains only a few nights.

15. Wintry air (Euctemon, Philippus, and Dositheus); changing wind (Egyptians); south wind (Hipparchus). This day is cold (Arabs), during which the coal is kindled. The Persians say: 'The Summer has put his hands into the water.' On this day the moisture of the wood is flowing from the lowest parts of the trees to the highest, and the frogs begin croaking.
16. A change in the winds and rain (Egyptians). People say that on this day the interior of the earth is getting warm. In Syria the mushrooms are coming forth; those which stand near the root of the olive-tree are deadly

---

[255] For an explanation of the *Coals*, see below, Shubāṭ 21.

poison, as people maintain. This may be true, for it is not approved of to take much of the mushroom and fungus, nor of that which is prepared from them. Its pharmacological treatment is mentioned in most of the medical compilations in the chapter of preparing poisons from these materials.

17. Nothing mentioned.
18. West wind, and hail falls, or rain (Egyptians).
19. Cold north wind (Hipparchus).
20. Winds (Egyptians).
21. Nothing mentioned. On this day the third Coal falls, called the great one. Between the falling of each of the two Coals there is an interval of one complete week. They were called Coals because they are days characterized by the spreading of the heat from the interior of the earth to the surface, according to those who hold this theory. According to those who hold the opposite view, this change is brought about by the air's receiving heat instead of cold from the body of the sun, for the body of the sun and the near approach of a column of rays are the first cause of the heat. With this subject also the question is connected why the earthen jars or pipes of which subterranean channels are formed, and the water of wells, are warm in the winter and cold in the summer.

Between 'Abū Bakr ibn Zakariyyā al-Rāzī and 'Abū Bakr Ḥusain al-Tammār several questions and answers, expostulations and refutations have been exchanged that will satisfy the curiosity of the reader and inform him of the truth.

The Arabs used these three days (the so-called Coals) in their months until they got into confusion, as we have mentioned, and these days no longer fell at their proper times. Thereupon they were transferred into (i.e., fixed on certain days of) the Greek months which keep always their proper places. On the first day, people say, the first and second Climata are getting warm, on the second the third and fourth, on the third the remaining Climata. Further, they say that on the Coal-days vapours are rising from the earth which warm the earth on the first Coal-day, the water on the second, and the trees on the third.

According to another view, they are days noticeable for the rising of Lunar Stations, or some special parts of them; whilst other subtle people maintain that they are the *termini* of the cold in winter, and serve to denote the differences in the beginning of heat and cold as known in the different countries. Some inconsiderate and over-zealous people of our ancestors have introduced these Coal-days into Khwārizm, so that the first fell on the twenty-first of Shubāṭ, the second a week later, and the third two weeks after the second one.

22. A cold north-east wind begins blowing and the swallows appear (Euctemon and Hipparchus).
23. Winds are blowing and the swallows appear (Callippus, Philippus, and the Egyptians); rain at the time of the appearance of the swallows, north-east wind during four days (Eudoxus, Conon, Callippus, and Philippus).
24. Cold north wind and west wind (Hipparchus); north-east wind with other winds (Egyptians); days with changeable air (Democritus).
25. Wintry air (Caesar and Dositheus).
26, 27. Nothing mentioned.
28. Cold north wind (Hipparchus).

In this month fall the Days of the Old Woman, i.e., seven consecutive days beginning with the twenty-sixth; if the year is a leap-year, four days fall into Shubāṭ and three into Adhār; if it is a common year, three fall into Shubāṭ and four into Adhār. They are called by the Arabs by special names; the first is called *al-Ṣinn*, that is, the severity of the cold, the second is called *al-Ṣinnabr*, that is, a man who leaves things as *ṣanbara*, that is, as something that is coarse and thick. The *nūn* in this word is not radical, the same as in *balanṣā*, the plural of *balaṣūs*. The third is their brother *al-Wabr*, so called from the verb *wabbara*, that is, *he followed the trace of these days*. The fourth is called *al-Āmir* (*commanding*), because he *commands* people to beware of him. The fifth is al-Mu'tamir, that is, he has an impulse of doing harm to mankind. The sixth is al-Mu'allil, that is, he diverts people by some relief which he affords.

The seventh is *Muṭfi' al-jamr* (the extinguisher of coals), the most severe of them, when the coals used to be extinguished. It is also called *Mukfi' al-qidr* (who turns the kettle upside down) in consequence of the cold wind of this day. Some poet has connected these names in a *versus memorialis* in this way:

The winter is closed by seven dusty (days),
Our *Old Woman's Days* of the month;
When her days come to an end,
Ṣinn, *Ṣinnabr*, and *Wabr*,
Āmir, and his brother *Mu'tamir*,
Mu'allil, and *Muṭfi' al-jamr*,
Then the cold retires, passing away with the end of the month,
And a burning (wind) comes to thee from the beginning of the next month.

The sixth day is also called *Shaibān,* and the seventh *Milḥān.* These days are scarcely ever free from cold and winds, the sky being dark and variously coloured. Mostly during these days the cold is most vehement, because it is about to *turn away* (i.e., to cease). And hence the Lunar Station *al-Ṣarfa* has got its name, because its setting occurs about this time.

Nobody need be astonished at the fact that the cold towards its end, when it is about to cease, is the most severe and vehement. Quite the same is the case with the heat, as we shall mention hereafter. Similar observations you may make in quite common physical appearances. For example, if the lamp is near the moment of extinction, because there is no more oil, it burns with an intense light, and flickers repeatedly, like the quivering (of human limbs). Sick people furnish another example, specially those who perish by hectic fever or consumption, or the disease of the belly, or similar diseases. For they regain power when they are near death; then those who are not familiar with these things gain new hope, whilst those who know them from experience despair.

I have seen a treatise of Yaʿqūb ibn Isḥāq al-Kindī on the cause of this appearance in these days (i.e., of the vehemence of the cold during them).

His whole argument comes to this, that the sun then reaches the quadrature of his apogee, which is the place of all changes, and that the sun's influence upon the atmosphere is greater than that of anything else. In that case it would be necessary that that change which the sun effects in its own sphere should be proportional to that one which it effects in the atmosphere, and that this effect should on an average continue as long as the moon stands in that quarter (of her own course) in which the effect commenced, and in that quarter of the sun in which the effect took place.

I have been told that ʿAbdallāh ibn ʿAlī, the mathematician, in Bukhārā, on having become acquainted with this treatise of al-Kindī, transferred these days into the calendar of his people in conformity with the amount of the progression of the apogee. Therefore they were called the Days of the Old Woman of ʿAbdallāh.

[Lacuna.]

Regarding the reason why these days were called the Days of the Old Woman, the ancients relate the following: They are the days which God mentions in his Book (Sūra LXIX.7), 'seven nights and eight days, unlucky ones', and the people of ʿĀd perished by their cold wind, their whirlwinds, and the other terrors which happened during them. Of all of them only one old woman remained, lamenting the fate of her nation. Her story is well known. Therefore these days are said to have been called the Days of the Old Woman.

People say that the wind which destroyed them was a west wind, for the prophet says: 'I have been assisted by the east wind – *viz.* on the *yaum al-khandaq* – and 'Ād has been destroyed by the west wind.' A poet says:

The west wind has destroyed the sandy tracts of 'Ād;
So they perished, thrown down like the trunks of palm-trees.

Further, people say that the Unlucky Days mentioned in the Koran (Sūra XLI.15) coincide, each set of four of them, with a day of the month in the date of which there is a four, that is, the fourth, or the fourteenth, or the twenty-fourth from beginning or end of the month.

Some people maintain that the Days of the Old Woman received their name from this, that an old woman, thinking that it was warm, threw off her *miḥsha'* (a sort of garment) and perished in the cold of these days.

Some Arabs maintain that the Days of the Old Woman (*al-'Ajūz*) were given this name because they are the *'ajuz*, that is, *pars postica*, of the winter.

We find that the Arabs have names for the five epagomenal days between Abān-Māh and Ādhār-Māh like those of the Days of the Old Woman. The first is called *Hinnabr*, the second *Hinzabr*, both words meaning the injury from cold; the third is called *Qālib al-fihr* (i.e., turning the braying stone upside down), *viz.* through the vehemence of the wind; the fourth, *Ḥāliq al-ḍufr* (i.e., cutting the nail), for they mean that the wind is so sharp as, for example, to cut the nail; the fifth is called *Mudaḥrij al-ba'r* (whirling about the dung), *viz.*, in the plains, so that the vehemence of the wind carries it to human habitations. Somebody has brought them into a verse in this way:

The first of them is *Hinnabr*, an excessive day,
After him comes *Hinzabr*, one who strikes with the fore-foot,
Striking till he comes who exercises justice.
And *Qālib al-fihr* is justly called thus;
And *Ḥāliq al-ḍufr* who evidently cuts
And splits the rocks by the cold.
After them the last of them, the fifth,
Mudaḥrij al-ba'r, the biting and licking one.
There is no sixth name after it.

Adhār

1. Nothing mentioned by the parapegmatists. People say that on this day the locusts and all creeping animals come forth, and that *the heat of heaven and the heat of the earth meet each other*. This is a somewhat

hyperbolical expression for the beginning of the heat, its increase and spreading, and for the air's preparing itself for the reception of the heat. For the heat is nothing but the rays of the sun detached from the body of the sun towards the earth or from the warm body which touches the inside of the Lunar sphere, which is called *Fire*.

Regarding the rays of the sun many theories have been brought forward. Some say that they are fiery particles similar to the essence of the sun, going out from his body. Others say that the air is getting warm by its being situated opposite to the sun, in the same way as the air is getting warm by being opposite to the fire. This is the theory of those who maintain that the sun is a hot, fiery substance.

Others, again, say that the air is getting warm by the rapid motion of the rays in the air, which is so rapid as to seem *timeless*, that is, without time (*zeitlos*). This is the theory of those who maintain that the nature of the sun has nothing in common with the natures of the four elements.

Further, there is a difference of opinion regarding the motion of the rays. Some say this motion is *timeless*, since the rays are not bodies. Others say this motion proceeds in very short time; that, however, there is nothing more rapid in existence by which you might measure the degree of its rapidity. For example, the motion of the sound in the air is not so fast as the motion of the rays; therefore the former has been compared with the latter, and thereby its time (i.e., the degree of its rapidity) has been determined.

As the reason of the heat which exists in the rays of the sun, people assign the acuteness of the angles of their reflection. This, however, is not the case. On the contrary, the heat exists in the rays (is inherent in them).

Regarding the body that touches the inside of the sphere, that is, the fire, people maintain that it is a simple element like earth, water, and air, and that it is of a globular form. According to my opinion, the warmth of the air is the result of the friction and violent contact between the sphere, moving rapidly, and his body, and that its shape is like a body which you get by making a crescent-like figure revolve around its chord. This explanation is in conformity with the theory, *viz.*, that none of the existing bodies is in its natural place, that all of them are where they are only in consequence of some force being employed, and that force must of necessity have had a beginning.

On this subject I have spoken in a more suitable place than this book is, especially in my correspondence with the youth Abū ʿAlī al-Ḥusain ibn ʿAbdallāh ibn Sīnā, consisting of discussions on this subject.

Both sorts of heat are brought to bear upon the earth in an equal manner during the four seasons. The heat of the earth consists either of the solar rays that are reflected from its surface, or of the vapours that are produced – according to one theory – by the heat of the interior of the earth, or – according to another – by that heat which accidentally comes to the earth from outside, for the motion of the vapours in the air causes them to get warm.

The heat of the fire (i.e., the body touching the inside of the sphere) remains always at the same distance (from us, i.e., is always of the same degree), because the rotation of the celestial sphere proceeds always at the same rate. And the reflected rays are not to be referred to the earth (i.e., the earth is not to be considered as their source), and the vapours reach only to a certain limit which they do not go beyond.

The author of this theory, I think, must believe that within the earth heat is contained which proceeds from the interior to the outside, whilst the air has become warm through the rays of the sun. *Thus the two sorts of heat meet each other.* This, at all events, *is* a theory, if there is any; one must accept it.

2. Cold north wind (Hipparchus); south wind and fall of hail (Egyptians).
3. Nothing mentioned.
4. Cold north wind (Euctemon). Sinān says that this is mostly true.
5. Wintry air (Egyptians). Beginning of the swallow-bringers (Caesar) they blow during ten days.
6. Troubled air (Egyptians). Beginning of the cold bird-bringers, which blow during nine days (Democritus).
7. Nothing mentioned. Some people say that a change of the violent winds takes place.
8. A change in the weather and cold north wind (Euctemon, Philippus, and Metrodorus); swallows and kites appear (Eudoxus). On the same day is the feast of the Small Lake of Alexandria.
9. North wind (Euctemon and Metrodorus); violent south wind (Hipparchus); light rain (Egyptians); the kites appear (Dositheus).
10. Nothing mentioned.
11. The ancients do not mention an apparent change on this day. Sinān says that there is frequently wintry air.
12. Moderate north wind (Callippus). People say that on this day the traces of the winter disappear, and that phlebotomy is advisable.
13. The bird-bringers begin blowing; the kite appears (Euctemon and Philippus).

14. Cold north wind (Euctemon and Hipparchus); west or south wind (Egyptians); the bird-bringers begin blowing (Eudoxus).
15. Cold north wind (Euctemon and Egyptians).
16. North wind (Callippus). This Sinān confirms from his experience.
17. Nothing mentioned. People say that on this day it is agreeable to go out on the sea. The snakes open their eyes, for during the cold season, as I have found them myself in Khwārizm, they gather in the interior of the earth and roll themselves up one round the other so that the greatest part of them is visible, and they look like a ball. In this condition they remain during the winter until this time.

On this day (the seventeenth) in a leap-year, and on the eighteenth in a common year, takes place the equinox, called the first equinox. It is the first day of the Persian spring and of the Chinese autumn, as we have mentioned. This, however, is impossible, for spring and autumn or winter and summer cannot at one and the same time alternately exchange their places except in countries north or south of the equator. And China, having only few degrees of latitude, does not lie south, but north of the equator, in the farthest end of the inhabited world towards the east.

The country south of the Line is not known, for the equatorial part of the earth is too much burned to be inhabitable. Parts of the inhabited world do not reach nearer the equator than to a distance of several days' journey. There the water of the sea is dense, because the sun so intensely vaporizes the small particles of the water, that fishes and other animals keep away from it. Neither we nor any of those who care for those things have ever heard that any one has reached the Line or even passed the Line to the south.

Some people have been beguiled by the expressions *aequator diei* and *linea aequitatis*, so as to think that there the air is *equal* (moderate), just as day and night there are equal. So they have made the equator the basis of their fictions, describing it as a sort of paradise and as being inhabited by creatures like angels.

As to the country beyond the Line, someone maintains that it is not inhabitable, because the sun, when reaching the perigee of his eccentric sphere, stands nearly in its utmost southern declination, and then burns all the countries over which he culminates, whilst all the countries of 65 degrees of southern latitude have the climate of the middle zone of the north. From that degree of latitude to the pole the world is again inhabitable. But the author of this theory must not represent this as necessary, because excessive heat and cold are not alone the causes which render a country uninhabitable, for they do not exist in the second quarter of the two northern quarters, and still that part of the world is not inhabited. So the matter is (and will be), because the apogee and perigee of the eccentric sphere, the sun's greater and less distance from the earth, are necessitated exclusively by the difference in the sun's rotation.

Abū Ja'far has designed a figure different from the eccentric sphere and the epicycle, in which the sun's distance from the earth, notwithstanding the difference of its rotation, is always identical. Thereby he gets two regions, a northern and a southern one, equal to each other in heat and cold.

The day of the equinox, as calculated by the Hindus according to their Canon – of which they are impudent enough to pretend that it is eternal, without beginning and end, whilst all the other Canons are derived therefrom – is their Naurōz, a great feast among them. In the first hour of the day they worship the sun and pray for happiness and bliss to the spirits (of the deceased). In the middle of the day they worship the sun again, and pray for the resurrection and the other world. At the end of the day they worship the sun again, and pray for health and happiness for their bodies. On the same day they make presents to each other, consisting of precious objects and domestic animals. They maintain that the winds blowing on this day are spiritual beings of great use for mankind. And the people in heaven and hell look at each other affectionately, and light and darkness are equal to each other. In the hour of the equinox they light fires in sacred places.

The omina of this day are the following, *viz.*: to rise from sleep lying on the back, the tree *salix aegyptia* and to fumigate with its wood before speaking. For he who performs this will be free from all sorts of pain.

People say that a man who has no children, on looking to the star *al-Suhā* in the night of this day and then having intercourse with his wife, will get children.

Muḥammad ibn Miṭyār maintains that in the hour when this day begins to decline (i.e., after noon,) the shadow of everything is half its size. This, however, is only partially the case, not in general. It is true only for such places of which the latitude is about 27 degrees.

On this day the crocodile in Egypt is thought to be dangerous. The crocodile is said to be the water-lizard when it has grown up. It is an obnoxious animal peculiar to the Nile, as the *thesking* is peculiar to other rivers. People say that in the mountains of Fusṭāṭ there was a talisman made for that district. Around this talisman the crocodile could not do any harm. On the contrary, when it came within its limits, it turned round and lay on its back, so that the children could play with it. But on reaching the frontier of the district it got up again and carried all it could get hold of away to the water. But this talisman, they say, has been broken and lost its power.

18. Wintry air and cold winds (Democritus and the Egyptians).
19. North wind (Hipparchus); winds, and cold in the morning (Egyptians).
20. North wind (Caesar).
21. North wind (Eudoxus).
22. Nothing mentioned.
23. North wind (Caesar); rain (Hipparchus).
24. Rain and mizzle (Callippus, Euctemon, and Philippus); a change in the weather (Hipparchus); thunder and a change in the weather (Egyptians). On this day people like to purify the children by circumcision. The fecundating winds are said to blow.
25. North wind (Eudoxus); a change in the weather (Meton, Conon, and the Egyptians).
26. Rain and snow-storm (Callippus); wind (Egyptians).
27. Rain (Callippus, Eudoxus, and Meton).

Of the rest of the month nothing is mentioned. Sinān says that the thirtieth frequently brings a change in the weather. God knows best!

Nisān

1. Rain (Callippus, Euctemon, Meton, and Metrodorus).
2. Nothing mentioned.
3. Wind (Eudoxus); rain (Egyptians and Conon).
4. West wind or south wind; hail falls. Sinān says that this is frequently the case.
5. South wind and changing winds (Hipparchus).
6. A change in the weather (Hipparchus and Dositheus). This is confirmed by Sinān.
7. Nothing mentioned.
8. Rain (Eudoxus); south wind (Egyptians).
9. Rain (Hipparchus); unmixed winds (Egyptians).
10. Unmixed winds (Euctemon and Philippus); rain (Hipparchus and Egyptians). The raining is confirmed by the experience of Sinān.
11. West wind and mizzle (Eudoxus).

12. Nothing mentioned.
13. Rain (Caesar and Dositheus).
14. South wind, rain, thunder, and mizzle (Egyptians). Sinān says that this is frequently the case.
15. Rain and hail (Euctemon and Eudoxus); unmixed winds (Egyptians).
16. West wind (Euctemon and Philippus); hail falling (Metrodorus).
17. West wind and rain (Eudoxus and Caesar); hail falling (Conon and Egyptians).
18. Winds and mizzle (Egyptians).
19. Nothing mentioned.
20. Wind, south wind or another one, the air unmixed (Ptolemy).
21. Cold south wind (Hipparchus). Sinān maintains that this is frequently the case. The water begins to increase.
22. Rain (Eudoxus); wintry air (Caesar and Egyptians). People fear for the ships at sea.
23. South wind and rain (Egyptians). People hold a fair at Dair Ayyūb. Abū Yaḥyā ibn Kunāsa says that the Pleiades disappear under the rays of the sun during forty days, and this fair is held when the Pleiades appear. So the Syrians make them rise fifteen days earlier than in reality they rise, because they are in a hurry to settle their affairs. This fair lasts seven days. Then they count seventy days until the fair of Buṣra. Through these fairs, that are held alternately in certain places, the commerce of the people of these countries has been promoted and their wealth been increased. They have proved profitable to the people, to both buyers and sellers.
24. Frequently hail falls (Callippus and Metrodorus); a change in the weather (Democritus); south wind, or a wind akin to it, and rain (Egyptians). The Euphrates begins to rise.
25. Mizzle and rain (Eudoxus and Egyptians).
26. Rain and frequently hail (Callippus and Euctemon); a change in the weather and west wind (Egyptians).
27. Dew and moisture (Caesar); winds (Egyptians).
28. Wind (Egyptians); rain (Eudoxus). Sinān confirms the rain from his own observations. On this, they say, the south wind blows, and then the streams and rivers begin to rise. This increase of the water, however, does not apply to all streams and rivers uniformly; on the contrary, they greatly differ from each other in this respect. For example, the Oxus has high water when there is little water in the Tigris, Euphrates, and other rivers. The fact is this, that those rivers the sources of which are situated in cold places, have more water in summer and less in winter.

For the greatest part of the ordinary volume of their water is gathered from springs, and an increase and decrease of them exclusively depend upon the fall of dew in those mountains where the rivers originate or through which they flow; thereupon the springs pour their volumes into the rivers. Now it is well known that dew-fall is more frequent in winter and beginning of spring than at any other season. In the countries far up to the north, where the cold is intense, this dew-water freezes at those seasons. But when the air is getting warm and the snow melts, at that time the Oxus rises.

As for the water of the Tigris and Euphrates, their sources are not so high up in the north. Therefore they have high water in winter and spring, because the dew that falls flows instantaneously into the rivers, and that portion of water that may have been frozen melts away in the beginning of spring.

The Nile, again, has high water when there is low water in both Tigris and Euphrates, because its source lies in the *mons lunae*, as has been said, beyond the Abyssinian city Assuan in the southern region, coming either exactly from the equator or from countries south of the equator. This is, however, a matter of doubt, because the equatorial zone is not inhabited, as we have before mentioned. It is evident that in those regions there is no freezing of moist substances at all. If, therefore, the high water of the Nile is caused by falling dew, it is evident that the dew does not stay where it has fallen, but that it directly flows off to the Nile. But if the high water is caused by the springs, these have the most abundant water in spring. Therefore the Nile has high water in summer, for when the sun is near us and our zenith, it is far distant from the zenith of those places whence the Nile originates, and which in consequence have winter.

As to the question why the springs have the most copious water in winter, we must observe: the all-wise and almighty Creator, in creating the mountains, destined them for various purposes and uses. Some of them have been mentioned by Thābit ibn Qurra in his book on the reason why the mountains were created. It is the same cause which renders complete the intention (of the Creator) which he had in making the sea-water salt.

Evidently more wet falls in winter than in summer, in the mountains more than in the plains. When, now, the wet falls and part of it flows away in the torrents, the remaining part sinks down into the channels in the mountain caves, and there it is stored up. Afterwards it begins to flow out through the holes, called springs. Therefore the springs have the most copious water in winter, because the substance by which they are nourished is then most copious. If, further, these mountain caves are clean and pure, the water flows out just as it is, that is, sweet. If that is not the case, the water acquires different qualities and peculiarities, the causes of which are not known to us.

The bubbling of the fountains and the rising of the water to a certain height are to be explained in this way, that their reservoirs lie higher than they themselves, as is the case with artificial well-springs, for this is the only reason why water rises upward.

Many people who attribute to God's wisdom all they do not know of physical sciences (i.e., who excuse their ignorance by saying 'Allah is all-wise!'), have argued with me on this subject. In support of their view they relate that they have observed the water rise in rivers and other watercourses, that the more the water flows away (from its source) the more it rises. This they assert in complete ignorance of the physical causes and because they do not sufficiently distinguish between the higher and lower situations (of the springs of rivers and of the rivers themselves). The matter is this, that they observed water flowing in mountain streamlets, the bed of which was going downward at the rate of 50–100 yards and more for the distance of one mile. If the peasants dig a channel somewhere in this terrain, and this channel is made to incline a little towards the country (i.e., if the channel is rising), at first the water flows only very little, until it rises to an enormous height above the water of the river; (then it commences to flow strongly).

If, now, a man who has no training in these things believes that the natural direction of the river is to flow in a horizontal line or with a small inclination (upwards), he must of necessity imagine that the river is rising in height. It is impossible to free their mind from this illusion unless they acquaint themselves with the instruments by which pieces of soil are weighed and determined, and by which rivers are dug and excavated – for if they weigh the earth through which the water flows, the reverse of what they believe becomes evident to them; or unless they study physical sciences, and learn that the water moves towards the centre of the earth and to any place which is nearest to the centre. There is no doubt that the water may rise to any place where you want to have it, even if it were to the tops of the mountains, if previously it descends to a place which is lower than its maximum of ascent (which it ultimately reaches), and if you keep away from it any substance which might occupy the place instead of the water when it finds the place empty. Now, the water in its natural function is only assisted by the co-operation of something forcible which acts like an instrument, and that is the air. This has frequently been carried out in canals, in the midst of which there were mountains which it was impossible to perforate.

An illustration of this principle is the instrument called Water-thief, *clepsydra*. For if you fill it with water and put both its ends into two vessels, in both of which the water reaches to the same level, then the water in the *clepsydra* stands still even for a long time, not flowing off into either of the two vessels. For the one vessel is not nearer (to the water) than the other, and it is impossible that the water should flow off equally into both vessels, for in that case the instrument would get empty. Now, emptiness is either a *non-ens*, as most philosophers suppose, or it is an *ens* which attracts bodies, as others believe. If, now, the *vacuum* cannot exist, the matter is impossible, or if it is something which attracts bodies, it keeps back the water and does not let it flow off, except its place be occupied by some other body. But if you then place the one end of the *clepsydra* a little lower (than the other), the water flows immediately off into that direction. For if its place has once become lower, it has come nearer to the centre of the earth, and so it flows towards it, flowing continually in consequence of the adhesion and connection of the water-atoms amongst each other. It flows so long until the water of that vessel, whence the water is drawn, is finished, or until the level of the water in the vessel where it flows is equal to the level of the water in the vessel whence it is drawn. So the question returns to its original condition. On this principle people have proceeded in the mountains.

Sometimes even the water rises in artificial fountains out of wells, in case they have got springing water. For one sort of well-water, which is gathered from droppings from the sides, does not rise at all; it is taken from neighbouring masses of water, and the level of the water which is gathered in this way is parallel to the level of those waters by which it is nourished. On the other hand there is one kind of water which bubbles (springs) already at the bottom. Of this water people hope that it may rise to the earth and flow on over its surface. This latter kind of water is mostly found in countries near to mountains, in the midst of which there are no lakes or rivers with deep water. If the source of such water is a reservoir much above the level of the earth, the water rises springing, if it is confined (to a narrow bed or channel); but if its reservoir be lower, the water does not succeed in rising to the earth. Frequently the reservoir is higher by thousands of yards in the mountains; in that case the water may rise up to the castles, and, for example, to the tops of the minarets.

I have been told that people in Yaman often dig until they come to a certain rock under which they know that there is water. Then they knock upon this rock, and by the sound of the knocking they ascertain the quantity of the water. Then they bore a small hole and examine it; if it is all right, they let the water bubble out and flow where it likes. But if they have some fear about the hole, they hasten to stop it up with gypsum and quicklime and to close it over repeatedly. For frequently they fear that from such a hole a spring similar to the Torrent of *al-'Arim* might originate.

As to the water on the top of the mountain between Abrashahr and Ṭūs, a small lake of one farsang in circumference, called Sabzarōd, one of the following three things must be the case:

1. Either its material is derived from a reservoir much higher than the lake itself, although it may be far distant, and the water flows into it in such a quantity as corresponds to that which the sun absorbs and vaporizes. Therefore the water of the lake remains in the same condition, quietly standing.
2. Or its material is derived from a reservoir which lies on the same level with the lake, and therefore the water of the lake does not rise above that of the reservoir.
3. Or, lastly, the condition of its sources in some way resembles that of the water of the instrument called *al-daḥj* and the Self-feeding Lamp. The case is this: You take a water-jug, or an oil-vase; in several places of the edge or lip of the vase you make fine splits, and you bore a narrow hole in it deeper than the mouth by as much as you wish the water to remain in the jug and the oil in the vase (i.e., the hole is to represent the line to which people wish the water or oil to rise). Thereupon you turn the jug upside down in the cup and the vase in the lamp. Then both water and oil flow out through the splits, until they reach the level of the hole. When, then, so much has been consumed as the hole allows to pass, then comes forth that which lies next to the hole. In this way both oil and water keep the same level.

   Similar to this little lake is a sweet-water well in the district of the Kīmāk in a mountain called Mankūr, as large as a great shield. The surface of its water is always on a level with its margin. Frequently a whole army drinks out of this well, and still it does not decrease as much as the breadth of a finger. Close to this well there are the traces of the foot, two hands with the fingers, and two knees of a man who had been worshipping there; also the traces of the foot of a child, and of the hoofs of an ass. The Ghuzzī Turks worship those traces when they see them.

   Moreover, similar to this is a small lake in the mountains of Bāmiyān, 1 mile square, on the top of the mountain. The water of the village which lies on the slope of the mountain comes down from that lake through a small hole in such a quantity as they require; but they are not able to make it flow more copiously.

Frequently the springing (rising of water) occurs also in a plain country which gets its water from a reservoir in a high situation. If the rising power of the water were kept down by an obstacle, and then this obstacle is removed, the water begins at once to spring (rise). For example, al-Jaihānī has mentioned a village between Bukhārā and al-Qarya al-Ḥadītha, where there is a hill that was perforated by diggers for hidden treasures. Suddenly they hit upon water which they were unable to keep back, and it has been flowing ever since till this day.

If you are inclined to wonder, you may well wonder at a place called Fīlawān (Failawān) in the neighbourhood of al-Mihrjān. This place is like a portico dug out in the mountain, from the roof of which water is always dropping. If the air gets cold, the water freezes and hangs down in long icicles. I have heard the people of al-Mihrjān maintain that they frequently knock the place with pickaxes, and that in consequence the spot which they knock becomes dry; but the water never increased, whilst reason would demand that it should always remain in the same condition if it does not increase.

More wonderful even than this is what al-Jaihānī relates in his *Kitab al-mamālik wa-l-Masālik* of the two columns in the grand mosque of Qairawān, the material of which people do not know. People maintain that on every Friday before sunrise they drop water. It is curious that this should take place just on a Friday. If it occurred on any week-day in general, it would be combined with the moon's reaching such and such a place of the sun's orbit, or with the like of it. This, however, is not admissible, since Friday is a *conditio sine qua non* of this occurrence. The Greek king is said to have sent to buy them. He said: 'It is better for the Muslims to utilize their prize than to have two stones in the mosque.' But the people of Qairawān refused, saying: 'We shall not let them pass out of the house of God into that of the devil.'

Still more marvellous than this is the self-moving column in al-Qairawān. For it inclines towards one side. People put something underneath when it inclines, and this you can no longer take away if the column again stands erect; if glass is put underneath, you hear the sound of breaking and crushing. This is no doubt a got-up piece of artifice, as also the place where the column stands seems to indicate.

We return to our subject, and say:

29. Wintry air (Caesar); winds, or moisture of the ground, and rain (Egyptians).
30. A change in the weather (Egyptians); winds and dew, moisture and mizzle (Callippus and Euctemon).

Ayyār

1. Mizzle (Egyptians).
2. Nothing mentioned.
3. Wind, mizzle, dew, moisture, and thunder (Egyptians). Rain (Eudoxus), mizzle (Egyptians).
4. Rain (Eudoxus), mizzle (Egyptians).
5. Rain (Dositheus). Sinān says that this is frequently the case and that it brings a strong change in the weather.
6. Wind (Egyptians), rain (Eudoxus), mizzle and a change in the weather. *[Lacuna.]*

   Some people extend the rainy season as far as this day. It is the time when the sun passes the (first) degrees of Leo. In this respect the matter stands as we have explained it at the beginning of the rainy season, when the sun moves in Cancer.
7. Winds (Egyptians). Sinān says that this is frequently the case, more particularly so if on the preceding day heaven has a rainy appearance.
8. Gushes of rain (Eudoxus and Dositheus), rain (Egyptians).
9. Rain (Egyptians).
10. A change in the weather and wind (Callippus and Euctemon), rain (Egyptians).
11. A change in the weather (Dositheus). Sinān says that it is true.
12. A change in the weather (Eudoxus, Metrodorus, and Hipparchus); rain (Caesar); west wind (Egyptians). People say that on this and the following day there is no fear of frost doing harm to the fruits. This remark can, however, only apply to one particular place; it cannot be meant in general.
13. Rain (Eudoxus); north wind and hail (Egyptians).
14. A change in the weather (Callippus, Euctemon, and Egyptians).
15. Rain (Caesar).
16. A change in the weather (Caesar). People say that on this day the first *Samūm* is blowing.
17. South wind or east wind and rain (Hipparchus and Egyptians).
18. A change in the weather (Eudoxus); rain and thunder (Egyptians).
19. A change in the weather and mizzle (Hipparchus and Egyptians).
20. Nothing mentioned.
21. A change in the weather (Caesar); south wind (Dositheus), west wind (Egyptians).
22, 23. Nothing mentioned.
24. A change in the weather (Callippus, Euctemon, and Philippus); winds (Egyptians).

25. A change in the weather (Euctemon, Philippus, and Hipparchus).
26. A change in the weather (Callippus and Euctemon); cold north wind (Egyptians).
27. Dew and moisture (Callippus and Euctemon); a change in the weather (Egyptians).
28. Rain (Metrodorus and Egyptians).
29. South wind or west wind (Hipparchus).
30. South wind (Caesar).
31. Nothing mentioned.

Ḥazīrān

1. Dew and moisture (Eudoxus and Dositheus); west wind (Egyptians).
2. West wind (Egyptians).
3. Wind and mizzle (Egyptians), and thunder.
4. Rain (Caesar).
5. Mizzle (Egyptians). Confirmed by Sinān.
6, 7, 8. Nothing mentioned.
9. West wind and thunder (Egyptians).
10, 11, 12. Nothing mentioned. The 11th is the Naurōz of the Khalif, when people in Baghdad splash in the water, strew about dust, and play other games, as is well known.
12. Sinān says that frequently a change of the weather takes place.
13. West wind and mizzle (Egyptians).
14. Nothing mentioned.
15. Mizzle (Egyptians).
16. Nothing mentioned. People say that on this day the water sinks into the earth, whilst the Nile begins to rise. The reason for this is, as we have mentioned before, the difference of their sources and of other circumstances, those of the Nile standing in direct opposition to those of all other rivers.

On this day in a leap-year, and on the 17th in a common year, the *plenitudo maxima* takes place, which is celebrated by Arabs and Persians. They call it *Mīrīn*, which means 'the Sun's getting full, that is, the summer-solstice. On this day light subdues darkness. The light of the sun is falling into the wells, as Muḥammad ibn Miṭyār mentions; but this is only possible in countries the latitude of which is like the greatest declination, over which, therefore, the sun culminates.

The *Ḥayawāniyya*-sect maintains that on this day the sun takes breath in the midst of heaven; that, therefore, the spirits recognize each other in the greatest heat. It is considered as a good omen to look into the intense heat. People eat pomegranates before having eaten anything else, and Hippocrates is said to have taught that he who eats a pomegranate on this day before having eaten anything else, enlightens his constitution and his humours are pure during forty mornings.

People relate, on the authority of Ḥanna the Hindu, that Kisra Parwīz has said: 'Sleeping in the shadow of a pomegranate cures a man of bad disease and makes him safe from the demons.'

It belongs to the *omina* of this day to rise in the morning from sleep on the left side, and to fumigate with saffron before speaking.

17. A change in the weather (Dositheus); heat (Egyptians).
18. West wind and heat (Egyptians).
19. Rain (Egyptians).
20. West wind, rain, and thunder (Egyptians).
21. Nothing mentioned.
22. A change in the weather (Democritus).
23. South wind or west wind (Hipparchus).
24. Nothing mentioned. People say that on this day the *Samūms* begin blowing during fifty-one days. The Oxus rises and frequently injures the shores and their inhabitants.
25. West wind and heat (Egyptians).
26. West wind (Democritus and Egyptians).
27. Nothing mentioned.
28. A change in the weather (Eudoxus); west wind and south wind and rain (Democritus); then the north wind begins to blow during seven days.
29. Nothing mentioned. People say that practical observers examine on this day the dew; if it is copious, the Nile rises; if it is not copious, the Nile does not rise, and they get a barren year.
30. Winds (Egyptians) and unmixed air.
31. Nothing mentioned.

Tammūz

1, 2. Nothing mentioned by our authorities.
3. South wind and heat (Caesar and Egyptians).
4. Wind (Egyptians); frequently it rains in their country.
5. South wind (Callippus, Metrodorus, and Hipparchus); west wind and thunder (Egyptians).
6. South wind (Callippus and Metrodorus); west wind and thunder (Egyptians).

7. A change in the weather (Ptolemy). According to Sinān the weather frequently changes.
8. Dew and moisture, according to Meton, in his country.
9. Dew (Euctemon and Philippus); west-by-west wind (Egyptians).
10. Bad air (Egyptians). On this day they begin to hold the fair of Buṣrā during twenty-five days; in the time of the Banū Umayya this fair used to last thirty to forty days.
11. Nothing mentioned.
12. West wind (Metrodorus); winds (Egyptians).
13. Unmixed winds (Hipparchus). According to Sinān the weather frequently changes.
14. Heavy wind (Caesar); the north wind begins to blow (Hipparchus); heat (Egyptians).
15. Nothing mentioned.
16. Frequently it rains in rainy countries (Ptolemy); rain and whirlwinds (Democritus); heavy wind (Egyptians).
17. Dew and heat (Dositheus and Egyptians).
18. The Etesian winds (ἐτησίαι) begin to blow (Hipparchus). According to the general consent of seamen and peasants, and all those who have experience in this subject, this is the first day of the dog-days, that is, seven consecutive days, the last of which is the twenty-fourth of this month. On each of these days they draw conclusions from certain changes of the weather regarding the months of the autumn and winter and part of spring; these changes mostly occur in the evening and morning. People maintain that these days are to the year what the *critical* days are to acute diseases, when their *criteria* appear, in consequence of which people conceive either hope or fear as to the end in which they will issue. Both words *bāḥūr* and *buḥrān* in the Greek and Syriac languages are derived from a word which means the decision of the rulers (κρί-σις and κρίσιμος ἡμέρα). According to another view, *buḥrān* is derived from *baḥr* (the Arabic for *sea)*, because the *critical* state of a sick person resembles the motion of the sea, called ebb and flow. This derivation is very likely correct, because of both appearances the motions of the moon, her cycles and phases, are the cause, whether the moon revolves in a Great Circle, as it is in the case of the flow, for the flow sets in when the moon reaches the western and eastern point of the horizon. The same is the case with the ebb, for it sets in when the moon reaches the sphere of the meridian of noon and midnight. Or whether it be that the moon revolves from one certain point of her cycle back to the same, or from the sun to that point. So the flow is the strongest in the first half

of the lunar month, the weakest in the second half. Besides, also, the sun has an influence upon this. It is curious what people relate of the Western Sea, *viz.* that there is flow from the side of Andalusia always at sunset, that then the sea decreases at the rate of about 5–6 farsang in one hour and then it ebbs. And this appearance takes place always precisely at this time.

If on the evening of the eighteenth there is a cloud on the horizon, people expect cold and rain at the beginning of Tishrīn I. If the same is the case at midnight, the cold and rain will come in the middle of Tishrīn I; and if it is the case towards morning, the same will come in the end of that month. The matter is the same, if you observe a cloud on the horizon during *daytime*; however, the changes of the sky in the night are more evident. And if you observe those changes on all four sides of the compass, the same, too, will occur in Tishrīn I. Herein the nights are counted after the days, as we have mentioned in the beginning of this book, in consequence of which those who count the nights before the days think that the night of the eighteenth is the nineteenth; therefore they consider the nineteenth as the first of the dog-days and the twenty-fifth as the last of them.

The first of these seven days serves to prognosticate the character of Tishrīn I, the second that of Tishrīn II, the third that of Kānūn I, etc. etc., and lastly, the seventh, that of Nīsān.

Practical observers prescribe the following: Take a plate some time before the dog-days, sow upon it all sorts of seeds and plants, and let it stand until the twenty-fifth night of Tammūz, that is, the last night of the dog-days; then put the plate somewhere outside at the time when the stars rise and set, and expose it uncovered to the open air. All seeds, then, that will grow in the year will be yellow in the morning, and all whose growth will not prosper will remain green. This experiment the Egyptians used to make.

Practical observers have produced many contrivances for the purpose of prognosticating the character of the year by help of these (the dog-) days; they have even gone as far as to use incantations and charms. So some people maintain that if you take the leaves of twelve different olive-trees, and write upon each leaf the name of some Syrian month, if you then put them, in the night we have mentioned, somewhere in a wet place, you will find that, if a leaf has dried up in this night, the month which was written upon it will be rainless.

According to others, you learn whether the year will have much rain or little, by this method: You look out for a level place, around which there is nothing that might keep off the dew, wind, and light rain; then you take 2 yards of a cotton dress, you weigh it and keep in mind its weight. Then you spread it over that place and leave it there during the first 4 hours of the night. Thereupon you weigh it a second time; then each *Mithqāl* which it weighs more the second time than the first time signifies one rainy day in that month which stands in relation with this particular dog-day of which we have heretofore spoken.

These dog-days are the time of the rising of Sirius *(Kalb al-Jabbār* or *al-Shi'rā al-Yamāniyya al-'Abūr).* Hippocrates, in his book of the seasons, forbids taking hot drugs and bleeding twenty days before and after the rising of this star, because it is the hottest time of summer and the heat reaches its maximum, and because summer time by itself warms, dissolves, and takes away all moist substances. However, Hippocrates does not forbid those things if you take but very little of them. Afterwards, when autumn comes with its cold and dryness, you cannot be sure whether the natural warmth may not be entirely extinguished.

Some people who have no practice in physical sciences and no knowledge of the *meteora,* think that the influence we have mentioned must be attributed to the body of this star, to its rising and revolution. They go even as far as to make people imagine that the air is warmed by its great mass; that, therefore, it is necessary to indicate and to explain its proper place and to determine the time of its rising. The same opinion is indicated by the verse of Abū Nu'ās:

Ilūl has gone and the hot night-wind passed away,
And Sirius has extinguished his fire.

For this reason 'Alī ibn 'Alī, the Christian secretary, maintains that the first of the dog-days is the twenty-second of Tammūz, suggesting that the dog-days have changed their place along with the star itself, whilst I maintain that Sirius always revolves during the whole year in one and the same orbit parallel to the equator. Hippocrates, however, meant by this time the central portion of the summer, the period when the heat is greatest in consequence of the sun's being near to our zenith, whilst he at the same time begins in his eccentric sphere to descend from the apogee of his orbit. And this event was in the time of Hippocrates contemporaneous with the rising of Sirius. Therefore he has only said in general *at the time when Sirius rises,* knowing that no scientific man could misunderstand the truth. For if Sirius changed its place so as to advance even as far as the beginning of Capricorn or Aries, the time during which he forbids taking drugs would not therefore advance in the same way.

Sinān says in his *Kitāb al-anwā'* that the shepherds have seven special days of their own, beginning with the first of Tammūz, which they use like the dog-days, drawing from them conclusions regarding the single winter months. They are known as 'the dog-days of the shepherds'. The weather of these days is always different from that of the time immediately preceding and following. During all or at least some of them heaven is never free from a speck of clouds.

19. West wind or heat (Egyptians). The Water Dogs are getting strong and do much damage.

20. West wind or a similar one (Egyptians). Practical observers say that on this day frequent cases of inflammation of the eyes occur.

21. The Etesian winds are blowing (Euctemon); the heat begins (Callippus, Euctemon, and Metrodorus).

22. Bad air (Euctemon); beginning of the heat (Hipparchus); west wind and heat (Egyptians).

23. Wintry air on sea, winds (Philippus and Metrodorus); beginning of the blowing of the Etesian winds (Egyptians). On this day Abū Jaʿfar al-Manṣūr began to build Baghdad, that part which is called *Manṣūr's-town*, on the western side of the Tigris in the present Baghdad. This was *a. Alexandri*, 1074. Astrologers are obliged to know dates like this, and must date from such an epoch by means of their knowledge of the *permutationes, terminationes, cycles*, and *directiones*, until they find the horoscopes of those people who were born at those times. It was Naubakht who determined the time (for the commencement of building). The constellation which heaven showed at the time, and the stations of the planets which appeared on heaven, were such as are indicated in Fig. Cat. 15.

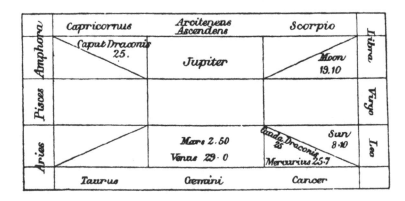

Fig. Cat. 15 The horoscope of Baghdad.

24. Winds (Philippus and Metrodorus); the Etesian winds blow (Hipparchus).
25. South wind (Eudoxus and Caesar); west or south wind (Egyptians). Sexual intercourse and all exertion are forbidden, because it is the time of the greatest heat. The river Oxus begins to rise.
26. South wind and heat (Philippus, Meton, Metrodorus, Democritus, and Hipparchus).
27. Dew and wet, and oppressive air (Euctemon and Dositheus). This oppressive air mostly occurs when heaven is covered and the air is in perfect repose. But often, too, this is peculiar to a place where this cause does not exist, for example, to the region beyond that bridge which, according to al-Jaihānī, was in old times built by the Chinese, reaching from the top of one mountain to that of another on the road that leads from Khotan to the region of the residence of the Khāqān. For those who pass this bridge come into an air which makes breathing difficult and the tongue heavy, in consequence of which many travellers perish there, whilst others are saved. The Tibetans call it the 'poison-mountain'.
28. Nothing mentioned.
29. Beginning of the Etesian winds (Dositheus); heat (Egyptians). They hold the fair of Buṣrā for a whole month, and that of Salamīya for two weeks.
30. The Etesian winds blow (Eudoxus); west wind and heat (Egyptians).
31. South wind (Caesar).

Āb

1. Heat (Hipparchus).
2. Nothing mentioned.
3. Dew falls (Eudoxus and Dositheus); a change in the weather (Caesar).
4. Great heat (Eudoxus).
5. Heat, still and oppressive air, then blowing of winds (Dositheus and Egyptians). They hold a fair at Adhri'āt during fifteen days, also in al-Urdunn, and in several districts of Palestine.
6, 7. Nothing mentioned.
8. The air is still and oppressive (Callippus); wind, and intense heat (Egyptians). According to Sinān, frequently there occurs a change of the air.
9. Heat and still air (Euctemon and Caesar); south wind and turbid air (Egyptians).
10. Heat and still air (Eudoxus, Metrodorus, and Dositheus); a change in the weather (Democritus). At this time the heat is very intense.

11. The northerly winds cease to blow (Callippus, Euctemon, and Philippus); heavy wind (Eudoxus); different winds blow together (Hipparchus); thunder (Egyptians). According to Sinān there is always a change of the weather on this day. He says: I do not know whether we, I and all those who make meteorological observations, are correct in describing a day like this. On this day there is almost always a change of the weather for the better. It is the first day when the air of Iraq begins to be agreeable. Sometimes this change is most evident, whilst at other times it is only slightly perceptible. But that the day should be free from such a change, almost never occurs.

    Some of the ancients consider this day as the beginning of the autumnal air, whilst others take as such the following day.

    Sinān says: Thābit used to say: If in a rare year that which we have described does not take place on this day, it is not likely to take place on the thirteenth or fourteenth, but rather in the middle of Āb. If it takes place on the eleventh, a season of agreeable air is sure to return about the middle of the month, though it may only be short.

12. Heat (Euctemon and Egyptians).

13. A change in the weather and still air (Caesar). Sinān says that on this day an irregular change of the air frequently occurs.

14, 15. Nothing mentioned.

16. A change in the weather (Caesar).

17. A change in the weather (Eudoxus).

18. Nothing mentioned. The *Samūms* are said to cease.

19. A change in the weather, rain, and wind (Democritus); west wind (Egyptians).

20. A change in the weather (Dositheus); heat and density in the air (Egyptians).

21. Nothing mentioned.

22. West wind and thunder (Eudoxus); a change in the weather and bad air (Caesar and Egyptians).

23. West wind (Egyptians).

24. A change in the weather (Eudoxus and Metrodorus). The heat relaxes a little at the time when the sun passes the first 6 degrees of Virgo.

25. A change in the weather (Eudoxus); south wind (Hipparchus); heat (Egyptians).

26. Rotating winds (Hipparchus). Between this day and the first of the Days of the Old Woman (i.e., 26 Shubāṭ) lies one half of a complete year. On this day the heat, at the time when it is about to disappear, returns once more with renewed force, as does also the cold at the time when it is

about to disappear. It is a time of seven days, the last of which is the first of Īlūl, called by the Arabs *Waqdat Suhail* (i.e., the burning of Suhail). It is the time of the winds that accompany the rising of *al-Jabha* (*frons leonis*, the tenth Lunar Station), but as Suhail rises in its neighbourhood, it has become the prevailing use to call the time by *Suhail* and not by *al-Jabha*. The heat of these days is more intense than at any time before or afterwards. But after this time the nights begin to be agreeable. This is an occurrence generally known among people, which scarcely ever fails. Muḥammad ibn ʿAbd al-Malik al-Zayyāt says:

The water had become cold and the night long,
And the wine was found to be sweet;
Ḥazīrān had left you, and Tammūz and Āb.

27. A change in the weather (Philippus).
28. West wind (Egyptians).
29. Rain and thunder; the Etesian winds are about to cease (Eudoxus and Hipparchus).
30. A change in the weather (Hipparchus).
31. The Etesian winds are about to cease (Ptolemy); changing winds (Eudoxus); winds, rain, and thunder (Caesar); east wind (Hipparchus).

Īlūl

1. A change in the weather and the Etesian winds are getting quiet (Callippus). A fair is held at Manbij (Mabbug).
2. Density in the air (Metrodorus). Conon says that on this day the Etesian winds cease.
3. Wind, thunder, and density in the air (Eudoxus); wet and dew (Hipparchus); fog, heat, rain, and thunder (Egyptians). On this day people begin to light their fires in cold countries.
4. Dense and changing air (Callippus, Euctemon, Philippus, and Metrodorus); rain, thunder, and changing wind (Eudoxus).
5. Changing winds and rain, and the Etesian winds are getting quiet (Caesar); rains and wintry air at sea, and south wind (Egyptians). On this day midsummer ends, and a time comes which is good for bleeding and for taking drugs during forty days.
6. West wind (Egyptians).
7. Density in the air (Philippus); a change in the weather (Dositheus).
8. West wind and a change in the weather (Egyptians).
9. Nothing mentioned.

10. The air is not troubled (mixed) (Dositheus).
11. The north winds are ceasing (Caesar).
12. South wind (Eudoxus).
13. A change in the weather (Callippus and Conon).
14. The north winds are ceasing (Eudoxus); a change in the weather (Democritus and Metrodorus). After this time no swallow is seen.
15. Wet and dew (Dositheus); rains and a change in the weather (Egyptians).
16. Density in the air, and rain at sea (Hipparchus).

On the sixteenth in a common year and on the seventeenth in a leap-year occurs the second equinox, which is the first day of the Persian autumn and the Chinese spring, as people maintain. But we have already explained that this is impossible.

The winds, now, blowing on this day are said to be of a psychical nature. To look towards the clouds that rise on this day emaciates the body and affects the soul with disease. I think the reason for this is that people conceive fear on account of the cold and the disappearance of the agreeable time of the year.

It is one of the *omina* of this day to rise from sleep in a worshipping attitude, and to fumigate with tamarisks before speaking.

People say that if a woman who is sterile looks on this day at the star *al-Suhā* and then has intercourse with her husband, she is sure to conceive.

Further, they say, that in the night of this day the waters are getting sweet. We have already heretofore shown the impossibility of such a thing.

This second equinox is, according to the *Canon Sindhind,* a great festival with the Hindus, like the Mihrjān with the Persians. People make each other presents of all sorts of valuable objects and of precious stones. They assemble in their temples and places of worship until noon. Then they go out to their pleasure-grounds, and there they assemble in parties, showing their devotion to the (Deity of) Time and humbling themselves before God Almighty.

17. Rain at sea and density in the air (Metrodorus).
18. West, then east wind (Egyptians).
19. Wet and dew (Eudoxus); west wind, mizzle, and rain (Egyptians). On this day the water returns from the upper parts of the trees to the roots.
20, 21. (Missing.)
22. Nothing mentioned.
23. Rain (Eudoxus); west wind or south wind (Hipparchus).

24. Nothing mentioned. On this day the fair of Thuʿāliba is held. Practical observers say that people mark on this day what wind is constantly blowing until night or until the time when the sun begins to decline; for this will be the most constant of all the winds of the year. This day they called the Turning of the Winds. The white-and-black crows appear on this day in most countries.
25. A change in the weather (Hipparchus and Eudoxus); west wind or south wind (Egyptians).
26, 27, 28. (Missing.)
29. A change in the weather (Euctemon and Eudoxus); west wind or south wind (Hipparchus).
30. Nothing mentioned by the ancients, either about the air or anything else.

This, now, is the calendar used by the Greeks, to which we have added all that Sinān has mentioned in his *Kitāb al-anwāʾ*. This is the concise summary of his book. We have not kept back anything which we have learned regarding the days of the calendar. We quote them by the names of the Syrians (i.e., as the first of Tishrīn, Kānūn, etc.) only, because they are generally known among people, and because this serves the same purpose (as if we were to call them by the Greek names).

Next we shall speak of the memorable days in the months of the Jews, if God Almighty permits!

A.xx. Codex Marcianus 335[256] parapegma

τοῦ πρωτοσπαθαρίου καὶ στρατηγοῦ τῶν Κιβυρραιωτῶν περὶ τῶν παρατηρουμένων ἀστέρων παρὰ τοῖς πλευστικοῖς, τῶν ποιούντων ζάλας καὶ ταραχὰς ἐν τῇ θαλάσσῃ.

τῇ ιδʹ τοῦ Σεπτεμβρίου μηνὸς ἄστρον τῆς Ὑψώσεως τοῦ τιμίου σταυροῦ καὶ γίνεται ταραχὴ τῆς θαλάσσης ἐξ ἀνέμου βορρᾶ εἴτε λιβὸς ἐπικρατοῦσα μέχρι ὡρῶν δύο· τῇ κʹ τοῦ αὐτοῦ μηνὸς ἄστρον τῆς ἁγίας Θέκλης καὶ γίνεται ταραχὴ τῆς θαλάσσης ἐν ὥρᾳ μιᾷ· τῇ ςʹ τοῦ Ὀκτωβρίου ἄστρον παρατηρούμενον τοῦ ἁγίου Θωμᾶ, κακοταρία γίνεται πολλὴ τῆς θαλάσσης καὶ οὐδέποτε ἄνευ ζάλης[257] διῆλθε τὸ τοιοῦτον ἄστρον· τῇ ιβʹ τοῦ αὐτοῦ μηνὸς ἄστρον τὸ καλούμενον Ταύρου οὐρὰ καὶ γίνεται κακοταρία

---

[256] Greek text following *CCAG* ii.214.   [257] Reading ζάλης for the MS τάλου.

πολλὴ καὶ ζάλη μεγάλη ἐν τῇ θαλάσσῃ· τῇ κϛ´ τοῦ αὐτοῦ μηνὸς ἄστρον τοῦ
ἁγίου Δημητρίου ποιοῦν ζάλην μεγάλην ἐν τῇ θαλάσσῃ. τῇ ια´ τοῦ **Νοεμβρίου** μηνὸς ἄστρον τοῦ ἁγίου Μηνᾶ καὶ γίνονται κακοταρίαι καὶ ταραχὴ
θαλάσσης πλέον τῶν λοιπῶν ἄστρων. τῇ ιδ´ τοῦ αὐτοῦ μηνὸς ἄστρον τοῦ
ἁγίου Φιλίππου καὶ αὐτὸ ποιοῦν θαλάσσης κακοταρίαν. ἀπὸ δὲ τῆς ιε´ τοῦ
αὐτοῦ μηνὸς ἄρχεται ἡ λεγομένη πλειοδυσία· καὶ κατὰ ἑβδόμην ἡμέραν
ἤγουν ἀπὸ τῆς ιε´ ἕως τῆς κβ´ δύνει ἄστρον· ὁμοίως καὶ καθεξῆς καὶ ἐν ταῖς
αὐταῖς ἡμέραις, ὅτε τὸ ἄστρον δύνει, γίνεται ἐν τῇ θαλάσσῃ κακοταρία·
ἐπικρατεῖ δὲ ἡ τοιαύτη Πλειὰς ἡμέρας μθ´, ἐν δὲ τῷ μέσῳ τούτων τῶν
ἡμερῶν ἐστιν ἄστρον τοῦ ἁγίου Νικολάου ποιοῦν καὶ αὐτὸ κακοταρίαν.
ἀπὸ δὲ τῆς αὐτῆς ϛ´ τοῦ **Δεκεβρίου** μηνὸς μέχρι τῆς ϛ´ τοῦ Ἰαννουαρίου
γίνονται κακοταρίαι πολλαὶ καὶ ζάλαι ἐν τῇ θαλάσσῃ· ἀπὸ δὲ τῆς αὐτῆς
ϛ´ μέχρι εἰκάδος τοῦ **Μαρτίου** μηνὸς πλέουσιν οἱ πλευστικοὶ μήτε κακοταρίαν πολλὴν ἔχοντες μήτε μὴν γαλήνην τελείαν· ἐν τῷ μέσῳ δὲ τῶν
ἡμερῶν τούτων ἀστέρων παρατήρησις οὐχ ὑπάρχει· ἀπὸ δὲ τῆς κε´ τοῦ
Μαρτίου μηνὸς ἡνίκα τὸ ἔαρ ἔρχεται, ἄρχονται οἱ παχνῖται· καὶ ὁ πρῶ-
τος παχνίτης ἐπικρατεῖ ἡμέρας ζ´, ὁμοίως καὶ οἱ λοιποί· εἰσὶ δὲ παχνῖται
ζ´ κατὰ τὴν ἀκολουθίαν τῆς πλειοδυσίας· καὶ εἰς τὴν πρώτην ἡμέραν τοῦ
πρώτου παχνίτου γίνεται ἄνεμος σφοδρὸς ὁ λεγόμενος εὖρος καὶ ἀστρα-
παὶ καὶ βρονταί, ὥστε μὴ δύνασθαι πλέειν ἐν τῇ τοιαύτῃ ἡμέρᾳ τινὰ καὶ
ἐπικρατεῖ ἡμερῶν δύο καὶ μετέπειτα πλατύνει καὶ δέχονται τὸν ἄνεμον
ἐξ εὐωνύμων καὶ ἀρμενίζουσι δεξιᾷ. οἱ δὲ ἐμπειρότεροι τῆς θαλάσσης ἐκ
τῶν κυμάτων καὶ τῶν ῥευμάτων διαγνώσονται, ποῦ πορεύονται, καὶ τῇ
τοι<αύ>τῃ ἡμέρᾳ ἀπολιβάζει, καὶ ἐπὶ τὴν αὔριον γίνεται κακοταρία· καὶ
ἐπικρατεῖ μέχρι τῆς α´ τοῦ β´ παχνίτου· ὁμοίως γίνεται καὶ αὐτὸ εἰς τὸν
δεύτερον παχνίτην· εἰς δὲ τὸν γ´ παχνίτην γίνεται πλέον κακοταρία παρὰ
τοὺς λοιποὺς παχνίτας· εἰς δὲ τὸν τέταρτον παχνίτην, γίνεται κακοταρία
ἐλαφροτέρα, ὡσαύτως καὶ εἰς τὸν πέμπτον καὶ εἰς τὸν ἕκτον καὶ εἰς τὸν
ἕβδομον. τῇ ια´ τοῦ **Μαΐου** μηνὸς ἐστι ἀστὴρ ὁ Σωρευτὴς καὶ γίνεται ἄνεμος
σφοδρός· ἀπὸ δὲ τῆς ιβ´ ἡμέρας τοῦ Μαΐου μέχρι τῆς κδ´ τοῦ Ἰουνίου μηνός,
ὅτε καὶ ἡ τροπὴ τοῦ ἡλίου γίνεται, πλέεται ἡ θάλασσα ἀπαρατηρήτως·
ἀστέρα γὰρ ἢ ταραχὴν θαλάσσης οὐ περιβλέπονται οἱ ναυτικοί· ἀπὸ δὲ
τῆς κδ´ τοῦ Ἰουνίου μηνὸς μέχρι τῆς ιε´ τοῦ **Αὐγούστου** φυσσῶιν οἱ ἄνεμοι
οἱ καλούμενοι ῥῆκται· ἐν δὲ τῇ Συρίᾳ φυσῶσι νότοι· δι᾽ αὐτοῦ γὰρ τοῦ
ἀνέμου ἐξέρχονται ἀπὸ Συρίας εἰς Ῥωμανίαν. ἐπικρατοῦσι δὲ οἱ τοιοῦ-
τοι ἄνεμοι ἡμέρας μθ´ καθὼς καὶ ἡ πλειοδυσία, καὶ κατὰ ἑβδόμην ἡμέραν
γίνεται ταραχὴ τῆς θαλάσσης· τοὺς δὲ λοιποὺς ἀστέρας οἱ πλευστικοὶ
οὐ παρατηροῦνται· καὶ αὐτοὺς γὰρ τοὺς εἰρημένους ἀστέρας οἱ πεῖραν
πολλὴν ἔχοντες, εἰ καὶ τάχα ταραχὴ καὶ κακοταρία γίνεται ἐν τῇ θαλάσσῃ,

οὐ παρατηροῦνται· ἀλλὰ καὶ ἐν ταῖς ἡμέραις τῶν τοιούτων ἀστέρων πλέουσι.

For the First Guardsman and the General of the Cibyrraeotes: Concerning the observed stars, those that cause squalls and disturbances at sea, according to sailors.

On the 14th of the month of **September:** the star of the Elevation of the Holy Cross,[258] and there is a disturbance of the sea from the north or the south-west wind dominating until the second hour. On the 20th of the same month: the star of St Thecla, and here is a disturbance of the sea in the first hour. On the 6th of **October:** the observed star of St Thomas, there is very foul weather on the sea, and this star has never passed without a squall. On the 12th of the same month: the star called the tail of Taurus, and there is very foul weather, and a large squall at sea. On the 26th of the same month: the star of St Demetrius causes a large squall at sea. On the 11th of the month of **November:** the star of St Mennas, and there is foul weather and a disturbance of the sea – more than the rest of the stars. On the 14th of the same month: the star of St Philip, and this causes foul weather on the sea. From the 15th of this month, what is called the *Pleiodysia* ('Setting-of-the-Pleiades') begins, and in the week, that is to say from the 15th to the 22nd, the star sets. Thus you will hold back, and there is foul weather at sea on these days, when the star is setting. This Pleiad is dominant for forty-nine days, and in the middle of these days the star of St Nicholas itself also causes foul weather. From the 6th of the month of **December** to the 6th of **January** there is also much foul weather and squalls at sea. From the 6th of this (month) to the 20th of the month of **March,** sailors sail since there is not much foul weather and the end of the month is not calm. The careful watch of the stars does not begin in the middle of these days, but from the 25th of the month of March, when spring comes, the frost-seasons begin,[259] and the first frost-season is dominant for seven days, and the rest are the same. There are seven frost-seasons in the wake of the *Pleiodysia*, and on the first day of the first frost-season there is the strong wind called *Eurus* (the east wind) and lightning and thunder, so people are not able to sail on this day, and this dominates for two days, but afterwards it widens and they catch the wind from the left and sail to the right. Those with more experience of the sea make distinctions from how the swells and flows move: on that day

---

[258] In this text the 'star of St Such-and-such' tends to fall on the feast day of that saint. Diverging dates are probably scribal errors.

[259] See note to the Antiochus parapegma, 22 April.

they drop off. The next day there is foul weather, and that dominates until the first (day) of the second frost-season. Similarly, the same thing happens in the second frost-season. In the third frost-season there is fouler weather than the remaining frost-seasons. In the fourth frost-season there is less foul weather, and similarly in the fifth and the sixth and the seventh. On the 11th of the month of **May** it is the star *Soreutes,* and there is a strong wind. From the 12th day of May until the 24th of the month of **June**, which is the summer solstice, the sea is sailed without worry, and sailors do not watch for stars or a disturbance of the sea. From the 24th of June until the 15th of **August**, the winds called the *Breakers*[260] blow gently. In Syria south winds blow gently, by which wind they travel from Syria to Romania. These winds dominate for forty-nine days, just like the *Pleiodysia*, and after seven days there is a disturbance of the sea. The sailors do not watch for the rest of the stars, and those with much experience do not watch for the aforementioned stars even. For they do not worry even if there is a sudden disturbance or foul weather at sea, and they sail even on the dates of those stars.

B.i. Thermae Traiani parapegma. See p. 168.

B.ii. Dura-Europus parapegma. See p. 170.

B.iii. Latium parapegma. See p. 171.

B.iv. Veleia inscription. See p. 172.

B.v. Neapolitan Museum 4072. See p. 173.

B.vi. Ostia inscription. See p. 173.

B.vii. Pompeii calendar. See p. 173.

B.viii. Pausilipum parapegma. See p. 174.

B.ix. Trier hebdomadal parapegma. See p. 176.

B.x. Trier parapegmatic mould. See p. 176.

B.xi. Soulousse hebdomadal parapegma. See p. 177.

[260] ῥῆκται.

B.xii. Arlon hebdomadal parapegma. See p. 177.

B.xiii. Rottweil parapegma. See p. 178.

B.xiv. Bad Rappenau hebdomadal parapegma. See p. 179.

C.i. P. Rylands 589[261]

(Column 9, fragment 4)

```
          .... παράπ[ηγμα τ]ῶν
κατ[ὰ σ]ελήνην νουμηνιῶ[ν ὡς εἰσι κ]α-
τὰ [τὰς ἡ]μέρας τῶ[ν] κατ᾽ Αἰγυπ[τίους δωδε]κα-
μή[νω]ν τεταγμέναι, οὗ ἐστὶν ἡ π[ερ]ίο-
δ[ος ἔ]τη μὲν εἴκ[ο]σι πέντε, μῆνες δ[ὲ σ]ὺν
ἐμβ[ο]λίμ[ο]ις τριακόσιοι ἐννέα, ἡμέρ[αι] δὲ
ἐν[ακι]σχ[ίλ]ιαι ἑκατὸν εἴκ[ο]σι πέντε. [σ]η-
μαί[νε]ι δὲ καὶ τοὺς κατὰ σελήνην μῆ-
νας καὶ τούτων τίνες ἦσὶ πλήρη[ς] κ[αὶ] τίνες
κοῖλ[οι κ]αὶ π[ο]ῖο[ι] αὐτῶν ἐμβόλιμοι κα[ὶ ἐ]ν τίνι
ζω[ιδί]ωι ἥλι[ο]ς καθ᾽ ἕκαστον μῆν[α στή]σεται.
ὅ[ταν] διέλθει τὰ εἴκοσι πέ[ν]τε ἔ[τη] πάλιν
ἐπ[ὶ τὴ]ν αὐτὴν ἀρχὴν ἥξει καὶ τὸ[ν α]ὐτὸν
τρόπ[ο]ν ἀλ[λάξ]εται. ἔστιν δὲ πρῶ[το]ν ἔτος
τῆς περιόδου [τ]ὸ αὐτὸ τῶι πρώτω[ι] ὡς
βασίλισσα Κλεοπάτ[ρ]α καὶ βασιλεὺ[ς Π]τολεμαῖος
ὁ υ[ἱ]ὸς θεοὶ Ἐπιφ[αν]εῖς ἄγουσιν²⁶² ἐν [ὧι] καὶ τὴν
βασιλείαν [π]αρ[ελ]άβοσαν. ὁ δὲ ἥλ[ιος] καθ[έστ]η
```

(Column 10, fragment 4)

```
μῆνα Ε[...
Θωὺθ Σκ[ορπίωι
Φαῶφι Τοξό[τηι
Ἀθὺρ Αἰγ[οκέρωι
Χοίαχ Ὑ[δ]ρ[οχόωι
Τῦβι Ἰχθύσι
```

---

261 Greek text from Turner and Neugebauer, 1949.

262 Turner and Neugebauer note that this is 'a formal and technical expression', which is true
enough, though its exact meaning is opaque to me.

Μεχεὶρ Κρί[ωι
Φαμενὼθ Τα[ύρωι
Φαρμοῦθι Δι[δύμοις
Παχὼν Κα[ρκίνωι
Παῦνι Λέ[οντι
Ἐπεὶφ Πα[ρθένωι
Μεσορὴ Χη[λαῖς

(Column 10, fragment 5)

αἱ δὲ κατὰ σ[ελήνην νουμη]νίαι
εἰσὶν ἔτους [πρώτου
Θωὺθ κ΄ [. . .[263]
Φ[α]ῶφι ιθ΄ [. . .
Ἀθὺρ ιθ΄ [. . .
Χοίαχ ιθ΄ [. . .

(Column 11, fragment 5)

[Τῦβι
[Μεχεὶρ
[Φαμενὼθ
[Φαρμοῦθι
Π[αχὼν
Π[αῦνι
Ἐπε[ὶφ
Με[σορὴ

(Column 11, fragment 6)

ἔτ[ους δευτέρου
Θω[ὺθ
Φαῶ[φι
Ἀθὺ[ρ

---

[263] Turner and Neugebauer think that day-counts for the lengths of the months would have been in the lacunae after each of these entries: Θωὺθ κ΄ [ἡμέραι κθ] // Φαῶφι ιθ΄ [ἡμέραι λ] // Ἀθὺρ ιθ΄ [ἡμέραι λ] // Χοίαχ ιθ΄ [ἡμέραι κθ]. If their reconstruction of column 12, fragment 6, is correct, then this is possible, but there the text is very badly damaged, and I am less confident of the reconstructibility here. Note also that in column 12 the day-counts precede the month names rather than following them as proposed here.

Χοία[χ
Τῦβ[ι
Μεχ[εὶρ
Φαμ[ενὼθ
Φαρμ[οῦθι
Παχ[ὼν

(Column 12, fragment 6)

ἡμέ]ραι κθ[
ἡμέρ]αι λ΄ Π[αχὼν
ἡμέρ]αι λ΄ [Παῦνι
ἡμέρ]αι κθ΄ [Ἐπεὶφ
ἡμέ]ραι λ΄ Μ[εσορὴ
ἡμέρ]αι κθ΄ ἔ[τους τετάρτου
ἡμέρ]αι λ΄ [
ἡμέρ]αι κθ΄ [
ἡμέρ]αι λ΄ [

(Column 9, fragment 4)

A parap[egma o]f the beginning[s of l]unar months (as they are ordered according to [the d]ays of the Egyp[tian ye]ar) of which the cy[cl]e is twe[n]ty-five years, (having) 309 months (including inter[ca]lary months), (or) 9,125 day[s]. It sh[o]ws the months according to the moon, and which of these will be [f]ull, which hollo[w], and w[h]i[c]h intercalary, an[d i]n which zo[dia]cal sign the sun [wi]ll be for each mont[h]. W[hen] twenty-five ye[ars] have passed, it will go t[o th]e same beginning, and it will change in th[e s]ame way. The fir[s]t year of the cycle is [t]he same as the firs[t] (year) that queen Cleopat[r]a and kin[g P]tolemy the y[ou]nger were taken as gods manifest and in [which] they [r]e[ce]ived the kingdom. The su[n] s[tand]s [. . .

(Column 10, fragment 4)

month [. . .
Thoth:	Sc[orpio
Phaophi:	Sagit[tarius
Athyr:	Cap[ricorn
Choiak:	A[q]u[arius
Tybi:	Pisces [
Mecheir:	Ari[es

Phamenoth:	Ta[urus
Pharmouthi:	Ge[mini
Pachon:	Ca[ncer
Payni:	Le[o
Epeiph:	Vi[rgo
Mesore:	Li[bra

(Column 10, fragment 5)

For the [first] year, the [begin]nings (of the months)
according to the m[oon] are:
Thoth: 20th [...
Ph[a]ophi: 19th [...
Athyr: 19th [...
Choiak: 19th [...

(Column 11, fragment 5)

[Tybi...
[Mecheir...
[Phamenoth...
[Pharmouthi...
P[achon...
P[ayni...
Epe[iph...
Me[sore...

(Column 11, fragment 6)   *(Column 12, fragment 6)*

For the second]		
ye[ar]:		
Th[oth...	...] 29 [da]ys	[Pharmouthi
Pha[ophi...	...] 30 [day]s	P[achon...
Athy[r...	...] 30 [day]s	[Payni...
Choia[k...	...] 29 [day]s	[Epeiph...
[Tybi...	...] 30 [da]ys	M[esore...
[Mecheir...	...] 29 [day]s	[For the fourth] y[ear]
[Phamenoth...	...] 30 [day]s	[...
[Pharmouthi...	...] 29 [day]s	[...
P[achon...	...] 30 [day]s	[...
[Payni...	[lines lost]	
[Epeiph...		

C.ii. Miletus I[264]

Inv. 456 B

1   ...]ΟΙΣΑ[...   1
                                        ··
                                         Λ
                          • ἐν Ὑδροχόω[ι ὁ] ἥλιος   [• ἐν Κριῶι ὁ ἥλιος?
 • ἐν Τοξότ]ηι ὁ ἥλιο[ς.   • [.....] ἑῶιος ἄρχεται δύνων   • [...
5 • ......] ἑῶιος δύνει καὶ Προ-   καὶ Λύρα δύνει.   ΤΕ[...   5

    κύων ἐ]ῶιος δύνει.   ··
                              • Ὄρνις ἀ[κ]ρόνυχος ἄρχεται δύν[ων.]   • ΜΕ[...
 • .....ἐ]ῶ[ιο]ς δύνει.   • • • • • • • • • •
 • Τοξότ]ης ἄρχεται ἑῶιος ἐ-   • Ἀνδρομέδα ἄρχεται ἑῶια ἐ[πι]   ΛΩ[...
                                   τέλλειν.   • Κ[...
    πιτέ]λλων καὶ Περσεὺς ὁ-   ··
                              • Ὑδροχόος μεσοῖ ἀνατέλλων.   Δ[...
10 λος ἑῶ]ιος δύνει.   •
 • Σκο]ρπίου τὸ κέντρον ἐπι-   • Ἵππος ἑῶιος ἄ[ρ]χ[ετ]αι ἐπι-   • ΚΕ[... ἐ-   10
    τέ]λλει ἑῶιον   τέλλει   πιτ[έλλει
    ··   •   • •
 • ....]ΕΥΜΑ ἑῶιον ἐπιτέλλει .   • Κένταυρος ὅλος ἑῶιος δύνει.   • ΟΕΝ[... ἀκρό-
                              • Ὑδρος ὅλος ἑῶιος δύνει.   νυχ[ος...
 • Ἰχ]θὺς ὁ νότιος ἄρχεται ἀκρό-   ··
                              • Κῆτος ἄρχετα[ι] ἀκρόνυχον   • Πλ[ειάδες
15 ν]υχος δύνειν.   δύνειν.   •
 • Ἀε]τὸς ἑῶιος ἐπιτέλλει.   Χ • Ὀιστὸς δύνει. ζ[ε]φύρων ὥ-   • ΚΑ[...   15
 • ?Δίδυμ]οι[? λήγ]ουσιν δυόμε-   ρα συνεχῶν   ΑΡ[...
                              • • • •   • [...
18 νοι]   • Ὄρνις ὅλος ἀκρόνυχος δύνει.   •   18
                              [•]

Inv. 456 B

                       ...]ΟΙΣΑ[...
                            ··
                       thirty (days)
                      • The sun is in Aquarius   [• The sun is in Aries]
•] The sun is [in Sagitta]rius.   • .....] begins setting in the morning   • [...
• ......] sets in the morning and   and Lyra sets.   ΤΕ[...
                              ••
  Pro[cyon] sets in the morning.   • Cygnus begins to set acronychally.   • <u>ΜΕ</u>[...
                              • • • • • • • • •

---

264 Greek text from Lehoux, 2005.

• . . . . .] sets in the [m]or[nin]g.
• Sagittar]ius begins ri[si]ng in

the morning and the w[hole] of
Perseus sets in [the
m]orning.
•] The middle of [Sco]rpio rises
in the morning.
    • •

• . . . .]EYMA rises in the morning.

•] The southern [f]ish begins
to set acro[n]ychally.
• Aq]uila rises in the morning.

• ?Gemin]i [?finishe]s setting.

• Andromeda begins rising in the
morning.
    • •
• Aquarius is in the middle of rising.

• Pegasus begins to rise in the
morning.
    •
• The whole of Centaurus sets in the
morning.
• The whole of Hydra sets in the
morning.
• Cetus begins to set acronychally.

X • Sagitta sets. The season of the
continuous west wind.
    • • • •
• The whole of Cygnus sets
acronychally.
[•]

ΛΩ[. . .
• K[. . .

Δ[. . .
•
• K<u>E</u>[. . .
[r]is[es . . .
    • •
• O<u>E</u>N[. . . acro-
nych[ally . . .
    • •
• The Pl[eiades
•
• KA[. . .
A<u>P</u>[. . .
• [. . .
•

Inv. no. 84

Current location unknown. Rehm treated this as part of Miletus I, but as I
argue in Lehoux, 2005, it is likely an independent inscription. I include it
here for the sake of completeness.

1   θ]ερινῆς τρο[π]ῆς [γε-   EXOM[
νομένης ἐπὶ Ἀψεύδους   KAIEΠ[. . .
Σκιροφοριῶνος ιγ΄, ἥ-   ΔEO[. . .
τις ἦν κατὰ τοὺς Αἰγυ-   KAIΣ[. . .
5   πτίους μία καὶ κ΄   EKKA[. . .
τ]οῦ Φαμενώθ, ἕως   THPIΔ[. . .
τῆ]ς γενομένης ἐπὶ   MEPA[. . .
Πολ]υκλείτου Σκι-   OΔΩ[. . .
ροφορι]ῶνος ιδ΄, κα-
10   τὰ δὲ τού]ς Αἰγυπτί-   ENNEA[. . .
ους τοῦ Παυ]νὶ τῆς ια΄,   KAITI[. . .
. . .]HΣION   ΔIΣT[. . .
EΛE

1   the s]ummer sol[s]tice [ha-
ppened in (the archonship of)   EXOM[
Apseudes on the 13th of Skirophorion,   KAIEΠ[. . .
which was, according to the Egy-   ΔEO[. . .
5   ptians the 21st   KAIΣ[. . .
o]f Phamenoth, but in   EKKA[. . .

(the archonship of)	ΤΗΡΙΔ[...
Pol]yclitus, it [h]appened	ΜΕΡΑ[...
on the 14th of Skirophorion,	ΟΔΩ[...
or acc[ording to th]e Egypt-	ΕΝΝΕΑ[...
[	ΚΑΙΤΙ[...
ians], the 11th of [Pay]ni.	
...]ΗΣΙΟΝ	ΔΙΣΤ[...
ΕΛΕ	

C.iii. Codex Vindobonensis 108, fo. 282ᵛ[265]

περὶ τῆς τοῦ ἐνιαυτοῦ διαιρέσεως.

διαιρεῖται ὁ ἐνιαυτὸς εἰς μέρη δ΄.

θέρους ἡμέραι ρκδ΄ ἀπὸ Πλειάδων ἐπιτολῆς εἰς Ἀρκτούρου ἐκφάνειαν καὶ
   Αἰγὸς ἐπιτολὴν.

μετοπώρου ἡμέραι νϛ΄ ἀπὸ Ἀρκτούρου ἐκφανείας καὶ Αἰγὸς ἐπιτολῆς εἰς
   Πλειάδων καὶ Ὠρίωνος δύσιν.

χειμῶνος ἡμέραι ρ΄ ἀπὸ Πλειάδων καὶ Ὠρίωνος δύσεως εἰς Ὀιστοῦ δύσιν
   καὶ Ἵππου ἐπιτολήν.

ἔαρος ἡμέραι πε΄ ἀπὸ Ὀιστοῦ δύσεως καὶ Ἵππου ἐπιτολῆς εἰς Πλειάδων
   ἐπιτολήν.

τῶν δὲ ἀπλανῶν ἀσέρων ἐπιτολή γίνεται τοῦτον τὸν τρόπον·
ἀνατολή[266] πρὸ ε΄ εἰδῶν Μαίων.

ἀπὸ Πλειάδων ἐπιτολῆς εἰς Αἰγὸς ἑσπερίας δύσιν ἡμέραι ιη΄.[267]

ἀπὸ Αἰγὸς δύσεως εἰς Ἀετοῦ ἐπιτολὴν ἡμέραι έ.

ἀπὸ δὲ Ἀετοῦ ἐπιτολῆς εἰς Ἀρκτούρου δύσιν ἡμέραι ιδ΄.

ἀπὸ Ἀρκτούρου δύσεως εἰς Ταύρου ἐπιτολὴν ἡμέραι δ΄.

ἀπὸ Ταύρου ἐπιτολῆς <εἰς> τροπὰς θερινὰς ἡμέραι λγ΄.

ἀπὸ τροπῶν θερινῶν εἰς Ὠρίωνος ἐκφάνειαν ἡμέραι ιη΄.

ἀπὸ Ὠρίωνος ἐκφανείας εἰς Ἀετοῦ δύσιν ἡμέραι ϛ΄.

ἀπὸ Ἀετοῦ δύσεως εἰς Κύνα ἡμέραι δ΄. ἐτησίαι ἄνεμοι ἄρχονται πνεῖν.

ἀπὸ Κυνὸς ἐκφανείας εἰς Λύρας δύσιν καὶ Ἵππου ἐπιτολὴν ἡμέραι ιγ΄.

---

[265] Greek text following Rehm, 1913. The format of Rehm's edition is very hard to follow. A new edition would be very desirable. In order to give this text a better fit with Euctemon citations in Geminus and Ptolemy, Rehm changes the beginning of the year to the summer solstice, and frequently queries the text. I have stuck as closely as possible to the MS here.

[266] Rehm restores a lost Πλειάδων here.

[267] Something has gone amiss between this entry and the following one. Here we have εἰς Αἰγὸς ἑσπερίας δύσιν and the next entry has ἀπὸ Αἰγόκερου δύσεως. Rehm emends both entries to Αἰγόκερω, Capricorn, but his reasons are opaque to me. The Madrid parapegma has the evening disappearance of Capella seventeen days after the rising of the Pleiades on the 5th of May.

ἀπὸ Λύρας δύσεως καὶ Ἵππου ἐπιτολῆς εἰς Προτρυγητοῦ ἐκφάνειαν καὶ
    Ἀρκτούρου ἐπιτολὴν καὶ Ὀϊστοῦ δύσιν[268] εἰς Ἀρκτούρου ἐκφάνειαν
    καὶ Αἰγὸς ἐπιτολὴν ἡμέραι ιʹ.

ἀπὸ Ἀρκτούρου καὶ Αἰγὸς ἐπιτολῆς εἰς ἰσημερίαν φθινοπωρινὴν ια ἡμέραι.

ἀπὸ δὲ ἰσημερίας φθινοπωρινῆς εἰς Ἐρίφων ἐπιτολὴν ἡμέραι βʹ.

ἀπὸ Ἐρίφων ἐπιτολῆς εἰς Πλειάδων ἑσπερίαν δύσιν ἡμέραι βʹ.

ἀπὸ Στέφανου ἐπιτολῆς εἰς Ὑάδων ἐπιτολὴν ἡμέραι ιγʹ.[269]

ἀπὸ Ὑάδων ἐπιτολῆς εἰς Ἀρκτούρου δύσιν ἡμέραι ιϛʹ.

ἀπὸ Ἀρκτούρου δύσεως εἰς Λύρας ἐπιτολὴν ἡμέραι εʹ.

ἀπὸ Πλειάδων δύσεως εἰς Ὠρίωνος ὅλου δύσιν ἡμέραι γʹ.

ἀπὸ Κύνος δύσεως εἰς Σκορπίου κέντρον ἡμέραι βʹ.

ἀπὸ Σκορπίου κέντρου εἰς Ἀετοῦ ἐπιτολὴν εἰς Αἰγὸς δύσιν ἡμέραι γʹ.

ἀπὸ Αἰγὸς δύσεως εἰς τροπὰς χειμερινὰς ἡμέραι ιβʹ.

ἀπὸ τροπῶν χειμερινῶν εἰς Δελφῖνος ἐπιτολὴν ἡμέρα αʹ.

ἀπὸ Δελφῖνος ἐπιτολῆς εἰς Ἀετοῦ δύσιν ἡμέραι δʹ.

ἀπὸ Ἀετοῦ δύσεως εἰς Δελφῖνος δύσιν ἡμέραι κʹ.

ἀπὸ Δελφῖνος δύσεως εἰς Λύρας δύσιν ἡμέραι ϛʹ.

ἀπὸ Ὀϊστοῦ δύσεως καὶ Ἵππου ἐπιτολῆς εἰς Προτρυγητοῦ καὶ Ἀρκτούρου
    καὶ Ἵππου δύσιν ἡμέραι ιϛʹ.[270]

ἀπὸ Ἵππου δύσεως εἰς ἰκτίνου ἐπιφάνειαν καὶ ἰσημερίαν ἐαρινὴν ἡμέραι ιβʹ.

ἀπὸ ἰσημερίας ἐαρινῆς εἰς Πλειάδων κρύψιν ἡμέραι ιδʹ.

ἀπὸ Ὑάδων κρύψεως εἰς Ὠρίωνος ὅλου κρύψιν εἰς Κύνος κρύψιν καὶ Λύρας
    ἐπιτολὴν ἡμέραι ιδʹ.

ἀπὸ Κυνὸς κρύψεως καὶ Λύρας ἐπιτολῆς εἰς Ἀετοῦ ἐπιτολὴν ἡμέραι γʹ.

ἀπὸ Ἀετοῦ ἐπιτολῆς εἰς Πλειάδων ἐπιτολὴν ἡμέραι ιγʹ.

Concerning the divisions of the year.

The year is divided into four parts:

Summer is 124 days, from the rising of the Pleiades to the appearance of
    Arcturus and the rising of Capella.

Autumn is fifty-six days, from the appearance of Arcturus and the rising of
    Capella to the setting of the Pleiades and Orion.

Winter is 100 days from the setting of the Pleiades and Orion to the setting
    of Sagitta and the rising of Pegasus.

---

[268] Rehm thinks, plausibly, that something has dropped out here. If we add up the total number
of days accounted for by this text we only reach 297, so there are definitely day-counts missing,
and the disruption of the usual formula of the text by an extra εἰς clause here is the giveaway.

[269] Here as elsewhere in this text, Rehm is right to suspect that an entry has dropped out between
this one and the one before it.

[270] Rehm emends the entry to εἰς Προτρυγητοῦ καὶ Ἀρκτούρου <ἐπιτολὴν> καὶ Ἵππου δύσιν.

Spring is eighty-five days from the setting of Sagitta and the rising of Pegasus to the rising of the Pleiades.

The rising of the fixed stars happens in this way:

The rising is on the fifth day before the Ides of May.

From the rising of the Pleiades to the evening setting of Capella, eighteen days.

From the setting of Capella to the rising of Aquila, five days.

From the rising of Aquila to the setting of Arcturus, fourteen days.

From the setting of Arcturus to the rising of Taurus, four days.

From the rising of Taurus to the summer solstice, thirty-three days.

From the summer solstice to the appearance of Orion, eighteen days.

From the appearance of Orion to the setting of Aquila, six days.

From the setting of Aquila to Sirius, four days. The Etesian winds begin to blow.

From the appearance of Sirius to the setting of Lyra and the rising of Pegasus, thirteen days.

From the setting of Lyra and the rising of Pegasus to the appearance of Vindemiatrix and the rising of Arcturus and the setting of Sagitta to the appearance of Arcturus and the rising of Capella, ten days.

From the rising of Arcturus and Capella to the autumnal equinox, eleven days.

From the autumnal equinox to the rising of the Haedi, two days.

From the rising of the Haedi to the evening setting of the Pleiades, two days.

From the rising of Corona to the rising of the Hyades, thirteen days.[271]

From the rising of the Hyades to the setting of Arcturus, sixteen days.

From the setting of Arcturus to the rising of Lyra, five days.

From the setting of the Pleiades to the setting of the whole of Orion, three days.

From the setting of Sirius to the middle of Scorpio, two days.

From the middle of Scorpio to the rising of Aquila[272] to the setting of Capella, three days.

From the setting of Capella to the winter solstice, twelve days.

From the winter solstice to the rising of Delphinus, one day.

From the rising of Delphinus to the setting of Aquila, four days.

From the setting of Aquila to the setting of Delphinus, twenty days.

---

[271] Here as elsewhere in this text, Rehm is right to suspect that an entry has dropped out between this entry and the one before it.

[272] Rehm thinks, plausibly, that something has dropped out here.

From the setting of Delphinus to the setting of Lyra, six days.

From the setting of Sagitta and the rising of Pegasus to the setting of Vindemiatrix, Arcturus, and Pegasus, sixteen days.

From the setting of Pegasus to the appearance of the kite and the vernal equinox, twelve days.

From the vernal equinox to the disappearance of the Pleiades, fourteen days.

From the disappearance of the Hyades to the disappearance of the whole of Orion[273] to the disappearance of Sirius and the rising of Lyra, fourteen days.

From the disappearance of Sirius and the rising of Lyra to the rising of Aquila, three days.

From the rising of Aquila to the rising of the Pleiades, thirteen days.

---

[273] Rehm thinks, plausibly, that something has dropped out here, and that an entry has dropped out just before this one.

E.iv. Vitruvius. See p. 202.

E.v. Diogenes Laertius. See p. 202.

E.vi. Proclus. See p. 203.

E.vii. Suda. See p. 203.

F.iv. Ṣafṭ el-Ḥenna naos. See p. 204.

F.v. The Letter of Diocles of Carystus (excerpt)[274]

γέγραφα δέ σοι καὶ περὶ τῶν τροπῶν τῶν εἰς τὸν ἐνιαυτόν, ἐν αἷς ἕκαστα γίγνεται, τίνα σε δεῖ προσφέρεσθαι καὶ τίνων ἀπέχεσθαι· ἄρξομαι δὲ ἀπὸ τροπῶν χειμερινῶν.

τροπὴ χειμερινή. αὕτη ἡ ὥρα αὔξει ἐν ἀνθρώποις κατάρρουν ὑγρασίαν ἕως ἰσημερίας ἐαρινῆς. . . . εἰσὶ δὲ εἰς ἰσημερίαν ἡμέραι Ϟ′.

ἰσημερία ἐαρινή. αὕτη ἡ ὥρα αὔξει ἐν ἀνθρώποις φλέγμα καὶ τοὺς γλυκεῖς ἰχῶρας τοῦ αἵμαλος ἕως πλειάδος ἐπιτολῆς. . . . εἰσὶ δὲ ἕως ηλειάδος ἐπιτολῆς ἡμέραι μϛ′.

πλειάδος ἐπιτολή. αὕτη ἡ ὥρα αὔξει ἐν ἀνθρώποις χολὴν πυρρὰν καὶ τοὺς πικροὺς ἰχῶρας ἕως θερινῶν τροπῶν. . . . εἰσὶ δὲ εἰς τροπὰς θερινὰς ἡμέραι μέ.

τροπαὶ θεριναί. αὕτη ἡ ὥρα αὔξει ἐν ἀνθρώποις χολὴν μέλαιναν ἕως ἰσημερίας φθινοπωρινῆς. . . . εἰσὶ δὲ εἰς ἰσημερίαν φθινοπωρινὴν ἡμέραι Ϟγ′.

ἰσημερία φθινοπωρινή. αὕτη ἡ ὥρα αὔξει ἐν ἀνθρώποις φλέγμα καὶ τὰ λεπτὰ ῥευμάτια ἕως πλειάδος δύσεως. . . . εἰσὶ δὲ εἰς πλειάδος δύσιν ἡμέραι μϛ′.

πλειάδοδ δύσις. αὕτη ἡ ὥρα αὔξει ἐν ἀνθρώποις φλέγμα ἕως τροπῶν χειμερινῶν. . . . εἰσὶ δὲ εἰς τροπὰς χειμερινὰς ἡμέραι μέ.

I have also written to you concerning the solstices of the year, in which each (disease) occurs, what things you should take and what things you should avoid. I will begin with the winter-solstice.

Winter-solstice: this season increases moist catarrhs in people until the vernal equinox . . . There are ninety days until the vernal equinox.

---

[274] Greek text from Paulus Aeginata, *Epitomae medicae* 1.100.1–6 (*CMG* ix.1, vol. 1).

Vernal equinox: this season increases phlegm in people and the sweet serums of the blood until the rising of the Pleiades . . . There are forty-six days until the rising of the Pleiades.

Rising of the Pleiades: this season increases yellow bile and the sharp serums in people until the summer-solstice . . . There are forty-five days until the summer-solstice.

Summer-solstice: this season increases black bile in people until the autumnal equinox . . . There are ninety-three days until the autumnal equinox.

Autumnal equinox: this season increases phlegm and the light rheums in people until the setting of the Pleiades . . . There are forty-six days until the setting of the Pleiades.

Setting of the Pleiades: this season increases phlegm in people until the winter-solstice . . . There are forty-five days until the winter-solstice.

F.vi. [Hippocrates] *Peri hebdomadon.* See p. 204.

F.vii. [Hippocrates] *On regimen.* See p. 205.

F.ix. Eudoxus (Leptines) papyrus, cols. xxi–xxiii[275]

(Col. xxi)

ἄστρων δι-
αστήμ[ατα]. ἀπὸ Ὠρίωνος εἰς
Κύνα ἡ[μέραι] . . . ἀπὸ Κυνὸς
εἰς Ἀρ[κτούρο]υ ἐπιτολὴν
ἡμέ[ραι] . . . –

(Col. xxii)

ἀπὸ τρ[οπῶν θερινῶ]ν <εἰς> ἰσημερίαν [μεθοπωρινὴν] ἡμέραι. –
ἀπ[ὸ. . . . . . εἰς Ἀρ]κτούρου
ἐπι[τολὴν ἡμέραι] μγ΄. –
ἀπὸ Πλε[ιάδων εἰς] Ὠρίωνος δύσιν
ἡμέραι κβ΄. –
ἀπὸ Ὠρίωνος εἰς Κυνὸς δύσιν

---

275   Greek text from Blass, 1887.

ἡμέραι δύο. –
ἀπὸ Κυνὸς ἐφ᾽ ἡλίου τροπὰς
ἡμέραι κδ. –
ἀπὸ τροπῶν χειμερινῶν εἰς
ζέφυρον ἡμέραι με. –
ἀπὸ ζεφύρου <εἰς> ἰσημερίαν
ἡμέραι μδ. –
ἀπὸ ἰσημερίας ἐαρινῆς εἰς
Πλειάδα ἡμέραι ν. –
ἀπὸ Πλειάδος ἐπιτολῆς εἰς τροπὰς θερινὰς ἡμέραι με. –
ἀπὸ τροπῶν θερινῶν εἰς ἰσημερίαν μεθοπωρινὴν ἡμέραι Ϟ α. –
Εὐδόξωι Δημοκρίτωι χειμε-
ριναὶ τροπαὶ Ἀθὺρ ὁτὲ μὲν κ´
ὁτὲ δὲ ιθ. –
Εὐδόξωι Δημοκρίτωι ἀπὸ τ[ρο]-
πῶν θερινῶν εἰς ἰσημ[ερί]αν [με]-
θοπωρινὴν ἡ[μέρα]ι . . , –

(*Col.* XXIII)

Εὐκτήμονι Ϟ´ – ,
Καλλίππωι Ϟ β´. –
ἀπὸ ἰσημερίας μεθοπωρινῆς
ἐπὶ χειμερινὰς τροπὰς Εὐ-
δόξωι ἡμέραι Ϟ β´ – ,
Δημοκρίτωι ἡμέραι Ϟ α´, –
Εὐκτήμονι Ϟ´, –
Καλλίππωι πθ. –
ἀπὸ τροπῶν χειμερινῶν
εἰς ἰσημερίαν ἐαρινὴν
Εὐδόξωι καὶ Δη[μ]οκρίτωι
ἡμέραι Ϟ α´ – ,
Εὐκτήμονι Ϟ β´, –
Καλλίππωι Ϟ´. –

Stellar Distan[ces.] From Orion to Sirius: [. . .] d[ays . . .] From Sirius to
the rising of Ar[cturu]s: [. . .] da[ys . . .] From the [summe]r-so[lstice] (to)
the [autumnal] equinox: [. . .] days. Fro[m . . . to the] ris[ing] of [Ar]cturus:
forty-three [days.] From the Ple[iades to] the setting of Orion: twenty-two

days. From Orion to the setting of Sirius: two days. From Sirius to the solstice: twenty-four days. From the winter-solstice to the west wind: forty-five days. From the west wind (to) the equinox: forty-four days. From the vernal equinox to the Pleiades: fifty days. From the rising of the Pleiades to the summer-solstice: forty-five days. From the summer-solstice to the autumnal equinox: ninety-one days.

According to Eudoxus and Democritus the winter-solstice (is on) the 20th or the 19th of Athyr. According to Eudoxus and Democritus from the summer-s[ol]stice to the [au]tumnal equ[in]ox is [. . .] d[ay]s; according to Euctemon: ninety; according to Callippus: ninety-two. From the autumnal equinox to the winter-solstice according to Eudoxus is ninety-two days; according to Democritus: ninety-one days; According to Euctemon: ninety; Callippus: eighty-nine. From the winter-solstice to the vernal equinox according to Eudoxus and De[m]ocritus is ninety-one days; Euctemon: ninety-two; Callippus: ninety.

F.x. Venusia *fasti*. See p. 207.

F.xii. Varro's season list. See p. 208.

F.xx. Florentinus season list (*Geoponica* I.1)[276]

περὶ διαιρέσεως ἐνιαυτοῦ, καὶ τῆς τῶν τροπῶν αὐτοῦ διαιρέσεως. Φλωρεντίνου.

τὸν ἐφεστῶτα τῇ τοῦ ἀγροῦ οἰκονομίᾳ ἀναγκαῖόν ἐστι καὶ τοὺς καιροὺς καὶ τὰς τῶν τροπῶν μεταβολὰς γινώσκειν. οὕτως γὰρ ἂν πρὸς τὴν ἁρμόζουσαν ἑκάστῳ καιρῷ ἐργασίαν τοὺς ἀγροίκους ἐφοδιάζων, πλεῖστα τὸν ἀγρὸν ὠφελήσειεν. οἱ τοίνυν πλεῖστοι, καὶ πρό γε πάντων Βάρων ὁ ῥωμαικὸς ἀρχὴν ἔαρος εἶπεν, ὅταν ὡς ἐπὶ τὸ πλεῖστον καὶ ὁ ζέφυρος ἄρχηται πνεῖν, ὅπερ ἐστὶ τῇ πρὸ ἑπτὰ εἰδῶν Φεβρουαρίων, ἡλίου ἐν Ὑδροχόῳ ὄντος καὶ τριῶν ἢ πέντε μοιρῶν γενομένου, τουτέστι, τρεῖς ἢ πέντε ἡμέρας ἐν τῷ ζωδίῳ ἔχοντος. συμπληροῦσθαι δὲ τὸ ἔαρ εἰς νόννας Μαίας. θέρος δὲ ἄρχεσθαι ἀπὸ τῆς πρὸ ὀκτὼ εἰδῶν Μαίων, τοῦ ἡλίου ἐν Ταύρῳ ὄντος· πληροῦσθαι δὲ εἰς τὴν πρὸ ἑπτὰ εἰδῶν Αὐγούστων. ἔτι δὲ μετόπωρον ἄρχεσθαι, ἀπὸ τῆς πρὸ ἓξ εἰδῶν Αὐγούστων, ἡλίου ὄντος ἐν Λέοντι· πληροῦσθαι δὲ εἰς τὴν πρὸ πέντε εἰδῶν Νοεμβρίων. χειμῶνα δὲ ἄρχεσθαι ἀπὸ τῆς πρὸ τεσσάρων εἰδῶν Νοεμβρίων, ἡλίου ὄντος ἐν Σκορπίῳ· πληροῦσθαι δὲ εἰς τὴν πρὸ ὀκτὼ εἰδῶν Φεβρουαρίων. τῶν δὲ τροπῶν

---

[276] Greek text from *Geoponica* I.1; also published in Wachsmuth, 1897, pp. 320–1.

ἡ μὲν χειμερινή <ἐστι τῇ πρὸ ὀκτὼ καλανδῶν Ἰαννουαρίων· ἡ δὲ θερινή>
ἐστι, τῇ πρὸ ὀκτὼ καλανδῶν Ἰουλίων, κἄν τινες αὐτὴν πρὸ ἓξ νόννων εἶναι
βούλωνται. καὶ τῶν ἰσημερινῶν ἡ μὲν ἐαρινή ἐστι τῇ πρὸ ὀκτὼ καλανδῶν
Ἀπριλλίων· τινὲς δὲ τῇ πρὸ ἐννέα. ἡ δὲ μετοπωρινὴ τῇ πρὸ ὀκτὼ καλανδῶν
Ὀκτωβρίων, ἤτοι πρὸ ἓξ. Πλειάδων δὲ ἐπιτολὴ τότε ἄρχεται γίγνεσθαι·
ἀπὸ τῆς πρὸ τεσσάρων εἰδῶν Ἰουνίων. δύσις δέ· ἀπὸ τῆς πρὸ τεσσάρων
νόννων Νοεμβρίων. ἡ δὲ τῶν Βρούμων ἑορτή ἐστι τῇ πρὸ ὀκτὼ καλανδῶν
Δεκεμβρίων.

On the division of the year and the division of its solstices, by Florentinus.

It is necessary to know what has been established in agriculture, both the
seasons and the changes of the solstices. Thus one can best maximize the
crops if the farmer is prepared for the tasks appropriate to each season.

The majority of people, and above all Varro the Roman, say that the
beginning of spring is for the most part when Zephyrus begins to blow,
which is the seventh day before the Ides of February, when the sun is in three
or five degrees of Aquarius, which is when it has been in the sign for three or
five days. Spring ends on the Nones of May. Summer begins on the eighth
day before the Ides of May, when the sun is in Taurus. It ends on the seventh
day before the Ides of August. Next, autumn begins on the sixth day before
the Ides of August, when the sun is in Leo. It ends on the fifth day before
the Ides of November. Winter begins on the fourth day before the Ides of
November, when the sun is in Scorpio. It ends on the eighth day before the
Ides of February.

Regarding the solstices, the winter one is on the eighth day before the
Kalends of January, the summer one on the eighth day before the Kalends of
July, although some think it to be on the sixth day before the Nones. As for
the equinoxes, the vernal is on the eighth day before the Kalends of April,
though some (think it) the ninth day before. The autumnal is on the eighth
day before the Kalends of October, or the sixth day before. The rising of the
Pleiades begins on the fourth day before the Ides of June, their setting on
the fourth day before the Nones of November. The festival of *Bruma* is on
the eighth day before the Kalends of December.

F.xxi. Geoponica phase list[277]

ἐπιτολὴ καὶ δύσις φανερῶν ἀστέρων. τῶν Κυντιλίων.

---

[277] Greek text from *Geoponica* 1.9.

ἐπειδὴ ἀναγκατόν ἐστιν εἰδέναι τοὺς γεωργοὺς φανερῶν ἀστέρων τὰς ἐπιτολὰς καὶ τὰς δύσεις, οὕτω τὰ περὶ τούτων συνέγραψα, ὥστε καὶ τοὺς παντελῶς ἀγραμμάτους ἀκούοντας ῥαδίως νοεῖν τοὺς καιροὺς τῆς τούτων ἐπιτολῆς τε καὶ δύσεως.

τῇ νεομηνίᾳ τοῦ Ἰαννουαρίου μηνὸς Δελφὶς ἐπιτέλλει. τῇ κϚ΄ τοῦ Φευρουαρίου Ἀρκτοῦρος ἑσπέριος ἐπιτέλλει. τῇ νεομηνίᾳ τοῦ Ἀπριλλίου, Πλειάδες ἀκρόνυχοι κρύπτονται. τῇ ιϚ΄ τοῦ Ἀπριλλίου, Πλειάδες ἑσπέριοι κρύπτονται. τῇ κγ΄ τοῦ Ἀπριλλίου, Πλειάδες ἅμα ἡλίου ἀνατολῇ ἐπιτέλλουσι. τῇ κθ΄ τοῦ Ἀπριλλίου, Ὠρίων ἑσπέριος κρύπτεται. τῇ λ΄ τοῦ Ἀπριλλίου, Ὑάδες ἅμα ἡλίου ἀνατολῇ ἀνατέλλουσι. τῇ ζ΄ τοῦ Μαΐου, Πλειάδες ἑωθινοὶ φαίνονται. τῇ ιθ΄ τοῦ Μαΐου, Ὑάδες ἑωθινοὶ φαίνονται. τῇ ἑβδόμῃ τοῦ Ἰουνίου, Ἀρκτοῦρος ἑῷος δύνει. τῇ κγ΄ τοῦ Ἰουνίου, Ὠρίων ἄρχεται ἐπιτέλλειν. τῇ δεκάτῃ τοῦ Ἰουλίου, Ὠρίων ἑῷος ἐπιτέλλει. τῇ ιγ΄ τοῦ Ἰουλίου, Προκύων ἑῷος ἐπιτέλλει. τῇ κδ΄ Κύων ἑῷος ἐπιτέλλει. τῇ κϚ΄ τοῦ Ἰουλίου, οἱ ἐτησίαι ἄνεμοι ἄρχονται πνέειν. τῇ λ΄ τοῦ Ἰουλίου, ὁ λαμπρὸς ἀστὴρ ὁ ἐν τῷ στήθει τοῦ Λέοντος ἐπιτέλλει. τῇ κε΄ τοῦ Αὐγούστου, Ὀιστὸς δύνει. τῇ ιε΄ τοῦ Σεπτεμβρίου, Ἀρκτοῦρος ἐπιτέλλει. τῇ τετάρτῃ τοῦ Ὀκτωβρίου, Στέφανος ἑῷος ἐπιτέλλει. τῇ κδ΄ τοῦ Ὀκτωβρίου Πλειάδες ἅμα ἡλίου ἀνατολῇ δύνουσι. τῇ ια΄ τοῦ Νοεμβρίου, Πλειάδες ἑῷοι δύνουσι, καὶ Ὠρίων ἄρχεται δύνειν. τῇ κβ΄ τοῦ Νοεμβρίου, Κύων ἑῷος δύνει.

Rising and setting of the visible stars, from Quintilius.

Because it is necessary for the farmer to know the risings and settings of the visible stars, I have written down these things, so that hearing them even completely illiterate people will easily know the seasons of their rising and setting.

On the 1st of the month of **January,** Delphinus rises.
On the 26th of **February,** Arcturus rises in the evening.
On the 1st of **April,** the Pleiades disappear acronychally.
On the 16th of April, the Pleiades disappear in the evening.
On the 23rd of April, the Pleiades rise at sunrise.
On the 29th of April, Orion disappears in the evening.
On the 30th of April, the Hyades rise at sunrise.
On the 7th of **May,** the Pleiades appear in the morning.
On the 19th of May, the Hyades appear in the morning.
On the 7th of **June,** Arcturus sets in the morning.
On the 23rd of June, Orion begins to rise.
On the 10th of **July,** Orion rises in the morning.

On the 13th of July, Procyon rises in the morning.

On the 24th, Sirius rises in the morning.

On the 26th of July, the Etesian winds begin to blow.

On the 30th of July, the bright star in the breast of Leo rises.

On the 25th of **August,** Sagitta sets.

On the 15th of **September,** Arcturus rises.

On the 4th of **October,** Corona rises in the morning.

On the 24th of October, the Pleiades set at sunrise.

On the 11th of **November,** the Pleiades set in the morning, and Orion begins to set.

On the 22nd of November, Sirius sets in the morning.

F.xxii. Byzantine season list[278]

τὸν ἐνιαυτὸν τοῦτον τὸν τρόπον διαιροῦμεν· ἀπὸ Πλειάδων εἰς ἡλίου τροπὰς ἡμέραι μβ'· ἀπὸ ἡλίου τροπῶν εἰς Ὠρίωνα ἡμέραι κ'· ἀπὸ δὲ Ὠρίωνος εἰς Κύνα ἡμέραι ια'· ἀπὸ δὲ Κυνός εἰς Ἀρκτοῦρον ἡμέραι να'· ἀπὸ δὲ Ἀρκτούρου εἰς ἰσημερίαν ἡμέραι ι'· ἀπὸ δὲ ἰσημερίας εἰς Πλειάδος δύσιν ἡμέραι με'· ἀπὸ δὲ Πλειάδος δύσεως εἰς Ὠρίωνα ἡμέραι κ'· ἀπὸ δὲ Ὠρίωνος εἰς Κύνα ἡμέραι ζ'· ἀπὸ δὲ Κυνὸς εἰς ἡλίου τροπὰς ἡμέραι κβ'· ἀπὸ δὲ ἡλίου τροπῆς εἰς ζέφυρον ἡμέραι με'· ἀπὸ δὲ ζεφύρου εἰς Ἀρκτοῦρον ἡμέραι ιε'· ἀπὸ δὲ Ἀρκτούρου εἰς ἰσημερίαν ἡμέραι λβ'· ἀπὸ δὲ ἰσημερίας εἰς Πλειάδος ἐπιτολὴν ἡμέραι με'· ὁμοῦ ἡμέραι τξε'.

καὶ αὖθις· διαιροῦμεν ἀπὸ Πλειάδος εἰς Λύρας δύσιν, θέρους ἡμέραι ϙδ'· ἀπὸ δὲ Λύρας δύσεως ἕως Ξίφους ἀρχομένου <δύεσθαι> φθινοπώρου ἡμέραι ϙ'· ἀπὸ δὲ Πλειάδος δύσεως ἕως ζεφύρου χειμῶνος ἡμέραι ϙα'· ἀπὸ δὲ ζεφύρου εἰς Αἰγὸς ἐπιτολὴν <ἔαρος> ἡμέραι ϙ'· ὁμοῦ γίνονται ἡμέραι τξε'.

τὰ ἄστρα ποσάκις δύνει καὶ ποσάκις ἀνατέλλει· Πλειάδες ἀνατέλλουσι δὶς πρωὶ καὶ ἑσπέρας, δύνουσι δὲ <δὶς>· Ἀρκτοῦρος ὁμοίως, Στέφανος ὁμοίως, Δελφὶς ὁμοίως. Λύρα Ὑάδες Ἀετὸς Κύων Ὠρίων ἀνατέλλει ἑῷος καὶ ἑσπέριος, Κύων δὲ ὁμοίως. Σκορπίος, Ἵππος ἐπιτέλλει ἅπαξ ἑσπέρας· Προτρυγητὴρ ἐπιτέλλει ἅπαξ ἡμέρας.

We divide the year this way: from the Pleiades to the solstice, forty-two days. From the solstice to Orion, twenty days. From Orion to Sirius, eleven

---

[278] Greek text from *CCAG* vii.162–3.

days. From Sirius to Arcturus, fifty-one days. From Arcturus to the equinox, ten days. From the equinox to the setting of the Pleiades, forty-five days. From the setting of the Pleiades to Orion, twenty days. From Orion to Sirius, seven days. From Sirius to the solstice, twenty-two days. From the solstice to the west wind, forty-five days. From the west wind to Arcturus, fifteen days. From Arcturus to the equinox, thirty-two days. From the equinox to the rising of the Pleiades, forty-five days. Altogether: 365 days.

Or alternately: we divide from the Pleiades to the setting of Lyra, summer is ninety-four days. From the setting of Lyra until the beginning of the sword, autumn is ninety days. From the setting of the Pleiades to the west wind, winter is ninety-one days. From the west wind to the rising of Capella is ninety days. Altogether: 365 days.

The stars, how often they rise and how often they set: the Pleiades rise twice, morning and evening, and they set twice. Arcturus similarly, Corona similarly, Delphinus similarly. Lyra, the Hyades, Aquila, Sirius, and Orion rise in the morning and the evening, Sirius similarly. Scorpio and Pegasus rise once, in the evening. Vindemiatrix rises once, in the day.

Appendix I: Authorities cited in parapegmata

Barron 'the Roman': is Marcus Terentius **Varro**, author of the *De re rustica*[1] and *De lingua latina*. He is mentioned in Lydus, Florentinus, and at the end of Clodius Tuscus, and is of course the author of the season list discussed in the catalogue, above.

Caesar: is mentioned in the parapegmata of Ptolemy, Lydus, Clodius Tuscus, and al-Bīrūnī. Pliny seems to associate the calendar reform of Julius Caesar and Sosigenes with the parapegmatic tradition,[2] and so it is possible that the Caesar cited in the parapegmata may well be Julius Caesar. Another possibility is Germanicus Caesar, who translated Aratus into Latin.[3]

Callaneus of the Indians: is mentioned only in Miletus I, and is otherwise unknown, although he may be the same person as the gymnosophist Callanus mentioned in the Alexander-history.[4] Pingree argues that Callaneus was using a Greek rather than an Indian method of astrometeorological prediction.[5]

Callippus: is known from Aristotle's *Metaphysics* Λ as having improved Eudoxus' concentric spheres model. He also inaugurated the 76-year cycle named for him. He was a contemporary of Aristotle. He is mentioned in Geminus, Pliny, Ptolemy, Lydus, al-Bīrūnī, and the Eudoxus papyrus.[6]

---

[1] See item F.xii in the Catalogue, above.

[2] Pliny, *HN* xviii.211.

[3] The question of whether the *Caesar* here refers to Julius or Germanicus is an open one. The earliest mention in a parapegma is Ptolemy's *Phaseis* (second century AD). This and later parapegmata give no information beyond the name 'Caesar'. Speculation ultimately rests on a judgment as to the weighting of one of two possibilities: either (a) Julius Caesar, in some kind of connection with his calendar reform, may have left some material that was later incorporated into parapegmata under his name (Pliny seems to hint as much at *HN* xviii.211); or (b) Germanicus Caesar's translation of Aratus (attested but now lost) may have included (or been related to) new material later incorporated into the parapegmatic tradition. On Germanicus' Aratus, see Gain, 1976; Gee, 2000.

[4] This suggestion was made by Diels and Rehm, 1904, p. 108, n. 1. For the story of Callanus, see Cic. *De div.* i.47.

[5] See Pingree, 1976, pp. 143–4.

[6] For the bibliography on Callippus, see Jones, 2000.

Chaldaeans: Columella mentions that they observe the winter solstice on
IX *K. Jan.* Pliny also mentions them. I argue in chapter 5 that the
attributions to them are probably not from original Babylonian sources.

Conon: fl. second half of the third century BC. He is mentioned by both
Archimedes and Apollonius as a mathematician, and by Seneca, Cal-
limachus, and Catullus as an astronomer. Ptolemy says he observed in
Italy and Sicily.[7] He is also mentioned in Pliny and al-Bīrūnī.

Crito: is mentioned once by Pliny. Rehm suggests he may be the historian
Crito of Naxus mentioned in the Suda.[8]

Democritus of Abdera: is the famous fifth-century presocratic philosopher.
He is mentioned in the Geminus, Pliny, Ptolemy, Clodius Tuscus, al-
Bīrūnī, and Lydus parapegmata, as well as in the Eudoxus papyrus.
Diogenes Laertius (IX.48) tells us he wrote 'a parapegma' called
The Great   Year, or Astronomy (Μέγας ἐνιαυτὸς ἢ ἀστρονομίη,
παράπηγμα), but by calling the book a parapegma Diogenes may
have meant nothing more than that it dealt with astrometeorology.[9]

Dositheus: pupil of Conon, late third century BC. Ptolemy says he observed
at Cos. He is also mentioned by Geminus, Pliny, al-Bīrūnī, and Lydus.

Egyptians: Neugebauer and Rehm thought this referred to Greco-Egyptian
observers, but I argue in chapter 6 that it may refer to native Egyptians.
They appear in Miletus II, Pliny, Antiochus, al-Bīrūnī, and Ptolemy.

Euctemon: is frequently associated with Meton (see below) in the ancient
sources. Together, they are known for a solstice observation at Athens
in 432, and Euctemon is also associated with the nineteen-year Metonic
cycle. He is mentioned in Miletus II, Geminus, Pliny, Ptolemy, Lydus,
al-Bīrūnī, and the Eudoxus papyrus.

Eudoxus of Cnidus: the astronomer, geographer, and mathematician. He
lived during the first half of the fourth century BC, and was the originator of the homocentric spheres model of the Cosmos used by Aristotle.
He is associated with an eight-year cycle. He appears in Miletus II,
Geminus, Columella, Pliny, Ptolemy, Clodius Tuscus, Lydus, al-Bīrūnī,
and the Eudoxus papyrus.

Hipparchus of Rhodes: very important astronomer, fl. late second century BC. Many of his observations are cited in Ptolemy's *Almagest*.

---

[7] This and the similar entries from this list ('Ptolemy says he observed at *x*') are all taken from the
Phaseis, p. 67.

[8] See Rehm, 1941, p. 37, n. 1.

[9] Sider, 2002, argues that the word parapegma is probably merely descriptive and a later addition
to the real title of Democritus' book, which was Μέγας ἐνιαυτὸς ἢ ἀστρονομίη. Nonetheless
Sider also thinks the Democritus text really was a parapegma.

Hipparchus discovered the precession of the equinoxes, and was also responsible for incorporating many Babylonian observation reports and numerical parameters into Greek astronomy.[10] He appears in the Columella, Ptolemy, Clodius Tuscus, al-Bīrūnī, and Lydus parapegmata, as well as being mentioned in connection with ἐπισημασίαι in *P. Oxy.* 4475.

Meton: fifth-century astronomer, frequently associated with Euctemon. He is mentioned in Geminus, Columella, Ptolemy, and al-Bīrūnī, and has the honour of being a character in Aristophanes' *Birds*.

Metrodorus of Chios: he is referred to as a pupil of Democritus.[11] Ptolemy tells us he observed in Italy and Sicily. He also turns up in Clodius Tuscus, al-Bīrūnī, and Lydus.

Parmeniscus: is mentioned once by Pliny.[12]

Philippus: is probably Philippus of Opus,[13] student of Plato. Ptolemy says he observed in the Hellespont. He appears in Miletus II, Pliny, Ptolemy, al-Bīrūnī, and Lydus.

Ptolemy: author of the *Almagest*, *Phaseis*, and *Tetrabiblos*. Mentioned as the authority for several weather predictions in the al-Bīrūnī parapegma. These entries are, interestingly, not taken from the *Phaseis*.

Tubero: is mentioned once by Pliny. He is referred to only as Tubero in the body of (and list of authorities for) book xviii, but in the list of authorities for books ii and xxxvi we find a Q. Tubero listed. Quintus Tubero was a Stoic and student of Panaetius, and an associate of Cicero's (he is one of the interlocutors in Cicero's *De republica*), and it seems plausible to associate the Tubero of book xviii with the Quintus Tubero of the other books.[14]

---

[10] See Toomer, 1988.

[11] But *HAMA*, p. 929, dates him to 'possibly the third century'.

[12] See Wendel, 'Parmensicus (3)' *RE*, Maass, 1892.   [13] *HAMA*, p. 740, 574.

[14] See also Bakhouche, 1996.

Appendix II: Tables of correspondence of parapegmata

Key

Lehoux: (A) Astrometeorological; (B) Astrological; (C) Astronomical; (D)
   Other; (E) Reports of Parapegmata (F) Related Texts and Instruments;
   (G) Dubia.
Rehm: (A) Inscriptional; (B) Literary; (III) Inauthentic.

Sources: Lehoux, Part II, Catalogue; Rehm (1949); Degrassi (1963).

Lehoux	Rehm	Degrassi
A.i	B 3	
A.ii	A 2	
A.iii	B 1	
A.iv	A 4	xii.2.57
A.v	B 4	
A.vi	B 5	
A.vii	B 20, B 6	
A.viii	B 2	
A.ix	B 12	xii.2.43
A.x	B 11	
A.xi	B 7b	
A.xii	B 13	
A.xiii	B 7a	
A.xiv	–	
A.xv	B 10	
A.xvi	B 9	
A.xvii	–	
A.xviii	–	
A.xix	–	
A.xx	B 14	
B.i	III 6	xii.2.56
B.ii	III 8	
B.iii	III 5	xii.2.49
B.iv	–	xii.2.59
B.v	III 3	xii.2.55
B.vi	–	xii.2.58

(*cont.*)

(*cont.*)

Lehoux	Rehm	Degrassi
B.vii	–	xɪɪ.2.53
B.viii	III 4	xɪɪ.2.52
B.ix	III 14	
B.x	III 15	
B.xi	III 16	
B.xii	–	
B.xiii	III 7	
B.xiv	–	
C.i	–	
C.ii	A 1	
C.iii	B 17	
C.iv	–	
D.i	A 3	
D.ii	III 10	xɪɪ.2.40
D.iii	–	xɪɪ.2.42
D.iv	–	
D.v	III 1	
D.vi	–	xɪɪ.2.54
D.vii	–	xɪɪ.2.51
D.viii	–	xɪɪ.2.50
E.i	–	
E.ii	III 2	
E.iii	B 15	
E.iv	–	
E.v	–	
E.vi	–	
E.vii	–	
F.i	–	
F.ii	–	
F.iii	–	
F.iv	–	
F.v	B 24	
F.vi	B 23	
F.vii	B 23	
F.viii	–	
F.ix	B 16	
F.x	B 8	xɪɪ.2.6
F.xi	–	xɪɪ.2.47, 48
F.xii	B 18	
F.xiii	B 19	
F.xiv	–	
F.xv	–	
F.xvi	B 25	

(*cont.*)

Lehoux	Rehm	Degrassi
F.xvii	–	
F.xviii	III 12, III 13	
F.xix	III 11	
F.xx	B 21	
F.xxi	–	
F.xxii	B 22	
G.i	–	
G.ii	B 17	
G.iii	–	
G.iv	–	
G.v	–	
G.vi	–	
G.vii	–	
G.viii	–	
G.ix	III 9	
G.x	–	
G.xi	–	

Rehm	Lehoux
A 1	C.ii
A 2	A.ii
A 3	D.i
A 4	A.iv
B 1	A.iii
B 2	A.viii
B 3	A.i
B 4	A.v
B 5	A.vi
B 6	A.vii
B 7a	A.xiii
B 7b	A.xi
B 8	F.x
B 9	A.xvi
B 10	A.xv
B 11	A.x
B 12	A.ix
B 13	A.xii
B 14	A.xx
B 15	E.iii

(*cont.*)

(*cont.*)

Rehm	Lehoux
B 16	F.ix
B 17	C.iii, G.ii
B 18	F.xii
B 19	F.xiii
B 20	A.vii
B 21	F.xx
B 22	F.xxii
B 23	F.vi, F.vii
B 24	F.v
B 25	F.xvi
III 1	D.v
III 2	E.ii
III 3	B.v
III 4	B.viii
III 5	B.iii
III 6	B.i
III 7	B.xiii
III 8	B.ii
III 9	G.ix
III 10	D.ii
III 11	F.xix
III 12	F.xviii
III 13	F.xviii
III 14	B.ix
III 15	B.x
III 16	B.xi

Bibliography

Aaboe, A. (1955) 'On the Babylonian Origin of Some Hipparchan Parameters,' *Centaurus*, 4.2, pp. 122–5

al-Biruni (1879) *The Chronology of Ancient Nations*, trans. C. E. Sachau, London

Allen, J. (2001) *Inference from Signs*, Oxford

Allen, T. W., ed. (1908) *Homer: Odyssea*, Oxford

Al-Rawi, F. N. H. and A. R. George (1991) 'Enuma Anu Enlil XIV and Other Early Astronomical Tables', *Archiv für Orientforschung*, 38/9, pp. 52–73

Andreau, J. (2000) 'Les marchés hebdomadaires de Latium et de Campanie au 1er siècle ap. J.-C.', in E. Lo Cascio, ed., *Mercati permanenti e mercati periodici nel mondo romano*, Bari, pp. 69–91

Ash, H. B., E. S. Forster, and E. H. Heffner, eds. (1948–55) *Columella: Rei rusticae*, Cambridge, MA

Aujac, G., ed. (1975) *Géminos: Introduction aux phénomènes*, Paris

Ax, W., ed. (1932) *Cicero: De natura deorum*, Leipzig

    ed. (1938) *Cicero: De divinatione*, Leipzig

Azkāyi, P., ed. (2001) *al-Biruni: al-Athār al-bāqiyah 'an al-qurūn al-khāliyah*, Tehran

Bakhouche, B. (1996) *Les textes latins d'astronomie*, Louvain

Bakir, A. M. (1966) *The Cairo Calendar No. 86637*, Cairo

Barnouw, J. (2002) *Propositional Perception: Phantasia, Predication, and Sign in Plato, Aristotle, and the Stoics*, Lanham, MD

Barton, T. (1994) *Ancient Astrology*, New York and London

Battstone, W. (1997) 'Virgilian Didaxis: Value and Meaning in the Georgics', in C. Martindale, ed., *Cambridge Companion to Virgil*, Cambridge

Bauer, J. (1998) 'Georgica Sumerica', *Orientalia*, 67, pp. 119–25

Beagon, M. (1992) *Roman Nature: The Thought of Pliny the Elder*, Oxford

Beattie, J. (1964) *Other Cultures*, London

Beaulieu, P. A. and J. P. Britton (1994) 'Rituals for an Eclipse Possibility in the 8th Year of Cyrus', *Journal of Cuneiform Studies*, 46, pp. 73–86

Becatti, G. (1954) 'I Mitrei', *Scavi di Ostia*, Rome

Beckh, H., ed. (1895) *Geoponica*, Leipzig

Bekker, I., ed. (1831) Aristotle, *Opera*, Berlin

Berggren, J. L. (1992) 'Review of W. Knorr's "Textual Studies in Ancient and Medieval Geometry",' *Ancient Philosophy*, 12, pp. 522–8

Bianchi, L. (1914) 'Der Kalender des sogenannten Clodius Tuscus', in F. Boll, ed., 'Griechische Kalender IV', *Sitzungsberichte der Heidelberger Akademie der Wissenschaften philosophisch-historische Klasse*, Heidelberg

Bickerman, E. J. (1980) *Chronology of the Ancient World*, 2nd edn., New York

Binsfeld, W. (1973) 'Römische Steckkalender in Trier', *Kurtrierisches Jahrbuch*, 13, pp. 186–9

Blass, F. (1887) 'Eudoxi ars astronomica qualis in charta aegyptiaca superest', reprinted in *Zeitschrift für Papyrologie und Epigraphik*, 115 (1997), pp. 79–101

Bloor, D. (1976) *Knowledge and Social Imagery*, London

Bobzien, S. (1998) *Determinism and Freedom in Stoic Philosophy*, Oxford

Böckh, A. (1863) *Über die vierjährigen Sonnenkreise der Alten, vorzüglich den Eudoxischen*, Berlin

Boll, F. (1905) 'Dodecaeteris', *Paulys Realencyclopädie der classischen Altertumswissenschaft*, vol. v, Stuttgart, pp. 1254–5

  (1909) 'Fixsterne', *Paulys Realencyclopädie der classischen Altertumswissenschaft*, vol. vi, Stuttgart, pp. 2407–2431

  (1910) 'Das Kalendarium des Antiochos', in F. Boll, ed., 'Griechische Kalender I', *Sitzungsberichte der Heidelberger Akademie der Wissenschaften philosophisch-historische Klasse*, Heidelberg

  (1911) 'Der Kalender der Quintilier und die Überlieferung der Geoponica', in F. Boll, ed., 'Griechische Kalender II', *Sitzungsberichte der Heidelberger Akademie der Wissenschaften, philosophisch-historische Klasse*, Heidelberg

Bottero, J. (1974) 'Symptômes, signes, écritures en Mésopotamie ancienne', in J. P. Vernant, ed., *Divination et rationalité*, Paris

Bouché-Leclercq, A. (1899) *L'astrologie grecque*, Paris

Bowen, A. and B. R. Goldstein (1988) 'Meton of Athens and Astronomy in the Fifth Century B. C.', in E. Leichty et al., eds., *A Scientific Humanist: Studies in Memory of Abraham Sachs*, Philadelphia

  (1989) 'On Early Hellenistic Astronomy: Timocharis and the First Callippic Calendar', *Centaurus*, 32, pp. 272–93

Brack-Bernsen, L. (1997) *Zur Entstehung der babylonischen Mondtheorie: Beobachtung und theoretische Berechnung von Mondphasen*, Stuttgart

  (2003) 'The Path of the Moon, the Rising Points of the Sun, and the Oblique Great Circle on the Celestial Sphere', in P. Barker *et al.*, eds., *Astronomy and Astrology from the Babylonians to Kepler: Essays Presented to Bernard R. Goldstein on the Occasion of his 65th Birthday: Part I*, Copenhagen

Bradley, R. S. (1999) *Paleoclimatology*, 2nd edn., San Diego

Brind'Amour, P. (1983) *Le calendrier romain*, Ottawa

Britton, J. (2002) 'Treatments of Annual Phenomena in Cuneiform Sources', in J. Steele and A. Imhausen, eds., *Under One Sky: Astronomy and Mathematics in the Ancient Near East*, Alter Orient und Altes Testament 297, Münster

Britton, J. and A. Jones (2000) 'A New Babylonian Planetary Model in a Greek Source', *Archive for History of Exact Sciences*, 54, pp. 349–73

Broughton, A. L. (1936) 'The Menologia Rustica', *Classical Philology*, 31, pp. 353–6

Brown, D. (2000) *Mesopotamian Planetary Astronomy-Astrology*, Groningen

Brown, M. (1997) 'Thinking About Magic', in S. D. Glazier, ed., *Anthropology of Religion*, London, pp. 121–36

Brown, N. (2001) *History and Climate Change*, New York

Brückner, A. (1931) 'Mitteilungen aus dem Kerameikos V', *Mitteilungen des Deutschen Archäologishe Instituts, Athenische Abteilung*, 56, pp. 1–97

Brugsch, H. K. (1883–91) *Thesaurus inscriptionum aegyptiacarum*, Leipzig

Bruns, I., ed. (1892) *Alexander of Aphrodisias*: *Praeter commentaria scripta minora*, Berlin, vol. ii.ii

Buchner, E. (1982) *Die Sonnenuhr des Augustus*, Mainz

Budyko, M. I. (1982) *The Earth's Climate, Past and Future*, New York

Buettner-Wobst, T., ed. (1889–1904), *Polybii historiae*, Stuttgart

Burnett, C. (1993) 'An Unknown Latin Version of an Ancient *Parapégma*: The Weather-Forecasting Stars in the *Iudicia* of Pseudo-Ptolemy', in R. G. W. Anderson, J. A. Burnett, and W. F. Ryan, eds., *Making Instruments Count*, Aldershot

Bury, R. B., ed. (1993) *Sextus Empiricus, Adversus mathematicos*, Cambridge, MA

Cagirgan, C. (1985) 'Three More Duplicates to Astrolabe B', *Türk Tarih Kurumu, Belleten*, 48, pp. 404–5

Caplice, R. I. (1974) *The Akkadian Namburbi Texts: An Introduction*, Los Angeles

Casaburi, M. C. (2003) *Tre-stelle-per-ciascun(-mese): L'Astrolabio B. Edizione filologica*, Naples

Christmann, E. (2003) 'Bermerkungen zu Autoren und ihrem Publikum in der römischen Landwirtschaftslehre', in M. Horster and C. Reitz, eds., *Antike Fachschriftsteller: Literarischer Diskurs und sozialer Kontext*, Stuttgart

Civil, M. (1994) *The Farmer's Instructions: A Sumerian Agricultural Manual*, Barcelona

Clagett, M. (1995) *Ancient Egyptian Science*, Philadelphia

Clermont-Ganneau, C. (1906) *Recueil d'archéologie orientale*, Paris

Coe, M. (1992) *Breaking the Maya Code*, London

Cohen, M. E. (1993) *The Cultic Calendars of the Ancient Near East*, Bethesda, MD

Colbert, E. (1869) *Astronomy Without a Telescope, being a Guide-Book to the Visible Heavens, with All Necessary Maps and Illustrations*, Chicago

Coles, R. A. (1974) *Location-list of the Oxyrhynchus Papyri and of other Greek Papyri Published by the Egypt Exploration Society*, London

Constans, L. A. and J. Bayet, eds. (1969) *Cicéron, Correspondance*, Paris

Corpus inscriptionum latinarum (1863), Berlin

Costa, A. (1762) *Raccolta di vari pezzi di antichità dissotterrati*, Rome

Cowley N. (1990) 'Africa in Plutarch', *Akroterion*, 35, pp. 80–3

Cramer, F. H. (1954) *Astrology in Roman Law and Politics*, Philadelphia

Csapo, E. (2005) *Theories of Mythology*, Oxford

Cumont, F. *et al.*, eds. (1898–1953) *Catalogus codicum astrologorum Graecorum*, Brussels

Curtis, H. D. and F. E. Robbins (1935) 'An Ephemeris of 467 A. D.', *Publications of the Observatory of the University of Michigan*, 6. 9, pp. 77–100

Dagron, G. (1990) 'Das Firmament soll christlich werden: Zu zwei Seefahrtskalendern des 10 Jahrhunderts', in G. Prinzing and D. Simon, eds., *Fest und Alltag in Byzanz*, Munich

Daressy, G. (1916) 'La statue d'un astronome', *Annales du Service des antiquités de l'Égypte*, 16, pp. 1–5

Davis, W. M. (1979) 'Plato on Egyptian Art', *Journal of Egyptian Archaeology*, 65, pp. 121–7

Degrassi, A. (1963) *Inscriptiones italiae*, Rome

De Lacy, P. H. and E. A. De Lacy, eds. (1978) *Philodemus, De signis*, Naples

della Corte, M. (1927) 'Pompei', *Notizie degli scavi di antichità*, 6.3, p. 98

Deman, A. (1974) 'Notes de chronologie romaine', *Historia*, 23, pp. 271–8

Depuydt, L. (1997) *Civil Calendar and Lunar Calendar in Ancient Egypt*, Leuven
    (1998) 'Ancient Egyptian Star Clocks and Their Theory', *Bibliotheca Orientalis*, pp. 5–44

de Romanis, A. (1822) *Le antiche camera esquiline dette comunemente delle Terme di Tito*, Rome

Dessau, H. (1904) 'Zu den Milesischen Kalenderfragmenten', *Sitzungsberichte der königlich preussischen Akadamie der Wissenschaften, philosophisch-historische Klasse*, 23, pp. 266–8

Dickie, M. W. (2001) *Magic and Magicians in the Greco-Roman World*, New York

Dicks, D. R. (1970) *Early Greek Astronomy to Aristotle*, London

Dictionary of Scientific Biography (1970–) C. C. Gillispie, ed., New York

Diels, H., ed. (1879) *Aetius: Doxographi graeci*, Berlin

Diels, H. and W. Krantz (1922) *Die Fragmente der Vorsokratiker*, Berlin

Diels, H. and A. Rehm (1904) 'Parapagmenfragmente aus Milet', *Sitzungsberichte der königlich preussischen Akadamie der Wissenschaften, philosophisch-historische Klasse*, 23, pp. 92–111

Dillon, J. (1996) *The Middle Platonists*, Ithaca, NY

Dilts, M., ed. (1974) *Claudii Aeliani varia historia*, Leipzig

Dindorf, L., ed. (1890–1928) *Dionis Cassii Cocceiani historia romana*, Leipzig

D'Isanto, G. (1984) 'Iscrizioni latine inedite dell'Antiquarium di S. Maria Capua Vetere', *Rendiconti della Accademia di archeologia, lettere e belle arti*, ns. 59, pp. 123–50

Dölger, F. J. (1950) 'Die Planetenwoche der griechisch-römischen Antike und der christliche Sonntag', *Antike und Christentum*, 6, pp. 202–38

Donahue, J. F. (1999) 'Euergetic Self-Representation and the Inscriptions at Satyricon 71.10', *Classical Philology*, 94, pp. 69–74

Donbaz, V. and J. Koch (1995) 'Ein Astrolab der dritten Generation: NV.10', *Journal of Cuneiform Studies*, 47, pp. 63–84

Douglas, M. (1966) *Purity and Danger*, London

Dozy, R. (1961) *Le calendrier de Cordue*, Leiden

Drower, M. S. et al. (1996) 'Europus', in S. Hornblower and A. Spawforth, eds., *The Oxford Classical Dictionary*, Oxford

Dunn, F. M. (1999) 'Tampering with the Calendar', *Zeitschrift für Papyrologie und Epigraphik*, 123, pp. 213–31

Duval, P. M. (1953) 'Les dieux de la semaine', *Gallia*, 11, pp. 282–93

Duval, P. M. and J. P. Boucher (1976) *Bronzes antiques du musée de la civilisation gallo-romaine à Lyon*, Lyon

Duval, P. M. and G. Pinault (1986) 'Les calendriers (Coligny, Villards d'Héria)', *Recueil des inscriptions gauloises*, Paris, vol. III

Ebeling, E. (1915) *Keilschrifttexte aus Assur religiösen Inhalts*, Leipzig

Eco, U. (1984) *Semiotica e filosofia del linguaggio*, Turin

Eijk, P. van der (2000) *Diocles of Carystus: A Collection of Fragments with Translation and Commentary*, Leiden

Elsner, J. (1997) 'Hagiographic Geography: Travel and Allegory in the Life of Apollonius of Tyana', *Journal of Hellenic Studies*, 117, pp. 22–37

Erbse, H. (1992) *Studien zum Verständnis Herodots*, Berlin

Erikkson, S. (1956) 'Wochentagsgötter, Mond und Tierkreis', *Studia Graeca et Latina Gothoburgensia*, vol. III

Ernesti, J. A., ed. (1810) *M. Tulli Ciceronis opera omnia*, Oxford

Erren, M. (1985) *P. Vergilius Maro, Georgica*, Heidelberg

Espérandieu, E. (1907–) *Recueil général des bas-reliefs, statues et bustes de la Gaule romaine*, Paris

   (1933) 'Séance du 13 Octobre', *Comptes rendus des séances de l'année de l'Académie des inscriptions et belles-lettres*, pp. 383–5

Evans, J. (1998) *The History and Practice of Ancient Astronomy*, Oxford

   (1999) 'The Material Culture of Greek Astronomy', *Journal for the History of Astronomy*, 30, pp. 237–307

   (2004) 'The Astrologer's Apparatus: A Picture of Professional Practice in Greco-Roman Egypt', *Journal for the History of Astronomy*, 35, pp. 1–44

Evans, J. and J. L. Berggren (2006) *Gemino's Introduction to the 'Phenomena'*, Princeton

Evans-Pritchard, E. E. (1937) *Witchcraft, Oracles and Magic among the Azande*, Oxford

Falconer, W. A., ed. (1923) *Cicero: De divinatione*, Cambridge, MA

Fantham, E., ed. (1998) *Ovid Fasti, Book IV*, Cambridge

Faraone, C. (1999) *Ancient Greek Love Magic*, Cambridge, MA

Farrell, J. (1991) *Vergil's Georgics and the Traditions of Ancient Epic*, New York

Fecht, R., ed. (1927) *Theodosius, De habitationibus liber, de diebus libri duo*, Göttingen

Fehling, D. (1989) *Herodotus and his Sources: Citation, Invention and the Narrative Art*, Leeds

Ferrua, A. (1985) 'Il giorno del mese', *Rivista di archeologia cristiana*, 61, pp. 61–75

Field, J. V. and M. T. Wright (1985) 'Gears from the Byzantines: A Portable Sundial with Calendrical Gearing', *Annals of Science*, 42, pp. 87–138

Fiorelli, G. (1867–8) *Catalogo del Museo nazionale di Napoli. Raccolta epigrafica*, Naples

Flammant, J. (1984) 'L'année lunaire aux origines du calendrier pre-julien', *Mélanges d'archéologie et d'histoire de l'École française de Rome*, 96, pp. 175–93

Flasch, D. (1996) *Marcus Terentius Varro: Gespräche über die Landwirtschaft*, Darmstadt

Fleming, J. R. (1998) *Historical Perspectives on Climate Change*, Oxford

Flobert, P., ed. (1985) *Varro, De lingua latina*, Paris

Fotheringham, J. K. (1924) 'The Metonic and Callippic Cycles', *Monthly Notices of the Royal Astronomical Society*, 84, pp. 383–92

Fowler, P. J. (2002) *Farming in the First Millenium A.D.: British Agriculture between Julius Caesar and William the Conqueror*, Cambridge

Frayn, J. M. (1993) *Markets and Fairs in Roman Italy*, Oxford

Frazer, J. G. (1900) *The Golden Bough*, London

Frede, M. (1980) 'The Original Notion of Cause', in M. Schofield, M. Burnyeat, and J. Barnes, eds., *Doubt and Dogmatism: Studies in Hellenistic Epistemology*, Oxford, pp. 217–49

French, R. (1994) *Ancient Natural History*, London

French, R. and F. Greenaway, eds. (1986) *Science in the Early Roman Empire: Pliny the Elder, his Sources and Influence*, London

Froehner, W. (1886) *Collection H. Hoffmann: catalogue des objets d'art antiques*, Paris

Fulvio, L. (1891) *Notizie degli scavi di antichità*, 15, p. 238

Furley, D. J. and J. T. Vallance (1996) 'Bolus', in S. Hornblower and A. Spawforth, eds., *The Oxford Classical Dictionary*, 3rd edn., Oxford

Gain, D. B. (1976) *The Aratus Ascribed to Germanicus Caesar*, London

Gardiner, A. (1957) *Egyptian Grammar*, Oxford

Gaspani, A. and S. Cernuti (1997) *L'astronomia dei celti*, Aosta

Gee, E. (2000) *Ovid, Aratus and Augustus*, Cambridge

  (2002) '*Vaga signa*: Orion and Sirius in Ovid's *Fasti*', in G. Herbert Brown, ed., *Ovid's Fasti*, Oxford

Germer-Durand, J. (1901) 'Inscriptions d'Abougoch, Esdoud, Naplouse, et Beisan', *Échos d'Orient*, 5, pp. 73–6

Gerstinger, H. and O. Neugebauer (1962) 'Eine Ephemeride für das Jahre 348 oder 424 n. Chr. in den PER, Pap. Graec. Vindob. 29370', *Sitzungsberichte Österreichische Akademie der Wissenschaften, philosophisch-historische Klasse*, 240.2, pp. 1–25

Gibbs, S. L. (1976) *Greek and Roman Sundials*, New Haven

Gibson, R. (1997) 'Didactic Poetry as "Popular" Form: A Study of Imperatival Expressions in Latin Didactic Verse and Prose', in C. Atherton, ed., *Form and Content in Didactic Poetry*, Nottingham

Gill, C. (1979) 'Plato's Atlantis Story and the Birth of Fiction', *Philosophy and Literature*, 3, pp. 64–78

Ginzel, F. K. (1914) *Handbuch der mathematischen und technischen Chronologie*, Leipzig

Gjerstad, E. (1961) 'Notes on the Early Roman Calendar', *Acta Archaeologica*, 32, pp. 193–214

Goessler, P. (1928) 'Ein gallorömischer Steckkalender aus Rottweil', *Germania*, 12, pp. 1–9

Goldstein, B. R. and A. C. Bowen (1991) 'The Introduction of Dated Observations and Precise Measurement in Greek Astronomy', *Archive for History of Exact Sciences*, 43, pp. 93–132

Gomme, A. W. (1956) *A Historical Commentary on Thucydides*, Oxford

Graf, F. (1997) *Magic in the Ancient World*, Cambridge, MA

(1999) 'Magic and Divination', in D. R. Jordan, H. Montgomery, and E. Thomassen, eds., *The World of Ancient Magic*, Bergen

(2002) 'Theories of Magic in the Antiquity', in Mirecki and Meyer, 2002

Grafton, A. T. and N. Swerdlow (1988) 'Calendar Dates and Ominous Days in Ancient Historiography', *Journal of the Warburg and Courtauld Institutes*, 51, pp. 14–42

Graßhof, G. (1993) 'The Babylonian Tradition of Celestial Phenomena and Ptolemy's Fixed Star Calendar', in H. D. Galter, ed., *Die Rolle der Astronomie in den Kulturen Mesopotamiens*, Grazer Morgenländische Studien 3, Graz, pp. 95–137

Green, C. M. C. (1997) 'Free as a Bird: Varro *De re rustica* 3', *American Journal of Philology*, 118, pp. 427–48

Green, S. J. (2004) *Ovid, Fasti I: A Commentary*, Leiden

Grenfell, B. P. and A. S. Hunt (1906–55) *The Hibeh Papyri*, London

Grenfell, B. P., A. S. Hunt, *et al.* (1898–) *The Oxyrhynchus Papyri*, London

Grenfell, B. P., A. S. Hunt, and J. G. Smyly (1902–76) *The Tebtunis Papyri*, London

Grigg, D. B. (1974) *The Agricultural Systems of the World: An Evolutionary Approach*, Cambridge

Grimal P. (1986) 'Pline et les philosophes', *Helmantica*, 37, pp. 239–49

Gruterus, J. (1707) *Inscriptiones antiquae totius orbis romani*, Amsterdam

Guattani, G. A. (1817) *Memorie enciclopediche di antichità e belle arti per l'anno 1816*, Rome

Habachi, B. and L. Habachi (1952) 'The Naos with the Decades (Louvre D 37) and the Discovery of Another Fragment', *Journal of Near Eastern Studies*, 11, pp. 251–63

Halma, N. B. (1819) *Table chronologique des règnes, prolongée jusqu'à la prise de Constantinople par les Turcs*, Paris

Hannah, R. (2001a) 'From Orality to Literacy? The Case of the Parapegma', in J. Watson, ed., *Speaking Volumes: Orality and Literacy in the Greek and Roman World*, Leiden

(2001b) 'The Moon, the Sun, and the Stars', in S. McCready, ed., *The Discovery of Time*, London, pp. 56–91

(2002) 'Euctemon's *Parapēgma*', in T. E. Rihll and C. J. Tuplin, eds., *Science and Mathematics in Ancient Greek Culture*, Oxford

(2005) *Greek and Roman Calendars*, London

Hartog, F. (1988) *The Mirror of Herodotus*, Berkeley, CA

Healy, J. (1999) *Pliny the Elder on Science and Technology*, Oxford

Heath, T. L. (1913) *Aristarchus of Samos, the Ancient Copernicus*, Oxford

Heiberg, J. L., ed. (1898) 'Ptolemy, *Almagest*', in *Claudii Ptolemaci opera quae exstant omnia*, vol. i, Leipzig

  (1907a) *Claudii Ptolemaei opera quae exstant omnia*, vol. ii, Leipzig

  (1907b) 'Ptolemy, *Phaseis*', in Heiberg, 1907a

Hellmann, G. (1916) 'Aegyptische Witterungsangaben im Kalender von Ptolemaios', *Sitzungberichte der Deutschen Akademie der Wissenschaften zu Berlin*, pp. 332–41

  (1917) 'Die Witterungsangaben in den griechischen und lateinischen Kalendern (Beiträge zur Geschichte der Meteorologie)', *Veröffentlichungen des Königlich Preußischen Meteorologischen Instituts*, 296, Berlin

Henzen, W. (1902) *Corpus inscriptionum latinarum*, vol. vi.4.2, Berlin

Herbert Brown, G., ed. (2002a) *Ovid's Fasti*, Oxford

  (2002b) 'Ovid and the Stellar Calendar', in G. Herbert Brown, 2002a

Herz, M. and C. Hosius, eds. (1903), *A. Gellii noctium atticarum libri xx*, Leipzig

Herz, P. and G. H. Waldherr, eds. (2001) *Landwirtschaft im Imperium Romanum*, St Katharinen

Heseltine, M. (1913) *Petronius, Satyricon*, London

Hitz, H. R. (1991) *Der gallo-lateinisch Mond- und Sonnen-Kalender von Coligny*, Dietikon

Hjelmslev, L. (1961) *Prolegomena to a Theory of Language*, trans. F. J. Whitfield, Madison, WI

Holleman, A. W. J. (1978) 'Les calendriers préjuliens à Rome', *L'antiqité classique*, 47, pp. 201–6

Hooper, W. D. and H. B. Ash, eds. (1935) *Varro, Rerum rusticarum*, Cambridge, MA

Höpfner, W. (1976) *Das Pompeion und seine Nachfolgerbauten. Kerameikos. Ergebnisse der Ausgrabungen X*, Berlin

Hornblower, S. and Spawforth, A., eds. (1996) *The Oxford Classical Dictionary*, 3rd edn., Oxford

Horowitz, W. (1989–90) 'Two Mul-Apin Fragments', *Archiv für Orientforschung*, 36/7, pp. 116–17

  (1996) 'The 360 and 364 Day Year in Ancient Mesopotamia', *Journal of the Ancient Near Eastern Society*, 24, pp. 35–44

Horsfall, N. (1989) ' "The Uses of Literacy" and the "Cena Trimalchionis": II', *Greece and Rome*, 2nd ser. 36, pp. 194–209

Hort, A., ed. (1926) Theophrastus, 'De signis', in *Inquiry into Plants*, Cambridge, MA, vol. ii

Høyrup, J. (1999) 'A Historian's History of Ancient Egyptian Science', *Physis*, 36, pp. 237–55

Huber, P. J. (1987) 'Astronomical Evidence for the Long and against the Short Chronologies', in P. Åström, ed., *High, Middle or Low? Acts of an International*

Colloquium on Absolute Chronology Held at the University of Gothenburg, 20th–22nd of August, 1987, Gothenburg

Huber, P. J., *et al.* (1982) *Astronomical Dating of Babylon I and Ur III*, Malibu

Hübner, W. (1998–9) 'De astrologia antiqua', *Acta classica universitatis scientarum Debreceniensis*, 34–5, pp. 171–86

Hunger, Herbert (1961–) *Katalog der griechischen Handschriften der Österreichischen Nationalbibliothek*, Vienna

Hunger, Hermann (1976) 'Astrologische Wettervorhersagen', *Zeitschrift für Assyriologie und vorderasiatische Archäologie*, 66, pp. 234–60

(1992) 'Astrological Reports to Assyrian Kings', *State Archives of Assyria*, vol. VIII, Helsinki

(1999) 'Non-mathematical Astronomical Texts and their Inter-relations', in N. M. Swerdlow, ed., *Ancient Astronomy and Celestial Divination*, Cambridge, MA

Hunger, H. and D. Pingree (1989) *MUL.APIN: An Astronomical Compendium in Cuneiform*, Horn

(1999) *Astral Sciences in Mesopotamia*, Leiden

Hunger, H. and E. Reiner (1975) 'A Scheme for Intercalary Months from Babylonia', *Wiener Zeitschrift für die Kunde des Morgenlandes*, 67, pp. 21–8

Ideler, J. (1825) 'Über den astronomischen Theil der *Fasti* des Ovid', *Abhandlungen der königlichen Akademie der Wissenschaften zu Berlin aus den Jahren 1822–3*, Berlin, pp. 137–69

Imhausen, A. (2002) 'The Algorithmic structure of the Egyptian Mathematical Problem Texts', in J. Steele and A. Imhausen, eds., *Under One Sky: Astronomy and Mathematics in the Ancient Near East*, Alter Orient und Altes Testament, Münster

(2003) 'Calculating the Daily Bread: Rations in Theory and Practice', *Historia Mathematica*, 30, pp. 3–16

Inscriptiones graecae (1863–) Berlin

Isager, S. and J. E. Skydsgaard (1992) *Ancient Greek Agriculture*, New York

Jacoby, K., ed. (1967) *Dionysius Halicarnassensis, Antiquitatum romanorum quae supersunt*, Stuttgart

Janowitz, N. (2001) *Magic in the Roman World: Pagans, Jews and Christians*, New York

Jenks, S. (1983) 'Astrometeorology in the Middle Ages', *Isis*, 74, pp. 185–210

Johnson, V. (1963) 'The Prehistoric Roman Calendar', *American Journal of Philology*, 84, pp. 28–35

Jones, A. (1991) 'The Adaptation of Babylonian Methods in Greek Numerical Astronomy', *Isis*, 82, pp. 441–53

(1993) 'Evidence for Babylonian Arithmetical Schemes in Greek Astronomy', in H. D. Galter, ed., *Die Rolle der Astronomie in den Kulturen Mesopotamiens*, Grazer Morgenländische Studien 3, Graz, pp. 77–94

(1994a) 'An Astronomical Ephemeris for A. D. 140: P. Harris I.60', *Zeitschrift für Papyrologie und Epigraphik*, 100, pp. 59–63

(1994b) 'Two Astronomical Papyri Revisited', *Analecta Papyrologica*, 6, pp. 111–26

(1996) 'Babylonian Astronomy and its Greek Metamorphoses', in F. J. Ragep and S. Ragep, eds., *Tradition, Transmission, Transformation*, Leiden, pp. 139–55

(1999a) *Astronomical Papyri from Oxyrhynchus*, Philadelphia

(1999b) 'Geminus and the Isia', *Harvard Studies in Classical Philology*, 99, pp. 255–67

(1999c) 'A Classification of Astronomical Tables on Papyrus', in N. Swerdlow, ed., *Ancient Astronomy and Celestial Divination*, Cambridge, MA, pp. 299–340

(2000) 'Calendrica I: New Callippic Dates', *Zeitschrift für Papyrologie und Epigraphik*, 129, pp. 141–58

(forthcoming) 'The Astrologers of Oxyrhynchus and Their Astronomy', in *Oxyrhynchus: A City and its Texts*

Junker, H. (1917) *Die Onurislegende*, Vienna

Kamal, A. B. (1906) 'Rapport sur quelques localités de la Basse-Égypte', *Annales du Service des Antiquités de l'Égypte*, 7, pp. 232–40

Kidd, D. A., ed. (1997) *Aratus, Phaenomena*, Cambridge

Klingner, F. (1967) *Virgil: Bucolica, Georgica, Aeneis*, Stuttgart

Koch, J. (1995–6) 'MUL.APIN II i 68–71', *Archiv für Orientforschung*, 42/3, pp. 155–62

Koch-Westenholz, U. (1995) *Mesopotamian Astrology*, Copenhagen

(2000) *Babylonian Liver Omens*, Copenhagen

Kolta, K. S. and D. Schwarzmann-Schafhauser (2000) *Die Heilkunde im alten Ägypten: Magie und Ratio in der Krankheitsvorstellung und therapeutischen Praxis*, Stuttgart

Krauss, R. (1999) 'Nähere Mitteilungen über Seth/Merkur und Horus-Horusauge/ Venus im grossen Tagewählkalender', *Studien zur Altägyptischen Kultur*, 27, pp. 233–54

(2002) 'The Eye of Horus and the Planet Venus: Astronomical and Mythological References', in J. Steele and A. Imhausen, eds., *Under One Sky: Astronomy and Mathematics in the Ancient Near East*, Alter Orient und Altes Testament 297, Münster

Krchnàk, A. (1963) 'Die Herkunft der astronomischen Handschriften und Instrumente des Nicholas von Kues', *Mitteilungen und Forsschungsbeiträge der Cusanus-Gesellschaft*, 3, pp. 109–80

Kroll, G. J., ed. (1899) *Procli Diadochi in Platonis rem publicam commentarii*, Leipzig

Kroll, W. (1930) *Die Kosmologie des Plinius*, Breslau

Kronenberg, L. J. (2000) 'The Poet's Fiction: Virgil's Praise of the Farmer, Philosopher, and Poet at the End of "Georgics 2" ', *Harvard Studies in Classical Philology*, 100, pp. 341–60

Kubitschek, W. (1928) *Grundriss der antiken Zeitrechnung*, Munich

(1932) 'Meton', *Paulys Realencyclopädie der classischen Altertumswissenschaft*, vol. xv, Stuttgart

Kugler, F. X. (1900) *Die Babylonische Mondrechnung*, Freiburg

Kühn, D. C. G., ed. (1821) *Claudii Galeni opera omnia*, Leipzig

Kuhn, T. S. (1970) *The Structure of Scientific Revolutions*, 2nd edn., Chicago

(1993) 'Afterwords', in P. Horwich, ed., *World Changes*, Cambridge, MA, pp. 311–41

Labat, R. (1939) *Hémérologies et ménologies d'Assur*, Paris

(1965) *Un calendrier babylonien des travaux des signes et des mois*, Paris

(1988) *Manuel d'épigraphie akkadienne*, 6th edn., Paris

Lacerenza, G. (1988) 'Il dio Dusares a Puteoli', *Puteoli*, 12, pp. 119–49

La'da, C. (2003) 'Encounters with Ancient Egypt: The Hellenistic Greek Experience', R. Matthews and C. Roemer, eds., *Ancient Perspectives on Egypt*, London, pp. 157–69

Lalonde, M. (1996) *La reconnaissance du temps*, Montreal

de Lama, P. (1818) *Iscrizioni antiche collocate ne' muri della scala Farnese*, Parma

Lamb, H. H. (1995) *Climate, History, and the Modern World*, 2nd edn., New York

Lanciani, R. (1878) 'Ostia', *Notizie degli scavi di antichità*, 3, p. 67

Langdon, S. (1935) *Babylonian Menologies and the Semitic Calendars*, London

Le Bouffle, A. (1989) *Le ciel des Romains*, Paris

Lehmann, P. (1930) 'Mitteilungen aus Handschriften II', *Sitzungsberichte der Bayerischen Akademie der Wissenschaften, philosophisch-historische Abteilung*, 2

Lehoux, D. (2002) 'The Historicity Question in Mesopotamian Divination', in J. Steele and A. Imhausen, eds., *Under One Sky: Astronomy and Mathematics in the Ancient Near East*, Alter Orient und Altes Testament 297, Münster

(2003) 'Tropes, Facts, and Empiricism', *Perspectives on Science*, 11, pp. 324–43

(2004a) 'Observation and Prediction in Ancient Astrology', *Studies in History and Philosophy of Science*, 35, pp. 227–46

(2004b) 'Weather, Why, and When', *Studies in History and Philosophy of Science*, 35, pp. 835–843

(2004c) 'Impersonal and Intransitive ἐπισημαίνει', *Classical Philology*, 99, pp. 78–85

(2004d) 'Logic, Physics, and Prediction in Hellenistic Philosophy: *x* Happens, but *y*?', in B. Löwe, V. Peckhaus, and T. Räsch, eds., *The History of the Concept of the Formal Sciences*

(2005) 'The Miletus Parapegma Fragments', *Zeitschrift für Papyrologie und Epigraphik*, 152, pp. 125–40

(2006a) 'Tomorrow's News Today: Astrology, Fate, and the Ways Out', *Representations*, 95, pp. 105–22

(2006b) 'Rethinking Parapegmata: The Puteoli Fragment', *Zeitschrift für Papyrologie und Epigraphik*, 157, pp. 95–104

(forthcoming) 'Image, Text, and Pattern: Reconstructing Parapegmata', in A. Jones, ed., *Reconstructing Ancient Texts*, Toronto

Lembke, J. (2005) *Virgil's Georgics*, New Haven

Le Roy Ladurie, E. (1967) *Histoire du climat depuis l'an mil*, Paris

Leitz, C. (1994) *Tagewählerei*, Wiesbaden

  (1995) *Altägyptische Sternühren*, Leuven

de Ligt, L. (1993) *Fairs and Markets in the Roman Empire*, Amsterdam

Linderski, J. (1996) 'Divination, Roman', in S. Hornblower and A. Spawforth, eds., *The Oxford Classical Dictionary*, 3rd edn., Oxford

Littré, E., ed. (1839–61) *Hippocrates: Œuvres complètes*, Paris

Lloyd, G. E. R. (1970) *Early Greek Science*, New York

  (1973) *Greek Science after Aristotle*, New York

  (1979) *Magic, Reason and Experience*, Cambridge

  (1983) *Science, Folklore, and Ideology*, Cambridge

  (1987) *Revolutions of Wisdom*, Berkeley, CA

  (1991a) *Methods and Problems in Greek Science*, Cambridge

  (1991b) 'The Invention of Nature', in Lloyd, 1991a

  (1996) *Adversaries and Authorities*, Cambridge

  (2002) *Ambitions of Curiosity*, Cambridge

  (2004) *Ancient Worlds, Modern Reflections*, Oxford

Lloyd, G. E. R. and N. Sivin (2002) *The Way and the Word*, New Haven

Long, C. R. (1992) 'Pompeii Calendar Medallions', *American Journal of Archaeology*, 96, pp. 477–501

Luck, G. (1985) *Arcana Mundi*, Baltimore

  (1990) *Magie und andere Geheimlehren in der Antike*, Stuttgart

Maass, E. (1892) *Aratea*, Berlin

  ed. (1893) *Arati phaenomena*, Berlin

  (1902) *Die Tagesgötter in Rom und den Provinzen*, Berlin

McCluskey, S. C. (1998) *Astronomies and Cultures in Early Medieval Europe*, Cambridge

MacCoull, L. S. B. (1991) 'Philoponus on Egypt', *Byzantische Forschungen*, 17, pp. 167–72

McEvoy, J. (1993) 'Platon et la sagesse de l'Égypte', *Kemos*, 6, pp. 245–75

MacMullen, R. (1970) 'Market-Days in the Roman Empire', *Phoenix*, 24, pp. 333–41

MacNeill, E. (1928) 'On the Notation and Chronology of the Calendar of Coligny', *Ériu*, 10, pp. 1–67

Malinowski, B. (1948) *Magic, Science and Religion*, New York

Manetti, G. (1987) *Le teorie del segno nell'antichità classica*, Milan

Manicoli, D. (1981) 'Un calendario astrologico al Museo della Civiltà Romana', *Bollettino dei Musei comunali di Roma*, 28–30, pp. 18–22

Manitius, C., ed. (1898) *Gemini elementa astronomiae*, Leipzig

Mansfeld, J. (1971) *The Pseudo-Hippocratic Tract 'Peri hebdomadon' Ch. 1–11 and Greek Philosophy*, Assen

Marcovich, M., ed. (1999) *Diogenes Laertius: Vitarum philosophorum libri*, Leipzig

Marganne M.-H. (1992) 'Les références à l'Égypte dans la matière médicale de Dioscoride', in *Serta Leodiensia secunda: mélanges publiés par les Classiques de Liège à l'occasion du 175e anniversaire de l'Université*, Liège, pp. 309–22

(1993) 'Links between Egyptian and Greek Medicine', *Forum: Trends in Experimental and Clinical Medicine*, 3, pp. 35–43

Marulli, T. (1813) *Sopra un'antica cappella cristiana. Scoperta di fresco in Roma nelle terme di Tito*, Rome

Matthaei, C. F., ed. (1802) *Nemesius, De natura hominis*, Halle

Maurach, G. (1978) *Germanicus und sein Arat*, Heidelberg

Mauss, M. (1902–3) 'Esquisse d'une théorie générale de la magie', *L'année sociologique*, 7, pp. 1–146

Maxwell-Stewart, P. G. (1995) *Studies in the Career of Pliny the Elder and the Composition of his 'Naturalis Historia'*, St Andrews

Mayhoff, C., ed. (1897–1906) *C. Plinii Secundi naturalis historiae libri* xxxvii, Leipzig

Mazzarino, A., ed. (1982) *M. Porcius Cato: De agri cultura*, Leipzig

Merkelbach, R. and M. L. West, eds. (1967) *Hesiod: Fragmenta Hesiodea*, Oxford

Meyboom, P. G. P. (1978) 'Un monument énigmatique "Dusari sacrum" à Pouzzoles', in M. B. d. Boer and T. A. Edridge, eds., *Hommages à Maarten J. Vermaseren*, Leiden, pp. 782–90

Michels, A. K. (1967) *The Calendar of the Roman Republic*, Princeton

Mikalson, J. D. (1996) 'Calendar, Greek', in S. Hornblower and A. Spawforth, eds., *The Oxford Classical Dictionary*, 3rd edn., Oxford

Mingazzini, P. (1928) 'Pozzuoli. – Frammento di calendario perpetuo', *Notizie degli scavi di antichità*, 6.4, pp. 202–5

Mirecki, P. A. and M. Meyer (2002) *Magic and Ritual in the Ancient World*, Leiden

Mommsen, T. (1852) *Inscriptiones regni Neapolitani latinae*, Leipzig

(1859) *Die römische Chronologie bis auf Caesar*, Berlin

Morgan, L. (1999) *Patterns of Redemption in Virgil's Georgics*, Cambridge

Müller, J. (1991) 'Intercalary Months in the Athenian Dark Age Period', *Schweizer Münzblätter* 164, pp. 85–9

(1994) 'Synchronization of the Late Athenian with the Julian Calendar', *Zeitschrift für Papyrologie und Epigraphik*, 103, pp. 128–38

Müller, K., ed. (1995) *Petronius Arbiter, Satyricon reliquiae*, Leipzig

Müller, K. and W. Ehlers, eds. (1965) *Petronius, Satyrica*, Munich

Mynors, R. A. B., ed., and comm. (1990) *Virgil, Georgics*, Oxford

Naas, V. (2002) *Le projet encyclopédique de Pline l'ancien*, Rome

Nelson, S. (1996) 'The Drama of Hesiod's Farm', *Classical Philology*, 91, pp. 45–53

(1998) *God and the Land: The Metaphysics of Farming in Hesiod and Vergil*, Oxford

Nesselrath, H.-G. (1996) 'Herodot und der griechische Mythos', *Poetica*, 28, pp. 275–96

Netz, R. (1999) *The Shaping of Deduction in Greek Mathematics*, Cambridge

Neugebauer, O. (1955) *Astronomical Cuneiform Texts*, London

(1969) *Exact Sciences in Antiquity*, New York

(1971) 'An Arabic Version of Ptolemy's Parapegma from the Phaseis', *Journal of the American Oriental Society*, 91, p. 506

(1975) *History of Ancient Mathematical Astronomy*, New York

Neugebauer, O. and H. B. van Hoesen (1959) *Greek Horoscopes*, Philadelphia

Neugebauer, O. and R. Parker (1969) *Egyptian Astronomical Texts*, Providence, RI

Newlands, C. (1995) *Playing with Time: Ovid and the Fasti*, Ithaca

Nilsson, M. P. (1920) *Primitive Time-Reckoning*, Oxford

(1952) 'Zur Frage von dem Alter des vorcäsarischen Kalenders', *Opuscula selecta*, Lund, vol. II, pp. 979–87

(1962) *Die Enstehung und religiöse Bedeutung des griechischen Kalenders*, Lund

Noè, E. (2002) *Il progetto di Columella*, Como

Nutton, V. (2004) *Ancient Medicine*, New York

Oates, J., et al. (2001) *Checklist of Editions of Greek, Latin, Demotic, and Coptic Papyri, Ostraca, and Tablets*, Oakville, CT

Oldfather, C. H., ed. (1933) *Diodorus Siculus: History*, Cambridge, MA

Olivieri, A., ed. (1935) *Aetii Amideni libri medicinales (Corpus medicorum graecorum VIII.1)*, Leipzig

Olmstead, G. (1992) *The Gaulish Calendar*, Bonn

Parisot, J. P. (1985) 'Les phases de la lune et les saisons dans le calendrier de Coligny', *Etudes indo-européens*, 13, pp. 1–18

Parker, R. (1950) *The Calendars of Ancient Egypt*, Chicago

(1959) *A Vienna Demotic Papyrus on Eclipse and Lunar-Omina*, Providence, RI

Parpola, S. (1993) *Letters from Assyrian and Babylonian Scholars: State Archives of Assyria*, vol. x, Helsinki

Paulys Realencyclopädie der classischen Altertumswissenschaft (1893–), Stuttgart

Pedersen, O. (1986) 'Some Astronomical Topics in Pliny', in R. French and F. Greenaway, eds., *Science in the Early Roman Empire: Pliny the Elder, his Sources and Influence*, Totowa, NJ

Pedroni, L. (1998) 'Ipotesi sull'evoluzione del calendario arcaico di Roma', *Papers of the British School at Rome*, 66, pp. 39–56

Peel, J. D. Y. (1969) 'Understanding Alien Belief Systems', *British Journal of Sociology*, 20, pp. 69–84

Pellat, C., ed. (1986) *Cinq calendriers Égyptiens*, Cairo

Pfeiffer, E. (1916) *Studien zum antiken Sternglauben*, Leipzig

Piale, S. (1817) 'Novembre', in Guattani, 1817, pp. 153–64

Pietri, C. (1984) 'Le temps de la semaine à Rome et dans l'Italie chrétienne (IV–VI s.)', *Le temps crétien de la fin de l'antiquité au Moyen-âge (III–XIII s.)*, Actes du colloque, *9–12 Mars, 1981*, Paris

Pingree, D. (1976) 'The Indian and Pseudo-Indian Passages in Greek and Latin Astronomical and Astrological Texts', *Viator*, 7, pp. 141–95

ed. (1986) *Vettius Valens, Anthologiae*, Leipzig

Pingree, D. and R. A. Gilbert (2002) 'Astrology', *Encyclopedia Britannica 2002, Deluxe Edition CD-ROM*

Plasberg, O., ed. (1922) *Cicero: Academica*, Leipzig

Popper, K. (1959) *Logic of Scientific Discovery*, London

Possanza, D. M. (2004) *Translating the Heavens: Aratus, Germanicus, and the Poetics of Latin Translation*, New York

Price, D. de Solla (1974) 'Gears from the Greeks', *Transactions of the American Philosophical Society Held at Philadelphia for Promoting Useful Knowledge*, ns 64.7

Pritchett, W. K. and O. Neugebauer (1947) *The Calendars of Athens*, Cambridge, MA

Pritchett, W. K. and B. L. van der Waerden (1961) 'Thucydidean Time-Reckoning and Euctemon's Seasonal Calendar', *Bulletin de correspondance Hellénique*, 85, pp. 17–52

Putnam, M. J. (1979) *Virgil's Poem of the Earth: Studies in the Georgics*, Princeton

Quack, J. F. (2003) 'Methoden und Möglichkeiten der Erforschung der Medizin im Alten Ägypten', *Medizinhistorisches Journal*, 38, pp. 3–15

Quine, W. V. (1953) 'Two Dogmas of Empiricism', *From a Logical Point of View*, Cambridge, MA

Radermacher, L., ed. (1959) *Quintilian, Institutionis oratoriae libri XII*, Leipzig

Radke, G. (1990) *Fasti romani*, Münster

Rampino, M. R., *et al.* (1987) *Climate, History, Periodicity, and Predictability*, New York

Reallexikon der Assyriologie und vorderasiatischen Archäologie (1928), Berlin

Rehm, A. (1904) 'Weiteres zu den milesischen Parapegmen', *Sitzungsberichte der königlich preussischen Akadamie der Wissenschaften, philosophisch-historische Klasse*, 23, pp. 752–9

    (1913) 'Das Parapegma des Euktemon', in F. Boll, 1911, pp. 2–38

    (1927) 'Der römische Bauernkalender und der Kalender Caesars', *Epitymbion Heinrich Swoboda dargebracht*, Reichenberg, pp. 214–28

    (1940) 'Episemasiai', *Paulys Realencyclopädie der classischen Altertumswissenschaft*, vol. suppl. VII, Stuttgart, pp. 175–98

    (1941) *Parapegmastudien*, Munich

    (1949) 'Parapegma', *Paulys Realencyclopädie der classischen Altertumswissenschaft*, vol. XVIII.4, Stuttgart, pp. 1295–1366

Reiner, E. (1999) 'Babylonian Celestial Divination', in N. M. Swerdlow, ed., *Ancient Astronomy and Celestial Divination*, Cambridge, MA

Reiner, E. and D. Pingree (1975–98) *Babylonian Planetary Omens*, Malibu and Groningen

Reinhardt, K. (1926) *Kosmos und Sympathie*, Munich

Reydellet, M. (1999) 'Une représentation du calendrier romain à la gloire d'Auguste et de sa maison: l'utilisation des phénomènes célestes dans les « Fastes » d'Ovide', *Revue des études latines*, 77, pp. 14–15

Ricci, S. de (1898) 'Le calendrier galois de Coligny', *Revue celtique*, 19, pp. 213–23, pls. I–VI

(1900) 'Le calendrier celtique de Coligny', *Revue celtique*, 21, pp. 10–27

Richards, E. G. (1998) *Mapping Time*, Oxford

Ritner, R. K. (1992) 'Implicit Models of Cross-Cultural Interaction: A Question of Noses, Soap, and Prejudice', in J. H. Johnson, ed., *Life in a Multi-Cultural Society: Egypt from Cambyses to Constantine and Beyond*, Chicago, pp. 283–90

(1993) *The Mechanics of Ancient Egyptian Magical Practice*, Chicago

(2000) 'Innovations and Adaptations in Ancient Egyptian Medicine', *Journal of Near Eastern Studies*, 59.2, pp. 107–17

Robson, E. (2003) 'Satisfaction, Subversion, and the Reluctant Reader: Some Thoughts on Writing Accessible History of Ancient Mathematics', *Studies in History and Philosophy of Science*, 34, pp. 423–9

Rocha-Pereira, M. H., ed. (1989–90) *Pausanias: Graeciae descriptio*, Leipzig

Rochberg, F. (1993) 'The Cultural Locus of Astronomy in Late Babylonia', in H. D. Galter, ed., *Die Rolle der Astronomie in den Kulturen Mesopotamiens*, Grazer Morgenländische Studien 3, Graz, pp. 31–45

(1998) *Babylonian Horoscopes*, Philadelphia

(1999) 'Babylonian Horoscopy: The Texts and their Relations', in N. M. Swerdlow, ed., *Ancient Astronomy and Celestial Divination*, Cambridge, MA

(2004) *The Heavenly Writing*, Cambridge

Rochberg-Halton, F. (1988) 'Aspects of Babylonian Celestial Divination: The Lunar Eclipse Tablets of Enuma Anu Enlil', *Archiv für Orientforschung*, Beiheft 22

(1992) 'Calendars, Ancient Near East', *Anchor Bible Dictionary*, New York

Rodgers, R. H., ed. (1975) *Palladius: Opus agriculturae*, Leipzig

Romilly, J. de, ed. (1962–4) *Thucydides, La guerre du Péloponnèse*, Paris

Rosén, H. B., ed. (1987–97) *Herodotus: Historiae*, Leipzig

Rostovtzeff, M. I., A. R. Bellinger, et al., eds. (1936) *The Excavations at Dura-Europus, Sixth Season, 1932–1933*, New Haven

Roughton, N. A., J. M. Steele, and C. B. F. Walker (2004) 'A Late Babylonian Normal and *Ziqpu* Star Text', *Archive for History of Exact Sciences*, 58, pp. 537–72

Rüpke, J. (1995) *Kalender und Öffentlichkeit*, Berlin

(2000a) 'Nundinae', *Der neue Pauly*, 8, Stuttgart, c. 1064

(2000b) 'Parapegma', *Der neue Pauly*, 9, Stuttgart, c. 322

Sachau, C. E., trans. (1879) *al-Biruni: The Chronology of Ancient Nations*, London

Sachs, A. (1948) 'A Classification of the Babylonian Astronomical Tablets of the Seleucid Period', *Journal of Cuneiform Studies*, 2, pp. 271–90

(1952) 'Sirius Dates in Babylonian Astronomical Texts of the Seleucid Period', *Journal of Cuneiform Studies*, 6, pp. 105–14

Sachs, A. and H. Hunger (1988) *Astronomical Diaries and Related Texts from Babylonia*, Vienna

Sadurska, A. (1979) 'Rzymskie kalendarze manipulowane oraz ich uwarunkowania historyczne', *Archeologia*, 30, pp. 69–85

Saint-Denis, E. de, ed. (1957) *Vergil, Georgica*, Paris

Sallmann, N., ed. (1983) *Censorinus: De die natali*, Leipzig

Salonen, A. (1968) *Agricultura mesopotamica*, Helsinki

Salzman, M. R. (1990) *On Roman Time: The Codex-Calendar of 354 and the Rhythms of Urban Life in Late Antiquity*, Berkeley

Samuel, A. E. (1972) *Greek and Roman Chronology*, Munich

Santini, C. (1975) 'Motivi astronomici e moduli didattici nei *Fasti* di Ovidio', *Giornale italiano di philologia*, ns. 6, pp. 1–26

Saunders, J. B. de C. M. (1963) *The Transitions from Ancient Egyptian to Greek Medicine*, Lawrence, KS

Saussure, F. de (1916) *Cours de linguistique générale*, Paris

Schaefer, B. E. (2000) 'The Heliacal Rise of Sirius and Ancient Egyptian Chronology', *Journal for the History of Astronomy*, 31, pp. 149–55

Schaumberger, J. (1952) 'Die Ziqpu-Gestirne nach neuen Keilschrifttexten', *Zeitschrift für Assyriologie*, 50, pp. 214–29

Schmitz, H.-G. (1993) 'Fiktion und Divination: Funf Thesen zur stoischen Semiotik des Orakels, wie Cicero sie entfaltet', *Philosophisches Jahrbuch*, 100, pp. 172–86

Schneider, J. G. (1794) *Scriptores rei rusticae veteres latini*, Leipzig

Schroeder, O. (1920) *Keilschrifttexte aus Assur verschiedenen Inhalts*, Leipzig

Sebeok, T. A., ed. (1986) *Encyclopedic Dictionary of Semiotics*, Amsterdam

Shackleton-Bailey, D. R. (1968) *Cicero's Letters to Atticus*, Cambridge

Sharrock, A. (1997) '*Haud mollia iussa*: A Response to Roy Gibson', in C. Atherton, ed., *Form and Content in Didactic Poetry*, Nottingham

Sider, D. (2002) 'Demokritos on the Weather', in A. Laks and C. Louguet, eds., *Qu'est que la philosophie présocratique?* Villeneuve d'Ascq

Sirago, V. A. (1995) *Storia agraria romana*, Naples

Slater, N. W. (1990) *Reading Petronius*, Baltimore

Slotsky, A. L. (1993) 'The Uruk Solstice Scheme Revisited', in H. D. Galter, ed., *Die Rolle der Astronomie in den Kulturen Mesopotamiens*, Grazer Morgenländische Studien 3, Graz, pp. 359–65

(1997) *The Bourse of Babylon*, Bethesda, MD

Smith, M. S., ed. (1975) *Petroni Arbitri Cena Trimalchionis*, Oxford

Smith, J. Z. (2002) 'Great Scott! Thought and Action One More Time', in Mirecki and Meyer, 2002, vol. 141, pp. 73–91

Snyder, W. F. (1936) 'Quinto nundinas Pompeis', *Journal of Roman Studies*, 26, pp. 12–18

Soldt, W. H. van (1995) *Solar Omens of Enuma Anu Enlil: Tablets 23(24)–19(30)*, Istanbul

Souriban, J. (1969) *Vitruve, DE l'architecture, livre* IX, Paris

Spaeth, O. von (2000) 'Dating the Oldest Egyptian Star Map', *Centaurus*, 42, pp. 159–79

Spalinger, A. (1991) 'Remarks on an Egyptian Feast Calendar of Foreign Origin', *Studien zur Altägyptischen Kultur*, 18, pp. 349–73

Spek, R. J. van der (1985) 'The Babylonian Temple during the Macedonian and Parthian Domination', *Bibliotheca Orientalis*, 50, pp. 542–62

Sperber, D. (1985) *On Anthropological Knowledge*, Cambridge

Spurr, M. S. (1986) 'Agriculture and the Georgics', *Greece & Rome*, 33, pp. 164–87

Staden, H. von (1989) *Herophilus: The Art of Medicine in Early Alexandria*, Cambridge

(1992) 'Affinities and Elisions: Helen and Hellenocentrism', *Isis*, 83, pp. 578–95

Steele, J. M. and A. Imhausen, eds. (2002) *Under One Sky: Astronomy and Mathematics in the Ancient Near East*, Alter Orient und Altes Testament 297, Münster

Stern, H. (1953) *Le calendrier de 354*, Paris

Storchi Marino, A. (2000) 'Reti interregionali integrate e circuiti di mercato periodico negli *indices nundinarii* del Lazio e della Campania', in E. Lo Cascio, ed., *Mercati permanenti e mercati periodici nel mondo romano*, Bari, pp. 93–130

Swerdlow, N. M. (1989) 'Ptolemy's Theory of the Inner Planets', *Journal for the History of Astronomy*, 20, pp. 29–60

(1998) *The Babylonian Theory of the Planets*, Princeton

(1999) 'Introduction', in Swerdlow, ed., *Ancient Astronomy and Celestial Divination*, Cambridge, MA, pp. 1–20

Symons, S. (2002) 'Two Fragments of Diagonal Star Clocks in the British Museum', *Journal for the History of Astronomy*, 33, pp. 257–60

Tambiah, S. J. (1990) *Magic, Science, Religion, and the Scope of Rationality*, Cambridge

Taub, L. (1993) *Ptolemy's Universe: The Natural Philosophical and Ethical Foundations of Ptolemy's Astronomy*, Chicago

(1998) 'Meteorology in the Ancient World', in G. A. Good, ed., *Sciences of the Earth: An Historical Encyclopedia*, New York, pp. 535–8

(2003) *Ancient Meteorology*, London and New York

Tavenner, E. (1918) 'The Roman Farmer and the Moon', *Transactions and Proceedings of the American Philological Association*, 49, pp. 67–82

Thomas, R. F. (1987) 'Prose into Poetry: Tradition and Meaning in Vergil's Georgics', *Harvard Studies in Classical Philology*, 91, pp. 229–60

ed. and comm. (1988) *Virgil, Georgica*, Cambridge

(1995) 'Vestigia Ruris: Urbane Rusticity in Virgil's Georgics', *Harvard Studies in Classical Philology*, 97, pp. 197–214

Thureau-Dangin, R. (1922) 'Tablettes d'Uruk', *Textes Cunéiformes*, vol. VI, Paris

Tibiletti, G. (1976–7) 'Qualche problema nundinario', *Rivista storica dell'antichità*, 6–7, pp. 27–34

Thompson, D. J. (1988) *Memphis under the Ptolemies*, Princeton

Thorndike, L. (1923) *History of Magic and Experimental Science*, New York

Toomer, G. J. (1974) 'Meton', *Dictionary of Scientific Biography*, New York

(1984) *Ptolemy's Almagest*, London

(1988) 'Hipparchus and Babylonian Astronomy', in E. Leichty, *et al.*, eds., *A Scientific Humanist: Studies in Memory of Abraham Sachs*, Philadelphia

(1996) 'Meton', in S. Hornblower and A. Spawforth, eds., *The Oxford Classical Dictionary*, 3rd edn., Oxford

Turner, E. G. and O. Neugebauer (1949) 'Gymnasium Debts and Full Moons', *Bulletin of the John Rylands Library*, 32, pp. 80–96

Tyler, E. B. (1871) *Primitive Culture*, London

Tyrell, R. Y. and L. C. Purser, eds. (1890) *The Correspondence of M. Tullius Cicero*, London

Urner-Astholz, H. (1942) 'Die römerzeitliche Keramik von Eschenz-Tasgetium', *Thurgauer Beiträge*, 78, p. 90f.

(1960) 'Der Wochensteckkalender von Eschenz-Tasgetium und die Verehrung der Wochengötter', *Jahrbuch der schweizerische Gesellschaft für Urgeschichte*, 48, pp. 44f.

Vasunia, P. (2001) *The Gift of the Nile: Hellenizing Egypt from Aeschylus to Alexander*, Berkeley, CA

Veith, I. (1965) *Hysteria: The History of a Disease*, Chicago

Verderame, L. (2002) *Le tavole I–VI della serie astrologica Enūma Anu Enlil*, Messina

Versnel, H. S. (1996) 'Magic', in S. Hornblower and A. Spawforth, eds., *Oxford Classical Dictionary*, 3rd edn., Oxford

Virolleaud, C. (1905–12) *L'astrologie chaldéenne*, Paris

Volkmann, R., ed. (1883) *Plotinus, Enneades*, Leipzig

Wachsmuth, C., ed. (1884–1912) *Ioannis Stobaei anthologiúm*, Berlin

(1897) *Ioannis Laurentii Lydi liber de ostentis et calendaria Graeca omnia*, Leipzig

Waeren, B. L. van der (1949) 'Babylonian Astronomy II: The Thirty-six Stars', *Journal of Near-Eastern Studies*, 8, pp. 6–26

(1960) 'Greek Astronomical Calendars and their Relation to the Athenian Civil Calendar', *Journal of Hellenic Studies*, 80, pp. 168–80

(1984a) 'Greek Astronomical Calendars I: The Parapegma of Euctemon', *Archive for History of Exact Sciences*, 29, pp. 101–14

(1984b) 'Greek Astronomical Calendars II: Callippus and his Calendar', *Archive for History of Exact Sciences*, 29, pp. 115–24

(1984c) 'Greek Astronomical Calendars III: The Calendar of Dionysius', *Archive for History of Exact Sciences*, 29, pp. 125–30

(1985) 'Greek Astronomical Calendars IV: The Parapegma of the Egyptians and their "Perpetual Tables" ', *Archive for History of Exact Sciences*, 32.2, pp. 95–104

Wagner-Roser, S. (1987) 'Ein Römischer Steckkalender aus Bad Rappenau', *Fundberichte aus Baden-Württemberg*, 12, pp. 431–8

Walsh, P. G. (1970) *The Roman Novel*, Cambridge

Waszink, J. H., ed. (1962) *Timaeus, a Calcidio translatus commentarioque instructus*, London

Weidner, E. F. (1915) *Handbuch der babylonische Astronomie*, Leipzig

(1928–) 'Fixsterne', *Reallexikon der Assyriologie und vorderasiatischen Archäologie*, Berlin

(1935–6) 'Aus den Tagen eines assyrischen Schattenkönigs', *Archiv für Orient-forschung*, 10, pp. 1–52

Weinstock, S. (1948) 'A New Greek Calendar and Festivals of the Sun', *Journal of Roman Studies*, 38, pp. 37–42

Wells, R. (1996) 'Astronomy in Egypt', in C. Walker, ed., *Astronomy Before the Telescope*, London

Wenskus, O. (1986) 'Les vrais et les faux calendriers agricoles romains chez Caton, Varron et Columelle', *Histoire et mesure*, 1.3/4, pp. 107–18

(1990) *Astronomische Zeitangaben von Homer bis Theophrast*, Stuttgart

(1998) 'Columellas Bauernkalender zwischen Mündlichkeit und Schriftlichkeit', in W. Kullmann, J. Althoff, and M. Asper, eds., *Gattungen wissenschaftlicher Literatur in der Antike*, Tübingen

West, M. L. (1971) *Early Greek Philosophy and the Orient*, Oxford

ed. (1978) *Hesiod: Works and Days*, Oxford

White, K. D. (1970) *Roman Farming*, London

Wiegand, T. (1904) 'Dritter vorläufiger Bericht über die von den Königlichen Museen begonnenen Ausgrabungen in Milet', *Sitzungsberichte der königlich preussischen Akadamie der Wissenschaften*, Berlin

Wiegand, T., et al., eds. (1919–) *Milet: Ergebnisse der Ausgrabungen und Untersuchen seit dem Jahre 1899*, Berlin

Wilkinson, L. P. (1982) 'The Georgics', in E. J. Kenney and W. V. Clausen, eds., *The Cambridge History of Classical Literature II: Latin Literature*, Cambridge

Willis, J., ed. (1963) *Macrobius: Saturnalia*, Leipzig

Wilson, J. A. (1970) *Herodotus in Egypt*, Leiden

Winstedt, E. O., ed. (1928) *Cicero, Letters to Atticus*, London

Witte, J. de (1877) 'Les divinités des sept jours de la semaine', *Gazette archéologique*, 3, pp. 50–7, pls. 8–9

Wormell, D. E. W., E. H. Alton, and E. Courtney, eds. (1997) *P. Ovidii Nasonis fastorum libri sex*, Leipzig

Worp, K. A. (1991) 'Remarks on Weekdays in Late Antiquity', *Tyche*, 6, pp. 221–30

Wünsch, R., ed. (1898) *Ioannis Laurentii Lydi liber de mensibus*, Leipzig

Yates, F. A. (1964) *Giordano Bruno and the Hermetic Tradition*, London

Yoyotte, J. (1989) 'Le nom égyptien du "ministre de l'économie" ', *Comptes rendus des séances de l'Académie des inscriptions et belles-lettres*, Paris, pp. 73–88

Astrometeorological index

Abbreviations

Calendrical

Ad.	Adhār		N.	Nones
Aq.	Aquarius		Nis.	Nisān
Ar.	Aries		Pach.	Pachon
Art.	Artemesion		Pay.	Payni
Ath.	Athyr		Pham.	Phamenoth
Ay.	Ayyār		Phao.	Phaophi
Can.	Cancer		Phar.	Pharmouthi
Cap.	Capricorn		Pis.	Pisces
Ch.	Choiak		pr.	*pridie*
Ep.	Epiphi		Sag.	Sagittarius
epag.	epagomenal days		Scor.	Scorpio
Gem.	Gemini		Shub.	Shubāt
Haz.	Hazīrān		Tau.	Taurus
Id.	Ides		Tam.	Tammūz
Īl.	Īlūl		Ti. I	Tishrīn I
K.	Kalends		Ti. II	Tishrīn II
Kān. I	Kānūn I		Th.	Thoth
Kān. II	Kānūn II		Tyb.	Tybi
Lib.	Libra		Vir.	Virgo
Mech.	Mecheir		Xan.	Xanthicos
Mes.	Mesore			

Sources

AB.	al-Bīrūnī		M. I	Miletus I
Aet.	Aëtius		M. II	Miletus II
Ant.	Antiochus		Mad.	Madrid
Byz.	Byzantine season list		Marc.	C. Marcianus 335
CT.	Clodius Tuscus		Ov.	Ovid
DC.	Diocles of Carystus		Ox.	Oxford
DM.	Lydus, *De mensibus*		Par.	Paris
EP.	Eudoxus (Leptines) papyrus		Plin.	Pliny
Flor.	Florentinus season list		PS.	Polemius Silvius, *Fasti*
Iud.	Iudicia		Ptol.	Ptolemy
Gem.	Geminus		Quin.	Quintilius
Geo.	*Geoponica*		Ry.	P. Rylands 589
Hib.	P. Hibeh 27		Vind.	*C. Vindob. philos. 108*

Note: source abbreviations appear in italics throughout to distinguish them from calendrical abbreviations.

Air

CT. 2 Oct., 27 Dec.; *Aet.* Intro.; *Quin.* Intro.; *Par.* 1 Dec.; *Iud.* 6 Nov., 7 May; *AB.* 7 Ti. II, 29 Kān. I, 19 Kān. II, 21 Shub., 1, 17 Ad., 28 Nis., 18, 27 Tam., 30 Īl.

bad: *Ptol.* 5, 6, 7, 10, 19, 20 Th., 4 Phao., 16, 28 Ep., 29 Mes.; *AB.* 10, 22 Tam., 22 Āb

blustery: *CT.* 2 Feb.; *Mad.* 2 Feb.

change in: *CT.* 3 Jan., 26 Apr., 8 Jul., 22 Sept., 8, 20 Oct., 1 Nov.; *Mad.* 4 May, 25 Jun., 19, 26 Sept., 1 Oct.; *Ox.* 6, 19 May; *Aet.* Intro., 4, 6, 19 May, 25 Jun., 14, 19, 25 Sept., 6 Oct.; *Quin.* Intro., 4 May, 13 K. Jun., 6 K. Jul., 14 Sept., 12 K. Oct.; *Par.* 5 Oct., 2 Nov., 3 Jan., 4, 18 May, 23 Jun.; *Iud.* 1, 7, 14, 19, 25 Sept., 6 Oct., 21, 27 Nov., 23 Dec., 29 Jan., 25 Feb., 27 Apr., 1, 4, 19 May, 2, 5, 9, 25 Jun., 4 Jul., 15 Aug.; *AB.* 25 Ti. I, 3, 18 Kān. II, 24 Shub., 8, 13 Āb, 4 Īl.

clear: *Ox.* 22 May; *AB.* 2 Ti. II

cloudy: *CT.* 8 Aug., 25 Sept., 28 Nov.; *Par.* 24 Sept.

cold: *AB.* 22 Ti. I

dense: *AB.* 7 Ti. II, 20 Āb, 2, 3, 4, 7, 16, 17 Īl.

dewy: *CT.* 16 Oct.

disturbed (confused/unsettled/unstable): *Ptol.* 7 Th., 23 Tyb., 15 Phar.; *CT.* 5 Feb., 26 Mar., 21 May, 2, 12, 25 Jun., 1, 3, 30 Jul., 25 Sept., 18 Nov., 1, 15 Dec.; *Mad.* 23 Mar., 23 May, 7, 10, 17, 27 Oct., 13, 21 Nov., 21 Dec., 17, 22, 26 Jan., 25 Feb.; *Ox.* 22 Feb., 23 Mar.; *Aet.* 23 Mar., 21 Apr., 24 May, 19, 25 Jul., 17, 23 Oct., 21 Nov., 1 Dec., 5, 25 Jan., 22 Feb.; *Quin.* K. Apr., 21 Art., 8 K. Jun., 25 Jul., 23 Sept., 15, 9 K. Nov., 10 K., K. Dec., 7 K. Mar.; *Par.* 23, 24, 25, 30 Sept., 17, 22 Oct., 15 Nov., 1 Dec., 11, 25 Jan., 21, 31 Mar., 24 May; *Iud.* 7, 23 Oct., 1, 21 Dec., 25 Jan., 21 Mar., 21 Apr., 24 May, 18, 25 Jul.; *AB.* 27 Ti. II, 6 Ad.

hot: *Col.* 7 K. Aug.

misty: *Ptol.* 16 Mes.; *Mad.* 13 Aug.; *Ox.* 1 May, 5 Aug.

mixed: *Ptol.* 28 Phao.

mixed, bad: *Ptol.* 13 Th., 6, 7, 25 Ath., 1, 29, 30 Ch., 16, 25 Phar., 6 Ep.

oppressive: *AB.* 5, 8 Āb

settled: *Mad.* 7 May; *Aet.* 7 May, 6 Nov.; *Quin.* 7 Id. May, 3 Id. Nov.; *Par.* 6 Nov.

south: *CT.* 2 Aug.; *Mad.* 24 May

still: *AB.* 5, 8, 9, 10, 13 Āb

stormy: *Gem.* 13 Scor.; *CT.* 15 Feb., 31 May, 24 Dec.

thundery: *CT.* 11 Jun.

turbid: *AB.* 1, 2 Ti. I, 13 Ti. II, 9 Āb

unmixed: *Ptol.* 27 Phao.; *AB.* 20 Nis., 30 Haz., 10 Īl.

wintry: *Ptol.* 27 Ath., 10, 12, 13, 15, 24 Ch., 20 Tyb.; *AB.* 2, 4, 5, 8, 27, 29, 31 Ti. I, 5, 6, 8, 10, 12, 13, 14, 16, 17, 19, 20, 21, 22, 23, 29 Ti. II, 1, 2, 3, 5, 6, 9, 10, 11, 12, 14, 16, 20, 24, 25, 28, 30, 31 Kān. I, 7, 8, 13, 18, 19, 20, 21, 24, 26 Kān. II, 4, 7, 11, 15, 25 Shub., 5, 11, 18 Ad., 22, 29 Nis., 23 Tam., 5 Īl.

Alferat

setting: *Iud.* 25 Sept.

Anaximander

Plin. NH xviii.57

Andromeda (see also: Pegasus and Andromeda, star shared by)

rising: *Ant.* 30 Aug.; *CT.* 31 Aug., 1 Sept.; *DM.* 1 Sept.; *Par.* 2 Sept.

rising, evening: *Col.* pr. K. Sept.; *Mad.* 2 Sept.; *Ox.* 1 Sept.

rising, morning: *M. I*

Anguifer (see also: Ophiucus)

setting, morning: *Col.* 11 K. Jul.

Antares

appearance: *Par.* 9 Nov.

disappearance: *Ptol.* 22, 29 Th., 6, 12, 17 Phao.

rising: *Ptol.* 25, 26, 27, 28, 29 Ath.

rising, evening: *Ptol.* 3, 4 Pach.

setting, morning: *Ptol.* 18, 19, 20, 21 Pach.; *Ant.* 17 May

Antevendemor

appearance: *Iud.* 28 Aug.

Antichanus

appearance: *Iud.* 4 Jul.

Apollo (see: Twin, northern)

Aquarius

Ov. 16 K., N. Feb., 15 K. Mar.; *Col.* 17 K. Feb.;
Plin. HN xviii.64 (16 K. Feb.); *CT.* 16
Jan.; *DM.* 11 K. Feb.; *Ox.* end Feb.; *Par.* 8
Mar.; *Ry.*; *M. I.*; *Flor.* 7 Id. Feb.
rising: *Col.* 15 K. Feb.; *Ant.* 22 Jan.; *CT.* 20, 21
Jan., 9, 23 Feb.; *Mad.* 24 Feb.; *DM.* 13, 12
K. Feb.; *Par.* 23 Feb.; *M. I.*
setting: *Col.* 8 K. Aug.; *Ant.* 21 Jul.; *CT.* 25 Jul.

Aquarius, bright star in

disappearance: *Ant.* 21 Jun.
rising: *Ant.* 21 Aug.
rising, evening: *Ptol.* 22, 24, 25, 26, 28 Tyb.
setting: *Mad.* 1 Aug.
setting, morning: *Ptol.* 14, 16, 19, 21, 23 Tyb.;
Ant. 15 Jan.

Aquarius, middle of

rising: *Gem.* 17 Aq.; *Col.* N. Feb.; *CT.* 5 Feb., 8
Aug.; *Par.* 5 Feb.
setting: *Col.* 7 Id. Aug.; *Ant.* 7 Aug.; *CT.* 7 Aug.

Aquila (see also: Ectos)

Hib. 4 Pay.; *Par.* 1 Jan.
appearance: *Iud.* 21 Dec.
appearance, evening: *Iud.* 2 Jun.
disappearance, evening: *CT.* 31 Oct.
rising: *Gem.* 15 Sag.; *Ov.* K. Jun.; *Col.* K, 4 N.
Jun., 6 K. Aug.; *Plin.* HN xviii.68 (13 K.

Jan.); *Ant.* 23 May, 1 Jun.; *CT.* 23, 27 May,
2, 5 Jun., 7, 20 Dec.; *Mad.* 25 Mar., 23
May, 19 Jun.; *Par.* 2 Jun.; *Vind.*
rising, acronychal (see also: 'rising, evening'):
Hib. 3 Phar.; *Gem.* 7 Gem.
rising, evening: *M. II*; *Gem.* 31 Tau.; *Plin.* HN
xviii.67 (4 N. Jun.), 69 (4 N. Jun.); *CT.* 22
Dec.; *?Mad.* 2 Jun.; *Ox.* 2 Jun.; *Aet.* 2 Jun.;
Quin. 3 N. Jun.; *Par.* 1 Jun.; *Byz.*
rising, morning: *Gem.* 16, 26 Sag.; *Col.* 7 Id.
Dec.; *Ant.* 9 Dec.; *M. I*; *Byz.*
setting: *Col.* 3 K. Aug.; *Plin.* HN xviii.69 (13 K.
Aug.); *CT.* 1, 5 Jan., 23, 26 Jul., 2 Aug., 29
Dec.; *Ox.* 20 Jul.; *Par.* 3, 5, 6, 22 Jan.; *Iud.*
25 Jan.; *Vind.*
setting, evening: *Gem.* 7 Cap., 25 Tau.; *Col.* 4 K.
Jan.; *Plin.* HN xviii.64 (3 K. Jan.); *CT.* 6
Jan.; *Mad.* 9 Jan.; *Aet.* 5 Jan.; *Quin.* 8 Id.
Jan (5 Jan.)
setting, morning: *Gem.* 28 Can., 5 Leo; *Plin.*
HN xviii.68 (13, 10 K. Aug.); *CT.* 30 Jul.,
1 Aug.; *Mad.* 25 Jul.; *Aet.* 25 Jul.; *Quin.* 25
Jul., 3 N. Aug. (25 Aug.)

Aquila, bright star in

disappearance: *Ptol.* 27 Ch.
rising: *Ptol.* 3 Tyb.; *Ant.* 30 Dec.
rising, evening: *Ptol.* 24, 27 Pach., 2, 6, 11 Pay.;
Ant. 28 May
rising, morning: *Ptol.* 23, 25, 27, 30 Ch.
setting, evening: *Ptol.* 30 Ch., 4, 7, 9 Tyb.
setting, morning: *Ptol.* 27 Ep., 2, 6, 10 Mes.

Aquila, middle of

setting: *Par.* 3 Jan.

Arabs

AB. Intro., 15, 21, 28 Shub.

Arctophylax (see also: Boötes)

setting: *Ov.* 3 N. Mar.

Arcturus (see also: Septemtrion)

Plin. HN xvIII.74; *Par.* 2 Nov., 11 Feb.; *Byz.*
appearance: *CT.* 8 Sept.; *Par.* 10 Sept.; *Iud.* 23
    Feb.; *Vind.*
appearance, morning: *Iud.* 14 Sept.
bright: *Par.* 25 Feb.
disappearance: *Ov.* 7 Id. Jun.
rising: *Gem.* 10, 17 Vir.; *Col.* N. Sept., 15 K.
    Oct.; *Ant.* 27 Feb., 5 Mar., 9 Sept.; *CT.* 11,
    25 Feb., 10 Mar., 4, 12, 13, 20 Sept.; *Mad.*
    24 Jan.; *DM.* 18 K. Oct.; *Ox.* 12 Mar.; *Aet.*
    14 Sept.; *Quin.* 14 Sept.; *Par.* 3, 8 Sept., 3
    Mar.; *Vind.*; *EP.*; *Geo.* 15 Sept.
rising, acronychal (see also: 'rising, evening'):
    Hib. 16 Ch.; *Gem.* 4. Pis.
rising, evening: *Gem.* 12 Pis.; *Col.* 9 K.
    Mar.; *Plin.* HN xvIII.65 (7 K. Mar.); *Ptol.*
    1, 5, 8, 12, 15 Pham.; *CT.* 26 Feb., 4 Mar.;
    Mad. 5 Mar., 25 Feb.; *Ox.* 25 Feb.; *Aet.* 25
    Feb.; *Quin.* 4 K. Mar.; *Geo.* 26 Feb.;
    Byz.
rising, morning: *Gem.* 19 Vir.; *Plin.* HN
    xvIII.74 (N. Sept.); *Ptol.* 23, 26, 29 Th., 3,
    6 Phao.; *Ant.* 28 Sept.; *CT.* 3 Mar.; *Par.* 9
    Sept.; *Byz.*
setting: *M. II*; *Col.* 7 Id. Jun., 6 K. Sept.; *Plin.*
    HN xvIII.69 (5 Id. May); *Ant.* 3 Dec.; *CT.*
    21, 26 May, 9 Jun., 29 Oct., 1 Nov.; *Mad.* 6
    Jun.; *Ox.* 1 Jun.; *Par.* 28 Oct., 21 Feb., 21
    May; *Iud.* 5 Jun.; *Vind.*
setting, acronychal: *Hib.* 4 epag.; *Gem.*
    8 Scor.
setting, evening: *Gem.* 5 Scor.; *Col.* 4 K. Nov.;
    Plin. HN xvIII.74 (pr. K. Nov.), 4 N. Nov.;
    Ptol. 18, 26 Phao., 4, 12, 21 Ath.; *CT.*
    21 Feb.
setting, morning: *M. II*; *Gem.* 32 Tau., 13 Gem.;
    Col. 11, 10 K. Jun.; *Plin.* HN xvIII.67 (3
    Id. May, 7 Id. Jun.), *post*-6 Id. May; *Ptol.*
    16, 26 Pach., 7, 18, 30 Pay.; *Ant.* 21 May;
    CT. 2 Mar., 8 Jun., 6 Nov.; *Ox.* 6 Jun.; *Aet.*
    6 Jun.; *Quin.* 7 Id. Jun.; *Par.* 6 Nov., 8 Jun.;
    Geo. 7 Jun.
visibility: *Gem.* 20 Vir.

Arcturus, middle of

appearance, morning: *CT.* 19 Sept.
setting: *Plin.* HN xvIII.68 (8 Id. Aug.)
rising, evening: *Plin.* HN xvIII.74 (pr. Id. Sept.)

Ares, feast of (see also: Mars)

Ox. 1 Mar.

Argo

rising: *Col.* pr. Id. Mar.
rising, evening: *CT.* 13 Mar.
setting: *Col.* 10 K. Oct.; *CT.* 22 Sept.; *Par.* 21
    Sept.

Ariadne, crown of (see also: Corona)

setting: *Mad.* 30 Jun.

Aries

Hib.?5 Tyb.; *Ov.* 10 K. Apr., 12 K. May; *Col.* 16
    K. Apr.; *Plin.* HN xvIII.59; *CT.* 17 Mar.;
    Ox. end April; *Par.* 17 Mar.; *Iud.* 3 Jul.; *AB.*
    18 Tam.; *Ry.*; *M. I*
disappearance: *Ov.* 7 K. May
rising: *Gem.* 3 Ar., 1 Tau.; *Col.* 10 K. Apr.; *CT.*
    22 Mar.; *Par.* 22 Mar.
rising, morning: *CT.* 26 Mar.
setting: *Gem.* 1 Lib.; *Col.* 11 K. Oct.

Aries, middle of

setting: *Col.* pr. N. Oct.; *CT.* 6 Oct.; *DM.* pr. N.
    Oct

Auriga

appearance: *Par.* 28 Sept.
rising: *Col.* 7 Id. Sept.
rising, morning: *Col.* 3 K. May, 8, 7, 6 K. Jun.
setting: *CT.* 3 Oct.; *Mad.* 20 Nov.; *Par.* 3 Oct.
setting, morning: *Col.* 4 N. Oct., 10 K. Jan.,
    Plin. HN xvIII.74 (5 N. Oct.)

Auriga, beak of

visibility: *Ov.* 8 K. Jun.

Auriga, bright star in (see also Capella)

rising: *Gem.* 20 Vir.

Auriga, star in the rear shoulder of

disappearance: *Ptol.* 23, 25 Pach.
rising: *Ptol.* 1 Pay.
rising, evening: *Ptol.* 21 Th., 8, 20, 28 Phao.;
    Ant. 19 Sept., 17 Oct.
rising, morning: *Ptol.* 6, 18, 24 Pach.
setting, evening: *Ptol.* 28 Pach., 1, 5 Pay.
setting, morning: *Ptol.* 13, 18, 23, 28 Ch.,
    5 Tyb.

Auriga, star on

rising: *Mad.* 20 Oct.

Autumn

Gem. 20 Vir.; *Col.* pr. Id. Aug.; *Plin.* HN
    xviii.59, 68 (3 Id. Aug.), 69 (14 K. Sept.),
    74; *Ptol.* 18, 19, 20, 21 Th., 19 Mes.; *PS.* 6
    K. Oct.; *CT.* 11, 17 Aug., 16 Sept., 15 Oct.;
    DM. pr. Id. Oct.; *AB.* 11 Kān. I, 6 Kān. II,
    Shub. Intro., 17 Ad., 18 Tam., 11 Āb, 16 Īl.;
    Vind.; *Flor.* 6 Id. Aug.; *Byz.*

Black (star?)

rising: *Mad.* 24 Nov.

Boötes (see also: Arctophlyax)

Ov. 3 Id. Feb.
appearance, morning: *Iud.* 1 Sept.
disappearance: *Ov.* 7 K. Jun.
setting: *Ov.* 3 N. Mar.

Breeze

Ov. N. Feb.
north-east: *Col.* N. Mar.; *PS.* pr. Id. Mar.

Caesar

Plin. HN xviii.57, 64 (pr. N. Jan.), 65 (14 K.
    Mar., Id. Mar.), 66 (K., N., 6 Id. Apr., 15, 8
    K., 6 N. May), 67 (11 K., 4 N. Jun., 11 K.
    Jul.), 68 (6 K. Jul., 10 K. Aug.), 74 (pr. Id.
    Aug., 11 K., 5 Id. Sept., 14, 8 K., 5, 4 N.
    Oct., pr. K. Nov.); *Ptol.* 3, 8, 14, 20 Th., 6,
    16, 26 Phao., 4, 6, 19, 20, 21, 25 Ath., 5, 7,
    8, 12, 13, 20, 28, 29 Ch., 15, 22, 24 Tyb.,
    12, 15 Mech., 1, 9, 24, 27 Pham., 18, 22, 27
    Phar., 4, 17, 20, 21, 26 Pach., 5, 10, 11 Pay.,
    9, 20, 27 Ep., 1, 7, 10, 20, 21, 23, 29 Mes., 2
    epag.; *PS.* Intro. Jul., 8 Id. Jul.; *DM.* 11 K.
    Feb., 12 K. Oct.; *AB.* 23, 31 Ti. I, 14, 16, 17
    Ti. II, 1, 3, 16, 24 Kān. I, 10, 17, 19 Kān. II,
    6, 25 Shub., 5, 20, 23 Ad., 13, 22, 27, 29
    Nis., 12, 15, 16, 21, 30 Ay., 4 Haz., 3, 14,
    25, 31 Tam., 3, 9, 13, 16, 22, 31 Āb, 5, 11
    Īl.

Callaneus

M. II

Callippus

Gem. 1, 27, 30 Can., 12, 29 Leo, 5, 17, 24 Vir., 1,
    5, 17, 28 Lib., 4, 9, 16, 28 Scor., 7, 16 Sag.,
    1, 15, 27 Cap., 2, 17 Aq., 2, 17, 30 Pis., 1, 3,
    23 Ar., 1, 4, 13, 32 Tau., 2 Gem.; *Plin.* HN
    xviii.71 (4 K. Oct.); *Ptol.* 4, 7, 16, 23 Th.,
    4, 16, 27, 28 Phao., 4, 6, 8, 15, 27 Ath., 5, 6,
    12, 13, 14, 15, 29 Ch., 13, 25, 29 Tyb., 14,
    17, 29 Mech., 14, 16, 21, 28, 30 Pham., 1,
    2, 21, 29 Phar., 2, 19 Pach., 1, 2 Pay., 4, 11,
    29, 30 Ep., 2 Mes., 4 epag.; *DM.* 13 K. Feb.,
    15 K. Nov.; *AB.* 1, 2, 13, 24, 25, 31 Ti. I, 11
    Ti. II, 1, 9, 10, 11, 29 Kān. I, 8, 24 Kān. II,
    8, 23 Shub., 12, 16, 24, 26, 27 Ad., 1, 24,
    26, 30 Nis., 10, 14, 24, 26, 27 Ay., 5, 6, 21
    Tam., 8, 11 Āb, 1, 4, 13 Īl.;
EP.

Cancer

Hib.?3 Phar.; *Ov.* 13 K. Jul.; *Col.* 13 K. Jul.; *Plin.*
    HN xviii.69, 65; *CT.* 19 Jun.; *Ox.* end July;
    AB. 7 Ti. II, 6 Ay.; *Ry.*

appearance: *Iud.* 5 Mar.
disappearance: *Ov.* 3 N. Jan.; *CT.* 24 Jul.
rising: *Gem.* 1, 27 Can.; *Plin.* HN xviii.65; *CT.* 15 May, 23 Jul.
setting: *Gem.* 27 Cap.; *Col.* 16 K. Feb., 3 N. Jan.; *CT.* 3, 22 Jan.; *DM.* 13 K. Feb.

Cancer, arms of

Ov. 3 N. Jan.

Cancer, middle of

rising: *CT.* 5 Jul.
setting: *Col.* pr. N. Jul.; *CT.* 2, 20 Jan.

Canopus

disappearance: *Ptol.* 2, 20 Phar., 5 Pach.; *Ant.* 14 Apr.
rising: *Ptol.* 14 Th., 2 epag.
rising, evening: *Ptol.* 22 Tyb., 6, 23 Mech.; *Ant.* 31 Jan.
setting, morning: *Ptol.* 24 Phao., 10, 23 Ath.; *Ant.* 8 Nov.

Capella (see also: Auriga, bright star in)

Ov. K. May; *Ant.* 2 Oct.
appearance: *Par.* 15 Sept.
disappearance, evening: *Mad.* 24 May; *Ox.* 24 May; *Aet.* 24 May; *Quin.* 8 K. Jun.; *Par.* 24 May
rising: *Gem.* 20 Vir.; *Ant.* 11 Sept.; *CT.* 7, 15, 30 Sept., 19 Dec.; *Par.* 7, 30 Sept., 18 Dec.; *Vind.*; *Byz.*
rising, acronychal (see also: 'rising, evening'): *Gem.* 4 Lib.
rising, evening: *Plin.* HN xviii.74 (5 Id. Sept.); *Ptol.* 3, 23 Th., 7, 21 Phao., 10 Mes.; *Ant.* 18 May; *Mad.* 6 Sept.; *Ox.* 6 Sept.; *Aet.* 7 Sept.; *Quin.* 7 Sept., 4 Id. Sept.; *Par.* 13 Mar.
rising, morning: *Gem.* 8, 9 Tau.; *Plin.* HN xviii.66 (8 Id. May), 74 (4 K. Oct.); *Ptol.*

29 Pham., 18 Phar., 2, 9, 12 Pach.; *Ox.* 6 May; *Aet.* 6 May; *Quin.* N. May (5 May)
setting: *CT.* 14 Dec.; *Ox.* 5 Dec.; *Vind.*
setting, acronychal: *M. II*
setting, evening: *M. II*; *Plin.* HN xviii.67; *Ptol.* 17, 20, 24, 28 Pach., 5 Pay.; *Ant.* 4 May; *CT.* 26 Dec.; *Mad.* 8 May, 14 Jun., 13 Dec.; *Vind.*
setting, morning: *Gem.* 23 Sag.; *Ptol.* 5, 9, 14, 19, 26 Ch.; *Ant.* 7 Dec.; *CT.* 23, 25 Dec.; *Aet.* 21 Dec.; *Quin.* 11 K. Jan. (21 Jan.); *Par.* 11, 23 Dec.

Capra

rising: *Iud.* 23 Dec.

Capricorn

Ov. 16 K. Feb.; *Col.* 16 K. Jan.; *Plin.* HN xviii.69 (8 K. Jan.); *CT.* 8 Jan., 17 Dec.; *Par.* 17 Dec.; *AB.* 20 Ti. II, 18 Tam.; *Ry.*
rising: *Gem.* 15 Cap.; *CT.* 20 Dec.; *Mad.* 14 Dec.; *Ox.* 21 Dec.

Capricorn, middle of

setting: *Col.* 8 Id. July

Cassiopeia

setting: *Col.* 3 K., pr. K. Nov.; *CT.* 30 Oct.

Castor (see: Twin[s])

Centaurus

Ov. 5, 3 N. May; *Par.* 24 Sept.
appearance: *Col.* 5 N. May; *CT.* 3 May; *Par.* 3 May
rising: *Par.* 23 Sept.
rising, morning: *Col.* 9 K. Oct., 8 K. Nov.
setting: *CT.* 24 Sept.
setting, morning: *CT.* 25 Oct.; *M. I*

Centaurus, star in right front hoof of

disappearance: *Ptol.* 27 Pay., 29 Ep., 23 Mes.
rising: *Ptol.* 24 Ath., 6, 23 Ch.
rising, evening: *Ptol.* 6, 17 Pach., 5 Pay.
setting, morning: *Ptol.* 22 Tyb., 19 Mech., 11 Pham.

Centaurus, star on right arm of

rising, evening: *Ant.* 13 May
setting, morning: *Ant.* 13 Feb.

Cepheus

rising: *Ant.* 9 Jul.; *CT.* 8 Jul.; *Ox.* 24 Feb.
rising, evening: *Col.* 7 Id. Jul.
setting: *Par.* 8 Jul.

Cetus

setting, acronychal: *M. I*

Chaldaeans

Col. 9 K. Jan; *Plin.* HN xvIII.57, 66 (N. Apr.), 68 (4 N. Jul.)

Cicero

Plin. HN xvIII.60; *PS.* 3 N. Jan.

Cion

Iud. 28 Aug.
appearance: *Iud.* 1 Dec., 18 Jul.

Claw, northern, bright star in

disappearance: *Ptol.* 2, 4, 6, 7, 8 Phao.
rising: *Ptol.* 3, 4, 5 Ath.
rising, evening: *Ptol.* 4, 8, 9, 10, 11 Phar.
setting, morning: *Ptol.* 10, 14, 25 Pach., 1 Pay.

Claw, southern, bright star in

disappearance: *Ptol.* 6, 12, 17, 21, 25 Th.
rising, evening: *Ptol.* 6, 7 Phar.
setting, morning: *Ptol.* 27, 29 Phar., 1, 5, 8 Pach.

Claws (see also: Libra)

setting: *Gem.* 23 Ar.

Cloud

Ov. N. Jan.; *Plin.* HN xvIII.60, 66; *CT.* 9, 30 Jan., 21 Feb., 22 Mar., 1, 2 Apr., 8 Aug., 25 Sept., 18 Oct., 6, 28 Nov.; *Par.* 24 Sept., 30 Jan.; *AB.* 7 Ti. II, 17 Kān. I, 18 Tam., 16 Īl.

Conon

Plin. HN xvIII.74 (4 K. Oct.); *Ptol.* 5, 16, 30 Th., 6, 15, 27 Ath., 4, 7, 29 Ch., 29 Mech., 29 Pham., 4, 8, 9, 22 Phar., 18 Pach., 2 Mes.; *AB.* 1, 11, 23, 29 Ti. II, 3 Kān. I, 23 Shub., 25 Ad., 3, 17 Nis., 2, 13 Īl.

Corona (see also: Ariadne, crown of; Stephanus)

Par. 1 Jan.; *Byz.*
appearance: *Iud.* 6 Oct.
rising: *Gem.* 7 Lib.; *Col.* 3 N. Oct.; *Plin.* HN xvIII.60, 74 (4 N., Id. Oct.); *Ant.* 9 Oct.; *CT.* 5, 8, 13 Oct.; *Mad.* 2, 8 Oct., 27 Nov.; *Ox.* 5 Oct.; *Par.* 4, 8 Oct.; *Vind.*
rising, acronychal (see also: 'rising, evening'): *Hib.* 26 Ch; *Gem.* 21 Pis.
rising, evening: *Mad.* 18 Mar.; *Byz.*
rising, morning: *Hib.* 4 Mes.; *Col.* 3 Id., pr. Id. Oct.; *Plin.* HN xvIII.74 (6 N. Oct.); *CT.* 11 Oct.; *Aet.* 6 Oct.; *Quin.* 9 N. Oct.; *Par.* 5, 11 Oct.; *Geo.* 4 Oct.; *Byz.*
setting: *Gem.* 10 Leo; *CT.* 1 Jan., 4 Jul., 5 Aug.; *Ox.* 24 Jun.; *Aet.* 27 Nov.; *Quin.* 4 K. Dec. (27 Nov.); *Par.* 1 Aug.; *Iud.* 27 Nov.
setting, acronychal: *Gem.* 9 Cap.

setting, morning: *Gem.* 16 Can.; *Col.* 4 N. Jul.;
Plin. HN xviii.68 (4 N. Jul.); *CT.* 7 Mar., 7
Jul.; *DM.* K. Jan., N. Mar.; *Par.* 7 Jul.
visibility: *Ov.* 8 Id. Mar.

Corona, bright star in

rising: *Col.* 8 Id. Oct.; *Plin.* HN xviii.74 (8 Id.
Oct.); *Ptol.* 2 Phar.; *Ant.* 5 Oct.
rising, evening: *Ptol.* 9, 14, 20, 26 Pham.
rising, morning: *Ptol.* 6, 10, 16, 22, 27 Phao.
setting, evening: *Ptol.* 15, 23 Ath., 2, 10,
19 Ch.
setting, morning: *Ptol.* 16, 27 Pay., 7, 18,
28 Ep.

Corona, star of

rising: *CT.* 8 Oct.

Corvus

Ov. 16 K., Id. Mar., 17 K. Apr.
setting: *Par.* 5 Jan.

Crane

CT. 5 Aug.

Crater (see also: Cup)

Ov. 16 K., Id. Mar., 17 K. Apr.
appearance, morning: *Col.* 13 K. Oct.
rising, evening: *Col.* 16 K. Mar.

Crito

Plin. HN xviii.74 (4 K. Oct)

Cup (see also: Crater)

appearance: *CT.* 19 Sept.; *Par.* 20 Sept.
rising: *Par.* 14 Feb.
rising, evening: *CT.* 14 Feb.

Cygnus

rising, evening: *Mad.* 16 May
setting, acronychal: *M. I*
setting, evening: *Mad.* 12 Mar.

Cygnus, bright star in

rising, evening: *Ptol.* 26 Phar., 8, 18, 30 Pach.,
10 Pay.; *Ant.* 6 Feb.
rising, morning: *Ptol.* 27 Ath., 7, 16 Ch., 4 Tyb.;
Ant. 21 Dec.
setting: *Ox.* 3 Feb.
setting, evening: *Ptol.* 4, 12, 21, 29 Mech., 7
Pham.
setting, morning: *Ptol.* 9, 17, 25 Th., 3 Phao., 5
epag.; *Ant.* 5 Sept.

Delfinus

appearance, evening: *Iud.* 9 Jun.

Delphinus

Ov. 5 Id. Jan., 4 Id. Jun.; *Quin.* 19 Aug.
appearance: *Iud.* 4 Jan.
appearance, evening: *Iud.* 28 Jan.
disappearance: *Ov.* 3 N. Feb.
rising: *Gem.* 2 Cap.; *Ant.* 5 Jan., 26 Dec.; *CT.* 5
Jan., 9, 14 Jun., 26 Aug., 30, 31 Dec.; *Mad.*
9 Jun., 13 Aug.; *Ox.* 5 Aug.; *Aet.* 4 Jan.;
Quin. N. Jan.; *Par.* 30, 31 Dec., 19 Jan.;
Vind.; *Geo.* 1 Jan.
rising, acronychal (see also: 'rising, evening'):
Hib. 11 Phar.; *Gem.* 18 Gem.
rising, evening: *Col.* 4 Id. Jun.; *Plin.* HN
xviii.67 (4 Id. Jun.); *Ox.* 9 Jun.; *Aet.* 9
Jun.; *Quin.* 4 Id. Jun.; *Par.* 9 Jun.; *Byz.*
rising, morning: *Col.* 6 K. Jan.; *Plin.* HN
xviii.64 (pr. N. Jan.); *CT.* 4 Jan., 27 Dec.;
Par. 4 Jan., 1 Aug.; *Byz.*
setting: *Col.* 3 K. Feb., Id. Aug.; *Plin.* HN
xviii.74 (pr. Id. Aug.); *Ant.* 17 Aug.; *CT.*
18, 29 Jan., 4 Feb., 13, 19 Aug.; *DM.* 15 K.
Feb.; *Par.* 8, 29 Jan.; *Iud.* 19 Aug.;
Vind.; *Byz.*
setting, acronychal: *Gem.* 4 Aq.
setting, evening: *Gem.* 27 Cap.; *Plin.* HN
xviii.64; *Mad.* 28 Jan.; *Aet.* 28 Jan.; *Quin.*
27 Jan. (4 K. Feb.); *Par.* 27 Jan., 4 Feb.; *Byz.*

4, 6, 7, 8, 9, 10, 12, 13, 14, 17, 18, 19, 21, 24, 26, 27, 28 Ay., 1, 2, 3, 5, 9, 13, 15, 17, 18, 19, 20, 25, 26, 30 Haz., 3, 4, 5, 6, 9, 10, 12, 14, 16, 17, 18, 19, 20, 22, 23, 25, 29, 30 Tam., 5, 8, 9, 11, 12, 19, 20, 22, 23, 25, 28 Āb, 3, 5, 6, 8, 15, 18, 19, 25 Īl.

Elevation of the Holy Cross, star of

Marc. 14 Sept.

Equinox, autumnal

Hib. 23 Ep.; *Gem.* 1 Lib.; *Col.* 8, 7, 6 K. Oct.; *Plin.* HN xviii.62, 59, 60; *Ptol.* 28 Th.; *PS.* 8 K. Oct.; *Ant.* 25 Sept.; *CT.* 21 Sept.; *Mad.* 26 Sept.; *Aet.* 25 Sept.; *Quin.* 6 K. Oct.; *Par.* 19 Sept.; *Vind.*; *DC.*; *EP.*; *Flor.* 8/6 K. Oct.; *Byz.*

Equinox, vernal

Hib. 20 Tyb.; *Gem.* 22 Pis., 1, 6 Ar.; *Col.* 9, 8 K. Apr.; *Plin.* HN xviii.59, 65, 66, 69; *Ptol.* 26, 29 Pham.; *PS.* 8 K. Apr.; *Ant.* 22 Mar.; *CT.* 19, 24, 27 Mar.; *Mad.* 21 Mar.; *DM.* 8 K. Apr.; *Ox.* 23 Mar.; *Aet.* 23 Mar.; *Quin.* 9 K. Apr.; *Vind.*; *DC.*; *EP.*; *Flor.* 9/8 K. Apr.; *Byz.*

Eridanus (see also: Esion)

appearance: *Ox.* 19 Feb.; *Iud.* 18 Mar.
appearance, morning: *Iud.* 7 May
setting: *Iud.* 21 Dec.

Eridanus and foot of Orion, star shared by

disappearance: *Ptol.* 17, 21, 24, 28 Phar., 3 Pach.
rising: *Ptol.* 28 Pay., 5, 12, 18, 24 Ep.
rising, evening: *Ptol.* 2, 7, 12, 16, 21 Ch.
setting, morning: *Ptol.* 9, 12, 14, 17, 20 Ath.

Eridanus, rearmost star of

appearance: *Iud.* 18 Mar.
disappearance: *Ptol.* 12, 25 Mech., 6, 16 Pham.
rising: *Ptol.* 21 Pay., 6, 22 Ep., 11 Mes.; *Ant.* 25 Feb., 3 Mar.
rising, evening: *Ptol.* 26 Ath., 9, 24 Ch., 13 Tyb.
setting: *Iud.* 21 Dec.
setting, morning: *Ptol.* 4, 17, 27 Th., 6 Phao.

Erifi

appearance, evening: *Iud.* 7 Oct.
setting: *Iud.* 10 Dec.

Esion

setting: *Iud.* 21 Dec.

Euctemon

M. II; *Gem.* 13, 27, 28 Can., 1, 14, 17 Leo, 10, 20 Vir., 1, 3, 5, 7, 30 Lib., 5, 10, 15, 27 Scor., 7, 10, 15, 19 Sag., 1, 2, 7, 14, 16, 27 Cap., 3, 17, 25 Aq., 12, 14, 22, 29 Pis., 1, 10, 23 Ar., 2, 4, 8, 13, 25, 30, 31, 32 Tau., 24 Gem.; *Plin.* HN xviii.57; *Ptol.* 7, 18, 29, 30 Th., 2, 4, 5, 27, 28 Phao., 3, 4, 7, 8, 13, 14, 16, 25 Ath., 5, 6, 8, 12, 13, 14 Ch., 3, 4, 11, 13, 23, 24, 25, 29 Tyb., 21, 28 Mech., 8, 12, 13, 17, 18, 19, 28 Pham., 1, 15, 20, 21 Phar., 1, 5, 15, 29, 30 Pach., 1 Pay., 15, 27, 28, 29 Ep., 3 Mes.; *DM.* 13 K. Feb., 15 K. Apr., pr. Id. Oct., 3 N. Dec.; *AB.* 1, 2, 25, 30, 31 Ti. I, 3, 4, 8, 10, 21 Ti. II, 2, 9, 10, 29, 31 Kān I., 8, 18, 20, 24 Kān II, 15, 22 Shub., 4, 8, 9, 13, 14, 15, 24 Ad., 1, 10, 15, 16, 26, 30 Nis., 10, 14, 24, 25, 26, 27 Ay., 9, 21, 22, 27 Tam., 9, 11, 12 Āb, 4, 29 Īl.; *EP.*

Eudoxus

M. II.; *Gem.* 9, 11, 27, 31 Can., 5, 10, 16, 18, 22, 29 Leo; 5, 10, 19 Vir., 4, 8, 10, 12, 17, 19, 22, 29 Lib., 8, 12, 14, 18, 19, 21, 29 Scor., 8, 12, 14, 16, 19, 21, 23, 26 Sag., 4, 9 Cap., 4, 11, 14 Aq., 4, 17, 21 Pis., 6, 13, 21, 27 Ar., 1, 4, 7, 9, 11, 21, 22 Tau., 5, 7, 13, 18, 24

Gem.; *Plin.* HN xviii.57, 74 (4 K. Oct.); *Ptol.* 1, 3, 6, 7, 15, 17, 21, 22, 26, 28 Th., 2, 5, 7, 12, 14, 16, 17, 19, 26 Phao., 6, 7, 18, 21, 23, 24, 26, 27 Ath., 1, 5, 8, 10, 11, 12, 13, 15, 18, 23, 24, 30 Ch., 1, 12, 14, 16, 26 Tyb., 1, 6, 9, 13, 14, 17, 29 Mech., 12, 25, 29 Pham., 1, 4, 6, 8, 13, 16, 20, 22, 27, 30 Phar., 3, 9, 13, 17, 18, 19, 22, 23 Pach., 7 Pay., 5, 30 Ep., 1, 5, 6, 10, 11, 18, 24, 29 Mes., 1, 2 epag.; *DM.* 12, 11 K. Feb., 4 N. Mar., 17 K. Apr., 8 Id. Sept., 6 N., pr. N. Oct., pr. Id. Nov., 6 N. Dec.; *AB.* 2, 4, 9, 11, 13, 14, 15, 23 Ti. I, 1, 3, 12, 13, 17, 19, 20, 22, 23, 29 Ti. II, 1, 6, 9, 11, 14, 19, 20 Kān. I, 7, 9, 11, 19, 21, 26 Kān. II, 1, 3, 7, 8, 11, 23 Shub., 8, 14, 21, 25, 27 Ad., 3, 8, 11, 15, 17, 22, 25, 28 Nis., 3, 4, 6, 8, 12, 13, 18 Ay., 1, 28 Haz., 25, 30 Tam., 3, 4, 10, 11, 17, 22, 24, 25, 29, 31 Āb, 3, 4, 12, 14, 19, 23, 25, 29 Īl.; *EP.*

Exeon

setting, evening: *Iud.* 24 May

Facule

setting: *Iud.* 21 Nov.

Fish, northern (see also: Pisces)

rising: *Gem.* 30 Pis.; *Col.* 3 Id. Mar.; *Plin.* HN xviii.65 (8 Id. Mar.)
setting: *Col.* 7 Id. Sept.

Fish, northern, bright star in (see also: Pisces)

rising, evening: *Ant.* 19 Oct.

Fish, southern (see also: Pisces)

Par. 12 Mar.
disappearance, morning: *CT.* 9 Mar.
rising: *Gem.* 17 Pis.; *Ant.* 25 Feb., 3 Mar., 1 Sept.; *CT.* 12 Mar.; *Ox.* 18 Oct.

setting: *Col.* 4 N. Sept.; *CT.* 2 Sept.
setting, acronychal: *M. I*

Fish, southern, bright star in (see also: Pisces)

disappearance: *Ptol.* 28 Ch., 4, 8, 13, 17 Tyb.
rising: *Ptol.* 11, 20 Pham., 3, 18 Phar., 9 Pach.
rising, evening: *Ptol.* 19 Th., 19, 26 Mes., 2 epag.; *Ant.* 23 Aug.
setting, morning: *Ptol.* 27 Ep., 2, 6, 9, 12 Mes.

Friccos

setting: *Iud.* 19 Aug.

Frost

Gem. 4 Scor.; *Col.* 5 K. Nov.; *Ptol.* 2 Ath.; *CT.* 13 Apr., 1 Nov.; *Ox.* 8 May; *AB.* 29 Ti. I, 12 Ay.

Frost-season (παχνίτη)

Ant. 22 Apr., 25 May; *CT.* 13, 24, 25 Apr.; *Mad.* 26 Apr., 20, 24 May; *Ox.* 21 Apr., 22 May; *Marc.* May

Galen

AB. Intro.

Gemini (see also: Twin[s])

Hib. 4 Pham.; *CT.* 18 May; *Ox.* end Jun.; *Ry.*
rising: *Gem.* 2 Gem.; *CT.* 23 May; *Mad.* 10 Feb.
setting: *Gem.* 16 Sag.; *M. I*

Geminus

Ov. 13 K. Jul.; *Col.* 14 K. Jun.

Golden boat

Hib. 26 Ch.

Greeks

Col. 6 N. Mar., 11 K. Jul.; *Plin.* HN xviii.57, 66; *CT.* 27 Feb.; *AB.* Intro., 30 Īl.

Haedi

Par. 27 Sept.
appearance: *Par.* 9 Oct.
appearance, evening: *Iud.* 7 Oct.
rising: *Col.* 5 K. Oct.; *Plin.* HN xviii.66 (7 K. May); *Ant.* 10 Oct.; *CT.* 29 Apr., 26 Sept., 4, 8, 9 Oct.; *DM.* pr. N. Oct.; *Par.* 25 Sept., 4, 8 Oct.; *Vind.*
rising, evening: *Gem.* 3 Lib.; *Col.* pr. N. Oct.; *Mad.* 7 Oct.; *Aet.* 7 Oct.; *Quin.* 8 Id. Oct.; *Par.* 6 Oct.
rising, morning: *Plin.* HN xviii.74 (3 K. Oct.); *CT.* 25 May, 27 Sept.; *Par.* 26 Sept.
setting: *DM.* 18 K. Feb.; *Iud.* 10 Dec.
setting, evening: *Plin.* HN xviii.74 (3 N. Oct.)
setting, morning: *Mad.* 10 Dec.; *Aet.* 10 Dec.; *Quin.* 3 Id. Dec. (10 Dec.), 11 K. Jan.

Hail

M. II; *Gem.* 23 Ar., 1, 2 Tau.; *Ov.* 18 K. May; *Col.* K. Feb., 13, 12 K., K. Mar., N. Apr., 11 K. May; *Plin.* HN xviii.57; *Ptol.* 8, 24 Mech., 6 Pham., 8, 9, 20, 21, 22, 29 Phar., 1 Pach.; *PS.* K. Feb., K. Mar., N. Apr., 11 K. May; *AB.* 29 Ti. I, 2, 18 Shub., 2 Ad., 4, 15, 16, 17, 24, 26 Nis., 13 Ay.

Halycon, days

Gem. 4 Pis.; *Col.* 8 K. Mar.; *Ptol.* 30 Mech.; *CT.* 22 Feb.; *Mad.* 23 Feb.

Halcyon, birds

PS. 6 Id. Apr.

Hare

setting: *CT.* 13 Aug.
setting, morning: *CT.* 21 Nov.

Heat

Gem. 1, 12, 14 Leo; *Col.* Id. Jun., 8, 7, 6, K., K., pr. N. Jul., 7 Id. Aug., K., 4 N. Sept.; *Plin.* HN xviii.68, 69; *Ptol.* 6 Th., 23, 24 Pay., 1, 9, 12, 20, 23, 25, 28 Ep., 2, 3, 5, 6, 8, 11, 12, 15, 17, 19, 27 Mes., 2 epag.; *PS.* 8 Id. Aug.; *CT.* 24 Jun., 27, 28, 29 Jul., 1, 6, 7, 8, 9, 10, 19 Aug.; *Ox.* 6, 24 Jun., 20 July; *Iud.* 2 Jun., 3 Jul.; *AB.* Intro., 22 Ti. I, 21, 28 Shub., 1, 17 Ad., 16, 17, 18, 25 Haz., 3, 14, 17, 18, 19, 21, 22, 25, 26, 29, 30 Tam., 1, 4, 5, 8, 9, 10, 12, 20, 24, 25, 26 Āb, 3 Īl.

Hera

Hib. 6 Mech., 11 Phar.

Heracles (see: Twin, southern)

Hesiod

Plin. HN xviii.57

Hindus

AB. 17 Ad., 16 Īl.

Hipparchus

Col. 16 K. Jan.; *Ptol.* 1, 2, 3, 6, 19, 26 Th., 1, 2, 10, 12, 20 Phao., 2, 6, 7, 11, 14, 15, 27 Ath., 11, 17, 22, 28, 30 Ch., 12, 16, 19, 21, 25, 28, 30 Tyb., 4, 14, 17, 18, 21, 25, 28, 30 Mech., 5, 6, 12, 13, 18, 23, 27, 28 Pham., 10, 11, 14, 15, 26 Phar., 17, 22, 24, 30 Pach., 4, 29 Pay., 11, 13, 19, 20, 24 Ep., 2, 8, 18 Mes., 2, 3 epag.; *DM.* 13 K. Feb.; *AB.* 7, 9, 29 Ti. I, 3, 7, 10, 23 Ti. II, 6, 13, 24 Kān. I, 7, 11, 13, 15, 20, 22, 25, 30 Kān. II, 8, 12, 15,

19, 22, 24, 28 Shub., 2, 9, 14, 19, 23, 24
Ad., 5, 6, 9, 10, 21 Nis., 12, 17, 19, 25,
29 Ay., 23 Haz., 5, 13, 14, 18, 22, 24,
26 Tam., 1, 11, 26, 29, 30, 31 Āb, 3, 16, 23,
25, 29 Īl.

Hippocrates

AB. 16 Haz., 18 Tam.

Hyades (see also: Facule; Succule)

M. II; *Ov.* 7 K., 4 N. Jun., 17 K. Jul.; *Mad.* 30
    Oct.; *Par.* 13 Apr.
appearance, acronychal: *Mad.* 2 Oct.
appearance, morning: *Mad.* 19 May; *Ox.* 19
    May; *Aet.* 19 May; *Quin.* 13 K. Jun.; *Iud.* 1
    May; *Geo.* 19 May
disappearance: *Gem.* 23 Ar.; *Col.* pr. Id. Apr.;
    CT. 12 Apr.; *Vind.*
disappearance, evening: *Ov.* 6 N. May; *Col.* 14
    K. May
rising: *Col.* 12 K. Jun.; *Ant.* 16 Mar., 1, 23 May,
    12 Nov.; *CT.* 6 Apr., 13, 20, 24 May, 1 Jun.;
    Mad. 26 May; *DM.* 9 K. Apr.; *Ox.* 25 Sept.;
    Par. 5 Apr., 20, 30 May; *Vind.*
rising, acronychal (see also: 'rising, evening'):
    Hib. 17 Mes.; *Gem.* 22 Lib.
rising, evening: *Plin.* HN xviii.74 (17 K. Nov.);
    Mad. 17 Oct.; *DM.* 14 K. Nov.; *Aet.* 17
    Oct.; *Quin.* 15 K. Nov.; *Par.* 17 Oct.;
    Byz.
rising, morning: *Gem.* 32 Tau., 5 Gem.; *Col.* K.
    May, 14 K. Dec.; *Plin.* HN xviii.66 (6 N.
    May), 74 (pr. K. Nov.); *CT.* 2 May; *Mad.* 1
    May, 29 Sept.; *Ox.* 1 May; *Aet.* 1 May; *Par.*
    1 May; *Geo.* 30 Apr.; *Byz.*
setting: *Gem.* 27, 29 Scor.; *Col.* pr. K. Dec.; *Ant.*
    5, 6 Nov.; *CT.* 7, 15, 16, 17, 19, 26 Apr., 10
    May, 27 Oct.; *Mad.* 1 Dec.; *DM.* 9 K. Apr.,
    K. Dec.; *Ox.* 1 Dec.; *Par.* 20 Nov., 15, 19
    Apr., 9, 13 May; *Iud.* 21 Nov.
setting, acronychal: *Hib.* 6 Mech.; *Gem.*
    21 Ar.
setting, evening: *Ov.* 15 K. May; *Plin.* HN
    xviii.66 (14 K. May)
setting, morning: *Col.* 11 K. Dec.; *CT.* 21 Nov.;
    Mad. 21 Nov.; *Aet.* 21 Nov.; *Quin.* 10 K.
    Dec.; *Par.* 10 Dec.
visibility: *Ov.* 6 N. May

Hyades, bright star in

disappearance: *Ptol.* 21, 23, 24, 26, 27 Phar.
rising: *Ptol.* 3, 7, 14, 17, 22 Pay.; *Ant.* 1 Jun.
rising, evening: *Ptol.* 7, 8 Ath.; *Ant.* 3 Nov.
setting: *Mad.* 15 Dec.
setting, morning: *Ptol.* 15, 16 Ath.; *CT.* 18 Jan.

Hydra

setting, evening: *Ox.* 1 Aug.
setting, morning: *M. I*

Hydra, bright star in

disappearance: *Ptol.* 9, 15, 20, 25, 30 Pay.
rising: *Ptol.* 22, 24, 27, 29 Mes., 1 epag.

Hydra, first part of (see also: Oridus)

appearance: *Iud.* 19 Aug.

Hyginus

Plin. HN xviii.63

Ipos

Iud. 21 Mar.
setting, evening: *Iud.* 22 Feb.

Isis, festival of launching of ship of

Ox. 9 Mar.

Jupiter

Ov. K. May

Kite

Gem. 17, 22, 30 Pis.; *Plin.* HN xviii.65 (8 Id. Mar., 15 K. Apr.); *Ptol.* 12, 13, 17, 21, 24, 25 Pham.; *CT.* 9, 10, 18 Mar.; *Par.* 9, 18 Mar., 1 Aug.; *AB.* 30 Ti. I, 8, 9, 13 Ad.; *Vind.*

Kite's star

Ov. 16 K. Apr.

Knee, star on

Ox. 26 Feb.

Lantern

rising: *Ox.* 1 Nov.

Lapsidis

appearance: *Iud.* 19 May
setting: *Iud.* 6 Nov.

Leo

Hib. 6 Pach.; *Col.* 13 K. Aug.; *Plin.* HN xviii.68 (13 K. Aug.); *CT.* 20 Jul.; *AB.* 6 Ay.; *Ry.*; *Flor.* 6 Id. Aug.
rising: *Gem.* 30 Can.; *Ant.* 1 Aug.; *CT.* 26 Jul., 8 Aug.
rising, morning: *CT.* 24 Jul.
setting: *Gem.* 2 Aq., 2 Pis.; *Col.* 10 K. Mar.; *CT.* 14, 18 Jan., 20 Feb.; *Mad.* 3 Feb.
setting, morning: *Col.* 17 K. Feb.; *CT.* 18 Jan.

Leo, back of

disappearance: *Ov.* 4 N. Feb.

Leo, bright star in

setting: *Mad.* 26 Jan.; *Aet.* 25 Jan.

Leo, (bright/Royal) star in breast of (see also: Leo, star in heart of)

disappearance, evening: *Ov.* 9 K. Feb.
rising: *Col.* 4 K. Aug.; *CT.* 29 Jul.; *Quin.* 3 K. Aug.; *Geo.* 30 Jul.
rising, morning: *Plin.* HN xviii.68 (3 K. Aug.)
setting: *Col.* 6 K. Feb.; *Ant.* 26 Jan.; *CT.* 27 Jan.; *Quin.* 7 K. Feb.
setting, morning: *Plin.* HN xviii.64 (8 K. Feb.); *Iud.* 25 Jan.

Leo, heart of

Ov. 9 K. Feb.

Leo, middle of

rising: *Gem.* 12 Leo; *Col.* pr. N. Aug.; *CT.* 4, 5 Aug.
setting: *Col.* 3 N. Feb.; *CT.* 3 Feb.; *Par.* 20 Feb.

Leo, star on head of

rising, evening: *Ant.* 18 Jan.

Leo, star in heart of (see also: Leo, star in breast of)

disappearance: *Ptol.* 10, 13, 16, 18, 21 Ep.; *Ant.* 11 Jul.
rising: *Ptol.* 18, 19, 20 Mes.; *Ant.* 11 Aug.; *Mad.* 7 Aug.; *Ox.* 17 Aug.
rising, evening: *Ptol.* 21, 22 Tyb.
setting, morning: *Ptol.* 6, 7, 8, 9, 11 Mech.; *Ant.* 1 Feb.

Leo, star on tail of

CT. 13 Mar.
disappearance: *Ptol.* 22, 23, 25 Mes.; *Ant.* 17 Aug.
rising: *Ptol.* 1, 2, 3 Th., 3, 4 epag.; *Ant.* 28 Aug.; *Mad.* 30 Jul.
rising, evening: *Ptol.* 6, 7, 8, 10, 13 Mech.
setting: *Ox.* 19 Mar.

setting, morning: *Ptol.* 13, 18, 25 Pham., 2, 12 Phar.; *Ant.* 14 Mar.

Leo, tail of

setting: *Par.* 22 Oct.

Lepus

setting, morning: *Col.* 10 K. Dec.; *Par.* 20 Nov., 1 Aug.

Libra (see also: Scorpio, claws of)

Ov. 8 Id. Apr.; *Col.* 13 K. Oct.; *Plin.* HN xviii.59; *CT.* 19 Sept.; *Par.* 18 Sept.; *Ry.*
appearance: *Iud.* 25 Feb.
rising: *CT.* 10 Oct.; *Par.* 10 Oct.
setting: *Col.* Id. Apr.; *Plin.* HN xviii.66 (6 Id. Apr.)
setting, morning: *Col.* 4 Id. Apr.; *CT.* 8 Apr.

Lightning

Gem. 16 Sag.; *Ptol.* 6 Phao., 14 Ch.; *AB.* 10 Kān. I; *Marc.*

Lyra

M. II; Ov. 3 N. May; *Par.* 3, 15 Nov.
appearance: *CT.* 24 Apr.; *Iud.* 13 Nov., 15 Aug.
appearance, evening: *Col.* 9 K. May; *CT.* 23 Apr.; *Par.* 23 Apr.; *Iud.* 4 May
disappearance: *Ov.* 10 K. Feb., 4 N. Feb.
disappearance, morning: *Par.* 1 Aug.
rising: *Gem.* 4 Tau.; *Ov.* N. Jan.; *Plin.* 6 Id. May *CT.* 5 Jan., 10 May, 6 Aug., 31 Oct., 18 Nov.; *Mad.* 6 Nov.; *Quin.* 18 K. Dec.; *Par.* 30 Oct., 5 Nov., 9, 28 May; *Vind.; Byz.*
rising, acronychal (see also: 'rising, evening'): *Hib.* 19 Mech.; *Gem.* 27 Ar.
rising, evening: *Mad.* 4 May; *Aet.* 4 May; *Quin.* 6 N. May; *Par.* 4 May; *Byz.*
rising, morning: *Gem.* 10, 13, 21 Scor.; *Col.* 3 Id., Id. May, 3 N. Nov., 16 K. Dec., N. Jan.; *Plin.* HN xviii.64 (N. Jan.), 66 (6 K. May); *Ant.* 28 Oct.; *CT.* 5, 14, 28 May, 3, 5, 15

Nov.; *DM.* pr. K. Nov.; *Aet.* 13 Nov.; *Par.* 11, 13 Nov., 14 May; *Byz.*
setting: *Gem.* 17 Leo; *Col.* 3 K., K., 3 N. Feb., 13 K. Sept.; *Plin.* HN xviii.59, 68 (3 Id. Aug.), 69 (6 Id. Aug.), 74; *Ant.* 27 Jul., 11 Aug.; *CT.* 17, 22, 26, 27, 30 Jan., 1, 3, 6 Feb., 18 Aug.; *Mad.* 25 Jul., 1 Nov., 1, 3 Feb.; *Par.* 5, 22, 25, 30 Jan., 3, 6 Feb.; *M. I; Vind.; Byz.*
setting, acronychal: *Hib.* 27 Mech.; *Gem.* 11 Aq.
setting, evening: *Gem.* 3 Aq.; *Col.* 11 K. Feb.; *Plin.* HN xviii.64 (pr. N. Feb.); *Quin.* 3 N. Feb.; *Par.* 12, 26 Jan.; *Iud.* 29 Jan.
setting, morning: *Gem.* 22 Leo; *Col.* pr. Id. Aug., *Plin.* HN xviii.69 (14 K. Sept.); *CT.* 11, 20 Aug.; *Mad.* 15 Aug.; *Ox.* 15 Aug.; *Aet.* 15 Aug.; *Quin.* 15 Aug.; *Par.* 15 Jan.

Lyra, bright star in

rising, evening: *Ptol.* 10, 19, 28 Phar., 8, 17 Pach.; *Ant.* 11 Jan., 2 May; *Ox.* 1 May
rising, morning: *Ptol.* 3, 11, 19, 26 Ath., 4 Ch.
setting: *Mad.* 25 Aug.
setting, evening: *Ptol.* 9, 18, 25 Tyb., 4, 12 Mech.; *Ant.* 4 Feb.
setting, morning: *Ptol.* 5 Th., 4, 13 Mes., 1 epag.

Lyra, middle of

rising: *Par.* 7 Jan.

Mars (see also: Ares)

CT. 8 Jan.

Meton

Gem. 25 Gem.; *Ptol.* 7, 11 Ath., 28, 29 Ch., 29 Pham., 1, 2 Phar., 14 Ep.; *AB.* 7 Ti. II, 24 Kān I, 25, 27 Ad., 1 Nis., 8, 26 Tam.

Metrodorus

Ptol. 5, 7, 17, 20, 23 Th., 5, 28 Phao., 4, 7, 13, 15 Ath., 6, 14, 29, 30 Ch., 13, 23 Tyb., 1, 15, 17, 30 Mech., 12, 13 Pham., 21, 29 Phar., 2, 17 Pach., 2 Pay., 11, 18, 27, 29, 30 Ep., 2

Mes., 1 epag.; *DM.* Id. Mar., 9 K. Apr., 1
Sept., 14 K., 4, 3 Id. Nov.; *AB.* 2, 25, 31 Ti.
I, 8, 11 Ti. II, 2, 10 Kān. I, 8, 18, 26 Kān. II,
8, 11 Shub., 8, 9 Ad., 1, 16, 24 Nis., 12, 28
Ay., 5, 6, 12, 21, 23, 24, 26 Tam., 10, 24 Āb,
2, 4, 14, 17 Īl.

Mist (see also air, misty)

Ptol. 6 Th., 27 Mes.; *CT.* 26 Jul., 4, 7 Aug.

Mizzle

AB. 24 Ad., 11, 14, 18, 25, 30 Nis., 1, 3, 4, 6, 19
Ay., 3, 5, 13, 15 Haz., 19 Īl.

Moon

Col. XI.ii.85, iii.22; *Plin.* HN XVIII.61, 63, 68,
75; *PS.* Id. Jan.; *AB.* 28 Shub., 28 Nis., 18
Tam.; *Ry.*
eclipse: *CT.* 10 Aug., 24 Sept.
full: *Col.* XI.ii.85; *Plin.* HN XVIII.61, 69, 75
half: *Plin.* HN XVIII.75
new: *Plin.* HN XVIII.68, 75
waning: *Col.* XI.ii.11, 52; *Plin.* HN XVIII.75
waxing: *Plin.* HN XVIII.75

New Year

Ox. 23 Sept.
Egyptian: *Ox.* 20 Aug.

Ophiucus

Ov. 11 K. Jul
rising: *Par.* 21 Jun.
setting, morning: *Col.* 11 K. Jul.; *CT.* 21 Jun.

Oridus

appearance: *Iud.* 19 Aug.

Orion

M. II; *Par.* 15, 23 Nov.; *EP.*; *Byz.*

appearance: *CT.* 19 Jul.; *Mad.* 15 Jul.; *Iud.* 9
Mar., 25 Jun., 3 Jul.; *Vind.*
disappearance: *Ov.* 5 Id. Apr., 5 Id. May; *Plin.*
HN XVIII.59, 66 (N. Apr., 5 K. May); *CT.*
31 Oct.; *Par.* 30 Oct.; *Vind.*
disappearance, evening: *Mad.* 26 Apr.; *Aet.* 27
Xan.; *Quin.* 12 K. May; *Geo.* 29 Apr.
rising: *Hib.* 9 Pach.; *Gem.* 13 Can., 24, 29 Gem.;
Plin. HN XVIII.65 (7 Id. Mar.), 68 (6 K., 4
N., pr. Id. Jul.); *Ant.* 13 Jun.; *CT.* 23, 27
Jun., 4, 9, 16, 19 Jul., 16 Oct., 18 Nov.;
Mad. 4 Jul.; *Ox.* 25 Jun., 1 Jul.; *Aet.* 3 Jul.,
27 Nov.; *Quin.* 6 K. Jul., 3 Jul., 4 K. Dec.
(27 Nov.); *Par.* 16 Oct., 7, 23 Jun., 1, 10
Jul.; *Geo.* 23 Jun.
rising, acronychal: *Gem.* 12 Scor.
rising, morning: *Hib.* 17 Phar.; *Gem.* 11 Can.;
Ant. 14 Jul.; *CT.* 19 Jun., 12 Jul.; *Mad.* 25
Jun.; *Aet.* 25 Jun.; *Geo.* 10 Jul.; *Byz.*
setting: *Gem.* 15, 19 Scor., 7 Sag.; *Ant.* 6 Nov.;
CT. 28 Oct., 7, 13, 23, 30 Nov.; *Mad.* 15
Oct.; *Ox.* 19 Nov.; *Quin.* 3 Id. Nov.; *Par.* 7
Nov., 2 Jan.; *Vind.*; *EP.*; *Geo.* 11 Nov.
setting, acronychal: *Gem.* 1 Tau.
setting, evening: *Gem.* 13 Ar.; *Mad.* 27 Nov.;
Iud. 27 Apr.; *Byz.*
setting, morning: *Gem.* 8 Sag.; *CT.* 13 Nov., 1
Dec.; *Par.* 13 Nov.

Orion, belt of (see also: Orion, star on belt of)

Ov. 6 K. Jul.
appearance: *CT.* 30 Jun.
rising: *Ant.* 25 Jun.
rising, evening: *Col.* Id. Oct.

Orion, foot of, star shared with Eridanus (see also: Eridanus and foot of Orion, star shared by)

Orion, middle of belt of

disappearance: *Ptol.* 24, 27 Phar., 1, 4, 7 Pach.;
Ant. 28 Apr.
rising: *Ptol.* 1, 6, 11, 17, 23 Ep.
rising, evening: *Ptol.* 30 Ath., 4, 7, 10, 13 Ch.

Perseus

appearance: *Mad.* 19 Sept.
rising: *CT.* 15 Apr.; *Ox.* 15 Apr.
setting, morning: *M. I*

Perseus, bright star in

rising, evening: *Ptol.* 10 Th., 2, 23 Ep., 11, 28
    Mes.
rising, morning: *Ptol.* 12 Mech., 3, 21 Pham., 3,
    14 Phar.; *Ant.* 17 Nov.
setting: *Ant.* 20 Apr.
setting, evening: *Ptol.* 22, 26 Phar., 1, 6, 12 Pach.
setting, morning: *Ptol.* 15, 20, 25 Ath., 1, 8 Ch.;
    Ant. 30 Mar.

Perseus, middle of belt of

setting, morning: *Ant.* 23 Apr.

Persians

AB. Intro., 15 Shub., 16 Haz., 16 Īl.

Philippus

M. II; *Plin.* HN xvIII.74 (4 K. Oct.); *Ptol.* 7, 10,
    30 Th., 4 Phao., 3, 7, 14 Ath., 6 Ch., 3, 11,
    13, 23 Tyb., 21, 29 Mech., 12, 17, 28
    Pham., 15, 21 Phar., 1, 5, 15, 29, 30 Pach.,
    15, 27, 28 Ep.; *DM.* 18 K. Feb., 9 K. Apr.;
    AB. 1, 30 Ti. I, 3, 4, 10, 23 Ti. II, 2 Kān. I,
    8, 18 Kān II, 11, 15, 23 Shub., 8, 13, 24
    Ad., 10, 16 Nis., 24, 25 Ay., 9, 23, 24, 26
    Tam., 11, 27 Āb, 4, 7 Īl.

Phoebe

Ov. 7 Id. Jun.

Phoebus

Ov. 16 K. Feb.

Pisces (see also: Fish)

Gem. 1 Ar.; *Ov.* 15 K. Mar.; *Col.* 15 K. Mar.; *CT.*
    15 Feb., 8 Mar.; *Ox.* end Mar.; *Par.* 15 Feb.;
    Iud. 8 Mar.; *Ry.*
rising: *CT.* 26 Mar.; *DM.* 17 K. Apr.; *Par.* 26
    Mar.
setting: *Plin.* HN xvIII.74; *Ant.* 22 Sept.; *CT.*
    21, 23 Sept.; *DM.* 16 K. Apr.
setting, morning: *Col.* 11 K. Oct.; *Par.* 9 Mar.

Pisces, binding of (see also: Fish)

rising: *Gem.* 1 Ar.
setting: *Ox.* 26 Mar.

Pisces, one fish in (see also: Fish)

disappearance: *Ov.* 5 N. Mar.

Pleiades (see also: Vergilie)

Ov. 4 N. Apr.; *CT.* 2 Oct.; *Par.* 12 Nov., 7 May;
    M. I; *EP.*; *Byz.*
appearance: *Col.* 6 Id. May; *CT.* 11 May; *Par.* 9
    Oct., 11 May; *Iud.* 1, 27 Apr.; *AB.* 23 Nis.
appearance, acronychal: *Mad.* 1 Apr.; *Aet.* 1
    Apr.
appearance, evening: *Gem.* 5 Lib.; *CT.* 27 Sept.;
    Par. 29 Sept.
appearance, in east: *CT.* 1 Oct.
appearance, morning: *CT.* 29 Sept.; *Mad.* 7
    May; *Ox.* 7 May; *Aet.* 7 May; *Quin.* 7 Id.
    May; *Par.* 1 Oct.; *Iud.* 7 May; *Geo.* 7 May
disappearance: *Gem.* 10 Ar.; *Ant.* 1, 5 Apr.; *CT.*
    11 Nov.; *AB.* 23 Nis.; *Vind.*
disappearance, acronychal: *Quin.* K. Apr.; *Geo.*
    1 Apr.
disappearance, evening: *Col.* 8 Id. Apr.; *Plin.*
    HN xvIII.66 (3 N. Apr.); *Mad.* 19 Apr.;
    Aet. 19 Apr.; *Quin.* 12 K. May; *Par.* 1 Apr.;
    Geo. 16 Apr.
disappearance, morning: *Gem.* 13 Ar.
rising: *Hib.* 13 Pham., *M. II*; *Gem.* 8 Lib., 13, 22
    Tau.; *Plin.* HN xvIII.66 (6 Id. May), 67, 69
    (6 Id. May); *Ant.* 31 May; *CT.* 1, 22 Apr.,
    12, 30 May, 14 Sept., 9, 12 Oct.; *Mad.* 12
    May; *Quin.* 21 Art.; *Par.* 11 Sept., 24 Oct.,

7, 22 Apr.; *Vind.*; *DC.*; *EP.*; *Flor.* 4 Id. Jun.; *Byz.*

rising, acronychal: *Ox.* 1 Oct.; *Hib.* 2 Mes.

rising, evening: *Col.* 6 Id. Oct.; *Plin.* HN xviii.74 (6 Id. Oct.); *DM.* N. Oct.; *Byz.*

rising, morning: *Col.* 10 K., N. May; *Plin.* HN xviii.59, 66; *Ant.* 9 May; *CT.* 7 May; *DM.* K. Oct.; *Aet.* 21 Apr.; *Par.* 6 May; *Geo.* 23 Apr; *Byz.*

setting: *Gem.* 15, 16 Scor.; *Col.* 5 K. Nov.; *Plin.* HN xviii.60 (3 Id. Nov.), 74 (3 Id. Nov.); *Ant.* 10 Oct., 6, 11 Nov.; *CT.* 20, 24, 28 Oct., 1, 7 Nov.; *Mad.* 13, 23 Nov.; *Quin.* 9 K. Nov.; *Par.* 20, 22 Oct., 1, 7 Nov.; *Marc.*; *Vind.*; *DC.*; *Flor.* 4 N. Nov.; *Byz.*

setting, acronychal: *Hib.* 27 Tyb.; *Gem.* 13 Ar.

setting, evening: *CT.* 3 Apr., 30 Sept.; *Iud.* 19 Apr.; *Vind.*

setting, morning: *Gem.* 4, 19 Scor.; *Col.* 13, 12 K., 6 Id. Nov.; *Plin.* HN xviii.57, 59; *CT.* 13 Nov.; *Mad.* 6 Nov.; *Aet.* 23 Oct., 6 Nov.; *Quin.* 3. Id. Nov.; *Par.* 6, 13 Nov.; *Iud.* 23 Oct.; *Geo.* 24 Oct., 11 Nov.

visibility: *Ov.* 3 Id. May

Pollux (see: Twin[s])

Procyon (see also: Antichanus)

Plin. HN xviii.68

appearance: *Iud.* 4 Jul.

disappearance: *Ptol.* 27 Pach., 1, 3, 6 Pay.; *Ant.* 28 May

disappearance, evening: *Mad.* 30 May

rising: *Plin.* HN xviii.68 (16 K. Aug.); *Ptol.* 19, 22, 24, 26, 28 Ep.; *CT.* 15 Jul.; *Ox.* 14 May, 6 Jul.; *Par.* 14 Jul.

rising, evening: *Ptol.* 25, 27, 29 Ch., 1, 3 Tyb.

rising, morning: *Col.* Id. Jul.; *Aet.* 14 Jul.; *Geo.* 13 Jul.

setting: *Plin.* HN xviii.69

setting, morning: *Ptol.* 20, 22, 24, 25, 26 Ch.; *Ant.* 23 Dec.; *M. I*

Ptolemy

AB. Intro., 3 Ti. II, 13, 28 Kān. II, 20 Nis., 7 Tam, 31 Āb.

Rain/rainy

Gem. 17 Leo, 10, 24 Vir., 9, 10, 14, 27, 28 Scor., 14, 16 Sag., 2, 3, 11 Aq., 4 Pis., 3, 6, 13 Ar., 1, 4, 7, 8, 11 Tau.; *Ov.* N. Jan., 8 Id. Apr., 7 K. May, 4 N. Jun.; *Col.* 17, 11, 5 K. Feb., 13, 12, 9 K. Mar., 10 K. Apr., 18, 15, 14, 11, 4, 3 K., 3 N., 7, 6 Id. May, 16, 15, 12 K., K., 4 N. Jun., 10, 6 K. Sept., 11, 10, 9, 8, 5 K., 6, 3 Id. Oct., 11 K., K., 3 N. Nov., 15, 12 K., pr. K., 3 Id. Dec., pr. N., 5 Id. Jan.; *Plin.* HN xviii.57, 60, 66 (6 Id. Apr.), 74; *Ptol.* 1, 3, 6, 7, 8, 18, 19, 20, 23, 26, 29 Th., 4, 5, 6, 7, 15, 16, 22 Phao., 7, 8, 9, 11, 13, 14, 21, 25, 27 Ath., 1, 4, 11, 12, 13, 18, 19 Ch., 14, 24, 25, 28, 29 Tyb., 1, 6, 12, 14, 24, 29 Mech., 27, 28, 30 Pham., 1, 2, 4, 6, 8, 13, 14, 15, 18, 20, 22, 27, 28, 29, 30 Phar., 1, 3, 4, 9, 10, 13, 15, 17, 18, 20, 22, 23 Pach., 3, 11, 25 Pay., 10, 22 Ep., 26 Mes.; *PS.* 4 N., 6 Id. Jan., 18, 11 K. Feb., 9 K. Mar., 11 K. Apr., 15, 11, 4 K., 3 N., 8 Id. May, 16, 12, 4 K. Jun., 6 K. Sept., 11 K., 6, 3 Id. Oct., 4 N. Nov., 15 K., pr. K., 3 Id. Dec.; *Ant.* 14 Jul.; *CT.* 21, 22, 27, 30, 31 Jan., 25 Feb., 6 Mar., 10, 21, 28 Apr., 5, 7, 10, 11, 20 Jun., 20 Aug., 1, 3, 5, 7, 13, 15, 20, 22, 23, 29 Sept., 4, 21, 22 Oct., 4, 22, 29, 30 Nov., 3, 13, 29 Dec.; *Mad.* 25 Mar., 25, 30 Apr.; *Ox.* 19 Feb., 25 Sept., 6 Nov.; *AB.* Intro., 1, 2, 12, 13, 17, 19 Ti. I, 3, 4, 5, 7, 8, 10, 17, 21, 23, 24, 27, 29 Ti. II, 3, 9, 10, 11, 14, 15, 24 Kān. I, 9, 20, 22, 26 Kān. II, 1, 4, 6, 8, 16, 18, 23 Shub., 9, 23, 24, 26, 27 Ad., 1, 3, 8, 9, 10, 13, 14, 15, 17, 22, 23, 24, 25, 26, 28, 29 Nis., 3, 4, 5, 6, 7, 8, 9, 10, 12, 13, 15, 17, 18, 28 Ay., 4, 19, 20, 28 Haz., 4, 16, 18 Tam., 19, 29, 31 Āb, 3, 4, 5, 15, 16, 17, 19, 23 Īl.

Regulus

setting, evening: *Par.* 26 Jan.

Sagitta

Par. 13 Jan.

appearance: *Par.* 4 Sept.

disappearance: *CT.* 4 Sept.

disappearance, evening: *Plin.* HN xviii.64 (N. Jan.)

rising, evening: *Gem.* 30 Tau.; *Col.* 8 K. Mar.
setting: *Plin.* HN xviii.74 (5 K. Sept.); *CT.* 18,
   19 Feb.; *Mad.* 28 Aug., 27 Feb.; *Aet.* 28
   Aug.; *Quin.* 4 K. Sept.; *Par.* 18, 22, 27 Feb.,
   1 Aug.; *M. I*; *Vind.*; *Geo.* 25 Aug.
setting, evening: *CT.* 27 Feb.; *Ox.* 22 Feb.; *Aet.*
   22 Feb.; *Quin.* 7 K. Mar.
setting, morning: *Gem.* 10 Vir.; *Plin.* HN
   xviii.74 (N. Sept.)

Sagitta, first part of

setting: *CT.* 13 Jan.

Sagittarius

Col. 14 K. Dec.; *CT.* 18, 25 Nov.; *DM.* 7 K. Dec.;
   Ry.; *M. I*
rising: *Gem.* 7 Sag., 1 Cap.; *DM.* 6 N. Dec.
rising, morning: *M. I*
setting: *CT.* 22 May, 4 Dec.; *Mad.* 13 Feb.
setting, evening: *Col.* Id. Feb.; *CT.* 13 Feb.;
   Mad. 22 Jan.; *Par.* 12 Feb.

Sagittarius, middle of

rising: *CT.* 6 Dec.
setting: *Col.* 8 Id. Dec.

Sagittarius, star on knee of

disappearance: *Ptol.* 18 Th., 11, 27 Phao., 6, 13
   Ath.; *Ant.* 1 Nov.
rising: *Ptol.* 6, 12, 18, 25 Tyb., 1, 6 Mech.; *Ant.* 7
   Jan.
rising, evening: *Ptol.* 18, 20, 24, 29 Pay.
rising, morning: *Ptol.* 15 Pay.
setting, morning: *Ptol.* 29 Pach., 6, 9, 13, 15 Pay.

St Demetrius, star of

Marc. 26 Oct.

St Mennas, star of

Marc. 11 Nov.

St Nicholas, star of

Marc.

St Philip, star of

Marc. 14 Nov.

St Thecla, star of

Marc. 20 Sept.

St Thomas, star of

Marc. 6 Oct.

Saturn

Plin. HN xviii.57

Scorpio

Hib. 2, 9 Mes; *Ov.* K. Apr.; *Col.* 14 K. Nov.; *CT.*
   19 Oct.; *DM.* 15 K. Nov.; *Ry.*; *Flor.* 4 Id.
   Nov.
appearance: *Ox.* 6 Nov.
disappearance: *Mad.* 7 Nov.
rising: *CT.* 8, 13 Dec.; *Mad.* 27 Oct.; *Ox.* 29 Oct.
rising, evening: *Byz.*
rising, morning: *Gem.* 18 Scor., 21 Sag.; *Col.* Id.
   Dec.; *Ant.* 4 Dec.; *CT.* 4 May
setting: *Hib.* 12 Pham., 14 Mes.; *M. II*; *Col.* Id.
   Mar., 17 K. Apr.; *Plin.* HN xviii.65 (Id.
   Mar.); *CT.* 29, 30 Mar., 1, 25 Apr., 14 May,
   6, 23 Oct.
setting, acronychal: *Gem.* 12, 17 Lib.
setting, morning: *Hib.* 5 Pham.; *Gem.* 11, 21
   Tau.; *Col.* K. Apr.; *CT.* 14 Nov.

Scorpio, bright star in

CT. 9 Nov.
rising: *Gem.* 16 Scor.; *Col.* 5 Id. Nov.

Scorpio, bright star in northern claw of

rising: *Ant.* 7, 31 Oct.

Scorpio, bright star in southern claw of

disappearance: *Ant.* 16 Sept.
rising, evening: *Ant.* 1 Apr.

Scorpio, claws of (see also: Libra)

Hib. Ep.
rising: *Gem.* 17 Lib.

Scorpio, first part of

visibility, morning: *Ov.* 17 K. Apr.

Scorpio, first stars in

setting: *Gem.* 29 Pis.

Scorpio, forehead of

rising: *Gem.* 4 Scor.; *Col.* 7 K. Nov.
setting: *CT.* 26 Oct.

Scorpio, middle/centre of

Vind.
rising: *Gem.* 10 Sag.
rising, morning: *M. I*
setting: *Col.* pr. N. May; *CT.* 6 May
visibility: *Ov.* pr. N. May

Scorpio, middle star of

rising: *CT.* 12 Nov.

Sea

Ov. 8 K. Jun.; *Plin.* HN 13 K. Aug., pr. Id. Sept.;
    AB. 13 Ti. II, 17 Ad., 18 Tam.; *Marc.* Mar.
change in the weather at: *Ptol.* 15 Tyb.,
disturbance at: *Ptol.* 1 Ch.; *CT.* 23, 27 Sept., 13
    Oct.; *Ox.* 3 Feb., 12, 19, 26 Mar., 28 Nov.,
    21 Dec.; *AB.* 27 Ti. II; *Marc.* 14, 20 Sept.
foul weather at: *Marc.* 15–22 Nov.
rain at: *Ptol.* 19, 20 Th.; *AB.* 16, 17 Īl.
sailing on: *CT.* 17 Mar. *Marc.* 12 May–24 Jun.
ships at: *AB.* 13 Ti. II
showery: *CT.* 23, 27 Sept.
storm at: *Gem.* 28 Can.; *Ptol.* 8 Th., 4, 14 Tyb.,
    29 Ep.
waves: *AB.* 13 Ti. I, 28 Ti. II
wind at bottom of: *AB.* 13 Ti. II
wintry air at: *AB.* 13 Ti. II; 30 Kan. I, 8 Kan. II

Septemtrion

appearance: *Iud.* 23 Feb.
appearance, morning: *Iud.* 14 Sept.

Ship of Isis (see: Isis, festival of launching of ship of)

Showers

Ptol. 26 Pay; *CT.* 8, 11, 13, 14, 16, 18, 23 Jan., 4,
    10, 20 Feb., 3, 22, 24, 26, 27, 29, 30, 31
    Mar., 6, 11, 24 Apr., 24, 30 May, 4, 12, 27,
    28 Jun., 11 Jul., 23, 27 Sept., 2, 27 Nov., 5
    Dec.; *Mad.* 2 Apr., 25 Aug., 15 Sept., 15, 20
    Oct., 12 Nov., 1 Dec.; *DM.* 11 K. Feb., 4 N.
    Sept., 6 N. Oct.; *Par.* 30 Jan., 22 Mar., 5, 27
    Apr., 1, 2 Jun.

Sinān (ibn Thābit ibn Qurra)

AB. Intro., 8, 13 Ti. I, 6, 11 Ti. II, 5, 11 Kān. I, 4
    Kān. II, 2, 8 Shub., 4, 11, 16, 30 Ad., 4, 6,
    10, 21, 28 Nis., 5, 7, 11 Ay., 5, 12 Haz., 7,
    13 Tam., 8, 11, 13 Āb

Sirius (see also: Cion)

Ov. 11 K. Jun.; *CT.* 27 Jul.; *AB.* 18 Tam.; *Vind.*;
    EP.; *Byz.*
appearance: *Col.* 7 K. Aug.; *Iud.* 1 Dec., 18 Jul.;
    Vind.
disappearance: *Gem.* 2 Tau.; *Plin.* HN xviii.66
    (4 K. May); *Ptol.* 3, 7, 12, 17, 23 Pach.; *Ant.*
    30 Apr., 13 May; *CT.* 25 Apr., 1 May; *Par.*
    30 Apr.; *Iud.* 27 Aug.; *Vind.*
disappearance, evening: *Col.* pr. K. May; *Plin.*
    HN xviii.66 (6, 4 K., 8 Id. May); *CT.* 30
    Apr.; *Mad.* 3 May
rising: *Hib.* 18 Pach.; *Gem.* 27, 30 Can.; *Ov.* 7 K.
    May; *Plin.* HN xviii.67 (13 K. Aug.), 69,
    (4 N. Jul.); *Ptol.* 22, 28 Ep., 4, 9, 14 Mes.;
    Ant. 19, 25, 30 Jul.; *CT.* 28 Jun., 23 Jul.;
    Mad. 19 Jul., 28 Aug.; *Ox.* 18 Jul., 1 Nov.;
    Aet. 28 Aug.; *Quin.* 9 K. Sept.; *Par.* 1 Aug.;
    AB. 18 Tam.
rising, acronychal (see also: 'rising, evening'):
    Gem. 16 Sag.
rising, evening: *Ptol.* 26 Ch., 1, 6, 10, 14 Tyb.;
    Ant. 28 Dec.; *CT.* 9 Dec.; *Byz.*
rising, morning: *Gem.* 25, 27 Can.; *CT.* 4 Jan.,
    30 Jun., 18, 24, 25 Jul.; *Aet.* 19 Jul.; *Quin.* 4
    N. Jul.; *Geo.* 24 Jul.; *Byz.*
setting: *Gem.* 7 Sag.; *Plin.* HN xviii.67 (12 K.
    Jun.), 69; *CT.* 17 May, 24, 27, 28 Nov., 2,
    30 Dec.; *Mad.* 26 Nov.; *DM.* 3 N. Dec.; *Ox.*
    28 Nov.; *Par.* 25 Apr., 17 May; *Vind.*; *EP.*
setting, acronychal: *Gem.* 4 Tau.
setting, evening: *Col.* 3 K. Jan.; *Mad.* 19 May, 16
    Jun.; *Par.* 2 Dec.
setting, morning: *Gem.* 12 Sag.; *Col.* 7 K. Dec.;
    Plin. HN xviii.64 (3 K. Jan.); *Ptol.* 24, 27
    Ath., 1, 5, 9 Ch.; *Ant.* 25 Nov., 2 Dec.; *CT.*
    29 Nov.; *Aet.* 1 Dec.; *Quin.* K. Dec. (1
    Dec.); *Par.* 1 Dec.; *Geo.* 22 Nov.
visibility: *Gem.* 23 Can., 1 Leo

Sirius, forerunners of/*prodromoi*

Ptol. 13 Ep.

Snow

Gem. 3 Ar.; *Col.* 10 K. Apr.; *Ptol.* 30 Pham.; *PS.*
    11 K. Apr.; *CT.* 11, 28, 31 Jan., 6, 26 Mar.;
    Ox. 14 May; *AB.* 26 Ad., 28 Nis.

Solstice, summer

Hib. 24 Phar.; *Gem.* Intro., 1 Can.; *Ov.* 6 K. Jul.;
    Col. 8, 7, 6 K. Jul.; Plin. HN xviii.59, 62,
    67, 68, 69 (8 K. Jun.); *Ptol.* 1 Ep.; *PS.* 8 K.
    Jul.; *Ant.* 24 Jun.; *CT.* 25 Jun.; *Mad.* 21
    Jun.; *Ox.* 22 Jun.; *Aet.* 25 Jun.; *Quin.* 6 K.
    Jul.; *Par.* 23 Jun.; *Marc.* 24 Jun.; *Vind.*;
    DC.; *EP.*; *Flor.* 8 K./6 N. Jul.; *Byz.*

Solstice, winter

Gem. 1, 4 Cap., 16 Aq.; *Col.* 16, 9 K. Jan.; *Plin.*
    HN xviii.59, 62, 63, 64; *Ptol.* 26 Ch.; *PS.* 8
    K. Jan.; *Ant.* 22 Dec.; *CT.* 24 Nov., 23 Dec.;
    Ox. 20 Dec.; *Aet.* 23 Dec.; *Quin.* 9 K. Jan.;
    Par. 23 Dec.; *AB.* 17 Kān. I *Vind.*; *DC.*;
    EP.; *Flor.* 8 K. Jan.; *Byz.*

Soreutes

Marc. 11 May

Sosigenes

Plin. HN xviii.57

Spica (star in Virgo)

disappearance: *Ptol.* 2, 5 Th., 3 epag.; *Ant.* 31
    Aug.
appearance: *Iud.* 19 Sept.
rising: *Gem.* 24 Vir.; *Col.* 14 K. Oct.; *Ptol.* 7, 8, 9
    Phao.; *Ant.* 2 Oct.; *CT.* 18 Sept.
rising, evening: *Ptol.* 17 Pham.; *Ant.* 12 Mar.;
    Mad. 23 Mar.
rising, morning: *Plin.* HN xviii.74 (16 K.
    Oct.); *Mad.* 19 Sept.; *Aet.* 19 Sept.; *Quin.*
    12 K. Oct. (19 Sept.)
setting: *Mad.* 15 Mar., 15 Sept.; *Ox.* 15 Sept.
setting, morning: *Ptol.* 30 Pham., 1, 2, 5, 7
    Phar.; *Ant.* 27 Mar.

Spring

Ov. 5 Id. Feb., 7 K., 3 Id. May; *Col.* 11 K. May;
    Plin. HN xviii.59, 61; *Ptol.* 13, 15, 17

Swallow

Taurus

Taurus, forehead of

Taurus, head of

Taurus, horns of

Taurus, middle of

Taurus, tail of

Teros

Thales

Thunder (see also: Air, thundery)

Wind, north

Hib. 26 Ch.; *Gem.* 19, 29 Lib., 4, 12, 14, 17, 29, 30 Pis.; *Ov.* N. Feb.; *Col.* 3 N. Feb., 10 K., 6 N., 3 Id. Mar., 12 K. Apr., 6 N. May, 12, 8, 7, 6 K. Jun., 3 Id. Dec.; *Ptol.* 14, 17 Th., 9, 16, 17, 20, 26 Phao., 6, 7, 14, 21, 24 Ath., 11, 17, 19 Ch., 19, 25 Tyb., 18, 25, 29, 30, Mech., 5, 6, 8, 12, 13, 14, 16, 18, 19, 21, 23, 24, 25, 27, 29 Pham., 1 Pay., 20 Ep.; *PS.* 10 K., 6 N, 4 Id. Mar., 12 K. Apr., 6 N. May; *CT.* 4, 7, 14, 15, 18, 20, 23, 24, 26, 27, 30 Jan., 3, 8, 17, 20, 24 Feb., 2, 3, 8, 10, 11, 14, 15, 18, 20, 21, 25 Mar., 10, 13, 30 Apr., 4, 20, 25, 26 May, 7, 10, 17, 22 Jun., 6, 11, 14, 16, 29 Jul., 10, 24, 25 Aug., 8 Sept., 14, 23, 27 Oct., 3, 5, 7, 15, 19 Nov., 4, 5, 10, 12, 13, 15, 21 Dec.; *Mad.* 9, 12, 20, 30 Mar., 30 Jun., 30 Jul., 2 Oct., 7 Nov., 26 Dec.; 9 Feb.; *DM.* N. Mar., 17 K. Apr., pr. N. Oct.; *Par.* 19 Jan., 7, 24 Feb., 14 Mar., 10, 11, 13, 20 Apr., 20 May; *AB.* 6, 13, 14, 23 Ti. I, 2, 3, 10, 17, 20 Ti. II, 6, 13, 15 Kān. I, 13, 20 Kān. II, 12, 19, 24, 28 Shub., 2, 4, 8, 9, 12, 14, 15, 16, 19, 20, 21, 23, 25 Ad., 13, 26 Ay., 28 Haz., 14 Tam., 11 Āb, 11, 14 Īl.; *Marc.* 14 Sept.

Wind, north-east

Col. 9 K. Mar., pr. Id. Mar., pr. N. Oct., 16, 15, 12 K. Dec.; *PS.* 9 K., pr. N., pr. Id. Apr., pr. N. Oct.; *CT.* 11 Dec.; *AB.* 22, 23, 24 Shub.

Wind, north-west

Ov. 6 N. May; *Col.* 3 N. Feb., 9 K. Mar., 16 K. Apr., 7, 6 Id. May, 7 Id. Jun., 4 N. Sept., 14, 11 K. Oct., 3 Id. Dec.; *Ptol.* 9, 22 Th., 15 Phao., 11 Ath., 15 Ch., 4, 30 Mech., 18, 30 Pham., 25 Phar., 1, 8, 13, 17, 26 Pach., 8, 9, 15, 19 Pay., 10, 11, 12, 15, 18, 25, 26 Ep., 6, 15, 16, 30 Mes., 5 epag.; *PS.* 3 N. Feb., 16 K. Apr., 7 Id. Jun., 4 N. Sept., 14, 11 K. Oct.; *CT.* 24 Feb., 13 Jun., 19, 20 Jul.; *Mad.* 6 Jan.; *DM.* 8 Id. Sept.

Wind, south

Hib. 6, 19, 27 Mech.; *M. II*; *Gem.* 9, 30, 31 Can., 10 Vir., 19, 29 Lib., 15, 16 Sag., 12, 14, 15, 16, 18 Cap., 4, 8 Tau., 2 Gem.; *Col.* 17, 5 K., K. Feb., 13, 12 K., K., 4 N., pr. Id. Mar., N., 7, 6, 5 Id. Apr., 10, 4 K., Id. May, 16, 15, 12 K. Jun., K. Jul., 8 K. Aug., 11, 5 K., 3 Id., pr. Id. Oct., 11 K., 8, 4 Id. Nov., 16, 15 K., pr. K., 7, 3 Id. Dec., pr. N., 6, 5 Id. Jan.; *Ptol.* 6, 8, 15, 18, 22, 25, 26 Th., 1, 2, 10, 16, 20, 22 Phao., 2, 4, 6, 7, 10, 13, 14, 19, 27 Ath., 1, 4, 10, 12, 13, 15, 17, 19, 22, 23 Ch., 9, 13, 14, 15, 16, 19, 27 Tyb., 3, 4, 8, 21, 24 Mech., 5, 6, 13, 18 Pham., 9, 10, 13, 19, 25, 26, 28, 29 Phar., 2, 4, 5, 8, 22, 26 Pach., 2, 4, 5, 7, 24, 29 Pay., 4, 9, 11, 14, 15, 23 Ep., 1, 2, 3, 5, 7, 10, 16, 26 Mes., 2 epag.; *PS.* 4. N., 8, 6 Id. Jan., 18, 5 K., K. Feb., 13 K. Mar., N. Apr., 10, 4 K. May, 8 K. Aug., 3 Id. Oct., 16, 15, 13, pr. K., 7 Id. Dec.; *CT.* 4, 6, 8, 9, 10, 12, 18, 20 Jan., 1, 4, 14, 16, 17, 19 Feb., 1, 9, 12, 13, 19 Mar., 7, 24, 25, 27, 29 Apr., 12, 13, 15, 19, 22, 26, 28, 29 May, 1, 4, 5, 14, 16, 18, 20, 22, 27 Jun., 2, 6, 7, 9, 18, 25, 26, 31 Jul., 4, 5, 10, 15, 16, 18, 20, 26, 26 Aug., 19, 26, 28, 29 Sept., 1, 12, 15, 22 Oct., 3, 4, 15, 17, 30 Nov., 9, 12, 13, 15, 21 Dec.; *Mad.* 19 Mar., 4, 20 Apr., 22, 24, 30 May, 29 Sept., 18, 26 Oct., 1, 20 Nov.; *DM.* 9 K. Apr., 4, 3 Id. Nov.; *Ox.* 15, 25 Apr., 1 Sept., 1, 18 Oct., 19 Nov.; *Par.* 18, 25, 27 Sept., 4, 8 Jan., 11 Feb., 1 Mar., 27 Apr., 12, 13, 14, 18, 28 May, 2 Jun.; *AB.* 7, 13, 17, 19, 23, 29, 31 Ti. I, 2, 3, 6, 8, 10, 14, 23, 27, 29 Ti. II, 11, 13, 15, 19 Kān. I, 4, 8, 9, 10, 11, 13, 28, 30 Kān. II, 2, 15 Shub., 2, 9, 14 Ad., 4, 5, 8, 14, 20, 21, 23, 24, 28 Nis., 17, 21, 29, 30 Ay., 23, 28 Haz., 3, 5, 6, 25, 26, 31 Tam., 9, 25 Ād., 5, 12, 23, 25, 29 Īl.; *Marc.*

Wind, south-east (see also: Wind, *eurus* and Wind, *Vulturnus*)

Col. K., 3 Id. Feb., Id. May, 16, 15 K. Jun., 15 K. Oct., 5, 4 Id. Nov., 7 Id. Dec.; *PS.* 4 Id. Feb., 17 K. Oct.; *AB.* 8, 14 Ti. II

General index

Milton Keynes UK
Ingram Content Group UK Ltd.
UKHW050108171024
449665UK00005B/67